INTRODUCTION TO CALCULUS

INTRODUCTION TO CALCULUS

by S. W. HOCKEY, M.A.
Assistant Master, Marlborough College

PERGAMON PRESS
OXFORD · LONDON · EDINBURGH · NEW YORK
TORONTO · SYDNEY · PARIS · BRAUNSCHWEIG

PERGAMON PRESS LTD.,
Headington Hill Hall, Oxford
4 & 5 Fitzroy Square, London W.1
PERGAMON PRESS (SCOTLAND) LTD.,
2 & 3 Teviot Place, Edinburgh 1
PERGAMON PRESS INC.,
Maxwell House, Fairview Park, Elmsford, New York 10523
PERGAMON OF CANADA LTD.,
207 Queen's Quay West, Toronto 1
PERGAMON PRESS (AUST.) PTY. LTD.,
19a Boundary Street, Rushcutters Bay, N.S.W. 2011, Australia
PERGAMON PRESS S.A.R.L.,
24 rue des Écoles, Paris 5e
VIEWEG & SOHN GMBH,
Burgplatz 1, Braunschweig

First Edition 1969
Library of Congress Catalog Card No. 68–21103

Printed in Hungary

08 012579 4

Contents

Preface

THIS book provides an introduction to calculus suitable for pupils who have achieved a satisfactory standard in mathematics at 'O' level. The length is such that most students should be able to digest the contents in a year.

The first part of the book is concerned with laying a solid foundation on which to build and develop the later work. Some of this will be familiar to pupils who have been introduced to the more elementary ideas of modern mathematics, and they will be able to omit a number of the sections without handicap. For the sake of those to whom such ideas are new, and for the sake of completeness, a sufficiently thorough treatment of all the concepts and vocabulary which form the basis of the calculus to follow has been included.

The introduction of graphical work follows the recommendations of the 1949 M.A. Report on the Teaching of Calculus and the M.A. Report "Analysis Course I" (1957), and the notions that a graph is a pictorial representation of a functional relationship and that its geometrical properties can be put to good use have been preceded by a thorough analysis of time-motion graphs. The ideas of continuity are dealt with fully, the way having been prepared by describing each point in a graph as an abbreviation for the mapping between a pair of related variable values and by considering in some detail the continuum of the natural numbers. As recommended by the 1962 Report on the Modernisation of Mathematical Teaching in the Netherlands (written for the International Commission for Mathematical Instruction's Sub-Committee for the Netherlands), the introduction of continuity precedes that of limits, and is dealt with in terms of mapping neighbourhoods. A number of examples included in the exercises later in the book remind the reader of the peculiar properties of certain functions, and, it is hoped, will develop some sense of the necessity for rigour and precise thinking.

The second part deals with some of the ideas of the Differential Calculus: adequate definition and discussion of the derivative are made possible by the treatment of continuity and limits just described. As also in the third part, and in sympathy with the British school of thought, opportunity has been taken to include a number of examples taken from the fields of applied mathematics; when these demand some special knowledge, they are marked with an asterisk. It is hoped that the readers will be constantly informed and reminded of the relevance and importance of the calculus to everyday life; the inclusion of such applications and the need for rigour should not necessarily be mutually exclusive. In problems involving physical units, the recommendations of the British Standards Institution have been borne in mind.

Rolle's Theorem and the First Mean Value Theorem—with the treatment of turning-points that can be developed from them—have been omitted, the intention being that these would form a suitable basis for a brief revision of this first-year course at the start of the second year. There are probably enough new ideas in this part for a first reading as it is.

The third part completes the volume with a treatment of the Integral Calculus. The notion of an integral is introduced carefully through the ideas of the least value and greatest value step functions and the area on a unity scale graph—a line of exposition which is a development of that described by the American writer Apostol. The image of the integral thus developed should prove an advantage rather than a handicap, and there are many examples included in the exercises which make deliberate reference to the physical interpretations which can be put upon the integral. If any weaning from these associations is thought necessary or even desirable as the book moves towards its closing stages, it should not prove difficult, for there is a full quota of the usual abstract or analytical examples.

A modern textbook would not be complete without some reference to the numerical methods which have been developed for dealing with functions which have no expressible or easily manipulated algebraic form, and are currently of great importance in translating the calculus into a suitable form for computers. For this reason, an account of Weierstrass's theorem and the Gregory–Newton formula for interpolation is given in Part 1, their relevance to numerical differentiation shown in Part 2, and their application to numerical integration presented in Part 3. It is hoped that readers will cover the numerical sections just as thoroughly as they cover the more abstract ones, and thereby gain a better understanding of the relationships which exist between the two sides of the calculus coin.

My thanks are due to Mr. W. R. Gordon of the Mathematics Department of the University of California and Dr. E. A. Maxwell of Queen's College, Cambridge, for reading the whole of the text and making many valuable suggestions and corrections; to Mr. D. A. Quadling of Marlborough College for finding the time and energy to read and comment on some aspects of the first part of the book; to a number of Examination Boards who gave permission to reprint questions set in their G.C.E. papers; to Mrs. P. A. Tupman, who, with great care and efficiency, typed the reams of hieroglyphs necessary in a text of this sort; to such reforming bodies as S.M.P. who have encouraged us all to rethink our mathematics; and to the publishers, and Mr. M. S. Gale, who were of the greatest assistance in preparing the drawings and the material for publication, and who helped to put the units in the text and exercises into metric form.

S. W. HOCKEY.

Acknowledgements

A LARGE number of examples are presented in the exercises at the end of the chapter sections; a number of these demand some special knowledge from fields outside the realms of pure mathematics, and where this is so, the questions are marked by an asterisk (*). Some questions are rather harder than the rest; these are prefixed with a dagger (†). A number of questions are taken from the G.C.E. papers set by the various examining bodies; permission to print these is gratefully acknowledged, and the questions are marked individually according to the following key:

(OC) Oxford and Cambridge Schools Examination Board.
(C) University of Cambridge Local Examinations Syndicate.
(NUJMB) Northern Universities Joint Matriculation Board.
(SU) Southern Universities Joint Board for School Examinations.
(L) University of London.

At the end of each chapter, a number of questions are provided for the reader who would like further practice on the topics included in the chapter; these examples are collected together in the "Miscellaneous" exercises. At the end of each part of the book, a number of questions of a rather harder and more extending nature are given in further "Miscellaneous" exercises.

Answers to most of the examples are printed at the end of the book.

PART 1

IMPORTANT FUNDAMENTALS

CHAPTER 1

A Basic Vocabulary

1.1. SETS, MAPS AND RELATIONS

In mathematical usage the word *set* is taken to mean a collection. The collection might be composed of objects, or ideas, or words, or a mixture of these, and at the simplest level a set is defined by specifying its constituent *members* (or, as they are often called, *elements*). It is understood that all the members of any set are individually distinguishable, and, in many cases, they will have at least something in common. Thus a herd of cows could be regarded as a set, the elements being the individual cows: or, to take a more abstract example, the names of the positions occupied by the players in a football team could define another (see Fig. 1.1).

When the elements of a set possess a common quality or feature, the set can often be defined more easily by describing the feature or features which qualify the elements for inclusion in the set. It is useful to enclose the members or description of a set in the bracket pair { }; thus the expressions {spring, summer, autumn, winter} and {seasons of the year} symbolise and define the same set in alternative ways. The order in which the elements are listed does not matter.

If we select some or all of the members of a given set to form a second set, the latter is said to be a *subset* of the former, and if, further, it contains only some of the members of the first set, it is said to be a *proper subset* of the former (or *parent*). Thus a set whose members are some or all of the positions on a football field is a subset of one whose members are all the positions on a football field:

Goalkeeper
Left back
Right back
Left half
Centre half
Right half
Left wing
Inside left
Centre forward
Inside right
Right wing

FIG. 1.1. A set is a collection of distinguishable elements.

and a set containing only the forward positions is a proper subset of that containing all the positions (Fig. 1.2).

For the sake of abbreviation, it is usual to label sets to which reference has to be made with capital letters; and the corresponding small letters are used to represent any individual member of the set. Thus x would be understood to represent a member of the set X; and a further symbol $\in$ is used to stand for the words "is a member of the set". Thus $p \in P$ is to be read as "p is a member of the set P".

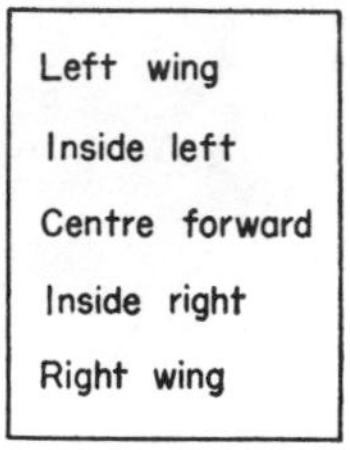

Left wing
Inside left
Centre forward
Inside right
Right wing

FIG. 1.2. A set consisting of the forward positions only is said to be a proper subset of that containing all the positions.

To illustrate the relationships which can exist between pairs of sets, let us consider the case of a car firm who have decided to produce three different models. One feature distinguishing the models is the capacity of their engine cylinders, and we will suppose that the firm have decided to make the three cylinder capacities 600 cm³, 1200 cm³ and 3200 cm³. The models are to be given different names, and after some research, the names Cyclops, Dandiprat and Gnat are chosen. The final problem is to decide which name to allocate to which model.

One way of indicating which name is to be associated with which cylinder capacity is to "pair off" the capacities with their corresponding names. Thus (600, Gnat), (1200, Dandiprat) and (3200, Cyclops) indicates quite clearly which name is to go with which model. Written in an ordered fashion such as this (where the first member of every pair is a cylinder capacity and the second a name), the three brackets and their contents can be referred to as a set of *ordered pairs*.

The set containing all the possible combinations of ordered pair elements whose first members are drawn from a set X and whose second members are drawn from a set Y is referred to as the *Cartesian product* of the sets X and Y, and the symbol denoting this set is written $X.Y$. For the example of the car models, the Cartesian product of the set X containing the numbers representing the cylinder

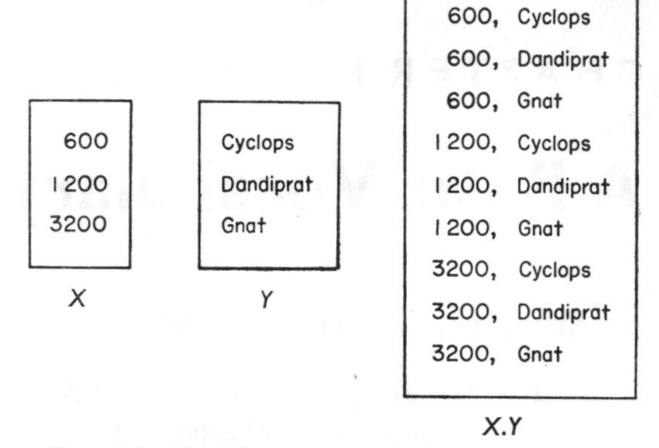

FIG. 1.3. The Cartesian product of two sets X and Y is denoted by the symbol $X.Y$ and contains all the possible combinations of ordered pair elements whose first members are drawn from X and whose second members are drawn from Y.

capacities and the set Y containing the three names is shown in Fig. 1.3. The set of ordered pairs indicating which name is to go with which cylinder capacity is a proper subset of this Cartesian product and is shown in Fig. 1.4.

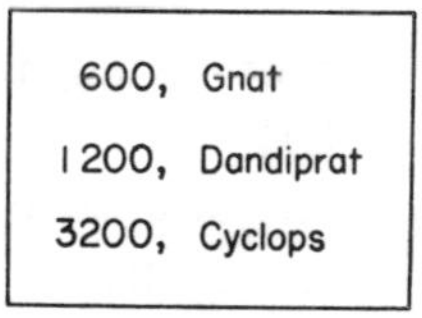

600, Gnat
1200, Dandiprat
3200, Cyclops

FIG. 1.4. A proper subset of a Cartesian product is called a relation.

For any two sets X and Y, a proper subset of the Cartesian product $X.Y$ is referred to as a *relation* from the set X to the set Y. Thus in the example just quoted, the set shown in Fig. 1.4 is a relation from the cylinder capacity set to the names set.

The set composed of the first elements of a set of ordered pairs forming a relation is referred to as the *domain* of the relation, and that composed of the second elements is referred to as the *range*. Thus together the domain and the range form the set of ordered pairs constituting the relation, and it should be noted that for a relation to exist between a pair of sets X and Y, there need not be a corresponding y for every x, but the domain of the relation must be built up of members of X which have corresponding members of Y. Similarly, the range of the relation must be built up of members of Y which have corresponding members of X. We can therefore say that the domain and range of a relationship from X to Y can be either proper subsets or just subsets of X and Y. It should be remembered that the Cartesian product $X.Y$ is not itself a relation; there must be some limitation on x or y to give a relation, since one or more members of $X.Y$ must be excluded to obtain a proper subset. Thus the set $X.Y$ in Fig. 1.3 is not a relation between the sets X and Y; the set shown in Fig. 1.4 is, and in this case the domain and the range of the relation are just subsets of X and Y. For the relation depicted in Fig. 1.5, both the domain and the range are proper subsets of the sets between which the relation exists.

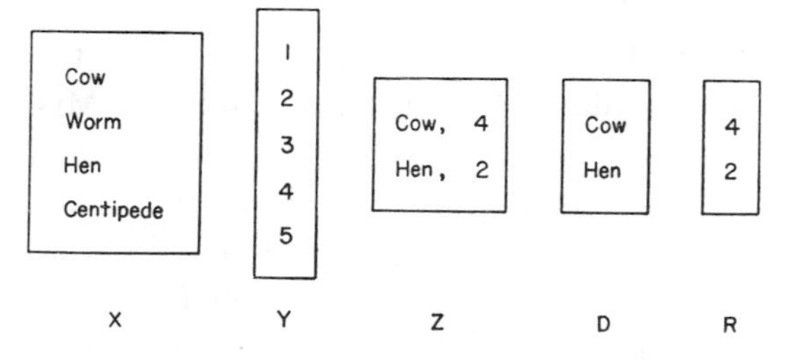

FIG. 1.5. The set Z defines the relation between the set X and the set Y. The domain of the relation Z is D; the range R. Note that in this case both D and R are proper subsets of X and Y respectively.

There are other ways of representing an association between the members of two sets: one is to link the corresponding members of the two sets by arrows (Fig. 1.6). Such a procedure is termed *mapping*, and the arrows are always seen to proceed from the members of the set containing the domain to the corresponding

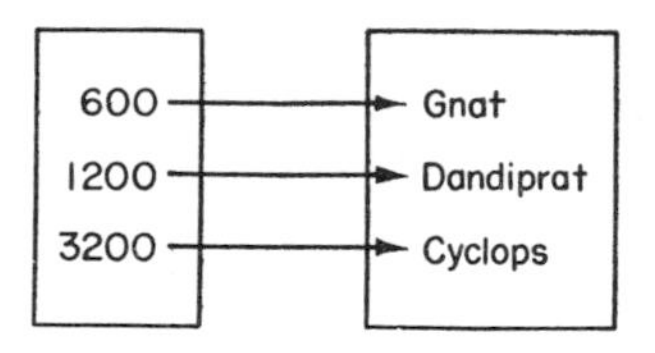

FIG. 1.6. A relation between two sets can also be depicted by arrows: this method can be described as "mapping".

images in the set containing the range (i.e. from the first members of the ordered pairs constituting the relation to the second members). The mapping between the sets X and Y is said to be from X "*onto*" Y if the range is the same as Y (i.e. if there is a link from an x for every y): if the range is a proper subset of Y (when some of the members of Y will have no associated member of X), the mapping is said to be from X "*into*" Y (Fig. 1.7).

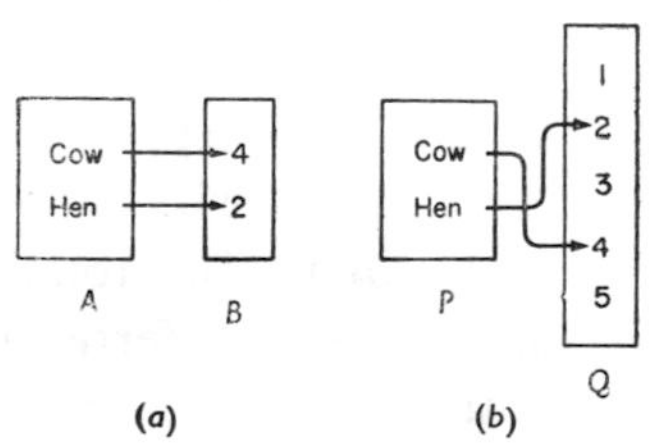

FIG. 1.7. The mapping in (a) is said to be from A "onto" B; in (b) it is said to be from P "into" Q. In this latter case, the range of the relation is a proper subset of Q.

A third way of representing a relation between two sets employs two perpendicular lines (called *axes*). Points are chosen in the horizontal line to represent the members of the domain, and points in the vertical line to represent the members of the range. A pair of axes suitable for the relation of Fig. 1.4 is shown in Fig. 1.8c. For the sake of neatness, the points are spaced evenly along the axes, and the mapping of Fig. 1.6 can be summarised by placing three points in the area between the axes as shown in Fig. 1.8d: each point is understood to map the domain element represented by the axial point vertically beneath it onto the range element represented by the point horizontally opposite it. Thus the point P maps 600 onto Gnat. The totality of all these "mapping" points is referred to as the *graph* of the relation.

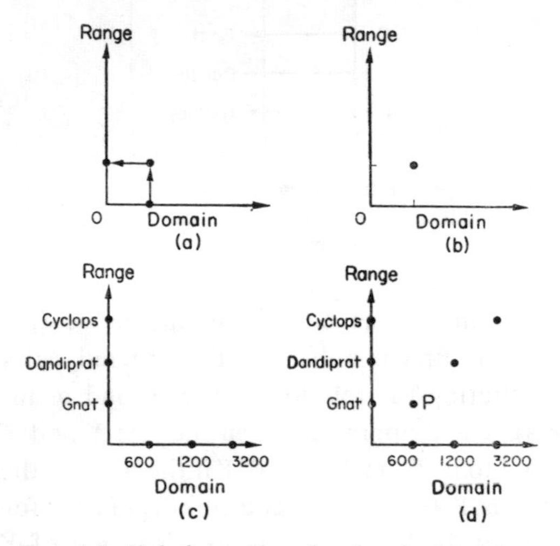

FIG. 1.8. Relations can also be depicted by a "graphical" map: the mapping represented by any point proceeds from the horizontal axis to the vertical axis (a), and the mapping arrows are—as it were—shrunk to a single point (b). The mapping of Fig. 1.6 is summarised in (d).

Example 1.1

Mr. and Mrs. Apathy were asked by Dr. Gallup whether they voted in the last Election or not, and if they did, whether they voted for the Conservative, Liberal or Labour candidate. Mr. Apathy said he voted for the Liberal candidate, and Mrs. Apathy said she voted for the Conservative candidate.

Write down a set N containing the names (separately) of Mr. and Mrs. Apathy, a second set V containing the four possible answers (Conservative, Liberal, Labour and "Didn't vote"), and represent the interest shown in the Election by Mr. and Mrs. Apathy by drawing a mapping between the sets. Is the mapping from N into V or onto V?

Write down

(i) the two ordered pairs constituting the relation R from N to V,
(ii) the domain X of the relation,
(iii) the range Y of the relation,
(iv) the Cartesian product $X.Y$.

Represent the relation R and the Cartesian product $X.Y$ graphically, and state whether the domain X and range Y are subsets or proper subsets of N and V respectively.

Writing down the sets and drawing in the mapping as instructed, we obtain

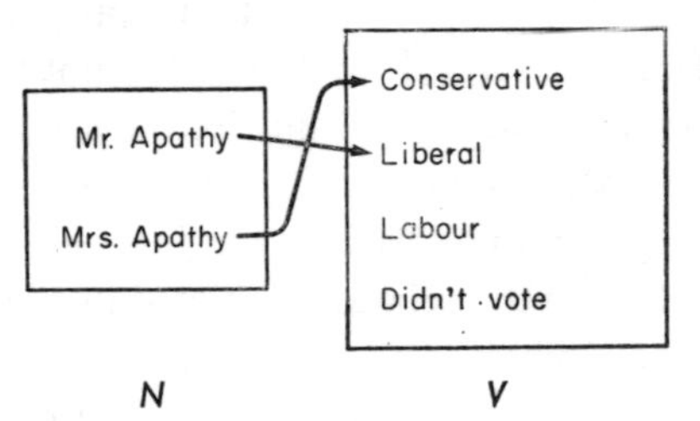

Since only two members of V are involved in the mapping, we can say that the mapping is from N into V.

(i) The relation R involves only the two ordered pairs (Mr. A, Liberal), (Mrs. A, Conservative). Note that the order is important: the ordered pairs (Liberal, Mr. A), (Conservative, Mrs. A) define the inverse relation R^{-1} which goes from V to N.
(ii) The domain X of the relation R must contain all the first elements of the ordered pairs. The set X is therefore {Mr. A, Mrs. A}.
(iii) The range Y must contain all the second elements of the ordered pairs constituting R, and Y is therefore {Liberal, Conservative}.

(iv) The Cartesian product $X.Y$ must contain all the ordered pairs which can be formed from the elements of X and Y, and is thus the set {(Mr. A, Liberal), (Mr. A, Conservative), (Mrs. A, Liberal), (Mrs. A, Conservative)}.

Choosing, by convention, the horizontal axis of our graph for the domain, we obtain the graphical representation of R shown in Fig. 1.9a, and that of $X.Y$ shown in Fig. 1.9b.

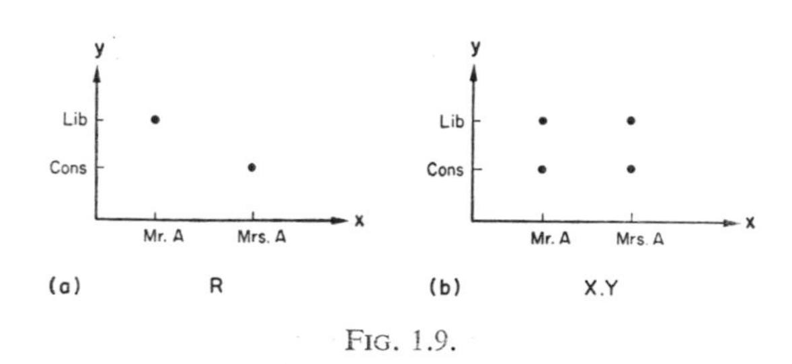

FIG. 1.9.

Since X contains both members of N it is a subset of N only: the range Y, however, contains only two of the four members of V, and it is therefore a proper subset of V.

The sense of order in a relation is very important. A relation R between two sets must always be directed from one set to the other, as must the Cartesian product of which it is a proper subset. Similarly the order in which the elements occur in the ordered pairs defining the relation must carry the elements from the first set of the relation in the first positions and the elements from the second set of the relation in the second positions. This then defines the domain as the subset or proper subset of the first set of the relation, and the range as the subset or proper subset of the second set of the relation.

It might be asked what happens if the sense of order between two related sets is reversed. Since a relation R from a set X to a set Y is a proper subset of the Cartesian product $X.Y$ (and therefore contains only some of the elements of the Cartesian product), this "reversed" relation will be a proper subset of the Cartesian product $Y.X$, and therefore another relation. This second relation is said to be the *inverse relation* of R, and can be conveniently represented by the symbol R^{-1} (Fig. 1.10).

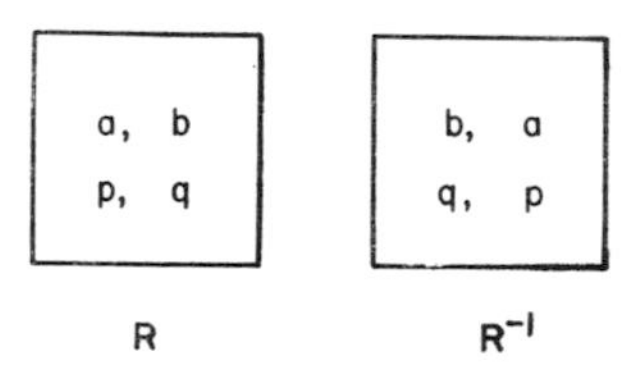

FIG. 1.10. R^{-1} is the inverse relation of R.

EXERCISE 1a

1. A certain school possessed four members of staff: Mr. Caneworthy (Headmaster), Mr. Pennygrab (Bursar), Mr. Chalkboard (the Senior—and only—Mathematics master) and Mr. Idlewild. Write down

(i) the set M containing the masters' names,
(ii) the set P containing the posts in the school,
(iii) the Cartesian product $M.P$,
(iv) the relation which exists between M and P.

Represent the relation of (iv) by mapping

(a) from M to P, (b) from P to M.

Which mapping is "into" the second set? Represent the mappings you have just drawn by a graph.

2. Three boys A, B and C take an examination and are given grades 1, 3 and 5 respectively. Write down the set of ordered pairs expressing their results, placing

(a) the letters describing the boys,
(b) the grade numbers,

in the first positions.

If there are only five grades altogether, depict the relation by mapping from N (a set constructed from the letters designating the boys) to G (a set constructed from all the possible grade numbers). Is the mapping from N into or onto G? Write down the relation from N to G and also the Cartesian product $N.G$. State whether the relation is a subset or a proper subset of $N.G$, and also whether the domain and range of the relation are subsets or proper subsets of N and G respectively.

3. Explain how the Cartesian product $A.B$ differs from the Cartesian product $B.A$.

4. Referring to the sets shown in Fig. 1.11, write down the letter of the set (i) $A.B$, (ii) which is a relation from A to B, (iii) which is a relation from B to A.

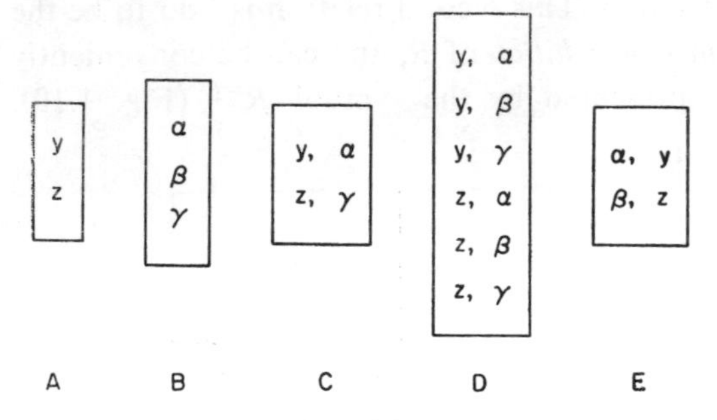

FIG. 1.11.

Is (a) C a subset or proper subset of D?
(b) E a subset or proper subset of D?
(c) the domain of C a subset or proper subset of A or B?
(d) the domain of E a subset or proper subset of A or B?
(e) the range of E a subset or proper subset of D?
(f) the range of E a subset or proper subset of A?

5. Add two members to the sets Y and Z of Fig. 1.5, so that the domain of the relation from X to Y contains all the members of the set X. Draw a mapping of the revised relation, and state whether the mapping from X to Y is "into" or "onto". What is the image of Hen?

6. Explain the meaning of the terms set, subset, proper subset, Cartesian product, relation, range, image.

7. A relationship exists between the sets X and Y depicted in Fig. 1.12. One mapping link is drawn. Deduce the nature of the relationship, complete the mapping, and state which set is the domain of the relation. Is the mapping from X into or onto Y?

8. Write down the set D containing the seven days of the week, and to the right of it another set N containing the ten numbers 0 to 9, inclusive. Draw in a set of mapping arrows to show how many school periods you have to attend on each of the days.

Is the mapping from D into or onto N?

Explain why the mapping you have just drawn represents a relation, and write down the domain and range of the relation. Is either a proper subset of its parent?

9. Represent the mapping you drew in question 8 by a graph.

10. Write down the set S containing all the subjects you study at school and the set N containing all the numbers from 1 to 9, inclusive. Map the subject you like best onto 1, the subject you like next best onto 2, and so on until all the members of S are linked to a member of N.

What is the image of mathematics?

Represent the relation existing from S to N by a graph, and state whether the range of the relation is a subset or a proper subset of N.

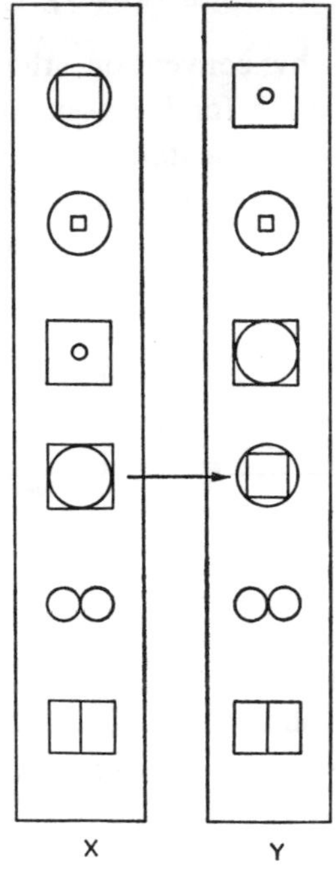

FIG. 1.12.

11. A relation exists between two sets X and Y. Which of the following statements are true?

(i) The relation is a set.
(ii) The relation is a set of ordered pairs.
(iii) The domain of the relation need not contain all the members of X.
(iv) The relation contains all the members of the Cartesian product $X.Y$.
(v) The Cartesian product $X.Y$ contains all the members of the relation.

12. Describe the appearance of a graph of a Cartesian product.

13. Set E consists of all the positive even numbers: set O consists of all the positive odd numbers. Which of the following statements are true?

(i) $2 \in E$ (v) $(7, 4) \in E.O$
(ii) $2 \in O$ (vi) $(7, 4) \in O.E$
(iii) $(2, 4) \in E.O$ (vii) $(6, 8) \in E.E$
(iv) $(2, 5) \in E.O$ (viii) $(12, 7) \in O.E$

†**14.** The sets P, Q and R have l, m and n members, respectively. How many members have the sets (i) $P.Q$,

(ii) $Q.R$? Are the sets $P.Y$ and $Z.R$ the same if Y and Z denote $Q.R$ and $P.O$, respectively? How many members have these sets?

†**15.** Can the set X, built up of n' members (p, q), be described as a relation if

(i) $n' < lm$, (ii) $n' \leqslant lm$, (iii) $n' = lm$,

given that the ordered pairs (p, q) are all members of the Cartesian product $P.Q$, where P has l members and Q has m?

1.2. FUNCTIONAL RELATIONS

The relation discussed on p. 4 is a particular type of relation: each member of the domain is linked with only one member of the range, and each member of the range is linked with only one member of the domain. Such a relation is called a "*one to one*" relation. If we now suppose that the car firm decides to make a more luxurious model of the Gnat, but to keep the engine size the same, we can revise the relation and include the new model—the Gadfly—in one of the following two forms:

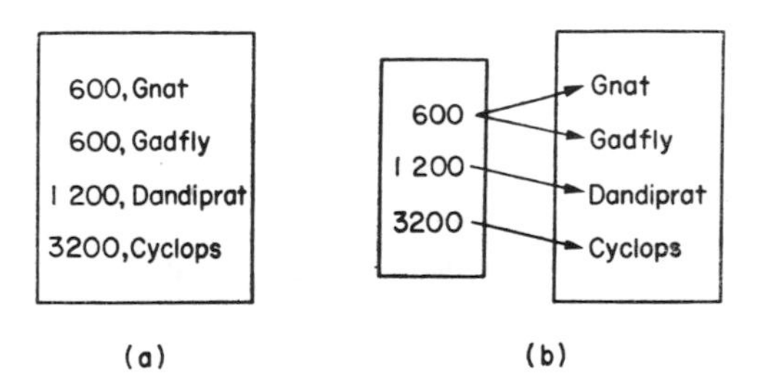

FIG. 1.13. A one to many relation.

Now two of the elements of the set of ordered pairs or relation of Fig. 1.13a have the same first member—viz. 600 in (600, Gnat) and (600, Gadfly). Such a relation is an example of a "*one to many*" relation: as seen again in Fig. 1.13b, in such a relation, one or more of the domain members can have "many" (i.e. two or more) range members or images associated with them, but each range member must be associated with only one domain member.

The order in which we have thought of the relation between the cylinder capacity and names' set can be easily reversed. That is to say, the relation can be rewritten in the form of Fig. 1.14, and although the actual ordered

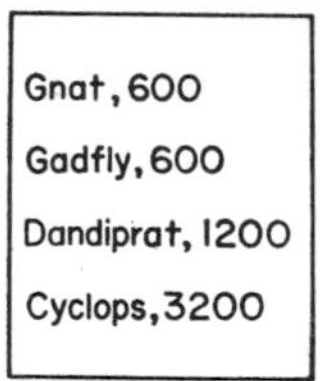

FIG. 1.14.

pairs are the same, the ordering of their component elements has been reversed. We can redraw the mapping as in Fig. 1.15, and re-

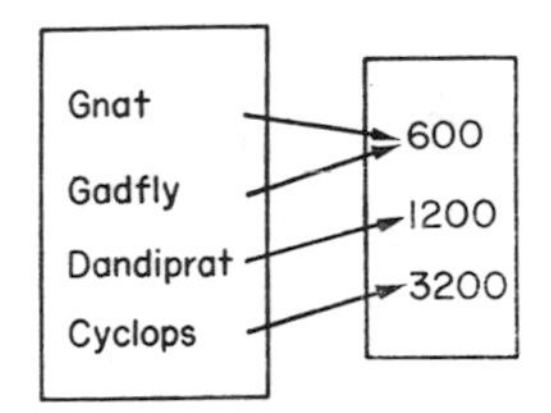

FIG. 1.15. A many to one relation, or "function".

ordering the elements in this way is equivalent to interchanging the domain and range sets: the names' set is now the domain, the cylinder capacity set the range. The relation is now a "*many to one*" relation: that is, each domain member is associated with only one range member, but some or all of the range members may each be associated with "many" (i.e. two or more) of the domain members. This type of relation is so common in practice that it is given a name of its own: it is referred to as a *function*.

The essential point about a function is that when the first element of any one of the ordered pairs is chosen, there is no ambiguity or choice about the associated second member. In our example, once the name of a model is chosen, the associated cylinder capacity is determined uniquely. It is the lack of this dition in the one to many relation which disqualifies it as a function; the half-filled pair (600,   ) can be the start of (600, Gnat) or (600, Gadfly), and we cannot tell which. Note that a one to one relation possesses this property of unambiguous association, and can therefore be classified as a function also.

A relation—defined as a set of ordered pairs—must have a sense of order about it, and so, therefore, must a function. Now if we reverse the ordering of the elements in a relation R, we always obtain another relation—the "inverse" relation R^{-1}—but if we reverse the ordering of the elements in a function, we may or may not obtain another function. Thus in

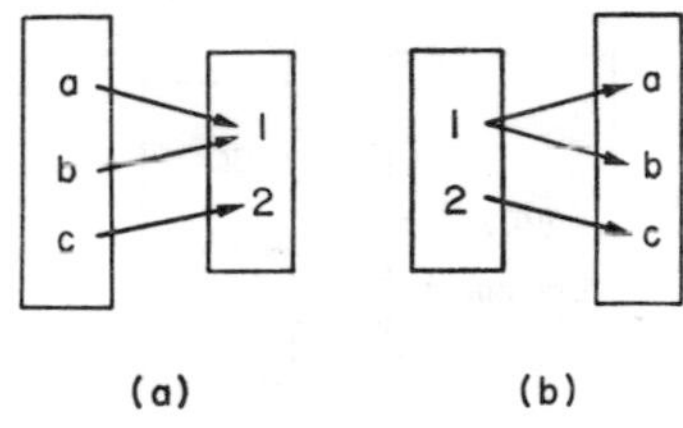

FIG. 1.16. Interchanging the elements in the ordered pairs of a function does not necessarily give another function.

Fig. 1.16a (where the functional relation is shown as a mapping), the reversal of the sense of the mapping does not lead to another function, for the map of 1.16b is that of a one to many relation. In Fig. 1.17a, however, the

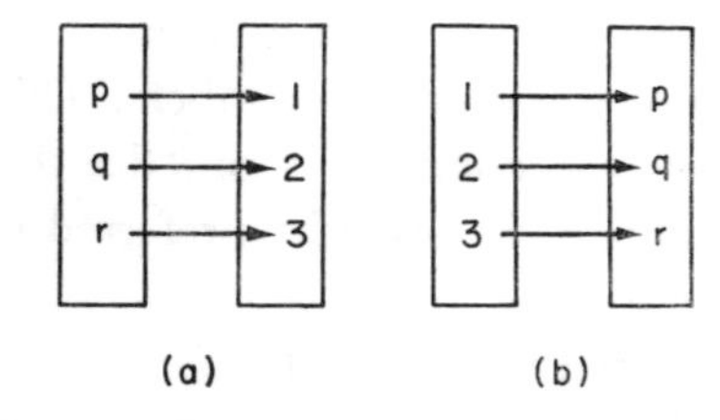

FIG. 1.17. An inverse function exists only if the relation between the sets forming the domain and range of a function is a one to one relation.

reversal of the mapping of the function leads to another one to one relation or function, and the functional relation of Fig. 1.17b is said to be the *inverse function* of f, and is denoted by the symbol f^{-1}. Thus a functional relation between two sets X and Y gives an inverse function if and only if it is a one to one relation. (It should be noted that f^{-1} does not represent $1/f$. The function symbol f is not an algebraic symbol of the usual kind, and $1/f$ has no meaning unless one is defined for it.)

There is one type of relation we have not yet mentioned. We have seen that if one or more of the domain elements is linked with more than one range element the relation can be described as one to many (Fig. 1.13b): if, in the same relation, one or more of the range elements is linked with more than one domain element, then it also possesses some many to one character. Such relations can be referred to as *many to many* relations (Fig. 1.18).

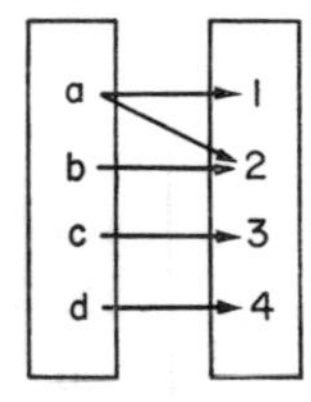

FIG. 1.18. A "many to many" relation.

Example 1.2

The languages studied at school can be classified as ancient or modern. Draw a mapping showing the classifications of Latin, French, German and Russian. State whether the relation R existing between the two sets is a one to many, many to one, many to many or one to one relation, and whether it defines a function F.

Into what category does the inverse relation R^{-1} fall? Is this relation functional?

The mapping showing the classification of the four languages is as follows:

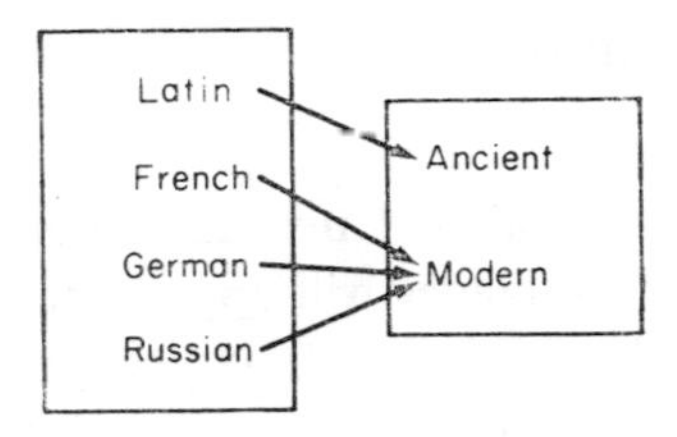

The relation existing between the two sets is a many to one relation, and can therefore define a function F (or, expressed another way, the relation is functional since each domain member is linked to only one range member).

The inverse relation R^{-1} can be visualised by thinking of the arrows in the reverse direction: clearly this is a one to many relation, since the domain (formerly range) element "Modern" is linked with more than one range element. Thus R^{-1} is not functional, and the inverse function F^{-1} of F does not exist.

Exercise 1b

1. The following sets are composed of numbers representing the mean air temperature in London for three different days in 1964 (the temperature being in °C), and the months in which the days occurred. Map the sets together, choosing the most likely mapping with

(a) set A as the domain,

(b) set B as the domain.

A	B
0	July
20	January
2	

Is the relation from A to B

(i) many to one?
(ii) one to many?
(iii) one to one?
(iv) functional?

Into which of these four categories does the relation from B to A fall?

2. Set A is composed of numbers representing the dates of various days in August 1988. What day of the week will August 4th, 1988, be? Map A on to B.

A	B
8	Tuesday
4	Wednesday
16	Thursday
17	Friday
5	Saturday
3	Sunday
2	
7	

Is the relation from A to B functional? Is there an inverse function?

3. Set B is the number of legs of various objects or animals. Map set A on to set B. Explain why the mapping represents a function, and draw the domain and range of the relation. Is there an inverse function?

A	B
Cow	2
Bird	3
Tripod	4
	5

4.

X	Y
6	12
5	10
4	8
3	6
2	4
1	2
0	0

The law enabling the mapping between the sets X and Y on the left is that each member of the set X is linked with the member of the set Y which has twice its numerical value. Copy the table, and draw in the mapping. Is the relation from X to Y a 1:1 relation? Is it functional? Is there an inverse function?

5. The numbers on a telephone occur in the same holes as groups of letters, as given by the table beneath:

1	2	3	4	5	6	7	8	9	0
	ABC	DEF	GHI	JKL	MN	PRS	TUV	WXY	OQ

Telephone numbers consisting of three letters and six digits are dialled by dialling nine digits, the three letters being replaced in succession by the digits appearing in the same hole of the dial. A set of three-letter elements and a set of three-digit elements is given beneath: copy the sets, and draw a mapping which depicts the association between the elements. Is the relation 1 : 1? Does the mapping represent a function if it is reversed in direction?

EUS	842
EAL	723
VIC	325
PAD	387
GER	437
SYD	793

6. The following table gives corresponding values of two related variables *A* and *B*.

A	8	17	24	56	89	95
B	2	8	49	81	83	83

Is the relation from (i) *A* to *B*
(ii) *B* to *A*
functional?

7. Describe the characteristic of the pattern which appears on the graph of a relation if the relation is

(i) many to one, (iii) one to one,
(ii) one to many, (iv) a function.

8. Redraw the graph of question 9 on p. 8, and state whether the relation from *D* to *N* is many to one, one to many or one to one. Is it also a function?

9. The graphs beneath represent relations existing between two sets. Which of the relations is

(i) many to one? (iv) many to many?
(ii) one to many? (v) functional?
(iii) one to one?

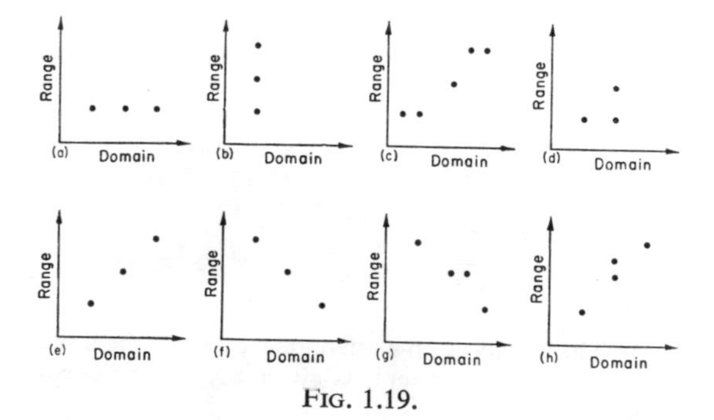

FIG. 1.19.

In which functional cases does an inverse function exist?

10. Is the mapping from a set whose elements are the dates of the month into or onto a set whose elements are the days?

1.3. NUMBERS

The elements forming the sets involved in this study are almost invariably numbers. In order that the reader will appreciate the later ideas in the book, a short account of the number systems in common use is given below.

Two sets which form a one to one relation with each other are said to have the same *cardinal* number. A set consisting of single objects only belongs to the whole family of sets containing single objects. Each can form a 1 : 1 relation with any other set in the family, and they all have the cardinal number we call *one*. Sets of triples belong to the family whose cardinal number we call *three*. From such ideas, we can construct a system of *counting* numbers; and if we count "one, two, three..." while laying out a number of objects, the number called out with the last object laid down will be the cardinal number of the set containing all the objects with which we started.

To represent these numbers, we employ a set of symbols. The symbols for the counting numbers and zero (i.e. 0, 1, 2, 3...) form a scheme for representing the so-called *natural* numbers. And for these numbers, we can construct definitions of the binary operations of addition and multiplication: thus, addition can be defined by the mapping of which the following are typical examples:

$$(1, 3) \xrightarrow{+} 4 \qquad\qquad 1+3 = 4$$
$$(2, 3) \xrightarrow{+} 5 \qquad \text{or} \qquad 2+3 = 5$$
$$(7, 4) \xrightarrow{+} 11 \qquad\qquad 7+4 = 11,$$

and multiplication by the following typical examples:

$$(1, 3) \xrightarrow{\times} 3 \qquad\qquad 1\times 3 = 3$$
$$(2, 3) \xrightarrow{\times} 6 \qquad \text{or} \qquad 2\times 3 = 6$$
$$(7, 4) \xrightarrow{\times} 28 \qquad\qquad 7\times 4 = 28.$$

From this natural number system, we can progress to the system of integers, defined in terms of the difference operation, represented by the symbol $\sim$. The difference $(a \sim b)$ between any two natural numbers a and b can be seen as a member of the family consisting of those differences $(u \sim v)$ for which $a+v = u+b$. These families of differences are called *integers*, and every difference $a \sim b$ belongs to only one family or integer. It is convenient to denote the family or integer of which $(a \sim b)$ is a member by the symbol $I(a \sim b)$; the symbol $I(p \sim q)$ represents the same integer if $a+q = p+b$. Now the integers whose second members are 0 have—to all intents and purposes—the same properties as the natural numbers, and the natural numbers can thus be used in place of this special class of integer—the so-called *positive integers*. Thus $I(0 \sim 0)$ can be replaced by 0, $I(2 \sim 0)$ by 2, and so on, and it is clearly more convenient to use the natural number symbols than the full integer form to represent them. The laws of addition and multiplication follow the laws of the natural numbers.

The integers whose first members are 0 are called *negative integers:* the abbreviated symbol in common use for $I(0 \sim a)$ is $-a$, and it is a simple matter to show that the sum of $-a$ and the "corresponding" positive integer a is 0. Inasmuch as any integer $I(p \sim q)$ can be written in one of the forms $I(x \sim 0)$ or $I(0 \sim x)$, we can allocate any integer to one of the two classes of positive and negative integers. $I(0 \sim 0)$ can be regarded as belonging to both.

The operation of division requires an extension of our number system: and the extension is made by way of defining a set of *rational* numbers. The quotient a/b is a member of the family consisting of all those quotients u/v for which $av = bu$: this family or rational number can be represented by the symbol $R(a/b)$, and this number is, of course, the same as $R(p/q)$ if $aq = pb$. For the sake of abbreviation it is usual to write a typical member a/b of the family $R(a/b)$ as representing the family.

Some of the rational numbers can be expressed in the form $R(a/1)$, where a is a natural number: the elements of this proper subset of the rational numbers again have the same properties as the natural numbers, and so we can represent a rational number of the form $R(a/1)$ by the symbol a. It also follows that the integers have the same properties as the members of this proper subset.

We can write that, if the rational number x/y is equivalent to p/q, the product of y and p/q is equal to x. Now if y is zero, any other form p/q of $x/0$ must be such that $0\,(p/q) = x$. It is an essential of any number system that the product of the zero element and any other member gives 0: and if this is to be so, the equality $0 \times p/q = x$ cannot be satisfied by any member p/q of the rational number system. Thus $x/0$ is sensibly excluded from the rational number system.

$0/0$ is also to be excluded: if $0/0$ has an alternative form p/q, then $0 = 0 \times (p/q)$. This time p/q can be *any* rational number or family of numbers, and this lack of identity or individuality leads to the exclusion of $0/0$ from the rational number system.

We can turn now to the graphical method of representing rational numbers by points in a line or axis. An example of the use of points in a line to represent the elements of a set has already been given: a similar approach enables us to extend our ideas to numbers. To do this we proceed as follows.

Draw a line, and choose any point in it to represent 0. Label this point P (Fig. 1.20), and choose another point to the right of it (say Q)

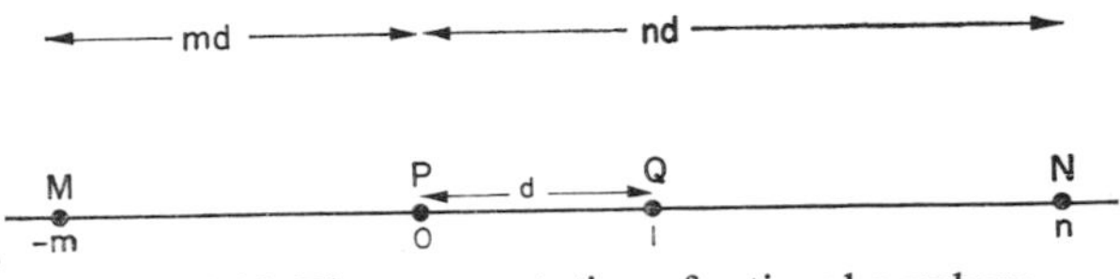

FIG. 1.20. The representation of rational numbers by points on the line.

to represent 1. Let the interval PQ be of distance d. Now let the point N situated at a distance $(n \times d)$ to the right of P represent the number n; and that point M at a distance $(m \times d)$ to the left of P the number $-m$ (where n and m are rational numbers). The following observations then apply:

1. each point in the line represents only one rational number (thought of as a family);
2. the points form a one to one relation with the rational numbers,* and by defining the operations of addition and multiplication geometrically, the points can be made to have the same properties as rational numbers;
3. the point representing the larger of the two numbers lies to the right of that representing the smaller. This is a direct consequence of the definition of "larger", where c is said to be larger than d if $\{c+(-d)\}$ is positive. Thinking of the addition of two numbers c and $-d$ as the process of measuring the distance PD' in the direction of $\overrightarrow{PD'}$ from C, where C and D' represent the numbers c and $-d$, respectively (Fig. 1.21), we can

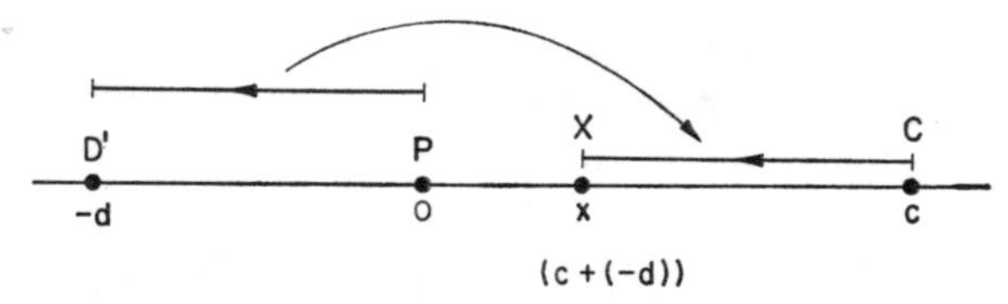

FIG. 1.21. The addition of c and $(-d)$ gives x, where C, D' and X represent c, $-d$ and x, respectively, and where $CX = PD'$.

see that the end of the measured distance (X) will lie to the right of P (and therefore represent a positive number) only if the length of PD' is less than the length of CP. Now if the point D represents the rational number d, the lengths of PD and PD' will be equal (Fig. 1.22). Thus

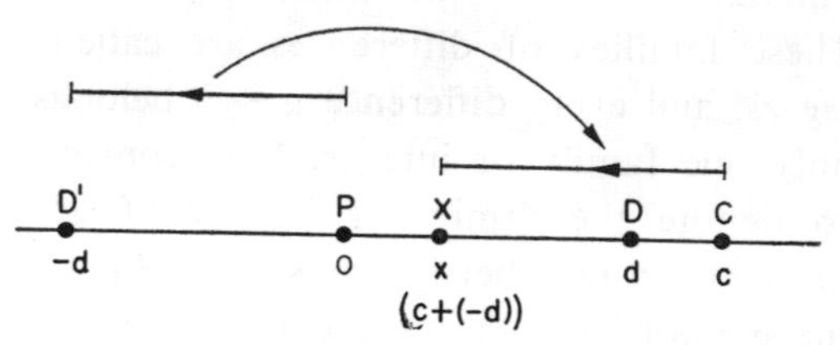

FIG. 1.22. If D represents d, $PD' = PD$, and $\{c+(-d)\}$ is positive only if C lies to the right of D.

X will lie to the right of P only if PD is less than PC, and for this to be so, C must lie further from P than D. So if c is greater than d, C must lie to the right of D.

For a negative number $-d$, we say that c is greater than $-d$ if $(c+d)$ is positive. For two negative numbers $-c$ and $-d$, we say that $-c$ is greater than $-d$ if $(-c+d)$ is positive. A little reflection will show that the rule that the greater number is represented by the point lying to the right holds for all three contingencies (Fig. 1.23). These principles are of fundamental importance, and although at first sight

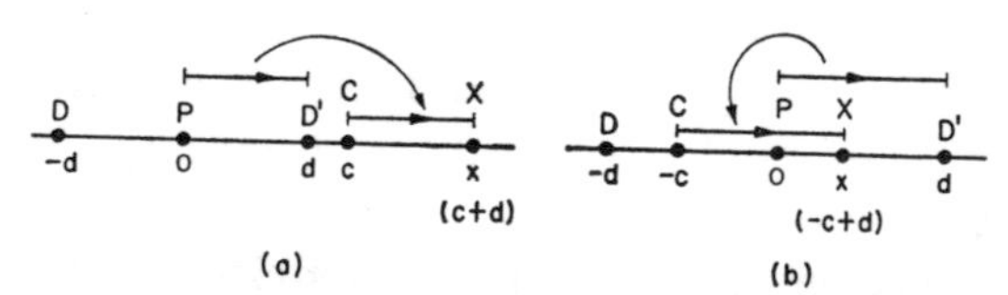

FIG. 1.23. (a) The number $-d$ is less than c if $(c+d)$ is positive. The point representing the sum $(c+d)$ is found by measuring a length equal to $\overrightarrow{PD'}$ from C, and, if c is positive, the end of this length will always be to the right of P and therefore represent a positive sum. (b) The number $-d$ is less than $-c$ if $(-c+d)$ is positive. Measure off $\overrightarrow{PD'}$ from C (which now lies to the left of P). The point X at the end of this measured length lies to the right of P only if C lies to the right of D.

* Or, as it is often expressed, the points and the rational numbers can be put into a 1:1 correspondence.

the statement that -3 is greater than -4 looks odd, it is a necessary part of the structure we have built up for our number system. The symbol for "is greater than" is $>$; and the complementary statement that c is less than d (defined as the equivalent of $d > c$) can be represented by the symbolism $c < d$. The graphical expression of this latter inequality is that the point C representing c lies to the left of D, where D represents d.

The set of rational numbers is unbounded, and the points representing the elements of the set are very close together in the line—so close, in fact, that if any two are chosen, another can be fitted between. (For example, if we consider the rational numbers 21/27 and 22/27, the point representing 43/54 will lie half-way between the points representing 21/27 and 22/27; and considering the points representing 43/54 and 22/27, the point representing 87/108 will lie half-way between these two, and so on). Surprisingly, there are more points in the line than can be accounted for by all the rational numbers, and the mapping of the rational numbers is "into" rather than "onto" the points in the line.

Consider, for example, the set of points representing the rational numbers x for which $x^2 > 2$, and the set representing the rational numbers x for which $x^2 < 2$ (Fig. 1.24). The

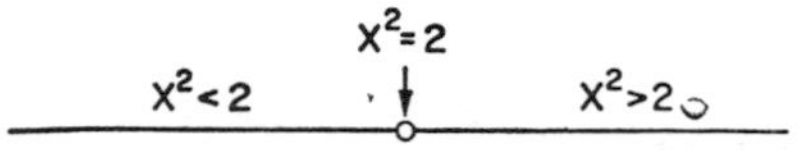

FIG. 1.24. The point dividing the points representing the numbers x for which $x^2 > 2$ from those representing the numbers x for which $x^2 < 2$ does not represent a rational number.

point dividing these two sets does not represent a rational number—for if it did, the number it represented could be expressed in the form p/q. If we take the quotient p/q in its lowest terms, with p and q both integers, we can write that $p^2/q^2 = 2$, since the point representing p/q divides the sets $x^2 > 2$, and $x^2 < 2$. Thus, $p^2 = 2q^2$, and this equation demands that p must be an even integer. Now if p is an even integer, it can be written in the form $2r$; and thus $p^2 = 4r^2$. Substituting this expression for p^2 in the equation $p^2 = 2q^2$, we obtain $4r^2 = 2q^2$. Thus, $q^2 = 2r^2$, and q must also be an even integer. But if p and q are both even, p/q will not be in its lowest terms, for both numerator and denominator can be divided by 2. This *reductio ad absurdum* tells us that the number whose square equals 2 cannot be written in rational form. Numbers which cannot be written in a rational form are termed *irrational*.

We have now completed the account of the number systems in common use, and our main conclusion allows us to employ each point in a line or axis to represent a number which is either rational or irrational. (The collection of all the rational numbers and all the irrational numbers is referred to as the "real" numbers system.) Each point represents only one number, and for any two points, that representing the larger number lies to the right of that representing the smaller. In such a manner we can represent graphically all the members of any set composed entirely of real numbers: and each number will be distinguishable as a point.

Two digressions are worth making. The first concerns the representation of rational numbers by decimals. If we wish to express any rational number by a decimal, one of three things will happen:

(i) the decimal terminates (e.g. 7, 0·25). This group contains all the integers and some rational numbers.

(ii) the decimal does not terminate but recurs (e.g. $0{\cdot}33\dot{3}\ldots$, $0{\cdot}0101\dot{0}\dot{1}\ldots$). This group contains further rational numbers, but no irrational numbers. (This can be shown by supposing, for example, that $N = .abc\, abc\, abc\ldots$, whence $1000N = abc.abc\,abc\,abc\ldots$. Subtracting, $999N = abc$, and thus $N = abc/999$. The argument can be extended to more complex recurrences.)

(iii) The decimal neither terminates nor recurs (e.g. 0·7070070007...). This group contains all the irrational numbers.

The second digression concerns the meaning of the word *infinite*. If we have a set containing all the real numbers, we could clearly never finish listing them, or, for that matter, counting them, and when there are as many elements or items as this, their number is said to be infinite. Similarly the representation of an irrational number demands a decimal so long that it could never be completed in writing (as we have just seen); the decimal has an infinite number of significant figures or places.

Mention of "infinite" provides a good occasion to discuss a misuse of the word "infinity" (symbol ∞). The term commonly arises when thinking of the value of a fraction of the form $1/x$ when x is small. If x is small, $1/x$ will be large, and if x grows smaller, then $1/x$ grows larger. As x approaches the value 0, $1/x$ becomes very large—even unmanageably large. We can say that it becomes infinite—meaning that it increases without bound, rather than being impossible to write down (although this is an unpleasant enough task!). But to follow this with the statement that "when x equals 0, $1/x$ equals infinity" is to descend into gibberish. There is no such number as "infinity" (or place for that matter), and even if there was, the fraction 1/0 has no meaning, as we have already seen. It is not logically possible to divide by zero.

We return to the main discussion of the use of points in a line to represent numbers by recalling the graphical representation of related variables described in Section 1.1. When all the elements of two related sets are numbers, it is necessary to use two axes or lines whose points can be used to represent the numbers occurring in the two sets, and the convention is to use the points above that representing 0 in the vertical (range) axis for the positive numbers (Fig. 1.25); and points below the zero to represent negative numbers. Our deductions about inequalities in the horizontal axis can then be translated into the following consequences for the vertical axis:

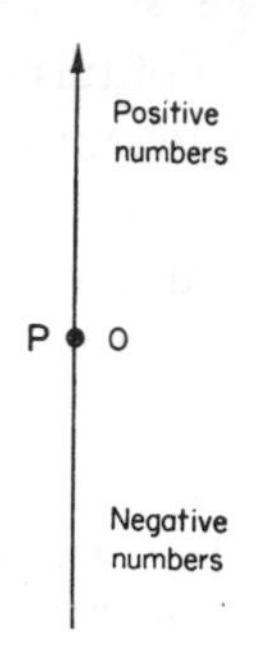

FIG. 1.25. In a vertical axis the points above the zero point P represent positive numbers, and the points below P represent negative numbers. This leads to the consequence that the larger of two numbers is represented by the point further up the axis.

(i) if $a > b$, the point A representing a is further up the axis than the point B representing b;

(ii) if $a < b$, A is below B.

1.4. VARIABLES

Consider an experiment which consists of heating an iron bar from 0°C to 600°C. As the bar is heated, its temperature changes progressively, and we can construct a set from the numbers representing all the temperatures (in °C) it passes through. This set will, in fact, contain all the real numbers between 0 and 600. Now for each particular temperature the iron bar has a certain length: and as the temperature is increased, the length will increase, and we can determine how, for example, the length of the bar increases progressively from (say) 100·0 cm to 122·0 cm. The totality of the numbers between 100·0 and 122·0 can be put together to form a second set—the set whose elements are the numbers representing the lengths through which the iron bar passes. Denoting the two sets we have now defined by T and L, we can write that for any $t \in T$,* $0 \leqslant t \leqslant 600$, and for any $l \in L$, $100{\cdot}0 \leqslant l \leqslant 122{\cdot}0$. The quantities t and l are called *variables*; and a suitable definition of a variable is that it is a symbol which can represent any one of the members of a specified set.

We shall be mainly concerned with situations where the elements of the sets with which we are dealing are members of the real number system, and the term variable will therefore generally imply a symbol which can represent any one of a specified set of real numbers. If at any moment the variable represents the number N, it can be said to "take the value N". If the set associated with a variable contains the totality of (real) numbers between two boundaries, the variable is said to be *continuous* throughout the range of values between the boundaries: if the set contains every real number imaginable, the variable can be said to be continuous throughout the whole range of real numbers.

We can now introduce a slightly different use of the word function. Let us suppose we have a functional relation involving two sets X and Y, the domain being X and the range being Y. If we select any $x \in X$, the associated $y \in Y$ is uniquely chosen. Thus the value which the range variable y takes at any instant is governed strictly by the value chosen at that instant for the domain variable x, or, as it is commonly put, "y is a function of x". The phrase "y is a function of x" is conveniently symbolised in the form $y = f(x)$, and it should now be understood that this implies that if any one of the domain values is assigned to x, the associated range value y is determined without ambiguity.

The symbol $f(3)$ is understood to represent the member of y associated with the member of x whose value is 3. Thus, if the mapping in Fig. 1.26 represents the functional relationship

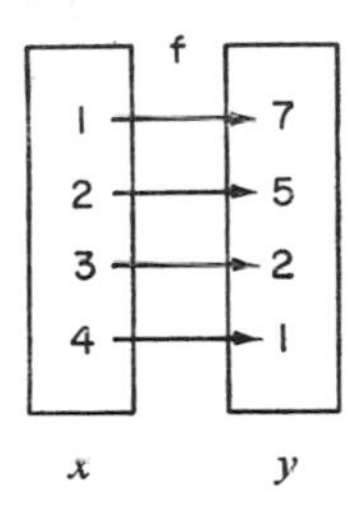

FIG. 1.26. The mapping of a functional relation f between two variables y and x is summarised by the equation $y = f(x)$.

f, we can see that $y = 2$ when $x = 3$, and thus $f(3) = 2$. Similarly if x takes the value 2, the associated value of y is 5, and we can write that $f(2) = 5$.

It is occasionally necessary to deal with two or more functional relations simultaneously. If the related pairs of variables are l and θ, and c and w, to write $l = f(\theta)$ and $c = f(w)$ implies that the number pairs defining the two functions are the same in both cases. For example, if $l = 3$ when $\theta = 5$, then it is suggested that $c = 3$ when $w = 5$. To avoid this implication, alternative abbreviations for "a function of" are used, and $g(w)$, $F(w)$, $G(w)$ and $\phi(w)$* are forms which are universally

* The $\in$ symbol is used here to mean "which is a member of". As we have seen, the members of two sets can be used to form a relation. If we are involved in a situation containing a relation, the variables representing the members of the two sets constituting the relation are said to be *related*. This statement does not imply any magic property of the variables, and should be interpreted to mean only that we are concerned with a situation which contains a set of number pairs or relation.

* ϕ is pronounced "phi".

recognised in their correct context as standing for a function of the variable inside the bracket.

Example 1.3

Draw horizontal and vertical lines, and mark points on the lines to represent the numbers 0, -4, 2 and -1 in the conventional manner.

State which is the larger in each of the pairs

(a) 2, 0; (b) 2, -4; (c) -1, 2;
(d) -4, 2; (e) -4, -1; (f) -1, 0.

The conventional representation of the numbers by points in the two lines is shown in Fig. 1.27; it will be recalled that positive

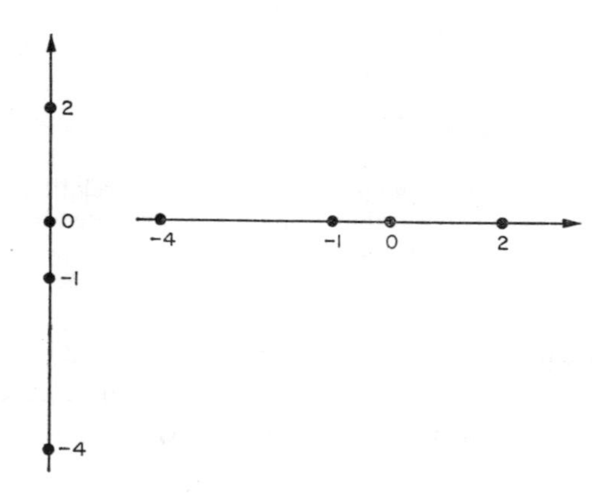

FIG. 1.27.

numbers are represented by points to the right of and above those representing zero.

The larger of the two numbers in any pair will be represented by the point lying higher in the vertical axis (or to the right in the horizontal axis). Considering the two points representing the pairs of numbers in turn, we obtain the answers to the second part of the question as: (a) 2, (b) 2, (c) 2, (d) 2, (e) -1, (f) 0.

Example 1.4

Two variables p and q are related by the law that q takes the value of the greatest integer less than or equal to p, and p can take any value. If the set P contains all values of p, and Q all values of q, classify the relation from (i) P to Q, (ii) Q to P. Is (iii) q a function of p, (iv) p a function of q?

Is Q a proper subset of P? Write down the values of q associated with the values (a) 5·714, (b) 3·0, (c) -1·7, of p.

(i) The set P contains all possible numbers, but Q contains only integers. Thus the mapping from P to Q is from all possible numbers (including integers) onto a set of integers: for example, all numbers between 1 and 1·9999... in P are mapped onto the member 1 of Q. Thus the relation from P to Q is many to one.

(ii) Conversely the mapping from Q to P is one to many, since each (integer) member of Q is mapped onto all the numbers whose value lies between its own value and that of the next larger integer.

(iii) Since each p value leads directly to a unique q value, q is a function of p (or, alternatively, since the relation from p to q is many to one, q is a function of p).

(iv) Since each q value is associated with more than one p value, selection of a value for q does not lead uniquely to an associated value of p (e.g. if q takes the value 5, p could take any value between 5 and 5·99999...). Thus q is not a function of p.

As stated in (i), P contains all numbers and Q contains only integers. Thus Q is a proper subset of P.

If (a) $p = 5{\cdot}714$, q takes the value of the greatest integer less than 5·714, namely 5.

(b) $p = 3{\cdot}0$, q takes the value 3, since this is the integer equal to 3·0.

(c) $p = -1{\cdot}7$, q takes the value -2 rather than -1, since -1 is greater than $-1{\cdot}7$ according to the conventions described on p. 14.

Example 1.5

Two variables p and q are related as indicated in the table beneath. If their relationship is denoted by the equation $p = f(q)$, write down the values of $f(17)$, $f(10)$ and $f(6)$.

q	0	3	6	10	12	15	17	20
p	1	2	2·5	2·9	3.2	3·4	3·5	3·6

The symbol $f(17)$ stands for the value of p when $q = 17$. This is 3·5, so $f(17) = 3{\cdot}5$. Similarly, $f(10) = 2{\cdot}9$, and $f(6) = 2{\cdot}5$.

Example 1.6

(i) If $f(x)$ stands for $(x^2-1)/(x-2)$, evaluate $f(3)$, $f(-1)$, $f(1)$, $f(2)$ and $f(3)/f(-3)$.

(ii) Derive an expression for $\{(x-2)f(x)\}$.

(iii) What is the value of $[f(3)]^2$?

(i) $f(3)$ stands for the value of the function f when x is put equal to 3, and is therefore numerically equivalent to

$$\frac{3^2-1}{3-2} = \frac{9-1}{1} = 8.$$

Similarly,

$$f(-1) = \frac{(-1)^2-1}{-1-2}$$

$$= \frac{1-1}{-3} = 0,$$

$$f(1) = \frac{1-1}{1-2} = 0,$$

$$f(2) = \frac{4-1}{2-2} = \text{an undefined value.}$$

Since $f(3) = 8$ and $f(-3) = \dfrac{9-1}{-3-2} = \dfrac{-8}{5}$ (substituting in the original expression), the ratio

$$\frac{f(3)}{f(-3)} = \frac{8}{-8/5}$$

$$= 8\left(-\frac{5}{8}\right)$$

$$= -5.$$

(ii) $\{(x-2)f(x)\} = (x-2)\dfrac{(x^2-1)}{x-2} = x^2-1.$

(iii) $[f(3)]^2 = 8^2 = 64$, since $f(3) = 8$ as shown above.

The answers are therefore 8; 0; 0; an undefined value; -5; (x^2-1); and 64.

Exercise 1c

1. By definition, the numerals representing the positive integers are subject to the laws of addition, some results of which are summarised in the table below (the operation of addition on two numerals giving the numeral occurring in the body of the table at the intersection of the original numeral column and row). Using the reverse procedure where necessary (i.e. splitting a numeral up into two component parts, e.g. $5 = 2+3$), show that

(i) $5+(-3) = 2$, (ii) $16+(-9) = 7$,
(iii) $2+(-1) = 1$, (iv) $-3+0 = -3$,
(v) $-4+2 = -2$, (vi) $-8+(-7) = -15$.

+	0	1	2	3	4
0	0	1	2	3	4
1	1	2	3	4	5
2	2	3	4	5	6
3	3	4	5	6	7
4	4	5	6	7	8

Assume that $a+(-a) = 0$.

2. Classify the following: 8, 0, 1, 7, 1/3, 8/5, 0·6˙, 0·7070707, 0·7070˙, π, $\sqrt{2}$, 1/11, -5, $-\frac{1}{4}$. Express the non-terminating repeating decimals as rational numbers.

3. Map the sets A and B together. Set A is composed of number pairs, the second of which is to be subtracted from the first [e.g. (7, 5) stands for $7-5$ or 2]. One link is drawn as an example.

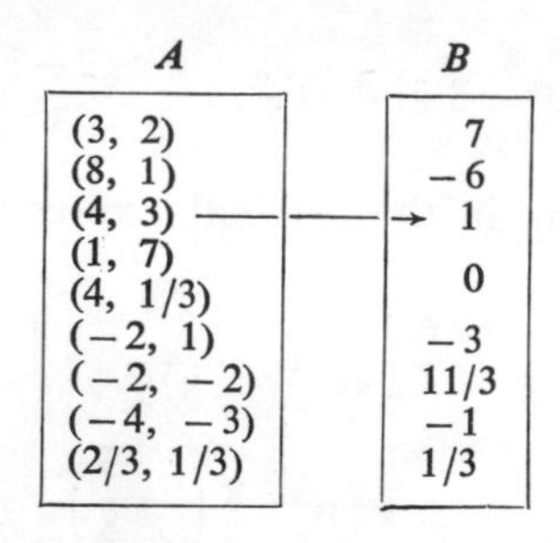

Is either set a function of the other?

4. Map the following sets together as in question 3, the brackets round the two numerals in the set A denoting multiplication instead of subtraction.

A	B
(2, 3)	6
(1, 7)	4
(4, $1\frac{1}{2}$)	7
(2, 2)	

Is either set a function of the other?

5. Points representing numbers in the line shown in Fig. 1.28 are labelled A, B, C, D, E. The numerals

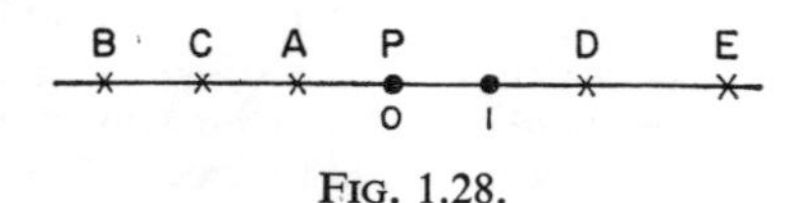

FIG. 1.28.

they represent define the set Y given below: map the letter set X with it. Are the numerals in set Y arranged in any particular order?

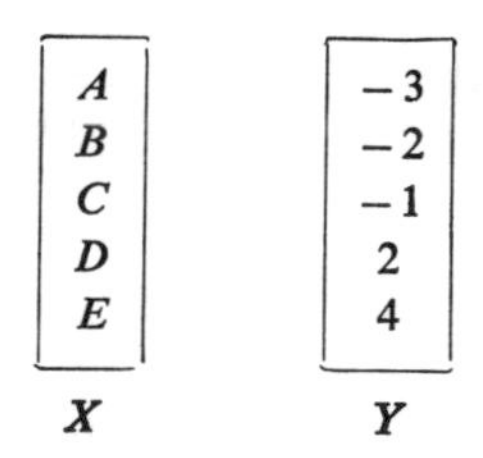

6. The set X contains all the positive integers. Is the variable x continuous? Would x be continuous if X contained

(i) all the negative integers?
(ii) all rational numbers?
(iii) all positive numbers?

7. The relationship existing between two variables x and y is that y is greater than x by 3. If the relation is symbolised by the equation $y = f(x)$, write down the values of $f(1)$, $f(-1)$, $f(7)$, and $f(-5)$.

8. Write down the larger of each of the following pairs of numbers:

(a)	$-3, -1$	(h)	$-0{\cdot}7, +0{\cdot}1$
(b)	$-3, +1$	(i)	$-0{\cdot}1, +0{\cdot}7$
(c)	$+8, +6$	(j)	$-0{\cdot}3, -0{\cdot}4$
(d)	$-9, -12$	(k)	$n, n-1$
(e)	$-6, -5$	(l)	$-n, -n-1$
(f)	$1{\cdot}1, 1{\cdot}101$	(m)	$-n, -n+1$
(g)	$-1{\cdot}1, -1{\cdot}101$		(n is a positive integer)

9. Which is the lesser of a/b and $(1-b/a)$ if

(i) $a > b$, (ii) $a < b$, (iii) $a = b$,

given that $a > 0$, $b > 0$?

10. Explain why the sum of a negative integer $-a$ and the corresponding positive integer a is zero.

11. For any positive number x, it is possible to determine how many positive even numbers there are less than it. Denoting this latter number by y, write down the values of y corresponding to the values of x: 2·8, 17·1, 15, 8, 0·7. (An example would be as follows: if $x = 7{\cdot}9$, the positive even numbers smaller than it are 6, 4 and 2, and are therefore 3 in number. Thus, for $x = 7{\cdot}9$, $y = 3$.)

Denoting the functional relationship defined by the above law by the symbol f, write down the values of $f(2{\cdot}8)$, $f(8)$, $f(9)$, $f(21{\cdot}8)$, $f(2n+1)$, $f(2n-3)$ and $f(2n)$ where n is any positive integer.

12. The symbol $g(p)$ stands for the number of primes less than p. Write down the values of $g(7)$, $g(6)$, $g(27)$ and $g(2{\cdot}8)$.

13. The modulus $|x|$* of a number x is defined by the equations:

$$|x| = x \text{ if } x \geqslant 0,$$
$$|x| = -x \text{ if } x < 0$$

(e.g. $|-3| = 3$; $|7| = 7$; $|-1{\cdot}1| = 1{\cdot}1$).

Write down the values of $|17|$, $|-8|$, $|7{\cdot}2|$ and $|-\frac{1}{2}|$.

14. State whether the first named number is a member of the following number group or not:

(i) 0, positive integers,
(ii) 0, negative integers,
(iii) $-1/3$, repeating non-terminating decimal,
(iv) $\sqrt{2}$, repeating non-terminating decimal,
(v) 100^{-100}, non-repeating, non-terminating decimal.

15. A variable y is zero whatever the value of the associated variable x. Is (i) y a function of x, (ii) x a function of y?

†**16.** Prove that

(i) $|y-x| = |x-y|$,
(ii) $|x|^2 = x^2$,
(iii) $|xy| = |x||y|$,
(iv) $|x+y| \leqslant |x|+|y|$.

†**17.** Denoting the greatest integer less than or equal to x by $[x]$, prove that

(i) $[-x] = -[x]$, if x is an integer,
(ii) $[-x] = -[x]-1$, if x is non-integer,
(iii) $[2x] = [x]+[\frac{1}{2}+x]$.

* This symbol is read as "mod x" or "the modulus of x".

†**18.** Two variables a and b are related. Given two corresponding pairs of values in the table below, find a possible expression for the relationship, and complete the table.

Values of a	3	5	7	−2			
Corresponding values of b	8	24			−1	3	0

†**19.** The non-terminating repeating decimal $0{\cdot}090909^{\cdot}$ can be written as $(9/10^2+9/10^4+9/10^6+\ldots)$. Write down an expression for the sum to n terms of this geometric progression, and discuss what happens to the value of this sum as n becomes very large.

†**20.** Convert the following denary fractions to "binary decimals", and discuss whether they are terminating (i.e. finite) or not, and repeating or not:

(i) $\frac{1}{4}$; (ii) $\frac{1}{5}$; (iii) $\frac{1}{3}$; (iv) $\frac{1}{7}$; (v) $\frac{1}{10}$.

†**21.** If three variables x, y and z are related functionally according to the equations $x = f(y)$, and $y = g(z)$, is it necessarily true to say that

(i) x is a function of z?
(ii) z is a function of x?

†**22.** If $\sqrt{(5-x)} = 1+\sqrt{x}$,

$$5-x = 1+2\sqrt{x}+x, \text{ by squaring.}$$

Rearranging and squaring to eliminate $\sqrt{x}$ gives

$$x^2-5x+4 = 0,$$

whence

$$x = 4 \text{ or } 1.$$

Now if $x = 4$, substitution into the original equation gives

$$\sqrt{(5-4)} = 1+\sqrt{4}, \quad \text{or}$$
$$1 = 3.$$

Discuss.

1.5. GRAPHICAL MAPPING

Let us return to consider further the bar heating experiment described in Section 1.4. In this experiment we investigate the relation between the length of the bar and its temperature, and we investigate such a relation by sampling. In reality the relationship exists between two sets of numbers—the one set containing all the length values which occur in the temperature range with which we are concerned, and the other containing all the temperature values in that range. We shall not be far from the truth if we say that both sets contain an infinite number of elements.

In such a situation we are neither able to determine by experiment all the number pairs of the relation, nor to represent each number pair of the relation by a point on a graph. The only practical solution is to determine or evaluate an evenly spaced sample of the number pairs and accept a smooth curve or line* drawn through their mapping points as an approximation to the infinite number of points needed to represent the whole relation.

The results of such a sampling experiment might give a set of number pairs such as that shown in Fig. 1.29. The mapping points re-

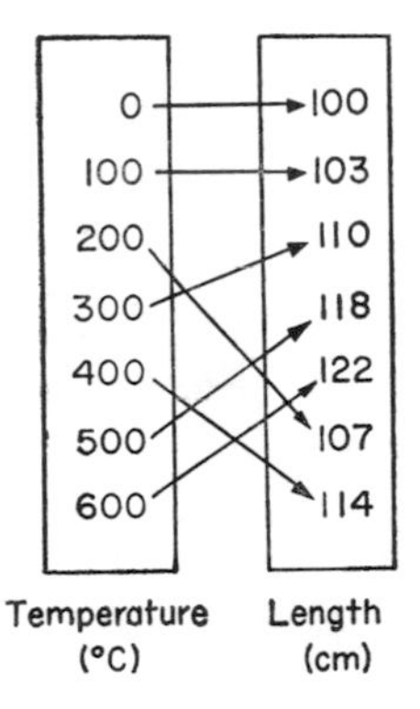

FIG. 1.29. In any experiment we are obliged to measure only a sample of the possible pairs of values of the variables involved.

presenting these number pairs can be plotted on a graphical map, and joined with a smooth curve (Fig. 1.30) which is then accepted as an approximate representation of the relation between the length of the bar and its temperature.

Now the map could equally well have been drawn with the length axis horizontal and the temperature axis vertical. It often happens, however, that the value of one of the variables involved in the physical situation is governed

* In practice, experimentally determined points will not lie on a smooth curve owing to errors in making the measurements or errors resulting from inaccuracies in the measuring instruments. In these cases, the best smooth curve through the points is drawn as in the example.

or even "caused" by the value of the other, and in such a case it is sensible to identify the variable whose value is consequential to the other with the range variable. We are then,

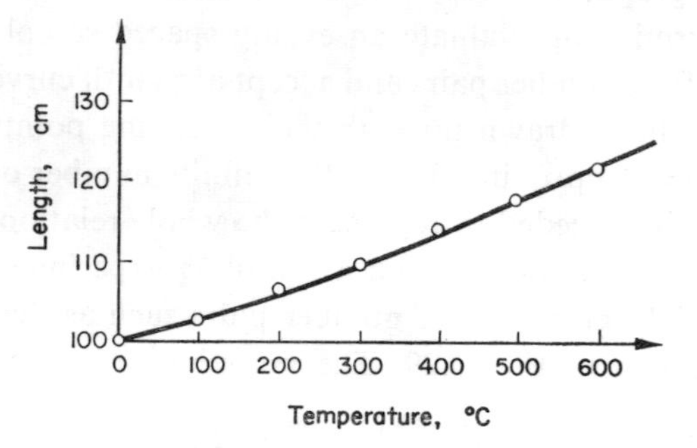

FIG. 1.30. The complete graphical map of the measured results.

in effect, associating the sense of direction from the first to the second member of an ordered pair or the direction of the mapping from the domain to the range with the direction from cause to effect in the physical relation. The variable whose value depends on the other can be referred to as the *dependent* variable: the domain (or cause) variable is then referred to as the *independent* variable, for its value can be selected by some independent decision or choice. Thus, in our bar expansion experiment, it does not make sense to regard the value of the length of the bar as something which can be arbitrarily fixed and from which the corresponding value of the temperature then results. The length is quite clearly the variable whose value depends on or is governed by the temperature, and it is therefore logical to regard the temperature as the independent variable, and the length as the dependent variable.

Since in our interpretation the expression "domain variable" is synonymous with "independent variable", and the expression "range variable" is synonymous with "dependent variable", we must, to be consistent, associate the vertical axis with the dependent variable and the horizontal axis with the independent variable (Fig. 1.31).

Readers will be familiar with the techniques used in plotting graphs, and our only purpose in dwelling on them is to instil the idea that a graph is merely a set of points. Each point represents a number pair, and the totality of the number pairs represent a relation which will generally be functional. For convenience we use graphs which are only approximations to the real situation, for it is impracticable to

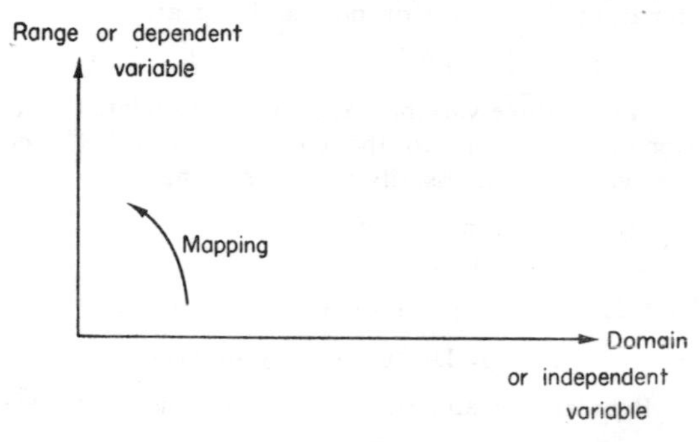

FIG. 1.31. It is usual to associate the dependent variable with the vertical axis and the independent variable with the horizontal axis.

plot more than a few points on a piece of graph paper. This fact need not hinder our study of the calculus, for being essentially an analytical subject, there will be little hindrance caused by the inaccuracies of practicality.

1.6. ALGEBRAIC EXPRESSION OF RELATIONSHIPS

Situations involving relations possessing an infinite number of elements do not only arise in experimental situations. We might, for example, have a relation existing between two sets X and Y, where the elements of X are all the real numbers between 2 and 7, inclusive, and the elements of Y all the real numbers between 4 and 14, inclusive, and where any member $x \in X$ is associated with the member $y \in Y$ whose value is twice its own. Such a relation would have an infinite number of elements, and, as we have seen, the full graph of the relation would accordingly have an infinite number of points. Although we need not bother to circumnavigate such an impracticable situation, we can none the less devise a scheme to compensate for the approximations we are ultimately forced to adopt. The scheme involves returning to the relationship itself, and replacing the infinity of number pairs by an algebraic expression of the statement summarising the nature of the relationship between the variables involved.

In many useful cases the nature of a relationship can be expressed in the form of an algebraic equation or set of equations. In the case just quoted, for example, the relation between any x and its associated partner y is summarised by the equations $y = 2x$, $2 \leqslant x \leqslant 7$. The expression of the nature of a relation by an equation or set of equations removes the hindrance often provided by the physical realities involved, and we shall see later that the ideas of the calculus are more easily handled if the relationships involved are represented by these more abstract expedients of the algebraic equation or graph.

Example 1.7

A car is travelling with a constant speed of 30 km/h. What is the expression relating its distance s from a fixed point t hours after it has passed a fixed point if s is measured in kilometres?

In 1 hour the car travels 30 km: in t hours it will therefore cover $(30t)$ km since the speed is constant. The relationship between s and t is thus $s = 30t$.

Example 1.8

A sum of money is invested in a concern which pays interest at the rate of 4% p.a. If the interest is reinvested in the concern, derive an expression for the relationship between the sum of money A accruing after T complete years and the original sum invested P.

Four per cent interest p.a. means that interest equal to four-hundredths of the sum invested at the beginning of the year is paid at the end of the year: the total sum owned is therefore $\{P+(4/100)P\}$ at the end of the first year. Writing this as $1{\cdot}04P$, it can be seen that the sum increases by a factor of 1·04 every year, so that after T years the amount accruing is $(1{\cdot}04)^T P$. The relationship between A, P and T is thus

$$A = (1{\cdot}04)^T P.$$

Example 1.9

The distance s of a train from a station at t minutes past noon is given by the equation $s = 1+t+t^2/20$, where s is measured in kilometres. Represent this graphically for the period 11.55 a.m. to 12.02 p.m.

First, a table of values is made: 11.55 a.m. is 5 minutes before noon and 12.02 p.m. is 2 minutes after noon. The value of t therefore ranges between -5 and $+2$.

t	-5	-4	-3	-2	-1	0	1	2
s	$-2{\cdot}75$	$-2{\cdot}2$	$-1{\cdot}55$	$-0{\cdot}8$	$0{\cdot}05$	$1{\cdot}0$	$2{\cdot}05$	$3{\cdot}2$

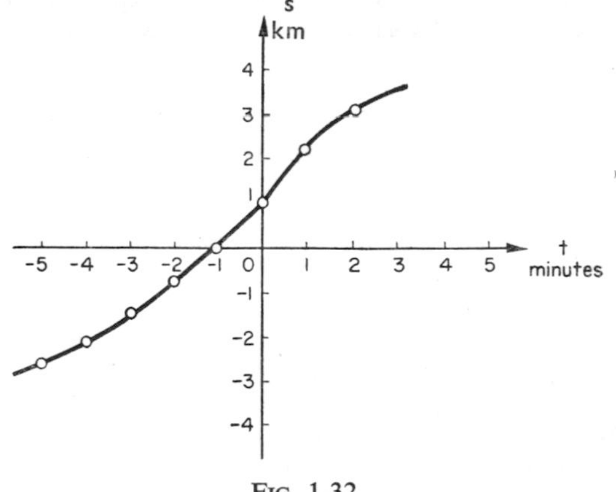

FIG. 1.32.

Suitable scales are chosen for the graphical axes, and the dependent variable (displacement) axis made vertical. The seven points are plotted and joined with a smooth curve (see Fig. 1.32).

EXERCISE 1d

1. Represent the following pairs of values graphically:

Independent variable	8	6	−1	10	12	−10
Dependent variable	−2	3	7	6	2	−3

2. The following pairs of variables are related: which is most sensibly chosen as the independent variable?

(i) Time of day and air temperature.
(ii) A sum of money invested and the amount of interest payable on it.
(iii) The height and the age of a growing boy.
(iv) The petrol consumption and the speed of a car.
(v) Cost of butter and the cost of living.
(vi) Volume of applause and audience appreciation.
(vii) The length of service and a man's salary.
(viii) The speed of a train and the interval of time since it started from rest.

3. The following table gives the length of a piece of elastic (l cm) when particular weights of value (w grammes) are hung on it. Represent the pairs of values graphically.

w (g)	0	10	20	30	40
l (cm)	20	20·5	21	21·5	22

Write down an algebraic equation expressing the relationship between l and w. Over what range of values does this hold?

4. The following table gives the profit earned by a firm in the years 1955–60. Represent the result graphically. Is it possible to express the relation between the two variables by an algebraic equation?

Year	1955	1956	1957	1958	1959	1960
Profit £000	175	283	394	397	421	356

5. A functional relationship between two variables f and t is expressed by the equation $s = f(t)$. Which is the independent variable?

6. Which is the independent variable in the graph of Fig. 1.32? Is this a reasonable choice?

***7.** Will a graph of the density of ice against temperature be upward pointing or downward pointing? Give reasons.

***8.** Graphs are to be drawn to depict the relationships between the following pairs of variables. Which variables should be represented along the horizontal axes?

(i) Population of the world and the years 1950–84.
(ii) The time of day and the air temperature.
(iii) The time of swing of a pendulum and its length.
(iv) The number of road accidents per year and the number of cars on the roads.

Make sketch graphs of the expected results. What are the dangers of extrapolating* the results?

9. Draw a graph to represent the following examination summary:

Mark in examination	20	21	22	23	24	25	26	27	28	29	30
Number of pupils gaining the mark	1	1	4	9	12	12	10	8	5	2	1

Is it sensible to join the points on the graph with a smooth curve or a straight line? Do the intermediate points have any meaning?

***10.** The specific heat capacity of a substance can be defined as the heat required to raise the temperature of 1 g of the substance by 1°C. Write down an equation which will give the amount of heat H (joules) needed to raise the temperature of m g of a substance by 1°C if the specific heat capacity of the substance is s joules g^{-1} °C^{-1}.

***11.** The coefficient of linear expansion of a substance is defined as the increase in length per unit length of a bar of that substance when its temperature is raised by 1°. Write down an expression relating the expanded length (l_1) of a bar in terms of its original length (l_0) when it is heated through t°, the coefficient of linear expansion of the substance of which it is made being α per degree.

***12.** Boyle's law for gases states that the volume of a fixed mass of gas is inversely proportional to its pressure, other physical conditions such as the temperature remaining constant. Denoting the pres-

* Extrapolation is the process by which an estimate of the value of one of the variables is made from the corresponding value of the other, one or both values being outside the domain or range of the relation.

sure by p and the volume of the fixed mass of gas by v, write an equation which expresses the law in symbolic form.

13. An amount of money £P is invested at $R\%$ simple interest. Write down an equation showing the amount of interest earned in T years.

14. The population P of a country increases each year by 2% of its value at the beginning of that year. Write down an expression for the population t years after the size of the population is P_0. Draw a sketch graph of the variation. How long will it take for the population to double?

15. The graphs in Fig. 1.33 describe the relationship between the two variables x and y. Is it true to say in any case that

(i) x is a function of y,
(ii) y is a function of x?

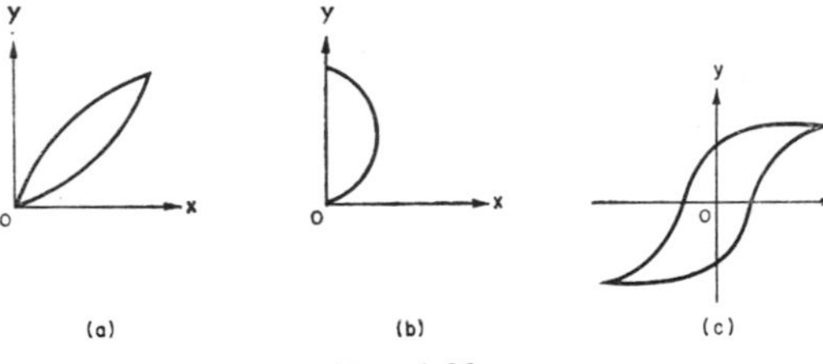

FIG. 1.33.

16. The cost of a certain metal is 25p per kg. If 1 cm³ of the metal weighs 0·005 kg, draw a graph to represent the cost of a square sheet of the metal over the range from 10 cm to 100 cm side, if the thickness is (i) 1 cm, (ii) 2 cm. How much is a 2 cm thick sheet of side 60 cm worth, and for each thickness, what is the length of the side of a sheet which can be bought for £1.80?

17. Plot a graph of the curve represented by the equation $y = x^3$ for x in the range -5 to $+5$, treating x as the independent variable. Find the value of $(2{\cdot}7)^3$ and $\sqrt{85}$ from your graph.

18. The distance of a train in kilometres from a station A is given by the expression $160t^2 - 80t^3$, where t is the time in hours after the train has left A. Draw a graph showing how far the train is from A at any instant during a period of $1\frac{1}{4}$ hours after leaving A.

19. A cylindrical metal can closed at both ends is to contain 200 cm³ of a liquid when full. Show that the area of metal needed to make the can is given by

$$S = 2\pi r^2 + \frac{400}{r},$$

where S is the area in square centimetres and r is the base radius in centimetres. Represent this graphically for

(i) $1 \leqslant r \leqslant 20$; (ii) $1 \leqslant r \leqslant 5$.

For what value of r is S a minimum? What is then the minimum value of S?

20. A piece of cardboard has an area of S square metres. If it is used to make an open carton of height y metres, with vertical sides and with a square base of side x metres, find its surface area in terms of x and y. What does this equal numerically? What is its volume V (in terms of x and y)?

Eliminate y between these equations, and represent the variation of S with x graphically, assuming V to be fixed and equal to 1 cubic metre.

21. Draw a graph which will enable the volume of a sphere up to 20 cm radius to be readily determined.

22. A car starts from rest and accelerates uniformly to a speed of $12t$ km per hour in t seconds. What is the average speed during the first t seconds, and how far will it have gone by then?

23. What is the definition of π? Write down an expression for the surface area of a closed cylindrical can whose height is three times its base radius. Explain the meaning of the letters you use.

24. A rectangular block has sides of length a, b and c. Write down an expression for the length D of the body diagonal. What is the length of the body diagonal of a cube of side a?

Rearrange both equations to make a the subject.

25. (i) y is the result of squaring x and adding 3. Is y a function of x?
(ii) y is the largest prime factor of x. Is y a function of x?
(iii) If $x \geqslant 1$, $y = 2x$. If $x \leqslant 1$, $y = 2$. Is y a function of x? Why would y not be a function of x if y was defined to take the value 3 for all $x \leqslant 1$?

26. (i) It is known that P varies inversely as the square of A. Corresponding pairs of values are shown in the following table.

P	t	$t+1$
Q	5	4

Calculate the value of t.

(ii) Illustrate on three separate sketch graphs the relationships between y and x in the cases (a), (b) and (c) described below. (Draw the usual x- and y-axes, each starting at 0 units. Graph paper and tables of numerical values are not necessary).

(a) x centimetres is the radius of a circle and y square centimetres is its area.
(b) x degrees is the temperature recorded on the Centigrade or Celsius scale, and y degrees is the same temperature measured on the Fahrenheit scale.
(c) x centimetres is the length of a rectangle of constant area and y centimetres is its breadth. (C)

27. It is known that y varies as a power of x. If $x = 2$ when $y = 8$ and $x = 3$ when $y = 40{\cdot}5$, find the simplest expression for y in terms of x. (SU)

28. If $f(x) = \log x$, show that $f(xy) = f(x)+f(y)$. Show also that $f(\sqrt{x}) = \frac{1}{2}f(x)$.

29. A stone is shot vertically into the air so that, t seconds after projection from a point P on the ground, the vertical height h m is given by $h = 40t - 5t^2$. Show that it returns to the ground 8 seconds after projection. Draw the graph of h for values of t between $t = 0$ and $t = 8$. Use your graph to find:

(i) the greatest height of the stone,
(ii) the time taken for the stone to pass between two points A and B on the vertical through P if $PB = 72$ and $PA = 45$. (SU)

†**30.** The relationship $y = \pm\sqrt{(9-x^2)}$ can give two values of y for every value of x numerically less than or equal to 3. Can y be said to be a function of x?

†**31.** Draw graphs of the so-called "step" functions given by the following relationships:

(i) $f(x) = [-x]$, (iii) $f(x) = [x+\frac{1}{4}]$,
(ii) $f(x) = 2[x]$, (iv) $f(x) = [x+a]$ where a is a constant.

{$[x]$ denotes the greatest integer less than or equal to x (e.g. $[7] = 7$, $[6{\cdot}5] = 6$, $[9{\cdot}1] = 9$. See also p. 20, q. 17)}.

†**32.** Would it be possible to represent three sets of numbers graphically? If all the points lay on a plane, how would it be possible to reduce the nature of the relationship between them to two dimensions?

What would be the nature of a graph representing the relationship between three sets of numbers if the points representing each kind of numbers were all equidistant from the zero point?

†**33.** The horizontal and vertical areas of a graph are interchanged, and the graph redrawn. By what operation or combination of geometrical operations (i.e. translation, rotation, reflection) can the second curve be obtained from the first?

†**34.** A certain firm makes a cylindrical soup can of height 6 cm and radius x cm. The cost of producing one can is $(S/1000)$p, where S is the surface area of the can in cm². Draw a graph showing the variation of the cost per can as a function of the volume. What is the cost of a can of radius (i) 6 cm, (ii) 8 cm, and what is the maximum radius of a can which costs not more than 0·5p to produce?

†**35.** Show that the graph of $y = ax^2$ is a 2 : 1 mapping with range $y \geqslant 0$. What happens if the domain is limited to $x \geqslant 0$?

1.7. CHANGES

It has been explained that a variable is a symbol which can represent any one of a set of numbers. There are many problems, however, in which the actual values of the variable are of little or no significance, the important factor being one of *change* in the value of the variable. To take an everyday example, the size of the task of walking to the top of the hill whose summit is 100 m above sea-level depends hardly at all upon the value of 100 m. What matters is the change in height above sea-level between the bottom and the top of the hill.

We can define the "change" in the value of the variable as its final value less its initial value. Thus if a variable x changes in value from x_1 to x_2, it can be said to have changed by (x_2-x_1). One has to deal with the amount by which a variable changes so frequently that the Greek letter delta (δ) is used as an abbreviation for the whole phrase "the change in the value of"; thus the equation $\delta x = 2$ is read as the statement "the change in the value of x is 2". Notice that no information is given concerning the actual value of x, and also that the δ symbols are not to be treated as algebraic quantities, e.g. they cannot be cancelled or squared. That is to say, $\delta a/\delta b$ is not the same as a/b, and $\delta^2 a$ has no meaning unless one is defined for it. The a and the δ are inseparable, and must always remain together as a single quantity, just as the prefix "sin" has to stay with the x in sin x. On the other hand, $(\delta a)^2$ is quite meaningful, being the square of (δa).

It will be understood that the sign of the change δx is positive only if $x_2 > x_1$. The variable x can be said to have increased in value if $x_2 > x_1$, and thus an increase is to be regarded as a positive change. Similarly the variable x can be said to have decreased in value if $x_2 < x_1$, and in this case the change as defined above is to be accorded a negative value. We can summarize these statements by the equations:

$$\delta x = x_2 - x_1;$$

$$\text{if} \quad \delta x > 0, \qquad x \text{ increases};$$

$$\text{if} \quad \delta x < 0, \qquad x \text{ decreases}^*.$$

* The statement $\delta x < 0$ is frequently used to express the fact that the value of δx is negative.

Now it is helpful to associate changes in the value of a variable with movements of a point representing the value of that variable on a graphical map. To develop this association, imagine a line whose points represent the totality of real numbers as described in Section 1.3, and that the position of a point X contained in that line indicates the value of a variable x at any instant. If x increases, X moves from the point representing the value x_1 to the point representing the larger number, i.e. to the right (Fig. 1.34a); and if x decreases, X moves from the point representing x_1 to a point representing the smaller number, i.e. to the left (Fig. 1.34b). Thus an increase or positive change can be associated with a rightward movement of the representative point in a line or axis; a decrease or negative change can be associated with a leftward movement of the point.

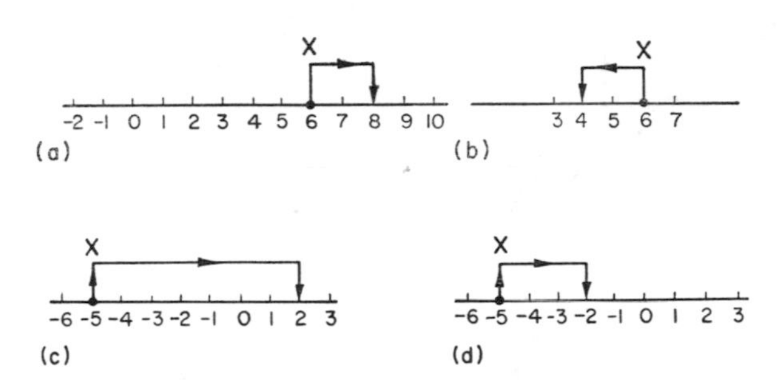

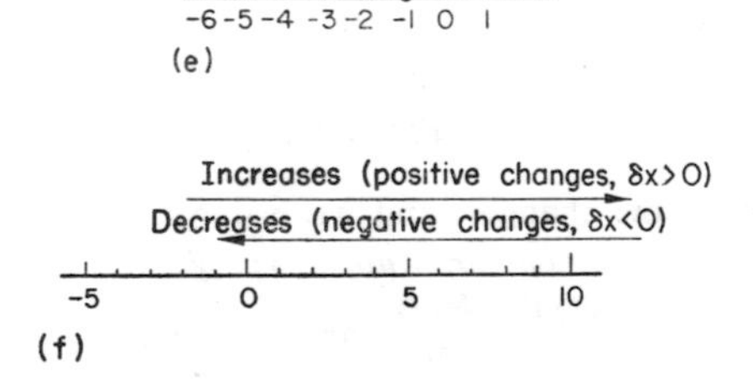

FIG. 1.34. (a) x increases from 6 to 8; (b) x decreases from 6 to 4; (c) x increases from -5 to 2; (d) x increases from -5 to -2; (e) x decreases from -1 to -3; (f) if x moves to the right, x has increased: if x moves to the left, x has decreased.

Similar reasoning can be applied to a vertical number axis, and a positive change or increase in the value of the variable represented by points in that axis can be associated with an upward movement of the point representing the instantaneous value of the variable, and a negative change or decrease can be associated with a downward movement (Fig. 1.35).

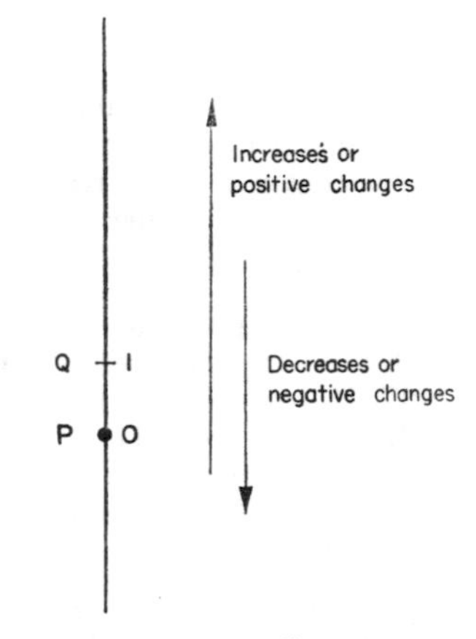

FIG. 1.35. The corresponding movements in a vertical axis are up for an increase, and down for a decrease.

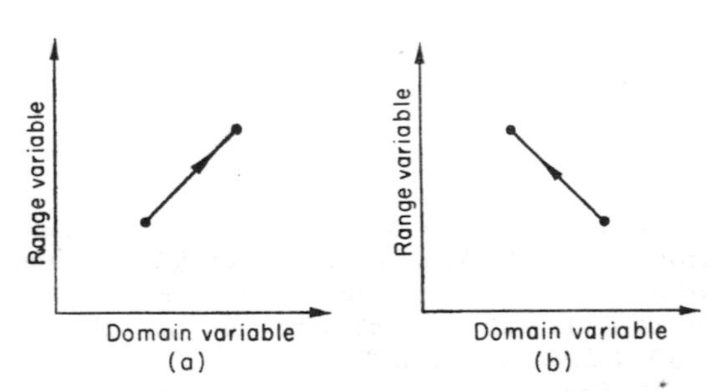

FIG. 1.36. The movements of a mapping point represent two simultaneous changes in a pair of related variables. A movement (a) up and to the right represents an increase in both variables involved and (b) up and to the left represents an increase in the range variable and a decrease in the domain variable.

A point moving in the mapping plane can represent a simultaneous change in the two variables it maps together: the sense in which it moves relative to the two axes defines the sense in which the two variables change. Thus a point moving up and to the right represents an increase in both of the related variables (Fig. 1.36a): a point moving upwards to the left represents an increase in the value of the vertical axis (range) variable and a decrease in the horizontal axis (domain) variable (Fig. 1.36b).

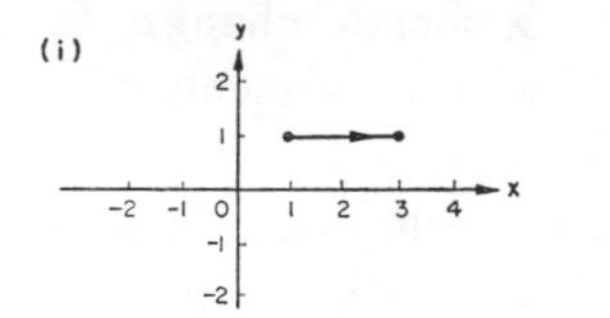

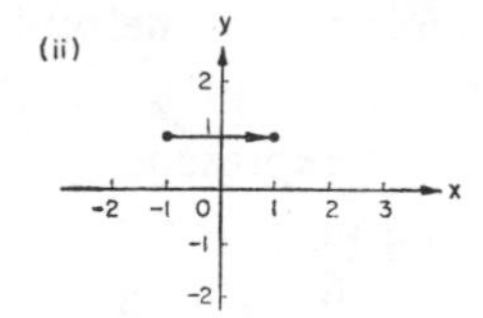

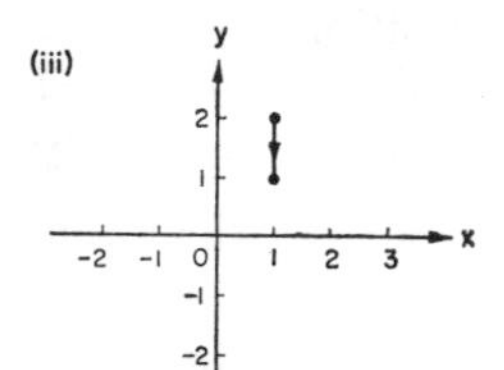

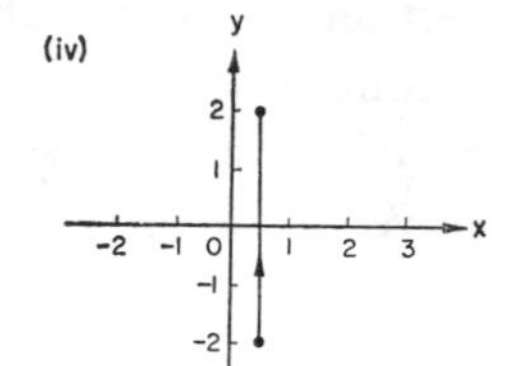

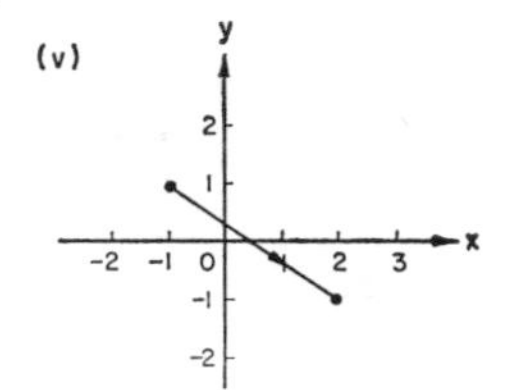

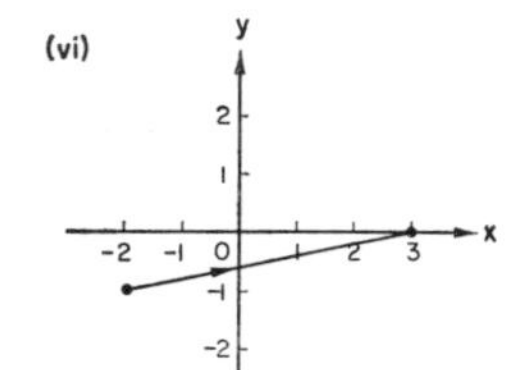

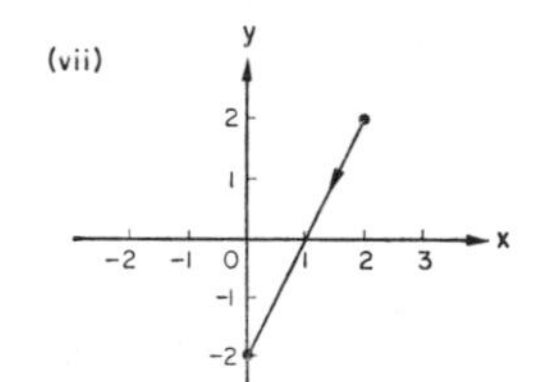

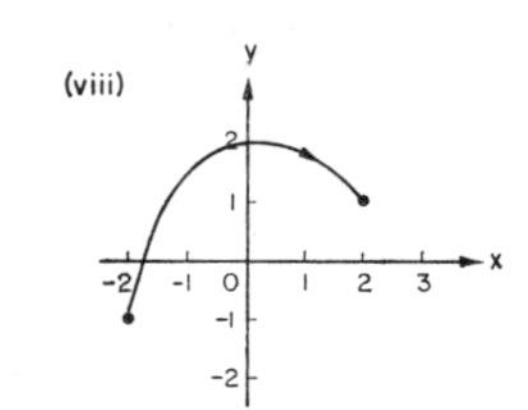

Fig. 1.37.

Exercise 1e

1. A variable y changes its value from the first of each pair of numbers beneath to the second. Give the value of the change with its correct mathematical sign, and state whether the change is an increase or decrease.

(i)	2, 5	(viii)	5, −9
(ii)	0, 11	(ix)	−17, −56
(iii)	17, 15	(x)	−11, −2
(iv)	4, 0	(xi)	11, −12
(v)	25, 28	(xii)	11, −11
(vi)	−3, 0	(xiii)	0, −8
(vii)	−2, −4		

2. The diagrams of Fig. 1.37 depict a point moving across a graphical map in the direction of the arrow. What are the changes in the values of x and y when the point moves from the tail of the arrow to the head?

3. Given that $(a)+(-a) = 0$, and that any positive number can be split into the sum of two positive numbers smaller than itself, show that

(i) $10-8 = 2$; (ii) $-10-11 = -21$; (iii) $2-7 = -5$.

4. The value of a variable denoted by g changes from 1·5 to 4·8. What is the value of δg? The value of another variable h changes from −3·1 to −1·4. What is the value of δh? What are the values of $(\delta g)^2$, $(\delta h)^2$ and $(\delta g)(\delta h)$?

5. Variables a and b have values of −5 and 2 respectively. If they change to −7 and −1 respectively, what are the values of

$$\delta a,\quad \delta b,\quad \frac{\delta a}{a}+\frac{\delta b}{b},\quad \frac{(\delta a)(\delta b)}{ab}\quad \text{and}\quad \left(\frac{\delta a}{\delta b}\right)?$$

6. A variable q has a value of 184. If it changes by δq, where δq equals (i) +107; (ii) −91; (iii) −256, what are its new values?

7. Variables x and y take the values −2 and 3 respectively. If they change to −5 and 4, what are the values of δx, δy, $\delta(x^2)$, $\delta(y^2)$, $(\delta x)^2$, $\delta(xy)$ and $\delta(x/y)$? Give a graphical picture of the changes.

8. Variables p and q change from values of 5 and −5 to −4 and −6 respectively. Draw a graphical map of the changes, and write down the values of

$$\delta p,\ \delta q,\ \delta(p/q),\ [\delta(pq)]^2,$$
$$\frac{\delta(p)^2}{p},\quad \frac{\delta(q^2)}{\delta q}\quad \text{and}\quad \left\{\frac{\delta p}{p}\right\}^2.$$

Why might an expression like $\delta p/p$ be ambiguous?

†9. The following set of numbers is composed of values of a variable y:

0 1 4 9 16 25 36 49 64

Copy the row, and beneath it—in the gaps—write down the values of δy for adjacent members. Beneath this write down the values of $\delta(\delta y)$—or, as it is often written, $\delta^2 y$—and finally $\delta(\delta^2 y)$ or $\delta^3 y$. What do you notice about the original set of numbers?

Now construct a row of numbers (z) which are the cubes of 1, 2, 3, 4, 5, 6, 7 and 8. Calculate and write down the values of δz, $\delta^2 z$, $\delta^3 z$, and $\delta^4 z$.

For what values of n would you expect $\delta^n x$ to be zero if:

(i) x equalled 1^4, 2^4, 3^4, 4^4, etc., in turn;
(ii) x equalled 1, 2, 3, 4, 5, etc., in turn;
(iii) x equalled 1^{18}, 2^{18}, 3^{18}, 4^{18}, etc., in turn?

†**10.** If $y = x^2$, write down an expression for $(y+\delta y)$ when x changes to $(x+\delta x)$. Find an expression for δy in terms of x and δx only.

SUMMARY

A *set* is a collection: the collection can be defined by listing or description. The members of a set are often referred to as its *elements* and they must be distinguishable from one another.

A set whose elements are drawn from another set is a *subset* of the latter: if the subset is constructed from some but not all of the elements of the parent, it is said to be a *proper subset* of the parent.

Sets can be referred to with capital letters: the corresponding small letter can be used to represent a typical member of the set. The expression $p \in P$ can be read "p is a member of P".

An *ordered pair* consists of a pair of elements, one drawn from each of two sets and written in a particular order. The set of all possible pairs formed by the elements of two sets X and Y is called the *Cartesian product* $(X.Y)$ of the two sets. Any proper subset of $X.Y$ is a *relation* from X to Y. The set constructed from all the first members of a relation (which is a set of ordered pairs) is called the *domain* of the relation: the set constructed from all the second members is called the *range*.

The relation obtained by reversing the order of each of the ordered pair elements of a parent relation R is called the inverse of R and is denoted by R^{-1}.

A relation can be depicted by listing two sets containing the domain and range in two vertical columns, and linking the paired members with arrows. This is called *mapping*. If all the members of the second set are situated at the head of an arrow, the mapping is said to be from the first set *onto* the second; if there are some members of the second set not situated at the head of an arrow, the mapping is said to be from the first set *into* the second.

If one or more domain elements are each associated with more than one range element, the relation is said to be a *one to many* relation (Fig. 1.13b); if one or more range element is associated with more than one domain element, the relation is a *many to one* relation (Fig. 1.15). If, simultaneously, both of these conditions hold, the relation is a *many to to many* relation (Fig. 1.18). If each domain element is associated with only one range element and each range element is associated with only one domain element (Fig. 1.6), the relation is a *one to one* relation. In this case, the elements of the domain and range are said to be in a 1 : 1 *correspondence* with each other.

A *function* is a many to one or a one to one relation: an *inverse function* exists only if the relation is a one to one relation.

Real numbers can be divided into four groups: the positive integers, the negative integers, the rational numbers and the irrational numbers. It is convenient to represent them by the points in a line or *axis*, and it is conventional to represent the positive numbers by using the points to the right of that chosen to represent zero in a horizontal axis and above that chosen to represent zero in a vertical axis. If d is the distance between the point (P) representing zero and that representing 1, the point representing N must be placed at a distance of Nd from P. This structure leads to the consequence that the larger of two numbers is represented by the point further to the right of (or above) that representing zero.

Variables are symbols which represent any member of the set associated with the variable: generally a variable is a symbol which can represent any member of a set composed entirely of numbers. If at any moment a variable represents the number N, it can be said to "take the value N".

If a relationship exists between two sets X and Y, the variables x and y associated with the sets are said to be related. If the relationship in one particular direction (say X to Y) is functional, the range or *dependent* variable y can be said to be a function f of the domain or *independent* variable x. The expression $y = f(x)$ represents the statement that y is a function of x, and the symbol $f(n)$ is understood to represent the value of y associated (uniquely) with $x = n$.

A relation or function can be represented graphically by points in a plane defined by two perpendicular axes: it is conventional to use the points in the horizontal axis to represent the values of the domain, independent or cause variable, and the points in the vertical axis to represent the values of the associated range, dependent or effect variable. Any point in the plane containing the two axes is understood to map the member of the domain set represented by the point in the horizontal axis directly beneath it onto the member of the range set represented by the point in the vertical axis directly opposite it. The entire collection of the mapping points is called a *graph* of the relation or function.

If a set contains all the real numbers, or all the real numbers between two boundaries, the associated variable is said to be *continuous*. When representing graphically any relation involving a variable which is continuous, we can plot only a sample of the points representing the ordered pairs of the relation: the rest of points can be approximately located by joining the sample points in straight lines or a sketched curve as appropriate.

If a variable p changes in value, the symbol δp is understood to stand for the value of the change. Increases are accorded a positive value and can be associated with rightward or upward movements on a graph; decreases are given a negative value, and are associated with downward or leftward movements.

Miscellaneous Exercise 1

1. Write down sets:

(i) A containing the names of the days of the week,
(ii) B containing the names of the months of the year,
(iii) C containing the colours of the rainbow,
(iv) D containing the continents of the world.

How many elements does each set contain? Display this answer in the form of a mapping from a set L containing the letters denoting the sets (i) to (iv) to a set N containing the numbers 1 to 12, inclusive. Would you say that the mapping is from L into N or from L onto N? Why?

2. A relationship exists between the sets L and N of question 1: write down the elements of the relation in the form of a set of ordered pairs. Does the domain of the relation contain all the members of L or not? Does the range contain all the elements of N?

Write down the Cartesian product $L.N$, and state whether the relation between L and N contains all the members of $L.N$ or not. Can a relation between two sets ever contain all the elements of the Cartesian product between the two sets?

3. Represent

(i) the members of the relation between L and N of question 1,
(ii) the members of the Cartesian product $L.N$,

by a graph.

Is the mapping in (i) (from L to N) one to many, many to one, or one to one?

4. Write down the inverse relation of that from L to N in question 1 as

(i) a mapping drawn from set N to set L,
(ii) a set of ordered pairs,
(iii) a graph.

Is this relation one to many, many to one, or one to one?

5. Is either of the relations between the sets L and N in questions 3 and 4 a function? What condition must be satisfied for a function to have an inverse function?

6. Rational numbers can be represented by terminating decimals, non-terminating, non-repeating decimals, or non-terminating, repeating decimals. Draw a set containing these three category names, and another set containing the fractions $\frac{1}{2}, \frac{1}{3}, \frac{1}{4}, \ldots, \frac{1}{9}$. Map each fraction onto the type of decimal which can represent it. What type of mapping is displayed by the relation between the two sets?

7. Represent the relation defined in question 6 by a graph.

8. Write down the inverse relation of that described in question 6, and represent it by a graph. What type of relation is this?

9. Classify the weather of the past three days into one of the groups o (overcast), b (bright periods), or s (mainly sunny).

Show the relationship between the day d and the weather w in the form of

(i) a mapping between the two sets D and W (such that $d \in D$, $w \in W$),
(ii) a set of ordered pairs,
(iii) a graph.

Into which category does the relationship fall? Is there an inverse relation? If so, show it in each of the three forms given for the original relation.

Is (i) w a function of d,
(ii) d a function of w?

10. Write down the names of the three major political parties and the names of their leaders in the form of two sets, and map the sets together. What type of relationship exists between the sets? Can the relation be described as a function? Is there an inverse function?

Represent the mapping by a graph, and explain how the graph displays the properties you have just attributed to the relation.

11. Set D is a set whose elements are dates, and set N is composed of numerals representing the number of complete years a girl had lived. The earliest date in set D is the date of her birth. Map the members of D onto those of N.

D	N
25.1.56	8
11.8.62	0
1.6.56	6
10.10.64	7
31.12.63	

Denoting a typical member of D by d, and one of N by n, is (i) n a function of d, (ii) d a function of n?

12. Sets A and B beneath contain further dates and completed years applying to the girl in question 11. Map the members of set A onto those of set B.

A	B
11.8.94	42
28.7.98	39
2.1.93	36
4.4.85	28

13. The relationship of the members in the two sets given beneath is that for any member in set A, the corresponding member in set B is 5 larger.

Copy the table, and complete the mapping.

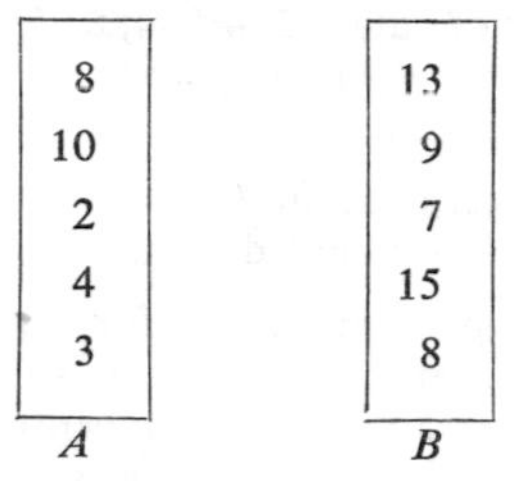

A	B
8	13
10	9
2	7
4	15
3	8

14. The relationships between the numerals in the pairs of sets below are that:

(i) x is greater than y by 3,
(ii) a is greater by 2 than the greatest integer less than b,
(iii) p is less than q by 3,
(iv) r is greater than s by 13.

Map the corresponding pairs of elements in the related sets together.

(i) X	(i) Y	(ii) A	(ii) B	(iii) P	(iii) Q	(iv) R	(iv) S
7	1	7	2·6	3	−1	0	−13
4	8	5	0	5	8	−27	−21
16	4	1	5·1	8	11	−8	−9
9	13	4	4·7	−4	6	4	−40
11	6	2	−1				
		6	3				

15. The members of set A below are number pairs, the first of which is to be divided by the second. The members of set B contain the result. Map the sets together.

A	B
(4, 2)	13
(3, 1)	3
(0, 2)	2
(2, 4)	0
(6, 3)	$\frac{1}{2}$
(169, 13)	

16. Write down the numbers which the points A–F in Fig. 1.38(a) and A–E in Fig. 1.38(b) represent.

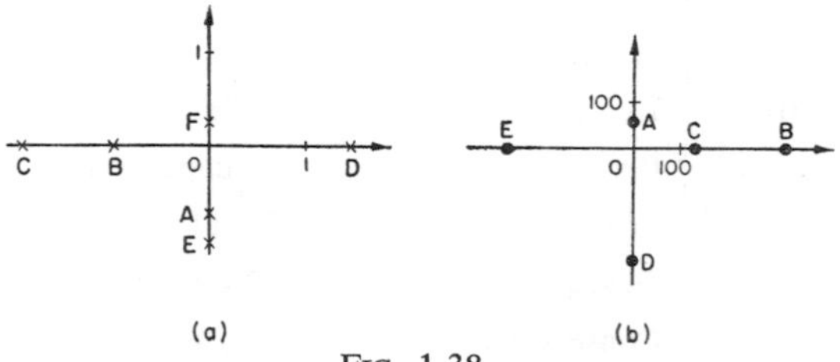

FIG. 1.38.

17. Two related variables p and q have corresponding pairs of values 3, 4·5; 4, 8; 5, 12·5; 6, 18; 7, 24·5. Estimate the value of q when $p = 5{\cdot}8$ by drawing a graph and using

(i) linear interpolation,
(ii) the curve you have drawn.

Find a simple expression for the relationship between p and q and calculate the value of q corresponding to $p = 5{\cdot}8$ from your expression.

18. $\theta(y)$ is a function of the variable y whose value is equal to the number of prime factors of y. Write down the values of:

(i) $\theta(3)$, (vi) $\theta(21)$,
(ii) $\theta(12)$, (vii) $\theta(22)$,
(iii) $\theta(16)$, (viii) $\theta(23)$,
(iv) $\theta(10)$, (ix) $\theta(2^n)$,
(v) $\theta(20$,

where n is an integer.

***19.** In each of the following pairs of sets, map set A onto set B. In which of the examples is (a) set A a function of set B, (b) set B a function of set A?

(i)
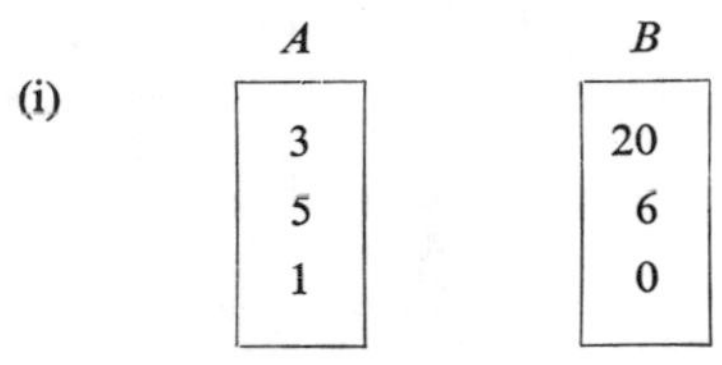

(Assume that the relationship can be expressed in the form $b = pa^2+qa+r$, where p, q and r are fixed.)

	A	B
(ii)	Orange Cherry Cucumber Sweetcorn Apricot	Yellow Orange Red Green
(iii)	Slings Sea Arrows Quality	Outrageous fortune Troubles Mercy
(iv)	Hot Black True Obsequious Conservative	White Independent Cold Progressive False
(v)	1066 1939 −4004 −44	Hitler Adam William Caesar

(vi)
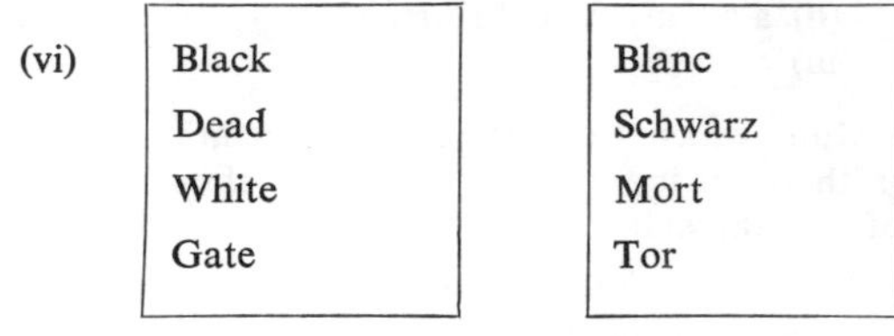

20. The points in two perpendicular axes are used to represent values of variables x and y. If x and y are related according to the following equations, and the related points of the two axes are joined by straight lines, what is the pattern of each mapping?

(a) $y = x$, (b) $y = 2x$, (c) $y = 10-x$.

21. The values of a variable y (where y is a function of another variable x) are denoted by the symbols y_0, y_2, y_4, y_6, etc., y_n representing the value of y corresponding to the x value n. The value of δy_n is defined by the equation

$$\delta y_n = y_{n+1} - y_{n-1},$$

and δy_n is called a "first order" difference. "Second order" differences are denoted by the symbol $\delta^2 y_p$, and are defined by the equation $\delta^2 y_p = \delta y_{p+1} - \delta y_{p-1}$. Higher order differences are defined and represented similarly. Construct a table showing the first, second and third order differences for the following values of y:

y_0	y_2	y_4	y_6	y_8	y_{10}	y_{12}	y_{14}
0	1	2	3	4	5	6	7

Repeat the example using the following values:

0	1	4	5	16	25	36	49
0	1	8	27	64	125	216	343
50	49	46	41	34	25	14	1

Comment on your results.

22. The relationship between two variables y and x is given by the equation $y = x^2+3x-7$. Denoting a pair of corresponding values of x and y by x_n and y_n, respectively, write down an expression for y_{n+1}, where y_{n+1} is the value of y when x takes the value x_n+1. Hence, write down an expression for $\delta y_{n+\frac{1}{2}}$ and by deriving an expression for y_{n-1}, and an expression for $\delta y_{n-\frac{1}{2}}$, show that $\delta^2 y_n = 2$. Comment on the result.

$$\left[\delta y_{n+\frac{1}{2}} = y_{n+1} - y_n, \quad \delta y_{n-\frac{1}{2}} = y_n - y_{n-1};\right.$$
$$\left.\delta^2 y_n = \delta y_{n+\frac{1}{2}} - \delta y_{n-\frac{1}{2}}\right]$$

23. Discuss the meaning of the statements that

(i) parallel lines meet at infinity,
(ii) the universe is finite.

24. Explain the terms integer and rational number. Why is it more correct to say that the integers have the same properties as the rational numbers of the form $R(a/1)$ than to say that the integers are a proper subset of the rational numbers?

25. Map together the following sets:

America
England
France
Wales
Russia

Washington
London
Cardiff
Paris
Moscow

26. The height and weight of twelve boys (A–L) were found to be as follows:

	A	B	C	D	E	F
Height (cm)	180	168	150	155	196	147
Weight (kg)	70·0	52·6	36·4	41·8	74·9	33·2

	G	H	I	J	K	L
Height (cm)	165	165	155	170	147	152
Weight (kg)	47·3	50·8	38·2	54·0	39·0	71·8

Represent these results graphically, labelling each point with the letter indicating whose vital statistics it represents. Which variable do you think is most sensibly regarded as the dependent one?

Indicate which regions on the graph contain points for those who are comparatively

(i) tall and thin, (ii) short and fat,
(iii) of average build, (iv) short and thin.

Can the relationship between the variables be said to be functional in either sense?

27. 1000 electric light bulbs are run until they burn out. The number remaining after t hours is as follows:

N	950	850	570	350	170	60	0
t	2800	3000	3200	3400	3600	3800	4000

Represent these results graphically, and plot another graph showing the percentage of the sample burning out between each determination.

28. Explain the meaning of the terms *interpolation* and *extrapolation*. Explain why the latter is often an unreliable procedure.

***29.** The time a pendulum takes to swing one complete to- and- fro oscillation is directly proportional to the square root of its length and inversely proportional to the acceleration of gravity. If the constant of proportionality is 2π, write down an expression for the time of one oscillation, explaining the meaning of the letters you use. Draw a graph showing the variation in the time of one oscillation of the pendulum for lengths between 50 mm and 500 mm, given that the acceleration due to gravity is 9·81 m/s^2.

How long must the pendulum be to have a period of (a) 1 second, (b) $\frac{1}{2}$ second?

***30.** The volume and pressure of a fixed mass of gas vary with each other according to the equation pv^{γ} = a constant if no heat is allowed to enter or leave the system while the changes are taking place. Draw a graph of the variation between p and v for values of p between 5×10^4 and 10^6 Nm^{-2}, given that the sample of gas under test occupies one cubic decimetre when the pressure is 1 atmosphere, and $\gamma = 1{\cdot}4$.

What would you expect the volume to be when the pressure is

(a) 2×10^5, (b) 7×10^5 Nm^{-2},

and what is the pressure when $v = 800$ cm^3? Would you expect the shape of the graph to change if p was measured in cm of mercury and v in cm^3?

31. A man pays £100 every year into a Society who pay him compound interest on his investment at 4% p.a. What is the sum he has invested at the end of

(a) 1 year, (b) 2 years, (c) 3 years, (d) T years?

Draw a graph of the amount he has invested after T years against values of T between 1 and 10.

32. At a certain moment, two concentric circles have radii of 2 cm and 3 cm, respectively. The radius of the outer circle then increases by 1 cm every second, while the radius of the inner circle increases by $\frac{1}{2}$ cm every second. Write down expressions for the radii of the two circles t seconds after they were 2 cm and 3 cm, respectively, and derive an expression for the area enclosed between them after t seconds. Represent the variation in this area with time graphically for $0 \leqslant t \leqslant 5$(s), and deduce how long it will be before the area enclosed between the circles is 100 cm^2. (Take $\pi = 3\frac{1}{7}$.)

33. The sum of the length and breadth of a rectangle always equals 6 cm. Draw a graph showing the variation in the area enclosed by the rectangle as the breadth varies, and show that the maximum area is enclosed when the rectangle is a square.

34. The sum of the radii of two circles always equals 10 cm. Draw a graph showing how the total area enclosed by both circles varies with the radius of one of them, and show that it is a maximum when the two circles are equal in size.

35. A trough whose length is 2 m has a triangular cross-section (see Fig. 1.39). Draw a graph to show

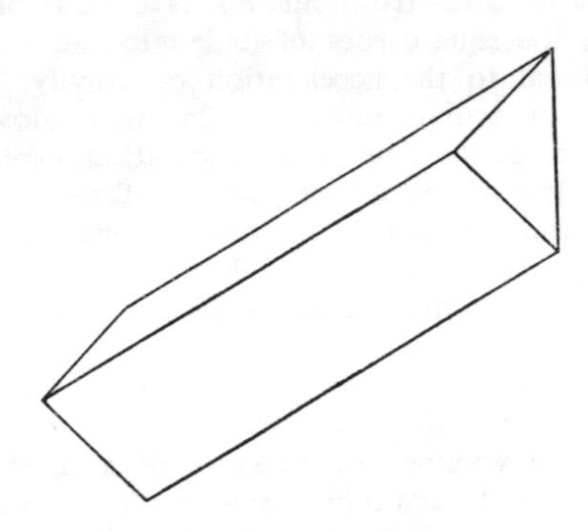

FIG. 1.39.

how much water is contained in the trough when the depth is x cm.

36. A piece of cardboard 12×10 cm is to be made into a rectangular, open container by cutting squares of side x cm from its four corners, and folding the cardboard along the dotted lines (Fig. 1.40). Show

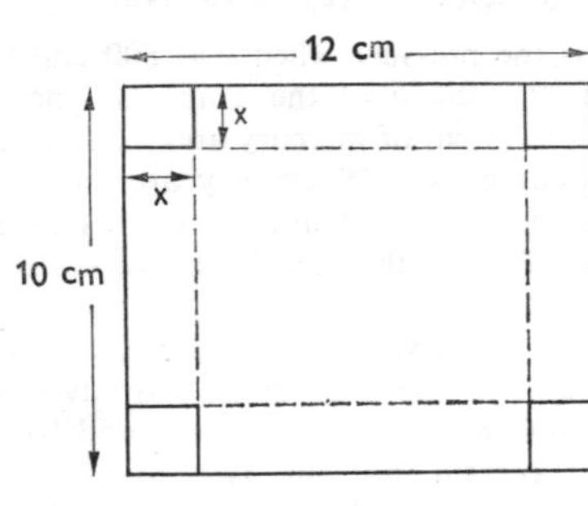

FIG. 1.40.

that the volume of the box is $(12-2x)(2x)$ cm³, and by plotting a graph of this function for $0 \leqslant x \leqslant 4$, find the maximum volume the box can have, and the value of x for which this volume is obtained.

37. A box with a square base and open at one end is to have a volume of 1000 cm³. What must be the dimensions if as little metal as possible is to be used in its construction?

***38.** Two points X and Y are situated 10 cm from the same side of a mirror, and 20 cm from each other. A ray of light from X strikes the mirror at a point P, P being x cm from the foot of the perpendicular from X to the mirror, and is then reflected to Y. Show that the length of its path from X to Y is $\sqrt{(10+x^2)}+\sqrt{(500-40x+x^2)}$.

By an adaptation of Fermat's Principle of Least Time, the actual path taken by the light is such that the distance it travels is a minimum. By plotting a graph of $\sqrt{(100+x^2)}+\sqrt{(500-40x+x^2)}$ for $0 \leqslant x \leqslant 20$, find the actual path taken by the light, and explain how this is in accordance with the second law of reflection.

39. It is believed that the quantity S which depends on t is given by the formula $S = a+bt^2$, where a and b are constants. By plotting S against t^2 for the table below, and drawing a straight line to pass approximately through the points, obtain approximate values for a and b. Note that a is given by the value S when $t = 0$.

t	2·0	2·4	3·0	3·6	4·0	4·5
S	10·4	12·7	16·5	20·0	23·1	28·0

(Take scales: 2 cm for 5 units of S, 2 cm for 4 units of t^2.) (OC)

40. If the value of a function $f(x)$ of a variable x is defined to be equal to the value of x^2, explain why:

(i) $f(-x) = f(x)$;
(ii) $f(3x) = 9f(x)$;
(iii) $f(a-x) = (a-x)(a+x)$, where a is a constant;
(iv) $f(x^2) = [f(x)]^2$;
(v) $f(a+x)-f(a) = x(2a+x)$.

CHAPTER 2

More about Graphs

2.1. A TIME-MOTION EXPERIMENT

Imagine a ball-bearing released from rest at the top of a tall column of oil contained in a glass cylinder. When it is released, it will fall under gravity through the oil, gaining speed until the viscous drag of the oil just balances the force of gravity and the ball-bearing travels with a constant speed (called the terminal velocity) to the bottom.

We can improve the precision of this description of the motion by using numbers. If we place a centimetre scale behind the cylinder so that its zero is at the base of the container, we shall be able to specify fairly accurately the position of the ball-bearing at any instant. And if we use a clock which is started at the instant the ball-bearing is released, we can say how long the ball-bearing takes to reach any particular position (see Fig. 2.1).

The results might well be like those shown in Table 1; and the relationship between the time values and the height values can be conveniently represented by plotting the results on a graph as described in Section 1.5.

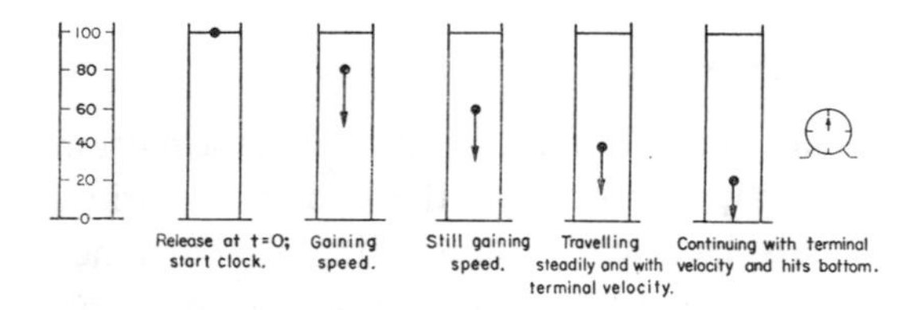

FIG. 2.1. Recording the position of a falling body at equal time intervals gives an accurate description of the motion.

TABLE 1

Time ofter release (s)	0	5	10	15	20	25	30	35
Height above base (cm)	100	95	87	75	60	40	20	0

Which variable should be regarded as the independent one? The height of the ball-bearing above the bench is determined by the time which has elapsed since its release, and it is not sensible to regard the time as consequential to the height. The horizontal axis should be allocated to the independent variable, which in this case is therefore the time variable, and the graph can be conveniently termed a height-time graph (by convention it is usual to refer to the dependent variable first).

The graph has the appearance of the downward sloping curve shown in Fig. 2.2, and if the curve is imagined to be the cross-section of a road running from left to right, it will be seen that the steepness increases at first and then takes on a steady value. That the speed of the ball-bearing also increases and then becomes steady is no coincidence, and this will be considered in more detail in the next section.

There are many situations in which the body in motion oscillates, that is, moves alternately

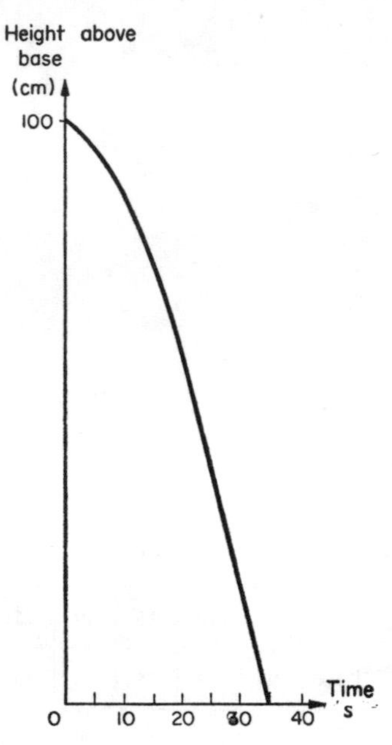

FIG. 2.2. The results can be represented by a graph.

to one side of a fixed reference point and then to the other. The motion of a cork bobbing up and down on water gives an example of an oscillatory movement. Yet another common type of motion is set into action by throwing a ball upwards and outwards from a cliff top (Fig. 2.3). In this case, the motion is not oscillatory, but the displacement relative to the point of projection is first in an upward direc-

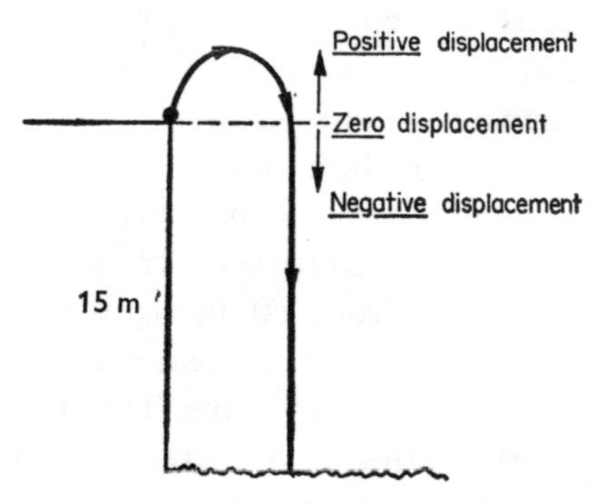

FIG. 2.3. A ball thrown upwards and outwards has displacements above and below cliff level.

tion, and then (after it has passed the level of the cliff top on the way down) in a downward direction. To represent these types of motion graphically we need to adopt some convention which enables us to distinguish between distances to one side of a fixed point and distances to the other. One convention commonly adopted uses positive numbers to represent the magnitudes of the distance of the body from the reference plane (e.g. the cliff top) in the one direction, and negative numbers for distances in the other direction. Thus, when the ball is 5 m above the cliff top it can be given a "displacement" value $+5$; and when it is 10 m below the cliff top, it can be given a displacement value -10. The graphical representation of a ball thrown up

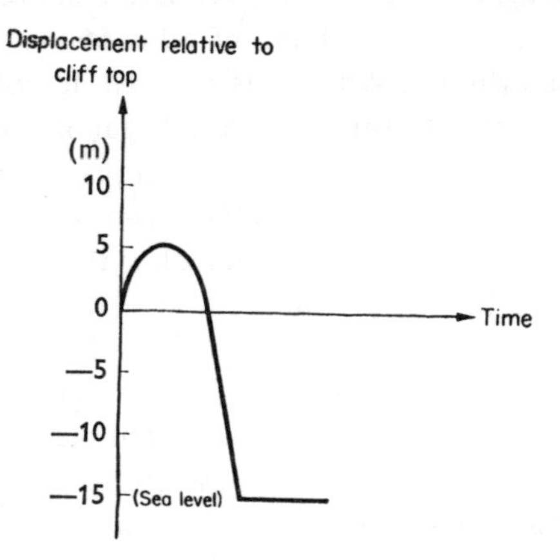

FIG. 2.4. The graph of the ball's motion. Positive values of displacement represent distances above the cliff top; negative values represent distances below the cliff top.

to a height of 5 m from a cliff top 15 m above sea-level then appears as shown in Fig. 2.4.

Example 2.1

The graph of Fig. 2.5b represents the motion of a lump of wood on the end of a spring, the displacement values being the distance of the wood from its rest position, distances below the rest position being registered as negative. Describe the motion.

At the start (time = 0) the wood is at its rest or normal position. The displacement then becomes positive as the wood moves up. Since the steepness of the graph decreases as time elapses, the speed decreases. Eventually (at the point marked X) the graph becomes flat and the distance of the wood above its rest position then decreases from its maximum value of 10 cm. After about 2·2 seconds it

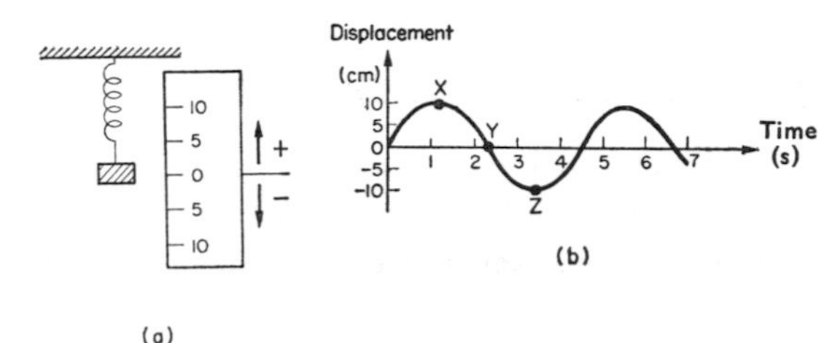

FIG. 2.5. The graph of the motion of the wood bob on a spring in (a) is given in (b).

passes through its rest position (Y), travelling fairly fast downwards (since the graph is fairly steep). The wood stops momentarily at its lowest point (10 cm below the rest position) after 3·3 seconds (Z), and then returns to the centre which it reaches after 4·4 seconds, travelling upwards. This is repeated every 4·4 seconds.

EXERCISE 2a

1. Draw the graph representing the motion of the ball-bearing to the oil column given in the text, the measured results being given on p. 35. Using the graph, answer the following questions:

(a) How far does the ball travel in

(i) the first 5 seconds,
(ii) the next 5 seconds?

Construct a table showing how far the ball travels in each 5-second period. How far does the ball travel in the fifth 5-second period?

(b) Represent on another graph the figures of the table constructed in (a).

(c) What is the average velocity of the ball-bearing in (i) the first 5-second period, (ii) the second 5-second period? What relation has the average velocity to the table of (a)? Draw a rough sketch graph showing the variation in the average velocity in a 5-second period with the time.

(d) Using the graph of (a), find how far the ball travelled in the periods

(i) 2 – 4 seconds,
(ii) 10 – 12 seconds,
(iii) 14 – 18 seconds,
(iv) 14 – 17 seconds,
(v) 14 – 15 seconds.

What were the average velocities in these periods? What would happen to the result of (v) if the interval of time was made progressively smaller?

2. The following cut-outs represent portions of graphs of the height of an object above a fixed level against the time. The height scale is up the page, and the time scale across the page; in each case the scales are the same. Describe the motion represented by the various cut-outs.

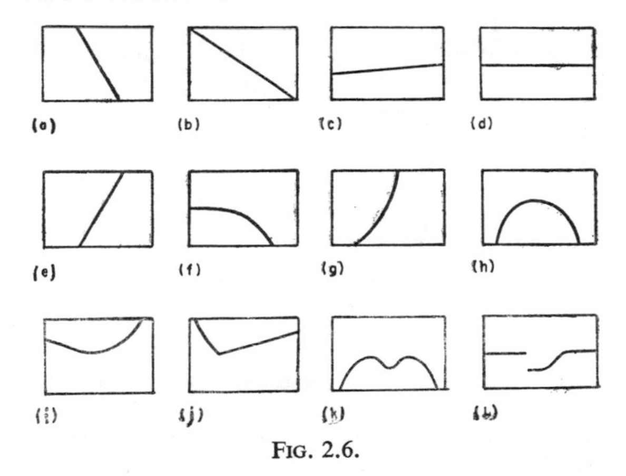

FIG. 2.6.

3. The picture shows a cam and lever which moves to the right. Under what conditions is the shape of the cam the same as the shape of the displacement graph for the lever?

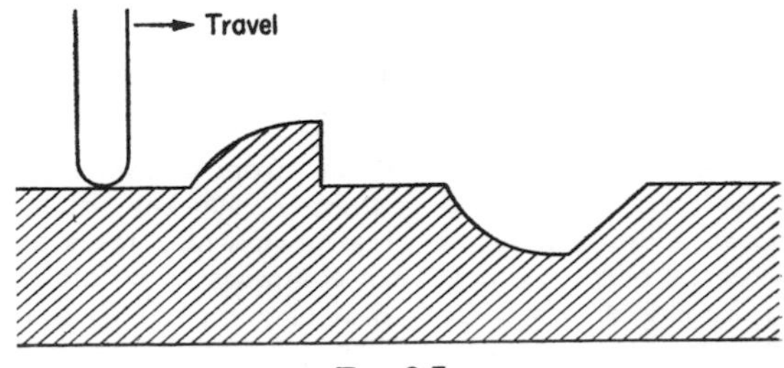

FIG. 2.7.

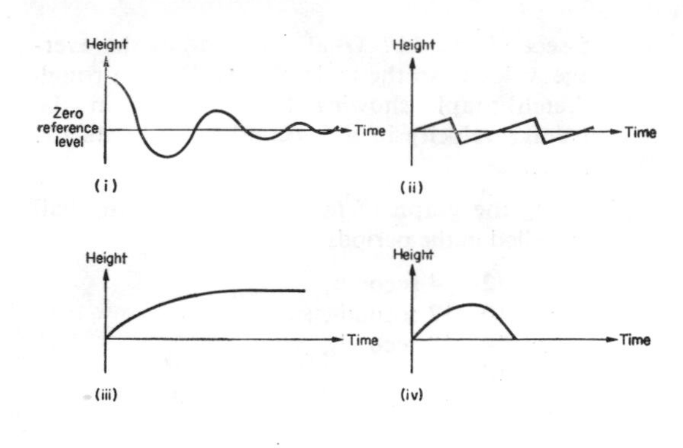

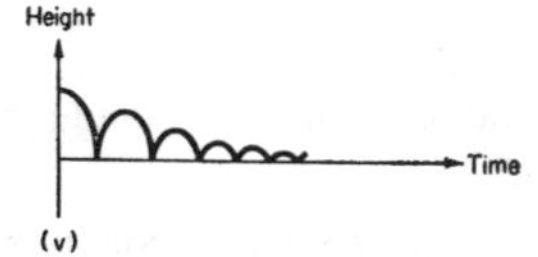

FIG. 2.8.

4. Describe the motion depicted in the following graphs:

5. If the ball whose motion is described by the graph of Fig. 2.4 has a constant horizontal velocity, explain why the shape of the graph will be unchanged if the time scale is replaced by a "horizontal distance" scale. Will the graph then be a scale drawing of the actual path of the ball?

What would happen to the gradient of the graph if:

(a) the horizontal scale was reduced,
(b) the horizontal scale was increased,
(c) the horizontal velocity of the ball was not constant, but decreased slowly due to wind resistance?

Illustrate your answers by sketches.

6. When a stone is projected into the air with a vertical speed of 26 metres/second, its height in metres above the ground after t seconds is given by the formula $h = 26t-5t^2$. Draw the graph of $h = 26t-5t^2$ for values of t from 0 to 6, taking 2 cm to represent single units of t, and 2 cm to represent 10 units of h.

From your graph estimate

(i) after how many seconds the stone will hit the ground,
(ii) its greatest height,
(iii) for how many seconds the stone is more than 30 metres above the ground. (OC)

7. A car sets out at 10 a.m. from a town A to travel at 65 km/h to a town B, 400 km distant. After 2 hours it has a breakdown which takes an hour to repair, and it then continues at 60 km/h. A second car starts from B at 11 a.m., travelling at 80 km/h to meet the first. Find by a graphical method the time when they meet and the number of minutes that the meeting is delayed as a result of the breakdown.

(Take 2 cm to represent 1 hour and 2 cm to represent 80 km.) (OC)

8. A car, running at 60 km/h, overtakes at 10 a.m. a cyclist going at 20 km/h, and at 10·15 a.m. meets a second cyclist going in the opposite direction at 17 km/h. Find by a graphical method the time at which the cyclists meet. (Take 2 cm to represent 10 minutes and 2 cm to represent 4 km. Start at 10 a.m.) (OC)

2.2 SPEED ON A GRAPH

The graph of Fig. 2.9 represents the motion of a rocket travelling upwards from a fixed reference level—say the earth's surface. The

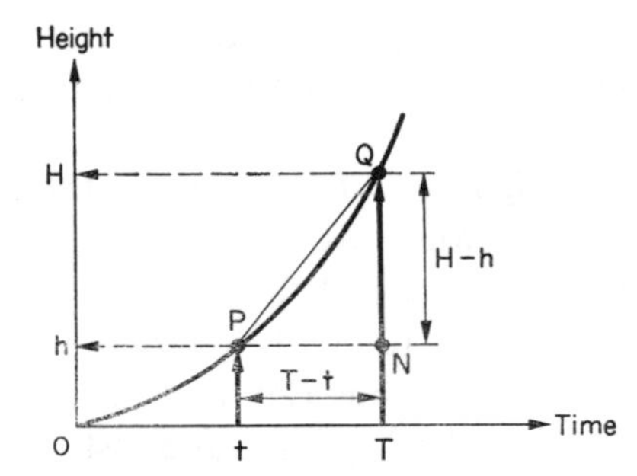

FIG. 2.9. The average speed between t and T is $(H-h)/(T-t)$, and the steepness of the graph is an indication of the speed.

point P maps together the height denoted by h and the time t at which the rocket is at that height: the point Q maps H and T. In the interval from t to T, the rocket moves from a height h to a height H, and the increase in height is thus $(H-h)$. And since the average speed in any interval is defined as the distance covered divided by the time taken, the average speed in the interval from t to T is $(H-h)/(T-t)$.

If the value of $H-h$ is expressed in cm and the value of $T-t$ in seconds, the ratio $(H-h)/(T-t)$ will give the speed in cm/s. The value of the speed can be calculated indirectly from the graph, and it can also be seen that for any fixed time interval $(T-t)$, the magnitude of the speed is directly related to the steepness of the graph. If the graph is very steep (in the sense of going from left to right), the value of $(H-h)$ will be comparatively large, and the distance covered in the time interval will be large. Thus, the speed is high. If the graph is nearly parallel to the time axis, the value of $(H-h)$, the distance covered in the same interval, and therefore the speed, will be comparatively low. Thus, steep graphs depict high speeds: flat ones depict low speeds.

The estimation of the speed of a body from the shape of a graph is, however, not easy, since the appearance of a graph can be altered so much by scaling (Fig. 2.10). The idea of

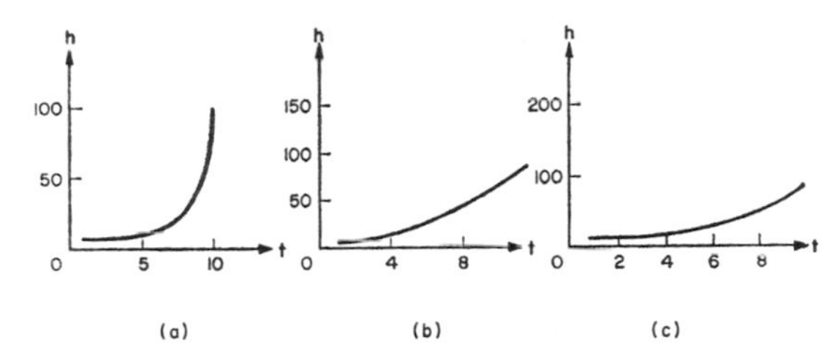

FIG. 2.10. The scale used along the axes has a great effect on the appearance of the graph.

a standard scale graph does not altogether solve the problem either, for the range of speeds that lie within human experience range from fractions of a centimetre per day to thousands of kilometres per second. Nevertheless, there are some advantages in employing a standard scale graph, as will be clear when the concepts of area and integration are studied in more detail, and the idea will be discussed further here.

Returning to Fig. 2.9, we can see that if 1 cm is used to represent 1 m along the distance axis, the length of QN in cm will be numerically equal to the distance $(H-h)$ travelled by the rocket between the moments represented by the points P and Q. Similarly, if 1 cm represents 1 second along the time axis, the length of PN in centimetres will be numerically equal to the value of $(T-t)$ in seconds. Thus, the average speed during the interval, $(H-h)/(T-t)$, will be given by the ratio QN/PN. Since the length of QN in centimetres represents the value of $(H-h)$ in metres, and PN the time in seconds, the value of this ratio gives the speed in m/s. We can also say that $QN/PN = \tan Q\hat{P}N$, and thus the tangent of the angle between the chord joining the points P and Q and the horizontal axis (or a line parallel to it) is numerically equal to the speed. In this case the apparent steepness of the graph is a true indication of the actual size of the speed. But, if other scaling is employed, a quick glance at the graph will only give a rough estimate of the variation in the speed.

We shall refer to a graph which employs 1 cm along both axes to represent a change of one unit in the value of the relevant variable as a "*unity scale*" graph: employing such a scaling enables absolute meaning to be attached to the geometrical properties of the graph. Here we have seen that the tangent of the angle between the chord PQ and the direction of the horizontal axis is numerically equal to the speed: if 1 cm along the vertical axis represents 1 km, and 1 cm along the horizontal axis 1 hour, the speed will be given in km/h. In most other cases, the geometrical properties of the graph may be useful only in a relative sense, and such quantities as speed should be calculated directly from the figures represented by the relevant points on the graph.

There is, however, another type of graph which does not give a misleading impression or useless geometrical dimensions as far as speed is concerned. If instead of using 1 cm on both axes to represent unit time and unit distance we use n cm, the dimensions of the graph will be scaled up by a factor n in both directions. Then the speed of the body whose motion the graph represents is properly given by

$$\frac{QN/n}{PN/n}$$

(Fig. 2.11), where QN and PN are measured in centimetres. But this ratio is the same as QN/PN, and the fact that n centimetres are used on both axes to represent unit change instead

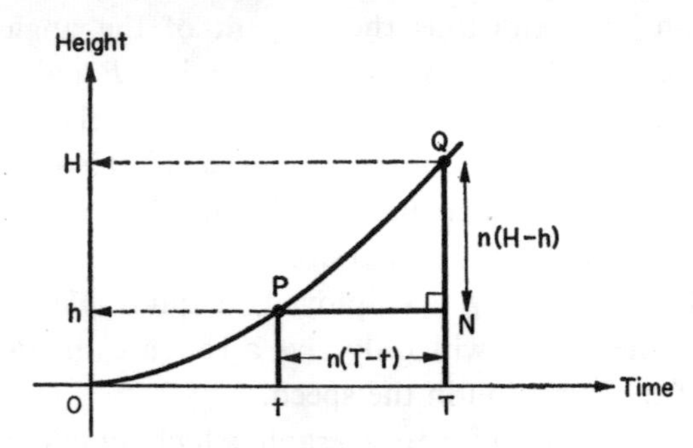

FIG. 2.11. The apparent steepness of a standard scale graph is the same as that of a unity scale graph.

of 1 centimetre can be ignored. A graph which employs the same distance along each axis to represent unit change in the value of the relevant variable will be referred to as a *standard scale* graph.

Although the geometrical property of steepness is the same on a standard scale graph as on a unity scale graph, the property of the area under the graph on the two types is not the same. Consider, for example, a unity scale graph representing the motion of a body travelling at a constant speed of 0·4 km/minute (Fig. 2.12). The distance travelled in, say, $\frac{3}{4}$ minute is simply $(0{\cdot}4\times\frac{3}{4})$ km, and the product of 0·4 and $\frac{3}{4}$ is also the product which gives the area between the straight line $v = 0{\cdot}4$ and a pair of ordinates separated by a distance representing $\frac{3}{4}$ minute. Similarly, the distance $(0{\cdot}4\times2{\cdot}5)$ km travelled in $2\frac{1}{2}$ minutes corresponds to the area "under" the graph between two ordinates $2\frac{1}{2}$ cm apart (Fig. 2.13). In general if we draw a unity scale velocity-time graph for a body travelling with constant velocity u (Fig. 2.14), the area which will be contained between the horizontal graph and any pair or ordinates t cm apart

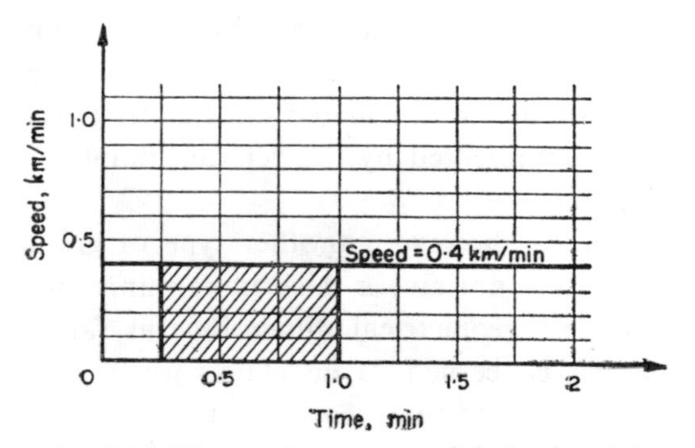

FIG. 2.12. The area between $v = 0{\cdot}4$, the time axis and the ordinates at 0·25 and 1·0 has the same numerical magnitude as the distance covered by the body travelling at 0·4 km/min in the interval from $t = 0{\cdot}25$ to $t = 1{\cdot}0$.

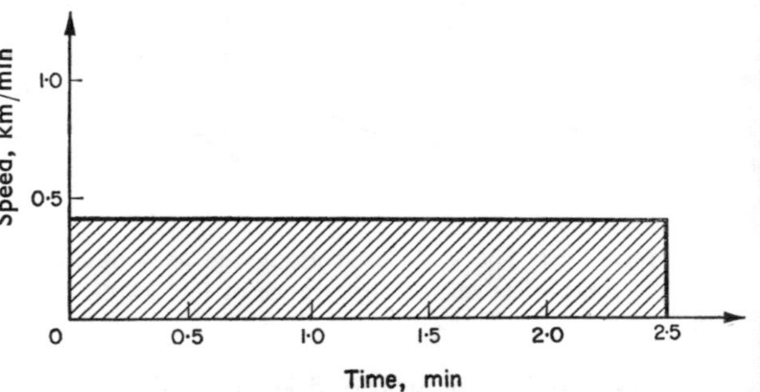

FIG. 2.13. The 1·0 cm² bounded by $v = 0{\cdot}4$, $v = 0$, $t = 0$ and $t = 2{\cdot}5$ represents a distance of 1·0 km...

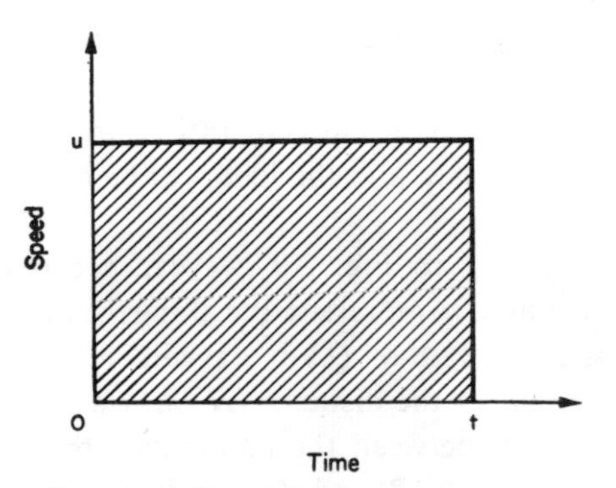

FIG. 2.14. ... and an area of ut cm² corresponds to a distance ut.

is (ut) cm²; and the numerical magnitude of this area gives exactly the distance travelled by that body during an interval whose length is represented by the distance between the ordinates.

If, however, we use a standard scale graph, the area contained between the graph and the time axis between any two ordinates will be n^2 times as great (Fig. 2.15). Thus, although we can still use the numerical value of the area on the graph to calculate or estimate distances travelled, we must beware of comparing distances on different graphs by comparing the representative areas directly.

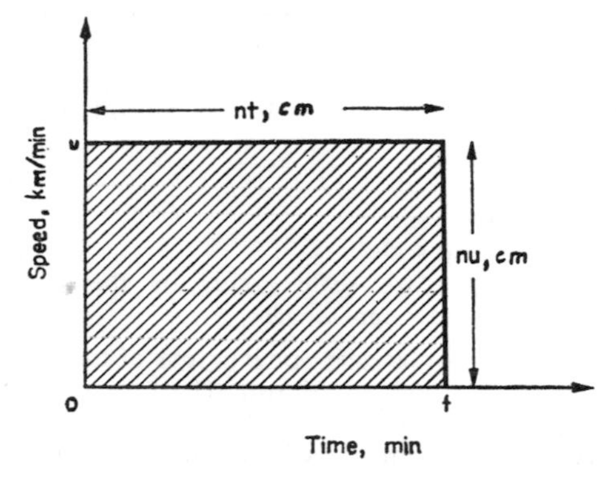

FIG. 2.15. On a standard scale graph, n^2ut cm² represent ut kilometres.

The examples which follow illustrate the use of the geometrical properties of graphs for these velocity–distance–time problems. A development of particular importance is made in Example 2.3, where the area on a graph which has neither a unity nor a standard scale is used to determine the distance travelled.

Example 2.2

The graphs of Fig. 2.16 represent the variation in the speed of a body, and are drawn with unity, standard and non-standard scaling, respectively. Calculate the distance travelled by the body between the times represented by the points P and Q on each of the graphs.

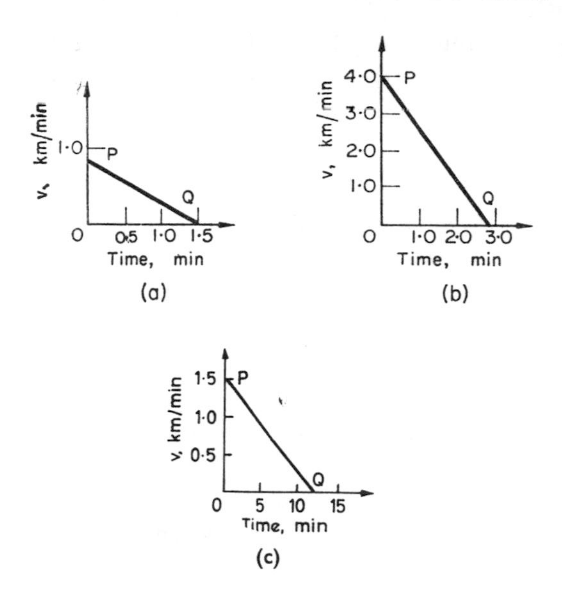

FIG. 2.16. (a) Unity scale. (b) Standard scale. (c) Non-standard scale.

(a) By direct measurement, $OP = 0{\cdot}8$ cm and $OQ = 1{\cdot}5$ cm, so the area OPQ can be calculated as $\frac{1}{2}(0{\cdot}8)(1{\cdot}5)$ or 0·6 cm². Since 1 cm² is the area of a square contained on the graph of a speed of 1 km/minute for 1 minute (Fig. 2.17a) and therefore represents 1 km, the distance travelled between the instance represented by P and Q is 0·6 kilometre.

(b) Again, by direct measurement, $OP =$ 2 cm and $OQ = 1{\cdot}4$ cm, so the area of the triangle OPQ can be calculated as $\frac{1}{2}(2)(1{\cdot}4)$ or 1·4 cm². In this case 1 cm

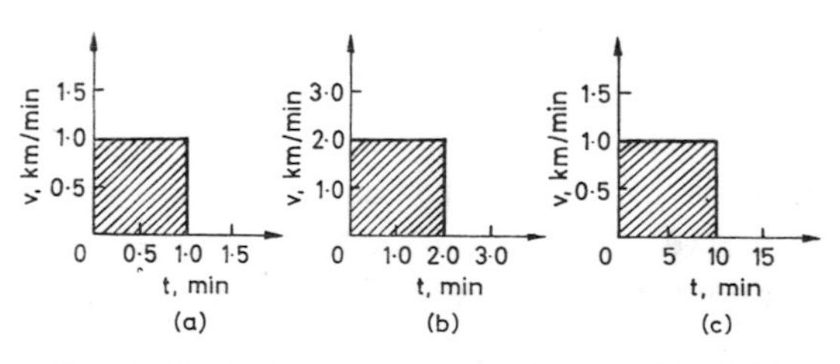

FIG. 2.17. (a) 1 cm² represents 1 km. (b) 1 cm² represents 2 km/min for 2 min, i.e. 4 km. (c) 1 cm² represents 1 km/min for 10 min, i.e. 10 kilometres.

represents 2 km/minute and 2 minutes, respectively, so a square of area 1 cm² represents 4 km (Fig. 2.17b). Thus, an area of 1·4 cm² will represent (4×1·4) or 5·6 kilometres.

(c) The area *OPQ* in Fig. 2.16c can be measured and calculated as $\frac{1}{2}(1{\cdot}5)(1{\cdot}2)$ or 0·9 cm². Now 1 cm represents 1 km/minute on the velocity axis and 10 minutes on the time axis, so an area of 1 cm² will represent 1 km/minute for 10 minutes or 10 kilometres (Fig. 2.17c). Thus, 0·9 cm² represents 9 kilometres, and the distance travelled by the body between *P* and *Q* is 9 kilometres.

Example 2.3

The graph shown in Fig. 2·18 represents the variation of the velocity of an aeroplane with time from the start of the flight. Estimate the total distance travelled during the flight.

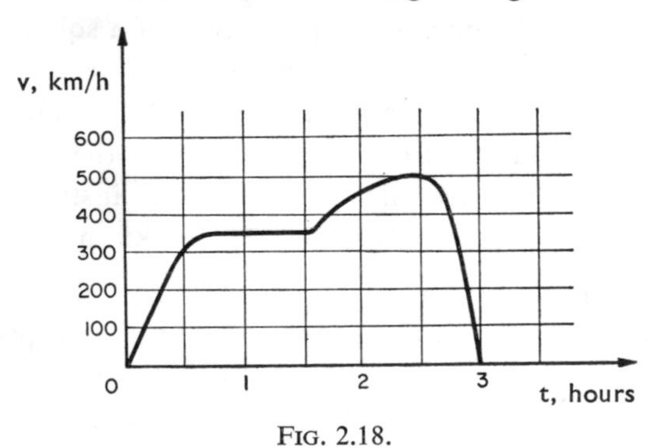

FIG. 2.18.

We can evaluate the area under the curve by counting squares, and it can be seen that the number of complete squares under the curve is 13. The part squares enclosed under the curve can be estimated to be the equivalent of approximately 9 squares,* so the total area under the curve is about 22 squares. By considering the 100 km/h horizontal line and the 1 hour ordinate line, it can be seen that a velocity of 100 km/h for 1 hour will enclose 2 squares. The distance travelled under these circumstances would be 100 km, and we can therefore equate two graph squares with 100 km. Thus, 22 graph squares represent a total distance of 1100 km. *Answer:* The estimated flight distance is 1100 km.

EXERCISE 2b

1. The distance *s* km of a car from a garage *t* minutes after leaving is given by the following table:

s (km)	0	1	2	3	4	5	6	7	8	9	10
t (min)	0	0·4	1·0	1·8	2·8	3·8	4·8	5·6	6·1	6·3	6·3

Draw
(i) a unity scale,
(ii) standard scale,
(iii) non-standard scale graph for the motion, and compare their appearances.

2. Three displacement-time graphs of the motion of a particular body are drawn. The first uses 1 cm along the axes to represent 1 km and 1 minute, respectively, the second uses 0·4 cm to represent 1 km along the displacement axis, and the third uses 1 cm to represent 1 km along the displacement axis (which is vertical) and 0·5 minute along the time axis. Describe the apparent steepness of the three graphs.

3. During a period of 2 minutes, the speed of a car decreases steadily from a value of 3 km per minute to zero. Draw a unity scale graph of the motion, and by calculating the area of the triangle formed by the graph and the axes, deduce how far the car travels while stopping.

4. A train decreases its speed uniformly from 3·8 km/minute to zero, and takes 6·2 minutes to do so. Draw a standard scale graph of the motion, and find the distance travelled by the train while stopping.

5. The speed of a jet aircraft varies with time after take-off as follows:

t (min)	0	5	10	15	20	25	30
v (km/min)	0	2	4	6	8	10	12

* When large numbers of squares are involved, an approximate estimate of the area of a number of part squares can be made by counting all squares of which half or more of their area is under the curve as one square, and ignoring all those with less than half their area included under the curve. Thus . . counts as one square, whereas . . is ignored.

Draw a

(i) unity standard scale graph
(ii) non-unity standard scale graph

to represent these figures, and find the areas between the two graphs and their respective time axes.

Hence write down the distance covered by the aeroplane in the

(a) first 4 minutes,
(b) fourth minute.

By what linear factor is your standard scale graph enlarged relative to the unity scale graph? By what factor are corresponding areas between the graphs and the time axes enlarged?

6. An aeroplane accelerates from rest to a speed of 4·5 km/minute during take-off, and takes 24 seconds to leave the ground. Given that the acceleration is uniform, draw a graph to represent the motion using 1 cm to represent 1 km/minute on the speed axis, and 1 cm to represent 0·1 minute on the time axis.

By considering the distance represented by an area of 1 cm² on such a graph, calculate the distance travelled by the aeroplane before taking off.

7. A standard scale velocity/time graph uses 1 cm to represent n km/minute and n minutes on the two axes. Write down the distance represented by an area of

(i) n^2 square centimetres, (ii) 1 square centimetre.

8. A velocity–time graph uses 1 cm to represent 50 m/h on the velocity axis and 1 cm to represent 5 minutes on the time axis. What distance does an area of 1 cm² on the graph represent?

9. A velocity–time graph uses p cm to represent q km/h on the velocity axis, and r cm to represent s minutes on the time axis. What distance does an area of 1 cm² on the graph represent?

10. Two men start out at 11.0 a.m. from Winborough to walk to Marlchester, a town 10 km away. A motor-cyclist takes a fourth man on the pillion, and upon reaching Marlchester, the fourth man gets off and the motor-cyclist returns to meet and pick up one of the pedastrians. The motor-cyclist takes him to Marlchester, and then returns to pick up the second pedestrian. He finally takes him to Marlchester. Given that the pedestrians' speeds are both 10 km/h and the motor-cyclist's 60 km/h, draw on a single diagram a graph which represents the journeys of the four men, and find at what time all four will be at Marlchester.

11. A train starting from a station X accelerates uniformly for 2 minutes and then has a speed of 90 km/h. It maintains this speed for 4 minutes and then is retarded uniformly for 3 minutes until it comes to rest at a station Y. Find the distance in kilometres between the two stations. (C)

12. An electric train travels between two stations 7 km apart in 5 minutes, starting from and finishing at rest. During the first $\frac{3}{4}$ minute, the acceleration is uniform, for the next $3\frac{3}{4}$ minutes the speed is constant, and for the last $\frac{1}{2}$ minute the retardation is uniform. Sketch a velocity–time graph of the motion, and hence or otherwise, find in km/h the maximum speed attained. (OC)

13. Two points A and B are a distance a apart. A particle moves from rest at A with an acceleration f until it acquires a speed V. It maintains this speed V for a time T, and then undergoes a retardation f which brings it to rest at B. Prove that

$$T = \frac{a}{V} - \frac{V}{f},$$

and show that the total time taken is

$$\frac{a}{V} + \frac{V}{f}.$$

Prove also that, if the average speed for the whole journey is $\frac{3}{4}V$, then the particle travels at speed V for two-thirds of the total distance. (OC)

14. Given that the maximum acceleration of a lift is 4 m per second squared, and the maximum retardation is 5 m per second squared, find the minimum time for a journey of 30 m

(i) if the maximum speed is 5 m/s,
(ii) if there is no restriction on speed. (SU)

2.3. DISPLACEMENT AND VELOCITY

The examples of motion we have considered so far have all involved height: but motion often takes place in a horizontal plane, and the distance of the moving body from a fixed point in the plane is the variable by which the movement can be described. Now distance is a *scalar* quantity—that is, it has only size—and direction in which the distance is measured must be specified if the motion is to be accurately described. The quantity specifying a distance and its direction is called the *displacement*, and is a *vector** quantity. A man who

* Vectors are not subject to the ordinary laws of algebra: in fact a vector can be defined as a quantity which has to be added to another vector by a "triangle" method. (A more general definition is that a vector is a $1 \times n$ or $n \times 1$ matrix, but we need not go into such generalities here.) This triangle method can be explained by the law that if two vectors are represented both in magnitude and direction by two

drives at 50 km/h due north for 12 minutes gives himself a displacement of 10 km due north relative to his position at the beginning of the 12-minute interval: the distance he is from his starting-point is simply 10 km (Fig. 2.19).

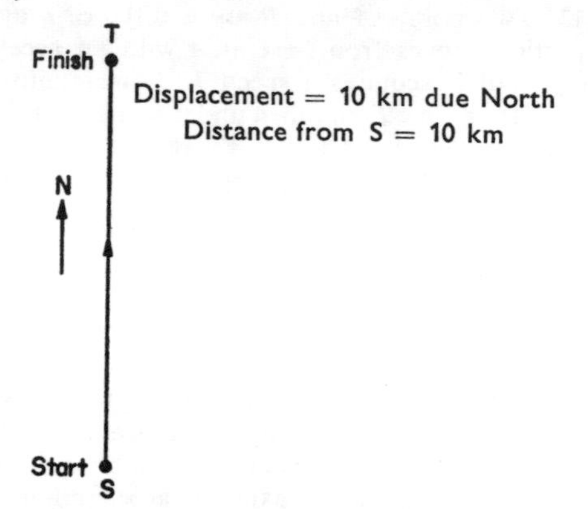

FIG. 2.19. Displacement is a vector, and has both magnitude and direction.

Fortunately, many motions met with in practice have displacements confined to a straight line: rather than having to specify the geographical direction, the only displacements involved are to one side or the other of a fixed reference mark. In these cases, the displacements to the one side can be represented by positive numbers, and the displacements to the other by negative numbers. Thus, in Fig. 2.20, the displacements to the left of the fixed reference point are represented by negative numbers (to compare with graphical axes), and those to the right of the reference point are represented by positive numbers. The direction of the displacement is thus implicit in the sign, and a quantity such as -17 cm is to be regarded as a displacement, and therefore a vector, rather than a distance, which cannot sensibly be negative. The letter generally used for the displacement variable is s.

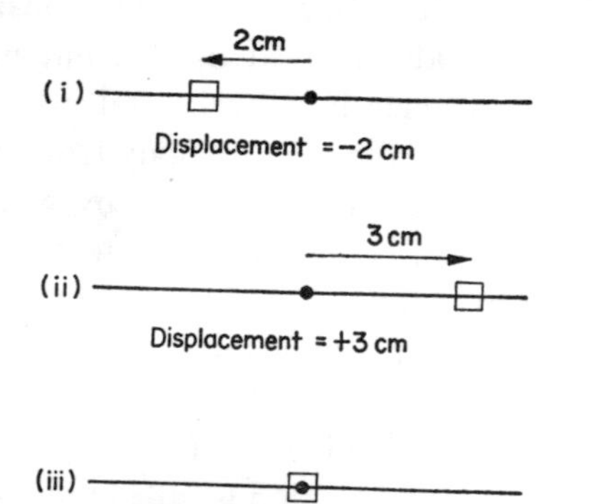

FIG. 2.20. Displacement in opposite directions can be represented by positive and negative numbers.

It is convenient to define a quantity equal to the ratio of the change of displacement occurring in an interval of time to the size of that interval of time. This ratio—as opposed to the ratio of the change in height or distance to the time interval—is called the *velocity*. Since displacement is a vector quantity, velocity also will be a vector quantity, and will thus have a direction associated with it. Very often the average speed of a body in an interval of time and its average velocity during the same interval are numerically equal, but a difference will occur whenever the direction in which the body travels during the interval changes. To take an example we can refer to example 2.1 on p. 36. If we consider the motion between the start and the moment represented by the point Y on the graph in Fig. 2.5b, we can see that the actual distance travelled by the wood is 20 cm—i.e. 10 cm out and 10 cm back—but the displacement at Y is zero. The average speed is thus 20/2·2 or 9·1 cm/s, whereas the average velocity is 0. Thus for motion along a line, the speed and velocity during an interval are not numerically equal if the motion reverses direction during the interval, and by considering a displacement–time graph with a highest or lowest point (Fig. 2.21), we can see that the ratio

---

sides of a triangle, their sum is represented both in magnitude and direction by the third side. It follows that a vector always possesses two independent properties which can be represented by a line magnitude and direction simultaneously.

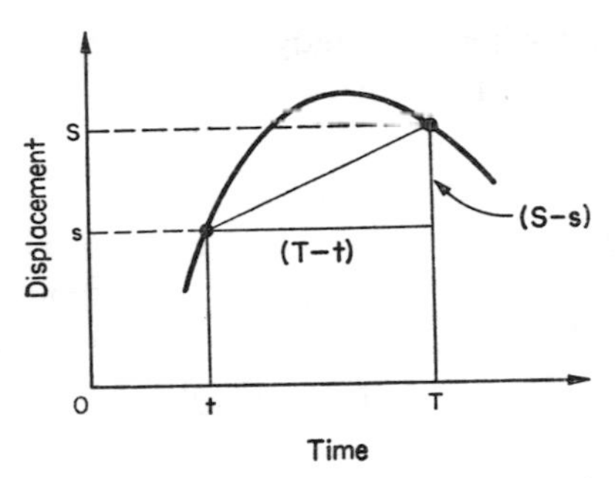

FIG. 2.21. The ratio $(S-s)/(T-t)$ gives a velocity rather than a speed.

$(S-s)/(T-t)$ gives the average velocity rather than the average speed in the interval between t and T. Accordingly it will be referred to as a velocity from now on. A more general example showing the difference between speed and velocity when the body in motion does not change direction along a straight line is given in example 2.4.

The expression $(S-s)$ represents the value of a change in displacement, and can be sensibly represented by the single symbol δs. If S represents a later value of the displacement than s, the value of T in the denominator of the expression for the velocity $(S-s)/(T-t)$ will be greater than that of t, and the denominator as a whole will be positive. The sign of the velocity will therefore be the same as the sign of δs, and if, for example, the displacement decreases, the velocity will take a negative value. We can generalise this deduction and say that if a body in motion moves in the direction of the negative displacement axis, its velocity will be negative, and if it moves in the direction of the positive displacement axis, its velocity will be positive. Thus, if we measure the vertical distance of a body from ground-level, using positive numbers for displacements above ground-level and negative numbers for displacements below ground-level, we can see that a velocity in an upward direction will be positive, and a velocity in a downward direction will be negative (Fig. 2.22). On a displacement–time graph, the upward direction of the vertical axis conventionally represents increasing displacement, and thus an upward-pointing curve represents a positive velocity (Fig. 2.23). Similarly, a downward-pointing curve represents a negative velocity.

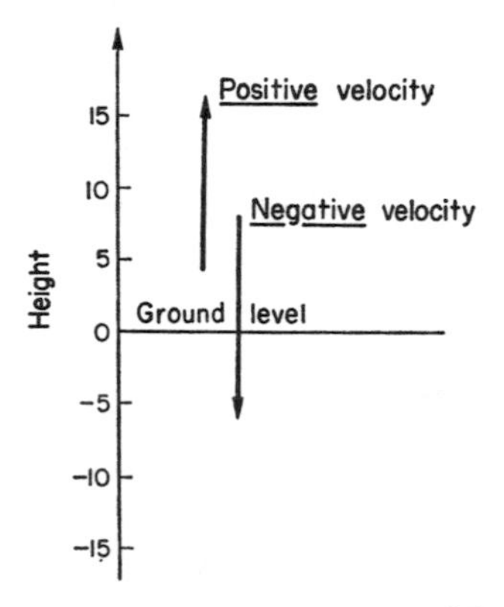

FIG. 2.22. If positive numbers are used for heights in an upward direction, upward velocities will be positive and downward velocities negative.

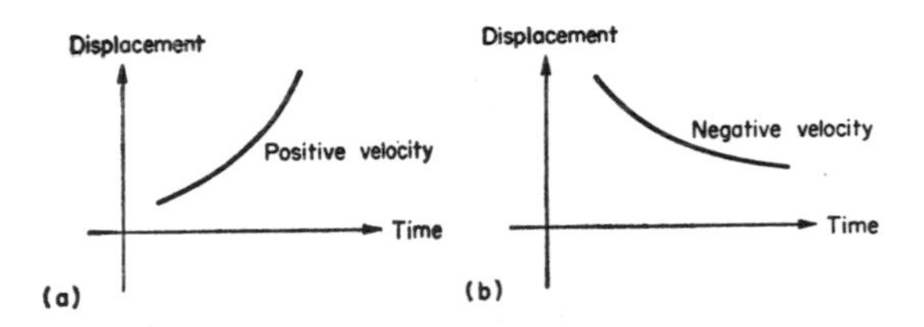

FIG. 2.23. The direction of the curve gives an indication of the sign of the velocity of the body whose motion it describes.

Example 2.4

The graph shown in Fig. 2.24 represents the motion of a body descending towards the

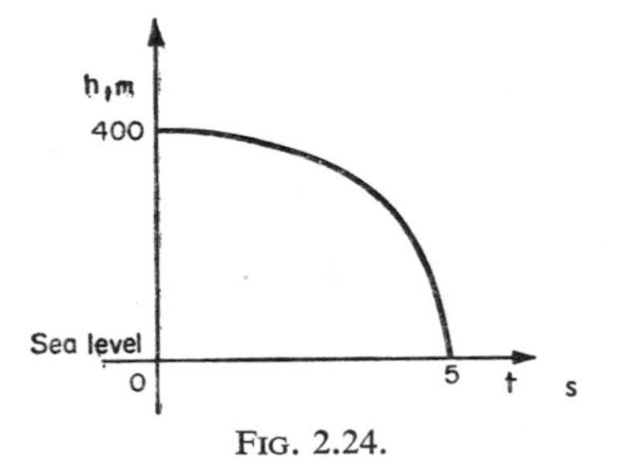

FIG. 2.24.

sea from a cliff. The displacement h is the height of the body in metres above sea-level at a time t seconds after the body commences its journey at the top of the cliff. Describe the motion, and calculate the average velocity during the third second.

Since the graph is initially horizontal, the body is released from rest at a height 125 m above sea-level. It descends with increasing speed (since the graph grows steeper). After 5 seconds, it strikes the sea and is brought to rest.

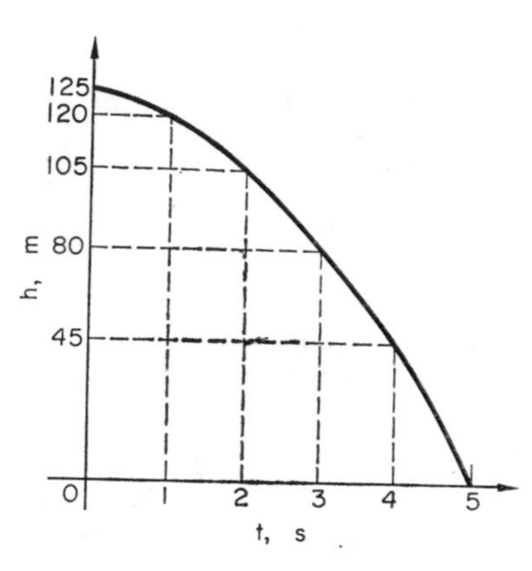

FIG. 2.25.

The value of h after 2 seconds is 105 m (Fig. 2.25), and after 3 seconds it is 80 m. The average velocity is thus $(105-80)/(3-2)$ or 25 m/s downwards.

Example 2.5

The graph of Fig. 2.26 represents the motion of a body moving about a fixed point, positive displacements representing distances to the right of the fixed point. Describe the motion, and calculate the average velocity and speed during the 2nd second, and between $t = 2$ and $t = 4$.

The body starts with a moderate velocity to the right, but this decreases and becomes zero at $t = 1$ when the body is 5 length units to the right of the fixed reference point. The velocity then becomes negative (i.e. the movement is directed towards the left), and at $t = 2$ the body passes through the fixed point. It then continues moving to the left, but slowing down, and at $t = 3$ it has stopped momentarily 5 length units to the left of the fixed point. It then returns with an increasing velocity to the fixed point, where it arrives after 4 seconds.

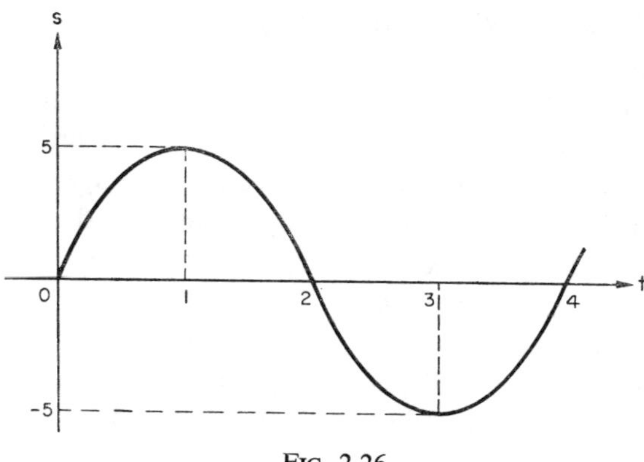

FIG. 2.26.

The distance covered during the second between $t = 1$ and $t = 2$ is 5 length units: the average speed is thus 5 length units/s. The change in displacement is -5 length units, and therefore the average velocity is -5 length units/s or 5 length units/s to the left.

The distance covered between $t = 2$ and $t = 4$ is 10 length units (5 units out and 5 units back), and therefore the average speed is 5 length units/s. The change in the displacement is zero (since it finishes where it started), and the average velocity during this interval is therefore zero.

Example 2.6

A fixed reference point P is set up as shown in Fig. 2.27. The direction of PN is due north. A man starts from P and walks 3 km due north to X, and then 4 km due east to Q. What is:

(i) his distance from P,
(ii) his displacement from P,
(iii) the distance he has covered?

If he took 2 hours to make the journey, what is his

(i) average speed,
(ii) average velocity?

Since PXQ is a right-angled triangle, PQ is equal to 5 km (by Pythagoras). The man's final distance from P is thus 5 km, and his displacement is 5 km N53°E. (Tan $Q\hat{P}X = 4/3$, and $Q\hat{P}X$ therefore equals 53° approximately). The

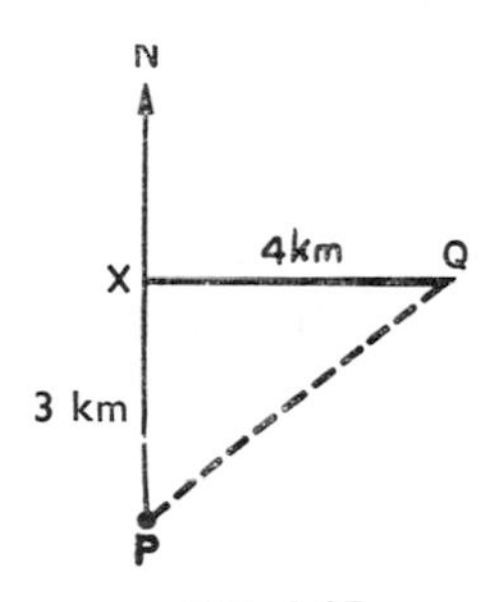

FIG. 2.27.

distance he covered is 7 km altogether (3+4), and therefore his average speed

$$= \frac{\text{distance covered}}{\text{time taken}} = 7/2 = 3\tfrac{1}{2}\ \text{km/h}$$

and his average velocity

$$= \frac{\text{displacement from starting point}}{\text{time}}$$

$$= 5/2 = 2\tfrac{1}{2}\ \text{km/h N53°E}.$$

EXERCISE 2c

1. Figure 2.28 represents a geographical map showing the positions of three points A, B and C relative to a fixed point P, the scale of the map being 1 cm to the kilometre. By measurement with ruler and protractor, write down the distances and displacements of the three points from P.

2. A man leaves a fixed reference point and walks 5 km due south, and then 5 km due east. What is his final displacement relative to the fixed reference point, and what is the total distance he has walked?

3. A man walks 100 metres in a direction 30° west of north. What further displacement will give him a position 100 metres 30° east of north?

4. Camford is 30 km N30°E of Oxbridge as the crow flies: the distance by road is 34 km. A car covers the distance between the two towns in 1 hour exactly. What is its

(i) average speed,
(ii) average velocity, for the journey?

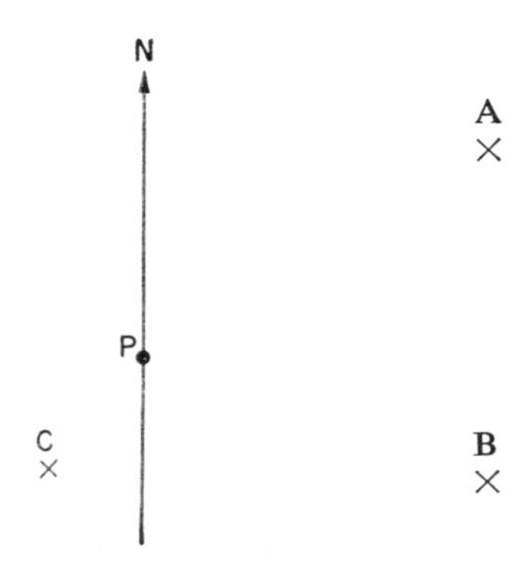

FIG. 2.28.

5. A ball is dropped from the top of a mine shaft, and its distance beneath the point from which it was dropped varies with the time after its release as shown in the table below. Represent its motion graphically, drawing three graphs, and using the following scales for the axes:

(i) 1 cm for 1 m and 1 second,
(ii) 1 cm for 1 m and 5 cm for 1 second,
(iii) 1 cm for 1 m and 10 cm for 1 second.

Distance of ball below top of mine (cm)	0	0·8	3·2	7·2	12·8	20·0
Time after release (s)	0	0·4	0·8	1·2	1·6	2·0

Join the points representing the positions of the ball at the instant of release and after half a second on each of the three graphs, and write down the value of the average speed of the ball during the first $\frac{1}{2}$ second. Measure the angles the lines drawn on each of the three graphs make with the positive directions of the time axes, and find the values of the tangents of these angles. What are the connections between the three tangent values and the speeds of the ball, and under what conditions does the value of the tangent of the chord joining two points on such a graph equal the speed during the interval represented by the points?

6. A car covers a distance of 12 km in 9 minutes, its distance from the starting-point varying with time as given in the table below Draw a graph describing the journey, and join the points describing the car's position $2\frac{1}{2}$ minutes and $6\frac{1}{2}$ minutes after the start of the journey with a straight line. Measure the angle this line makes with the positive direction of the time axis, and hence deduce the average speed of the car in the interval.

What is the maximum speed of the car?

Distance from starting point (km)	0	0·5	1·5	3·0	4·8	6·8	8·8	10·2	11·2	12·0
Time after start (min)	0	1	2	3	4	5	6	7	8	9

7. The table beneath represents the motion of an oscillating particle on a spring, displacements above its equilibrium position being accorded positive values. Draw a graph representing its motion, and calculate the average speed and the average velocity of the particle in the following intervals:

(i) $t = 0$ to $t = 0{\cdot}6$, (iv) $t = 1{\cdot}0$ to $t = 1{\cdot}6$,
(ii) $t = 0{\cdot}6$ to $t = 1{\cdot}0$, (v) $t = 1{\cdot}0$ to $t = 2{\cdot}0$,
(iii) $t = 0$ to $t = 1{\cdot}2$, (vi) $t = 0{\cdot}4$ to $t = 2{\cdot}0$.

Displacement (cm)	10·0	8·7	5·0	0	−5·0	−8·7	−10·0	−8·7	−5·0	0	5·0
Time (s)	0	0·2	0·4	0·6	0·8	1·0	1·2	1·4	1·6	1·8	2·0

8. A wild goose can fly at 80 km/h in still air. Aiming due north, it is blown 60 km east every hour by the wind. What will be its displacement at the end of an hour relative to its position at the beginning of the hour, and what is its velocity relative to the earth?

†**9.** A gun-boat is 20 km due north of a frigate at noon, and is travelling at 40 km/h in a direction W60°S. The frigate is travelling at 20 km/h due west. Make a scale drawing of their positions at noon and at 1 p.m. What is the displacement of the gun-boat relative to the frigate at 1 p.m.? What is the velocity of the gun-boat relative to the frigate?

†**10.** A boat can be rowed in still water at 6 knots: it is rowed straight across a stream which flows with a velocity of 8 knots. What is the velocity of the boat relative to the bank?

2.4. COORDINATE GEOMETRY

The ideas of calculus are more freely developed by removing the physical restrictions inevitably imposed by considering specific problems. This can be done by employing generalised variables which are denoted by the letters y and x, representing the dependent or range variable and the independent or domain variable, respectively. Attention can then be given more to the shape of a graph than to the nature of the problem it represents, and in doing this we become more conscious of the geometrical patterns resulting from different forms of relationship between x and y. The branch of mathematics which investigates the properties of curves through the algebraic expressions representing them is known as *Coordinate Geometry*.

To introduce the vocabulary of coordinate geometry, let us recapitulate one or two ideas we have already met, and as a basis for the recapitulation, let us consider a situation concerned with the marketing of silver.

We will suppose that the market price for gold is \$1·23 per gramme. The cost of x g will then be \$$(1{\cdot}23x)$, or, denoting the cost in dollars by y, we can write the relationship between x and y in the form $y = 1{\cdot}23x$. In accordance with the convention mentioned above, we are regarding the weight (x) as the domain or independent variable and the cost (y) as the range or dependent variable.

To plot a graph of this relationship, we plot a sample of the points representing pairs of values of y and x: one could be the point P

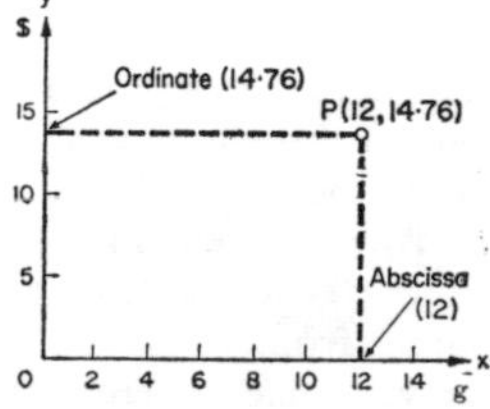

FIG. 2.29. The point P maps the weight 12 g with its cost \$ 14·76: its coordinates are 12 and 14·76; its abscissa is 12 and its ordinate 14·76.

shown in Fig. 2.29. The role of P can be regarded in two ways: it can be said to represent the ordered number pair (12, 14·76), or it can be said to map the weight value 12 g onto its cost \$14·76. In either case the two numbers it ties together are expressed solely in the position of P relative to the axes, and from the geometrical viewpoint, we can regard the position of P as precisely expressed by the ordered number pair (12, 14·76). The numbers 12 and 14·76 are called the *coordinates* of P, and the first number inside the bracket is always understood to be the value of the domain or independent variable represented by the point. Since the letter x is conventionally used for the domain variable, the number 12 can be said to be the *x-coordinate* of P, or, as it is frequently called, the ***abscissa*** of P. Thus in this case the abscissa of P is 12. Similarly, the second number inside the bracket containing the coordinates gives the *y-coordinate* or *ordinate*. It will be realised that the abscissa is a value associated with the horizontal axis, and the ordinate a value associated with the vertical axis (Fig. 2.29).

To complete the graphical representation of the equation $y = 1{\cdot}23x$, we must now construct some more points. Let us choose to plot those with abscissae 2, 6 and 16, and to label them A, B and C, respectively. By calculation from the equation $y = 1{\cdot}23x$, the corresponding ordinates can be found to be 2·46, 7·38 and 19·68, respectively, and the points are shown plotted in Fig. 2.30. It will be found possible to draw a straight line to pass through all the points, and we can therefore say that the equation $y = 1{\cdot}23x$ gives a straight line when represented graphically. For this reason, such a relationship or equation can be referred to *linear*, and $y = 1{\cdot}23x$ is said to be the *equation of the line*. It can also be seen from Fig. 2.30 that the line goes through the point (0, 0): this point is known as the *origin*.

The coordinates of any point lying in the line whose equation is $y = 1{\cdot}23x$ must "satisfy" the equation. This statement means that, if the value of the ordinate of any particular point is substituted for the letter y in the equation, and the corresponding value of the abscissa of that point substituted for x simultaneously, then the left-hand side of the equation must equal the right-hand side. For example, A is the point for which $x = 2$, and $y = 2{\cdot}46$; substituting these values into the equation $y = 1{\cdot}23x$, we find that the left-hand side equals 2·46, as does the right-hand side ($2 \times 1{\cdot}23$). The equation is thus satisfied. On the other hand, D (see Fig. 2.31a) is the point (6, 15); substituting its coordinates into the equation, we find that the left-hand side equals 15 and the right-hand side equals 7·38. The equation is not satisfied, and D does not lie in the line. It might be noted in passing that the ordinates y and the abscissae x of points which lie above the line satisfy the inequality $y > 1{\cdot}23x$

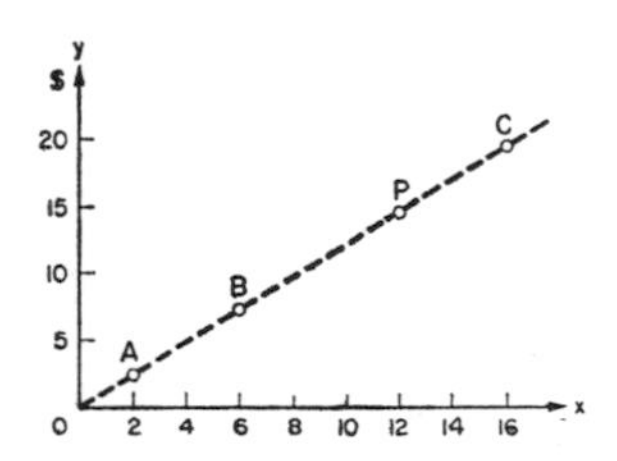

FIG. 2.30. The points $A(2,2{\cdot}46)$, $B(6,7{\cdot}38)$, $P(12,14{\cdot}76)$ and $C(16,19{\cdot}68)$ lie on a straight line.

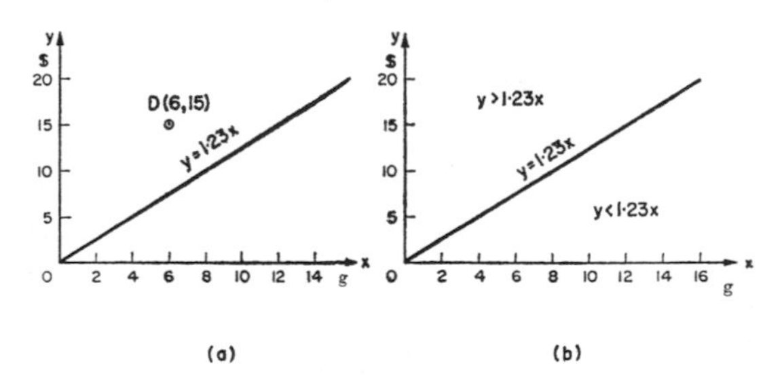

FIG. 2.31. (a) The coordinates (6, 15) of D do not satisfy the equation $y = 1{\cdot}23x$: and D does not lie in the line; (b) points for which $y > 1{\cdot}23x$ lie above the line, and those for which $y < 1{\cdot}23x$ lie below the line.

or $(y-1{\cdot}23x) > 0$. For points below the line, $(y-1{\cdot}23x) < 0$ and it is only for the points in the line that $(y-1{\cdot}23x) = 0$ (Fig. 2.31b).

More complicated relationships can also be represented graphically. The equations will not generally be simple, and the graphs not necessarily linear. Relationships or graphs which are not linear are logically referred to as *non-linear*.

If the law relating the two variables in any particular problem can be written as an algebraic equation, it will be possible only in the simpler cases to arrange the equation in the form with a single y on one side of an equation and some *expression* (i.e. a set of terms) involving only x on the other. When this is done, the relationship is said to be expressed in an *explicit* form, and relationships expressed in a form where the two variables are inseparable (e.g. $y^3+x^2y+7x^2 = 5$) are said to be *implicit*. The statement in Section 1.1 that that $y = f(x)$ indicates that y is a function of x can now be augmented to include the fact that the relationship can be expressed in an explicit form: an implicit form of relationship between x and y is better represented by the equation $g(x, y) = 0$. As explained in Section 1.1, $f(3)$ represents the value of the function $f(x)$ where $x = 3$: the corresponding "implicit" expression $g(2, 7)$ represents the value of the function $g(x, y)$ when $x = 2$ and $y = 7$.

The relationships we shall come across will be mostly expressible in an explicit form, and we can therefore generally take the word "function" to be synonymous with "a set of terms". Thus, in the problem concerning the value of given weights of gold, the relationship between y and x is explicit and expressible in the form $y = 1{\cdot}23x$. The particular function of x which y equals in this case is the single term $1{\cdot}23x$. Or if the law relating y to x is that y must be 7 more than $3x$, then $y = 3x+7$. In this case y is another function of x, and the form of the function is $3x+7$. [$f(x) \equiv 3x+7$.]

A final word about dependent and independent variables: if the relationship between the variables involved in the problem can be expressed explicitly, then it is usual to choose as the independent variable the one which appears on the more complicated side of the equation (usually the right), e.g. x in $y = x^2$, p in $V = 3+7p+p^2$, and n in $C = 6n^3-n$. To plot a point on the graph of the relationship, a value can be arbitrarily assigned to this variable, and the value of the dependent variable is then easily calculated; the reverse order is obviously much more difficult, and in some cases impractical.

Example 2.7

Arrange the equation $(y+13)/(x+4) = 5$ in an explicit form for y, and sketch its graph. Write down the abscissa of the point whose ordinate is -3, and the ordinate of the point whose abscissa is 2.

Shade in the area occupied by points which satisfy the inequality $y > 5x+7$, and state whether the point (1, 10) lies in the line $y = 5x+7$ or not.

Multiplying the equation throughout by $(x+4)$ gives $y+13 = 5(x+4)$. Expanding the right-hand side and rearranging then leads to the explicit equation for y, viz. $y = 5x+7$.

Plotting a number of sample points shows the graph to be linear; a sketch of $y = 5x+7$ is shown in Fig. 2.32a.

The value of x when $y = -3$ is obtained by substitution into the equation $y = 5x+7$,

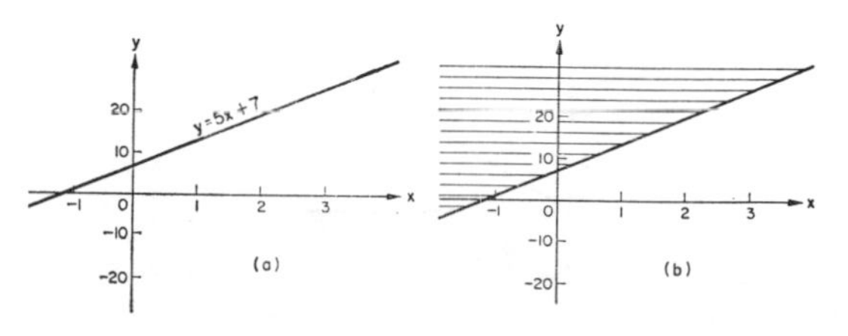

FIG. 2.32.

giving

$$-3 = 5x+7, \quad \text{or}$$
$$x = -2.$$

Thus, the x-coordinate or abscissa of the point whose y-coordinate or ordinate is -3 is -2.

Similarly, when $x = 2$, $y = 17$, so the ordinate of the point whose abscissa is 2 is 17.

Any point whose ordinate is greater than $5x+7$ lies above the line $y = 5x+7$, and so the region occupied by the points satisfying this inequality is the entire area above the line (shown shaded in Fig. 2.32b).

If $x = 1$, $y = 12$ for a point in the line: the point (1, 10) thus satisfies the inequality $y < 5x+7$, and therefore lies below the line

EXERCISE 2d

1. Draw an x-axis ranging from -2 to $+2$ and a y-axis ranging from -5 to $+5$. Plot the following points: (0, 0), (1, 2), $(-0{\cdot}5, -1)$, (2, 4), $(-1{\cdot}5, -3)$, $(-1, 1)$, $(2, -2)$, $(-0{\cdot}5, -0{\cdot}5)$ and $(1{\cdot}5, 1{\cdot}5)$.

2. Plot the curves $y = x^2$ and $y = 7x-4$ for $0 \leqslant x \leqslant 8$. What are the coordinates of their intersections? What are the abscissa of their intersections?

3. Plot the straight lines $y = 5x-3$, $y = 3x-3$, $y = -2x-3$, $y = 5x+1$, $y = 5x+2$ and $y = -5x+1$, all on the same graph. What can you say about a and b in the equation $y = ax+b$?

4. Four points, A (1, 5), B $(-1, 2)$, C $(-5, -5)$ and D $(2, -4)$, are plotted on a graph. What are the changes in the x- and y-coordinates in going from (i) A to B; (ii) B to C; (iii) C to D; (iv) A to C; (v) B to D?

Accord the changes their proper sign.

What is the abscissa of (a) A, (b) C, and what is the ordinate of (c) B, (d) D?

5. A point whose coordinates are (p, q) is used to represent the mapping between a value p of the variable P and the value q of the variable Q.

Which variable is

(i) the dependent one,
(ii) the domain variable,
(iii) the "cause" variable,

and which variable has its values represented by the points in the horizontal axis? What is the abscissa of the point representing the mapping?

6. Write down the abscissa of the intersection of the lines $y = 5x+7$ and $y = 8x$.

7. Draw the lines $y = 1-x$ and $2y = 2x+1$ on a single sheet of graph paper. Indicate the region in which a point whose coordinates obey the relations

(i) $y+2-1 > 0$
(ii) $y+x-1 = 0$, and
(iii) $2y-2x-1 > 0$ can move.

What are the restrictions on the coordinates of a point which moves in the remaining area? Are the lines included in this area? What conditions does the point at the intersection of the lines satisfy? What is the (a) abscissa, (b) ordinate of the intersection?

8. Draw a graph of $y = 2^x$ for values of x between -4 and $+4$. By considering the point whose abscissa is $\frac{1}{2}$, write down a value for $\sqrt{2}$. What is the value of $\sqrt[3]{2}$?

9. Rearrange the following equations in an explicit form for y:

(i) $3x^2y = 1$, (iii) $\sin y = 5x$,
(ii) $x^2+y-7 = 0$, (iv) $x-3/y+1 = 4$.

10. If $f(x)$ represents the function x^2, write down the values of

(i) $f(3)$, (ii) $f(-4)/f(1)$, (iii) $\dfrac{f(4)-f(-2)}{f(3)-f(1)}$.

11. If $f(x)$ represents x^2, find an expression for

$$\{f(x+\delta x)-f(x)\}.$$

12. If $g(x, y)$ represents the expression x^2+3xy, write down the values of

(i) $g(5, 0)$, (iii) $\dfrac{g(2, 2)}{g(4, 0)}$,

(ii) $g(-1, -3)$, (iv) $g(3, 1)-g(2, 1)$.

13. If $g(x, y)$ represents the expression $x^2-2xy+7$, write down an expression for

(i) $g(x+a, y)-g(x, y)$,
(ii) $g(x, y+b)-g(x, y)$.

14. If $f(x) \equiv 1/x$, find an expression for

$$\left\{\frac{f(x+\delta x)-f(x)}{\delta x}\right\}.$$

15. Explain why the equation $y^2 = a^2-x^2$ expresses a *relation* between x and y, but y cannot be said to be a *function* of x. How can the relation be resolved into two functions? What are the domain and range of the (i) relation, (ii) two functions?

16. How many functions does the relation $y^2 = x^3-x^2$ express? Give sketches to show the functions separately, and write down their domains and ranges.

17. Can the graph of a function be symmetrical with respect to the

(i) y-axis, (ii) x-axis?

18. Sketch the graph of the function $f(x)$ defined by the equations: $f(x) = 1$ if x is rational, $f(x) = 0$ if x is irrational. What is the

(i) domain, (ii) range of the function?

19. A function is defined only for integral values of x, and under these conditions it takes the value 2. Draw a sketch graph of the function, and describe its domain and range.

20. By considering the gradient of the line joining the point $(-1, -2)$ to $(1, 0)$ and that of the line joining $(1, 0)$ to $(3, 2)$, show that $(-1, -2)$, $(1, 0)$ and $(3, 2)$ are collinear.

21. The points $P(x_1, y_1)$ and $Q(x_2, y_2)$ are plotted on a graph, and a straight line drawn through them. What is the ordinate of the point on the line whose abscissa is a? Assume that $x_1 < a < x_2$, and give your answer in terms of the coordinates of P and Q and a.

22. Repeat question 21 for the point whose abscissa is a, but where $a < x_1 < x_2$.

†**23.** Sketch the graph of the function $\sqrt{(9-x^2)}$ and its inverse $g(x)$ obtained by the restriction that $-3 \leqslant g(x) \leqslant 0$. Why is it necessary to make a restriction?

†**24.** What geometrical relationship does the graph of a function bear to its inverse if the latter exists?

†**25.** Find the equation of the circle of which the line joining the point A (1, 3) and B (9, 9) is a diameter and show that the circle touches the y-axis. Find the gradient of the other tangent to the circle which passes through the origin. (NUJMB)

†**26.** Write down the domain and the range of the functions

(i) $y = \sqrt{\{1-\sqrt{(1+x)}\}}$,
(ii) $y = \sqrt{\{1-\sqrt{(1-x)}\}}$.

2.5. CONTINUITY

The word *continuous* has already been used in a semi-technical sense: the occasion was in Section 1.4, when a variable was defined as a symbol which can represent any one of a set of numbers, and a continuous variable a symbol which can represent any one of a set of numbers when the set contains all the real numbers (possibly between two limits). We must now extend the use of the word, and will consider under what conditions the graph of a function can be said to be continuous.

We have seen that the elements of a set can be represented by points in a line, or axis: and that a relation between two sets can be represented by a mapping consisting of arrows leading from the elements of the domain to the corresponding elements of the range. Let us now shrink the two related sets to a pair of parallel axes whose points represent the elements (Fig. 2.33), and let us further suppose

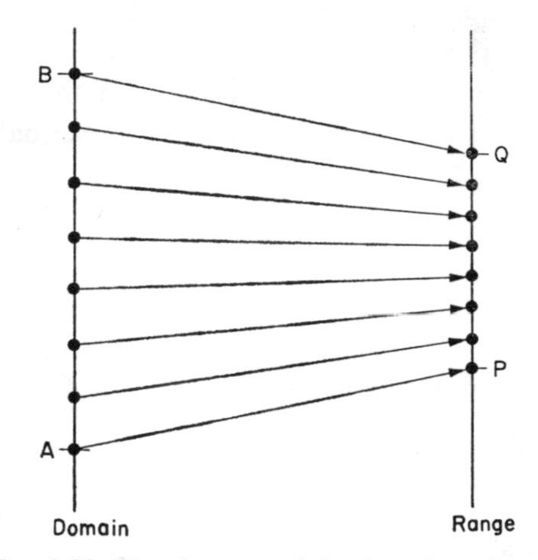

FIG. 2.33. The elements of the domain and range can be represented by points in two parallel axes.

that every single point in each axis represents a different element of the appropriate set (cf. Section 1.2). It will be impossible to draw in all the mapping arrows, but we can usefully think in terms of the mapping between (sample) points and the neighbourhoods around these points. By the *neighbourhood* around a point we will mean the set of points within a specified distance of the central point. Thus, if in Fig. 2.33 all the elements of the domain represented by points between A and B map onto (or into) the elements of the range represented by the points between P and Q, we can say that the neighbourhood AB maps onto (or into) the neighbourhood PQ.

If we now suppose that the point B maps onto the point Q, we sense that in general the image of a small neighbourhood BC which adjoins B will be a small neighbourhood adjoining Q (for example, the neighbourhood QR in Fig. 2.34). This idea springs from the intuitive

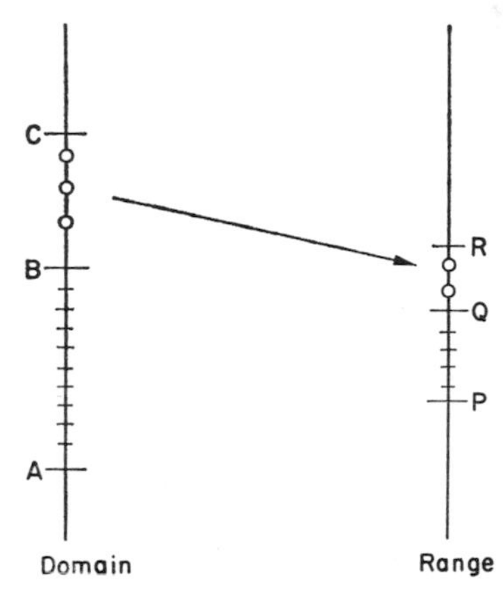

FIG. 2.34. A small neighbourhood BC adjoining B will, in general, have as its image a small neighbourhood such as QR adjoining Q.

notion of continuity, a notion derivable from the philosophy that small changes in cause lead to small changes in effect, or, more mathematically, that a small change in x leads to a small change in y or $f(x)$. To be more precise, let us consider more closely the point representing a in the domain axis. Associated with this point is the point representing the element $f(a)$ in the range axis (Fig. 2.35). Now, if we choose a neighbourhood of size α about $f(a)$ (i.e. think of all the elements represented by points contained within a distance of α of $f(a)$), we must find (if possible) a neighbourhood about a whose image (i.e. set of image points) lies entirely within the range neighbourhood chosen. Only if it is possible to do this for even vanishingly small values of α can the function $f(x)$ be said to be continuous. Or if y denotes the variable identified with $f(x)$, then under these circumstances the relationship between x and y can be said to be continuous at $x = a$.

Let us see how to apply this reasoning to test whether the relationship $y = 2x$ is continuous at $x = 3$. We first find the range point corresponding to $x = 3$ (namely $y = 6$), and then select a neighbourhood of size α about this (so that $6+\alpha \geqslant y \geqslant 6-\alpha$) (Fig. 2.36).

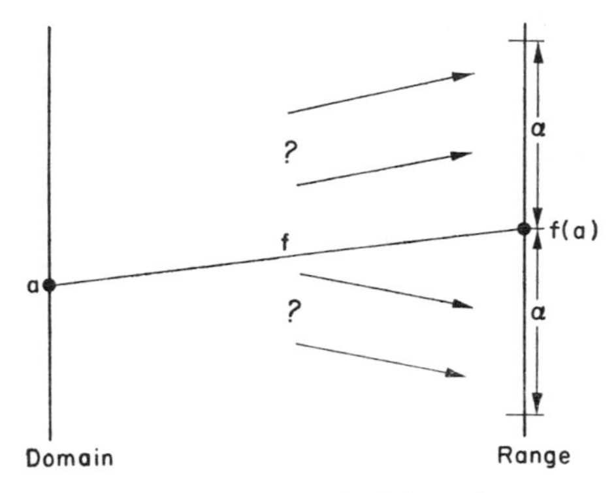

FIG. 2.35. If the image of a is $f(a)$, and the neighbourhood with boundaries α from $f(a)$ on either side is the image of a finite neighbourhood about a, no matter how small α, the functional relation f is continuous at a.

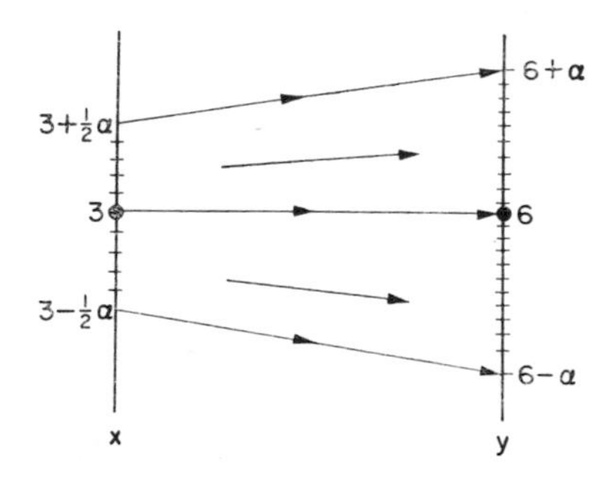

FIG. 2.36. The neighbourhood of size α about $y = 6$ is the image of any small neighbourhood lying within $\frac{1}{2}\alpha$ of $x = 3$.

Since $x = \frac{1}{2}(6+\alpha)$, or $(3+\frac{1}{2}\alpha)$, maps onto $y = 6+\alpha$, and $x = \frac{1}{2}(6-\alpha)$ or $(3-\frac{1}{2}\alpha)$ maps onto $y = 6-\alpha$, we can see that any neighbourhood within $\frac{1}{2}\alpha$ of $x = 3$ (i.e. $3+\frac{1}{2}\alpha > x > 3-\frac{1}{2}\alpha$) will have its image within the chosen range neighbourhood. And since this is true no matter how small α, we can say that the relationship is continuous at $x = 3$.

If this is continuity, what is discontinuity? Intuitively it is easy to realise that the image of an object seen through a bifocal lens has a discontinuity where the image is broken (Fig. 2.37); to prove it in formal language is perhaps a little more tricky, but a line of reasoning such as follows will suffice. Select the object point A which gives the image point P, and choose a neighbourhood of P, say QR (Fig. 2.38). We then have to ask whether we can find a neighbourhood of A whose image lies within QR. This cannot be done, for any neighbourhood BC of A will map partly on

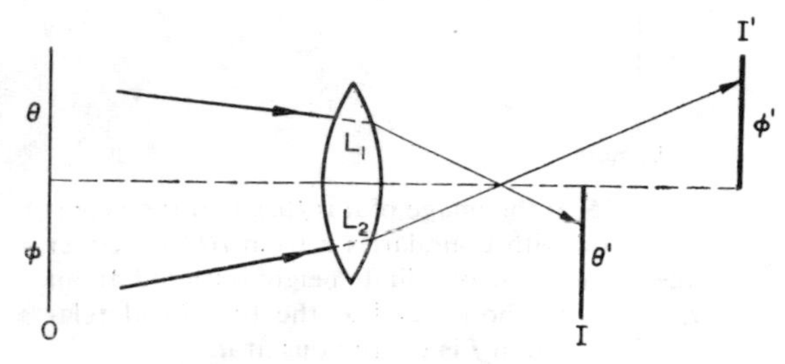

FIG. 2.37. The image II' of an object O formed by a bifocal lens is discontinuous ...

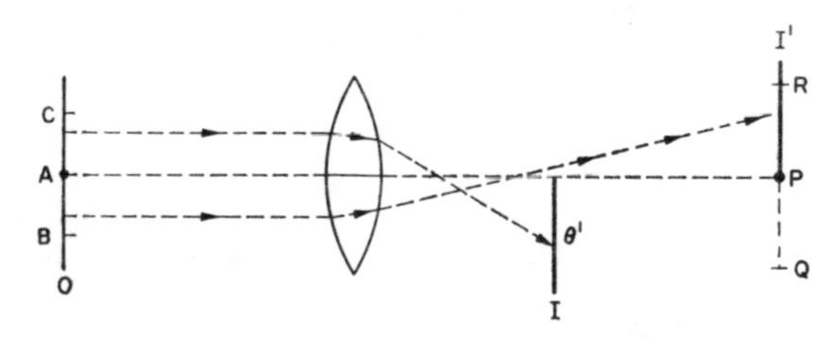

FIG. 2.38. ... because a neighbourhood QR on the image PI' is not the image of any small neighbourhood on the object about A.

to the further image I' (and therefore partly into the neighbourhood QR), but partly onto the nearer image I: as shown in Fig. 2.38 the region AC will map into θ', and the region BA into PR. This will happen no matter how small the neighbourhood of A, and so it is not possible to find a domain neighbourhood which maps into the chosen range neighbourhood P—that is, the image is discontinuous at P.

The occurrence of discontinuity at a break in the image can be used to apply our ideas to graphs. It is in fact true that whenever a break occurs in the graph, the graph can be

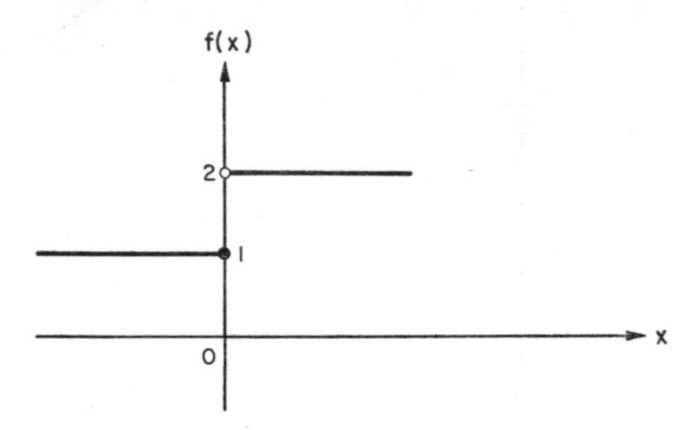

FIG. 2.39. $f(x)$ is discontinuous at $x = 0$.

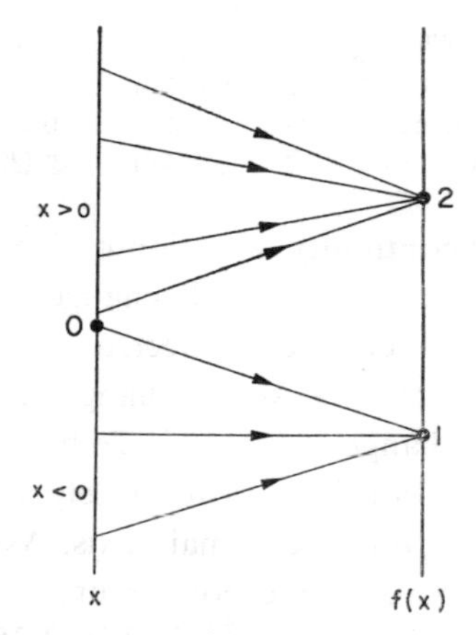

FIG. 2.40. All values of $x > 0$ have the image $f(x) = 2$; all $x \leqslant 0$ have the image $f(x) = 1$.

said to be discontinuous, and the functional relationship expressed by the graph is also discontinuous. Thus a function defined as $f(x) = 2$ for $x > 0$, $f(x) = 1$ for $x \leqslant 0$ (see Fig. 2.39), is discontinuous at $x = 0$. In formal language, if we represent the elements of the variable x and its function $f(x)$ by the points in two parallel lines (Fig. 2.40), we can easily see that all points above $x = 0$ map onto $f(x) = 2$, and all points below $x = 0$ (inclusive) map onto $f(x) = 1$. Now consider the point $x = 0$ and its image $f(x) = 1$ (Fig. 2.41); and choose a neighbourhood about $f(x) = 1$. Is there a domain neighbourhood about $x = 0$ whose image lies entirely within this neighbourhood? The answer is, of course, no, for any neighbourhood about $x = 0$ will map partly onto $f(x) = 1$ and partly onto $f(x) = 2$. Thus the function $f(x)$ is discontinuous at $x = 0$ or $f(x) = 1$.

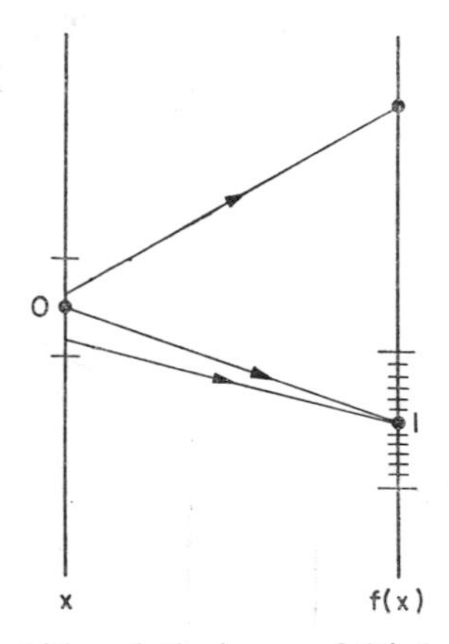

FIG. 2.41. Although the image of O is 1, no small neighbourhood about O will map into a small neighbourhood about 1.

One rather different but commonly occurring example should be mentioned. The graph of a function like $y = 1/x$ has an obvious break at $x = 0$: to prove its discontinuity leads to an immediate snag, for we cannot find an image point for $x = 0$. This in itself solves the problem, for a function cannot be continuous or discontinuous at a point where it is not defined or does not exist. But because in any case we cannot define a value for the function at the point $x = 0$ to make it continuous, it is often said to have a discontinuity at the point. Consider, for example, defining $f(x)$ as 10^{10} at $x = 0$. Then choosing a neighbourhood of 10^{10} (say between $(10^{10}+1)$ and $(10^{10}-1)$ as in Fig. 2.42), we find that any small neighbourhood of $x = 0$ contains values of x smaller than 10^{-10} (e.g. $x = 10^{-20}$ or 10^{-100}), and therefore maps into a neighbourhood beyond 10^{20} or 10^{100}. Thus a neighbourhood of $x = 0$ cannot be found so that its

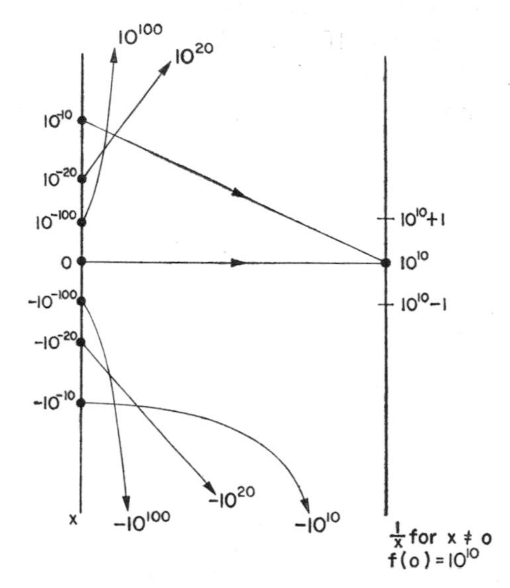

FIG. 2.42. No neighbourhood about $x = 0$ will map into a neighbourhood about $f(0)$.

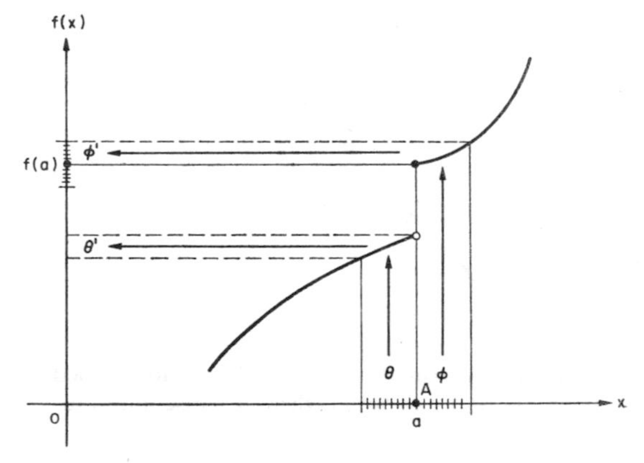

FIG. 2.43. The neighbourhood of θ maps into its image neighbourhood θ', and ϕ maps into its image neighbourhood ϕ'. No small neighbourhood about A will map entirely into a small neighbourhood about $f(a)$.

image lies within the chosen range neighbourhood of $10^{-10}+1 \geqslant f(x) \geqslant 10^{-10}-1$, and this will be true for any value chosen for $f(0)$, no matter how large. Such a discontinuity is sometimes called an infinite discontinuity.

We can express our ideas of continuity in terms of graphical representation. The graph of a function is the collection of points mapping the domain axis elements with their range axis elements. If a break occurs at $x = a$ (Fig. 2.43), then it is not possible to find any neighbourhood of a which maps into a small

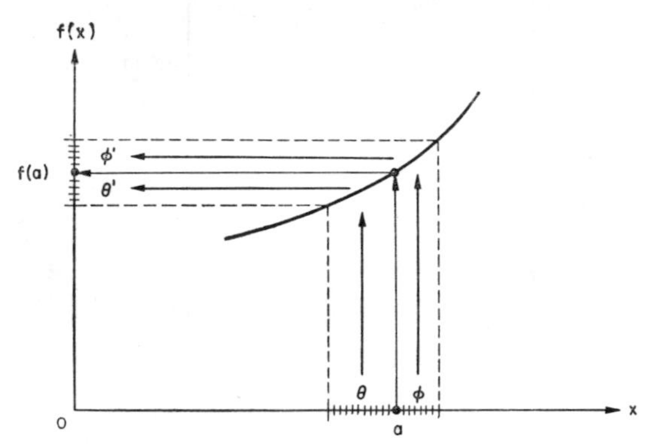

FIG. 2.44. The graph of a continuous function has no breaks in it.

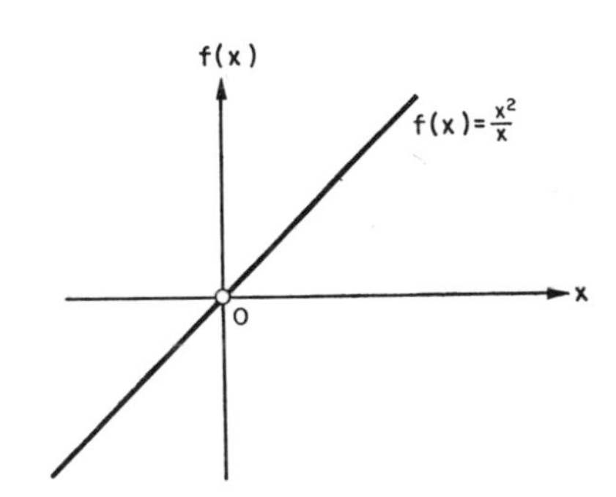

FIG. 2.45. The function $f(x) = x^2/x$ is not defined at $x = 0$, and its graph is therefore discontinuous at $x = 0$.

neighbourhood of $f(a)$, for, taking the graph of Fig. 2.43 as an example, the left-hand part θ of the neigbourhood of a maps into a neighbourhood θ' lying some way from $f(a)$, and this will happen no matter how small θ. Conversely, if there is no break in the curve (Fig. 2.44), the functional relationship it represents will be continuous. Remember though that the function must be defined at the point in question—it is possible to have a smooth-looking curve or line with one point missing (e.g. x^2/x at $x = 0$) (Fig. 2.45); and also that the visual appearance of a graph is not a proof of continuity. In proving the continuity of a graph or function it is in general safer to use the parallel axis approach described earlier in this section.

A few final comments:

(i) A function is said to be continuous throughout an interval if it is continuous at all points within that interval.

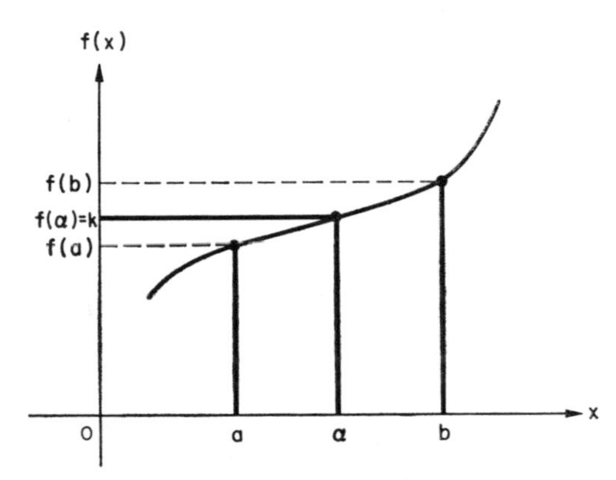

FIG. 2.46. If $f(x)$ is continuous throughout $a \leqslant x \leqslant b$, and if $f(a) \leqslant k \leqslant f(b)$, then there must be some number α between a and b such that $f(\alpha) = k$.

(ii) If a function $f(x)$ is continuous throughout an interval $a \leqslant x \leqslant b$, and if k is a number between $f(a)$ and $f(b)$, then there is a number α in the interval such that $f(\alpha) = k$. This can be seen to be a reasonable statement by considering a graph such as that in Fig. 2.46, and although such an intuitive approach gives no proof, it is beyond our scope to delve into the fully rigorous argument here.

(iii) If a function $f(x)$ is continuous in an interval $a \leqslant x \leqslant b$, then there is a number α in the interval such that there is no value of $f(x)$ greater than

$f(\alpha)$ in the interval (Fig. 2.47a). There is also another number β in the interval such that there is no value of $f(x)$ less than $f(\beta)$ in the interval (Fig. 2.47b). We can say this more concisely by stating that the function attains a maximum and a minimum within the interval throughout which it is continuous. Notice that either the α or β or both may be at a or b (Fig. 2.47c).

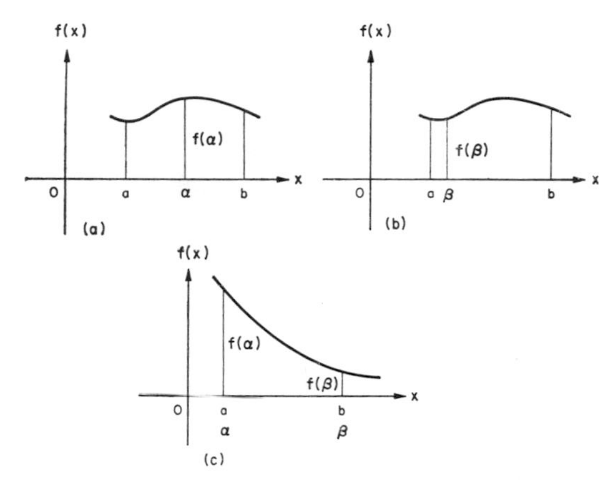

FIG. 2.47. If $f(x)$ is continuous throughout $a \leqslant x \leqslant b$, then: (a) there is a number α in the interval $a \leqslant \alpha \leqslant b$ such that $f(x) \leqslant f(\alpha)$ for all other values of x in the interval; (b) there is a number β in the interval such that $f(x) \geqslant f(\beta)$ for all other values of x in the interval (c) even if $\alpha = a$ and $\beta = b$.

Example 2.8

Is the function defined by the equation $y = |x|$ continuous at $x = 0$? Sketch its graph.

The nature of the relationship defined by the equation $y = |x|$ can be expressed in more detail by the pair of equations $y = x$ if $x \geqslant 0$, $y = -x$ if $x < 0$ (see p. 20, question 13). Thus, when $x = 0$, $y = 0$, and if the function is to be continuous, we must be able to find a neighbourhood about $x = 0$ whose image lies within any small neighbourhood about the associated range point $y = 0$. Using a pair of parallel axes to represent the mapping between x and y (Fig. 2.48), we can see that all points above 0 in the x-axis map onto points above 0′ in the y-axis, where we are supposing that the positive numbers in both axes are represented by the points above 0 and 0′, and where 0 and 0′ represent zero in the two axes. And, secondly, all points below 0 in the x-axis (i.e. the points representing negative values of x) map into the same points above 0′,

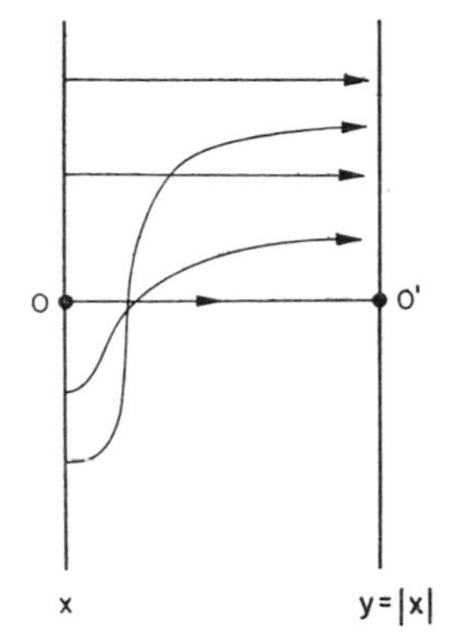

FIG. 2.48.

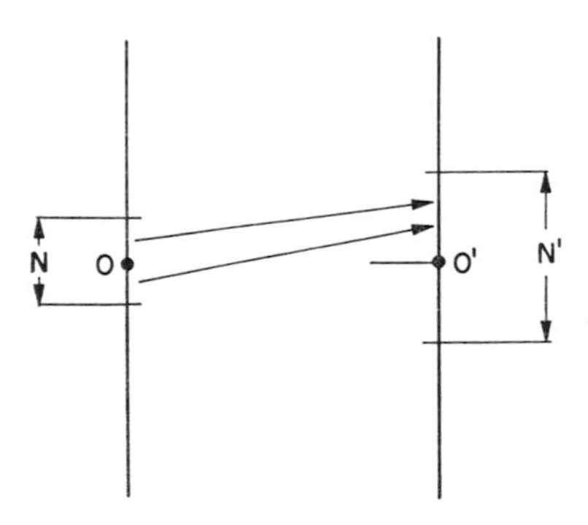

FIG. 2.49.

since for $x < 0$, $y = -x$, and y is therefore always positive.

Choosing a neighbourhood N' about 0′ (Fig. 2.49), we can see that any smaller neighbourhood N around 0 has its entire image in the upper half of N': this image is in fact produced twice—once by its upper half and once by its lower half. Thus, it is possible to find a domain neighbourhood whose image lies within any small-range neighbourhood about $0'$, and $y = |x|$ is therefore continuous at $x = 0$.

The sketch of the graph of $y = |x|$ is shown in Fig. 2.50: it is important to note that although the graph suffers a sharp turn at the origin, it is nevertheless continuous, as we have just proved.

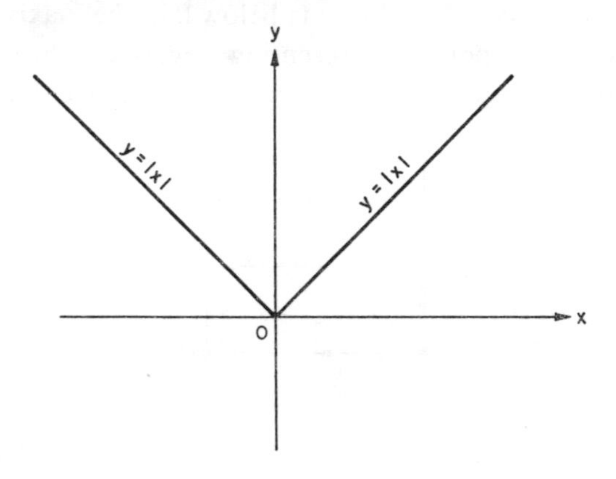

FIG. 2.50.

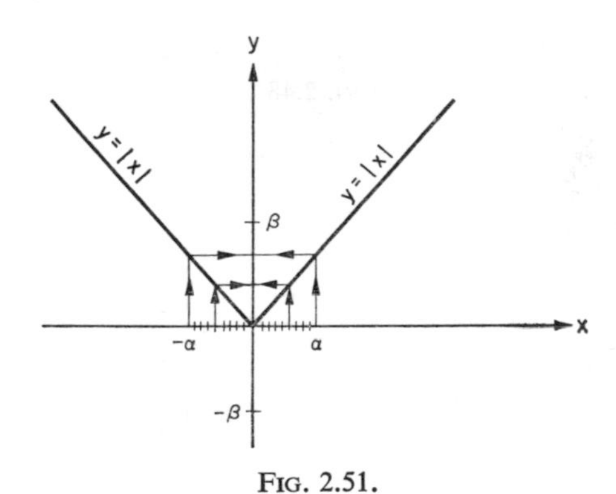

FIG. 2.51.

It is of interest to display the continuity of the graph using the perpendicular axes: redrawing Fig. 2.50 in Fig. 2.51, and selecting a neighbourhood between $y = \beta$ and $y = -\beta$, we can see that it is possible to find a neighbourhood on the x-axis whose image lies within this y-axis neighbourhood. In fact, any neighbourhood of size α on the x-axis (where $\alpha < \beta$) is entirely mapped into the region between $y = 0$ and $y = \beta$. Thus again the graph is continuous at $x = 0$.

Example 2.9

Discuss the continuity of a function defined by the equations:

$f(x) = x$ if x is an integer,

$f(x) = 0$ if x is not an integer.

Sketch the graph of the function $f(x)$.

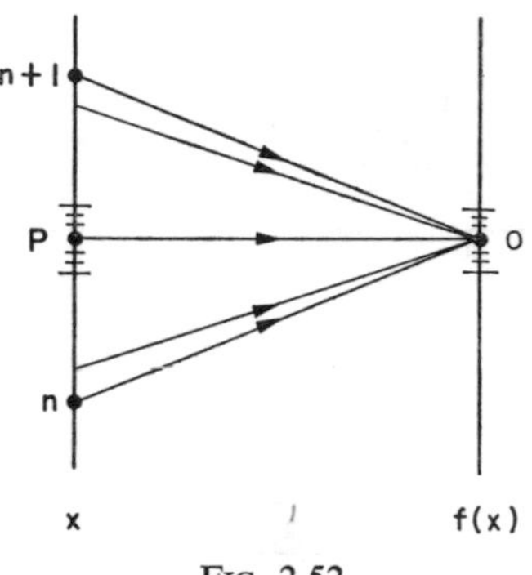

FIG. 2.52.

Confining our attention, for the moment, to the region between two integers $x = n$ and $x = n+1$, we can say from the definition that each point lying in the x-axis between these two integers maps onto O in the $f(x)$-axis (Fig. 2.52). If we therefore choose a non-integral value for x (say $x = p$, and where $n < p < n+1$), the point representing it will map onto O. It also follows that any small neighbourhood about p maps onto O, and thus a domain neighbourhood can be found to map into any small neighbourhood about O. The function $f(x)$ is therefore continuous at all non-integral values of x.

If, however, x takes the integral value n (where $n \neq 0$), the point representing it maps onto N', where N' represents the value n in the $f(x)$-axis (Fig. 2.53). But we cannot find a neighbourhood about n to map into a small neighbourhood about N', for the points adjacent to n map onto O, no matter how close to n

they are. Thus, the function $f(x)$ is discontinuous at all integral values of x other than 0. That $f(x)$ is continuous at $x = 0$ is left as an exercise for the reader.

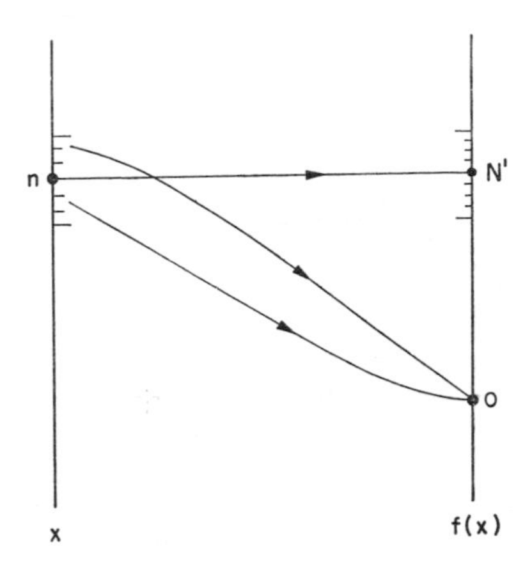

FIG. 2.53.

EXERCISE 2e

1. Prove that the geometrical operations of translation on a plane and reflection are continuous.

2. Prove that x as a function of x is continuous.

3. Prove that a constant is a continuous function of any variable.

4. Sketch the graphs of

(i) $y = \dfrac{x^2+1}{x+1}$, (ii) $y = \dfrac{x^2+1}{x-1}$,

and discuss the nature of any discontinuities arising in them.

5. Sketch the graphs of

(i) $\dfrac{x^2-1}{x-1}$, (ii) $\dfrac{x^2-1}{x+1}$,

and state where any discontinuities occur. Explain or prove why the graphs are discontinuous at the points you mention.

6. Draw sketch graphs of

(i) $y = 1-x^{2/3}$;
(ii) $y = x$ for $x \leqslant 2$,
$y = -x$ for $x > 2$;

and discuss whether or not they are continuous in the region $-5 < x \leqslant 5$.

7. Sketch the graph of the function $f(x) = 3+x$ for $x > 0$, $f(x) = 3-x$ for $x < 0$. Why is there a discontinuity at $x = 0$? Will this be removed if we define $f(0)$ as (i) 0, (ii) 3?

8. Sketch the graph of $y = 2^{-1/x^2}$ for $x \neq 0$. Why is there a discontinuity at $x = 0$? Will this be removed if we make $f(0) = 0$?

9. Sketch the graph of $y = \sec^2 x$, and write down a general expression for the positions of its discontinuities.

†10. Prove that the sum of two continuous functions is itself continuous.

†11. Prove that

(i) the product of two continuous functions,
(ii) the quotient (as far as defined) of two continuous functions,

is itself continuous.

†12. Prove that a continuous function of a continuous function is itself continuous.

†13. Discuss whether or not the function $f(x) = x$ for $x > 0$, $f(x) = 0$ for $x < 0$, is continuous at $x = 0$.

†14. A function $f(x)$ is defined by the following equations:

$$f(x) = x \text{ if } x \text{ is rational,}$$
$$f(x) = -x \text{ if } x \text{ is irrational.}$$

Sketch the graph of the function, and discuss whether or not it is continuous.

†15. State the positions of discontinuity in the graph of $[x]$, where the symbol $[x]$ denotes the greatest integer less than or equal to x. Explain why discontinuity occurs at these points. (The function $[x]$ is often referred as a "step" function.)

†16. Sketch the graph of a function showing the number of real roots of $ax^2+ax+1 = 0$ as a function of a. Discuss whether or not there are points at which the function is discontinuous.

†17. For what values of x (if any) are the following functions undefined?

(i) $\dfrac{1}{x-3}$, (v) $\dfrac{1}{\sin x}$,

(ii) $\dfrac{1}{2x+7}$, (vi) $\dfrac{1}{x^2-16}$,

(iii) $\tan x$, (vii) $\dfrac{x}{\sin x}$,

(iv) $\sin x$, (viii) $\dfrac{\sin x}{x}$.

†18. Discuss the continuity of the following functions:

(i) $y = \dfrac{1}{x}$, (vi) $\tan x$,

(ii) $y = [x]$, (vii) $\dfrac{\cos x}{2x-3}$,

(iii) $y = \dfrac{1}{x^2+4}$, (viii) $\dfrac{1}{1-\cos x}$,

(iv) $\dfrac{x^2-9}{x^2-4}$, (ix) $\dfrac{x^2}{x^2-5x-6}$.

(v) $\operatorname{cosec} x$,

2.6. LIMITS

Let us consider for a moment the graph of the function $y = (\sin x)/x$ for small values of x. We can construct a table of values for x equal to 0·1, 0·08, 0·06, 0·04 and 0·02 radians (see below), but if $x = 0$, we obtain the expression 0/0 for y, which therefore has no defined value when x is zero. For small negative values of x we can continue the table, and if we plot

X (radians)	−0·10	−0·08	−0·06	−0·04	−0·02
$\frac{\text{Sin X}}{\text{X}}$	·9983	·9989	·9994	·9998	·9999

X (radians)	0·02	0·04	0·06	0·08	0·10
$\frac{\text{Sin X}}{\text{X}}$	·9999	·9998	·9994	·9989	·9983

the points representing the values shown, we obtain the curve of Fig. 2.54. The graph is, of course, discontinuous at $x = 0$, since y is not defined there, but we can make it continuous by defining y to equal 1 when $x=0$ (Fig. 2.55). The value of y for which the graph can be made continuous at $x = 0$ is called the *limit* of $(\sin x)/x$ at $x = 0$.

Generalising, we can say that if we have a function $f(x)$ which is undefined at $x = a$, and if its graph can be made continuous at $x = a$ by defining $f(a)$ to equal b, then $f(x)$ possesses a limit at $x = a$, and the value of this limit is b.

As usual, mathematicians employ shorthand notation to represent these ideas: for the example just considered, we can write

$$\lim_{x=0} \left(\frac{\sin x}{x}\right) = 1$$

to mean that the limit at $x = 0$ of $\sin x/x$ is 1.

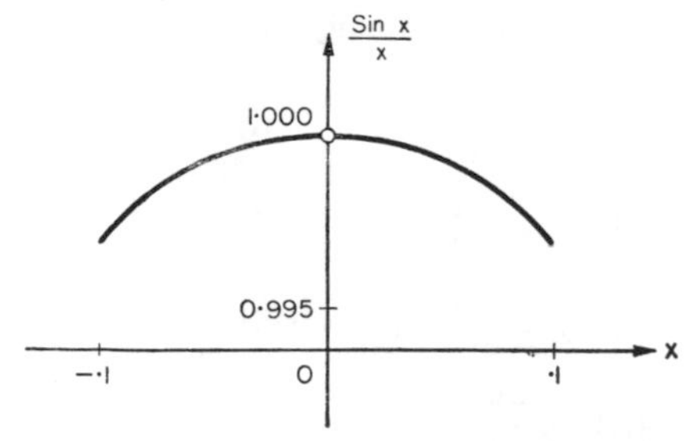

FIG. 2.54. A function $(\sin x)/x$ is undefined at $x = 0$, and is therefore discontinuous there.

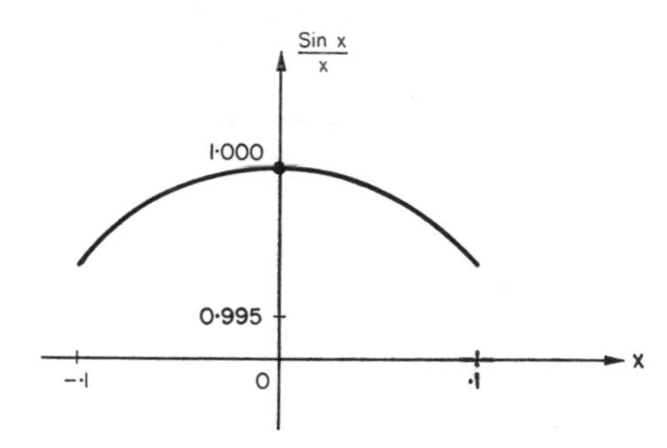

FIG. 2.55. By defining $f(0)=1$ and $f(x)=(\sin x)/x$ for $x \neq 0$, $f(x)$ becomes continuous at $x = 0$.

There is another property of the function $(\sin x)/x$ that is worthy of mention. From the graph of Fig. 2.54, it is clear that as x approaches the value 0 from above (and, for that matter, from beneath, but we will confine our attention to the positive side only), the value of $(\sin x)/x$ approaches 1: that is to say, the nearer the value of x is to 0, the nearer the value of $(\sin x)/x$ is to 1. We can express this symbolically by writing that

$$\text{as } x \to 0+, \quad \frac{\sin x}{x} \to 1,$$

and where "$x \to 0+$" means that x *decreases* towards the value 0. But expressed in this fashion, our ideas are intuitive rather than rigorous, and we must build our concept of "approach" on more sophisticated and solid bases.

Returning to our definitions of neighbourhoods and mapping, we can see that if we choose the point representing 1 in the $(\sin x)/x$ axis, and describe a neighbourhood N about it (Fig. 2.56), the image of any one-sided neighbourhood N' between $x = 0$ and $x = +\alpha$ on the x axis will map into N if α is sufficiently small. This will be true no matter how small

N, and so, for any small-range neighbourhood, we can find a one-sided domain neighbourhood whose image lies entirely within it. Since we can do this, we can say that the limit

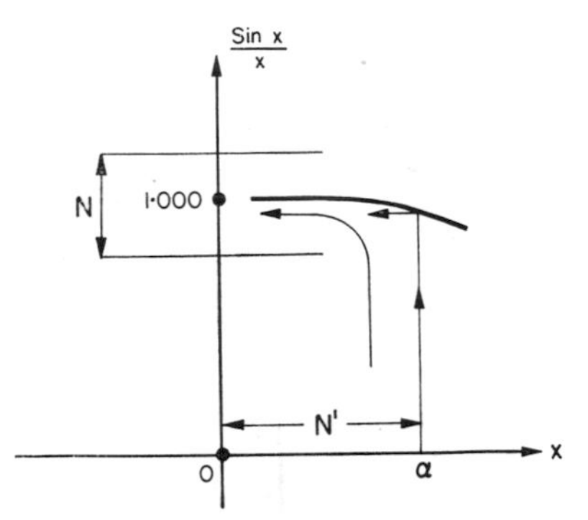

FIG. 2.56. Any one-sided neighbourhood N' will map into the neighbourhood N about the range value 1.

of $(\sin x)/x$ as x approaches 0 from above is 1, and writing this symbolically,

$$\lim_{x\to 0+}\left(\frac{\sin x}{x}\right) = 1.$$

(A more rigorous proof of this equality is given in the Appendix, p. 254.)

To generalise this dynamic, one-sided limit in which the equality sign is replaced by an arrow, we set up the convention that:

(i) if a function $f(x)$ is undefined at $x = a$,

and (ii) if for any small neighbourhood N about a range value b, a domain neighbourhood with a lower boundary at $x = a$ can be found whose image lies entirely within N,

then as x approaches a from above, the value of $f(x)$ "approaches" b, or the limit of $f(x)$ as x approaches a from above is b. Symbolically,

$$\lim_{x\to a+} \{f(x)\} = b.$$

We can similarly define a dynamic limit for x approaching a from below by specifying in (ii) that the domain neighbourhood must have an upper boundary $x = a$. We write this limit in the form

$$\lim_{x\to a-} \{f(x)\} = b$$

(see Fig. 2.57).

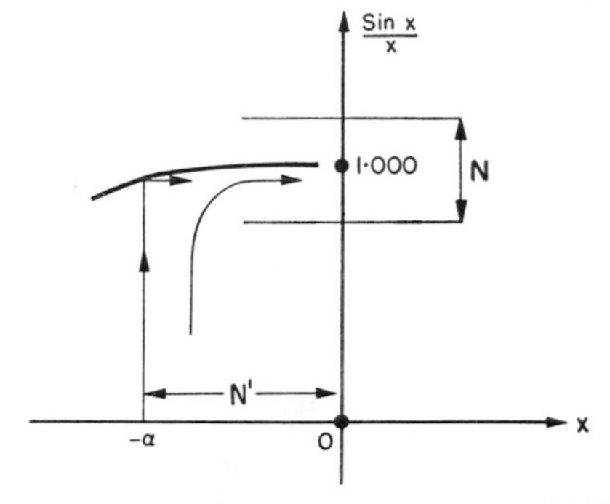

FIG. 2.57. If a one-sided neighbourhood N' can be found to the left of $x = 0$ such that its image lies entirely within N, we can say that $\lim_{x\to 0-} \{f(x)\} = 1$.

It will be appreciated now that the static limit equation

$$\lim_{x=a} \{f(x)\} = b$$

can be written only if

$$\lim_{x\to a+} \{f(x)\} = \lim_{x\to a-} \{f(x)\} = b,$$

that is, the dynamic limits from above and below must be equal to each other.

Consider now the behaviour of a function such as $1/x$ as x increases without bound. A graph of the function is shown in Fig. 2.58,

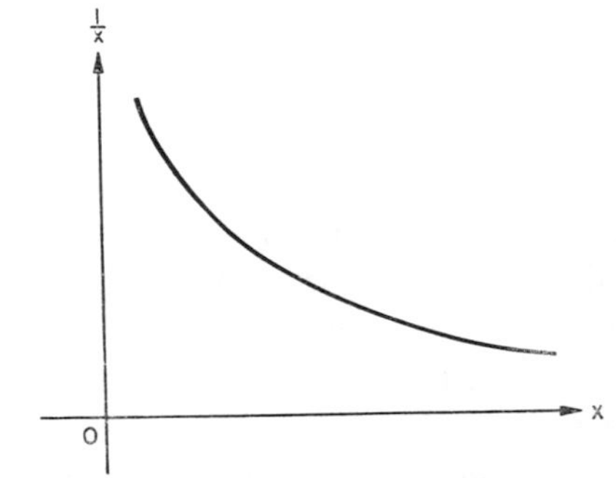

FIG. 2.58. As x approaches zero from above, $1/x$ increases without bound.

and it is very apparent from this that as x increases the value of $1/x$ gets closer and closer to zero. Numerically, if $x = 100$, $1/x = 0{\cdot}01$; if $x = 10^6$, $1/x = 0{\cdot}000001$, or if $x = 10^{100}$, $1/x = 10^{-100}$. We can say that as x increases without bound, $1/x$ approaches 0: or, in symbols

$$\text{as } x \to \infty, \quad \frac{1}{x} \to 0.$$

The use of the symbol $\to \infty$ is perhaps a little unfortunate, but conventional. We have defined the arrow $\to$ as representing the words "approaches the value", and the symbol ∞ is read as "infinity". To read $\to \infty$ as "approaches (the value) infinity" is therefore conventional, but in fact a preferable phrase is "increases without bound" (cf. Section 1.3).

Now previously we have defined the expressions

$$\lim_{x \to a+} f(x) = b$$

and "as $x \to a+$, $f(x) \to b$" to be equivalent. Logically, therefore, we should be able to write our statement about $1/x$ in the form

$$\lim_{x \to \infty} \frac{1}{x} = 0;$$

and this is, in fact, commonplace.

How can this be phrased in mapping terms? As before, we choose the value of the limit (0) as the centre of a small range neighbourhood N (Fig. 2.59a), and define the meaning of $x \to \infty$ as the statement that it must be possible to find a domain neighbourhood N' lying to the right of the point representing some x value p so that its range lies entirely within N. (In this case, of course, if N is the region between $y = \alpha$ and $y = -\alpha$, p must be greater than $1/\alpha$: but the value of p necessary for the image to lie within N is irrelevant, and all that matters is that a domain neighbourhood to the right of some specified point can be found.)

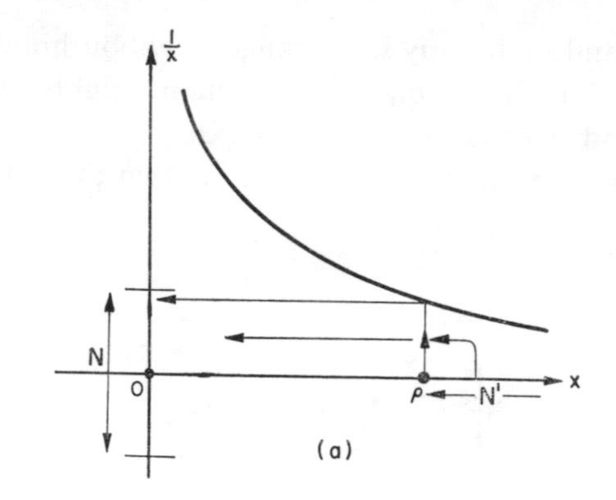

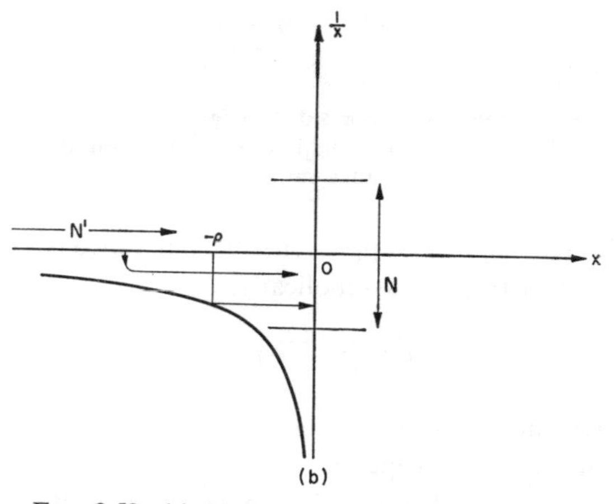

FIG. 2.59. (a) If for some value ϱ of x, the neighbourhood N' to its right maps into any small neighbourhood N, $\lim_{x \to \infty} (1/x) = 0$; and (b) if for some value $-\varrho$ of x, the neighbourhood N' to its left maps into any small neighbourhood N, $\lim_{x \to \infty} (1/x) = 0$.

The analogous equation

$$\lim_{x \to -\infty} \left(\frac{1}{x}\right) = 0$$

can be easily defined to mean that a domain neighbourhood to the left of some point can be found so that its image lies within any small neighbourhood N about $y = 0$ (Fig. 2.59b).

Consider finally the behaviour of a function such as $1/(x-2)$ for values of x near 2. If x is greater than 2 by 1/10 (i.e. $x = 2{\cdot}1$), y would equal 10; if x is greater than 2 by 1/100 (i.e. $x = 2{\cdot}01$), y would equal 100; and, generally, if x is greater than 2 by $1/p$, y would equal p. Thus, as $x \to 2+$, y increases without bound, or, to abbreviate in symbols even further,

$$\text{as } x \to 2+, \quad y \to \infty.$$

We must now define the meaning of this expression in rigorous mapping terms. This is not difficult, for by saying that the value of the limit is infinite, we can choose to mean that we are considering first a range neighbourhood N bounded only on its lower side, but by a fairly large positive value (Fig. 2.60a). Then, by writing that

$$\lim_{x\to 2+}\left(\frac{1}{x-2}\right) = \infty,$$

we mean that it is possible to find a domain neighbourhood N' with a lower boundary of $x = 2$ whose image lies entirely within N. Similarly, the expression that

$$\lim_{x\to 2-}\left(\frac{1}{x-2}\right) = -\infty$$

will mean that for any range neighbourhood N with an upper bound at a fairly large but negative value (see Fig. 2.60b), it is possible to find a domain neighbourhood N' with upper boundary at $x = 2$ whose image lies entirely within N.

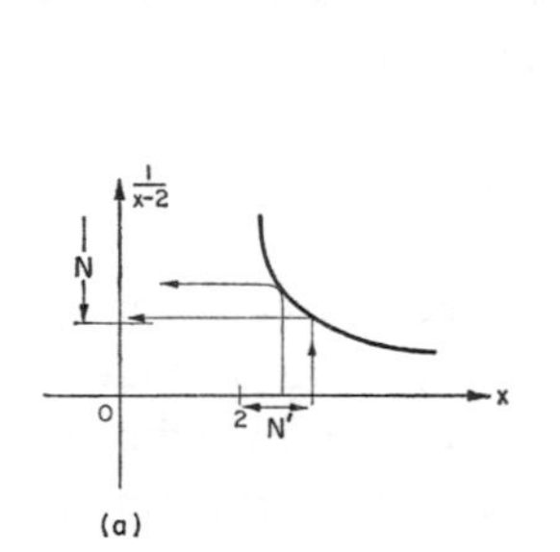

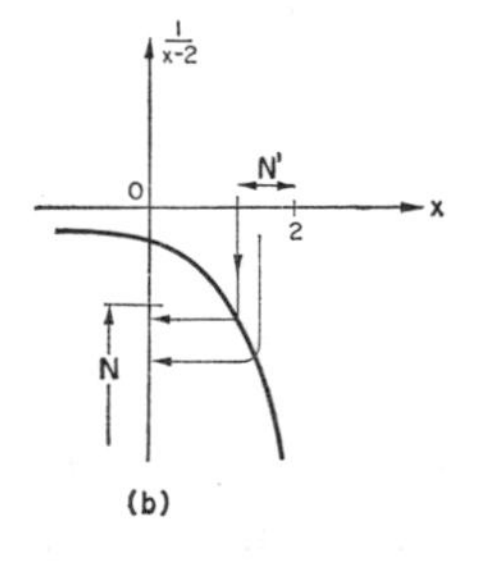

FIG. 2.60. (a) $\lim_{x\to 2+} \{1/(x-2)\} = \infty$; (b) $\lim_{x\to 2-} \{1/(x-2)\} = -\infty$.

Example 2.10

Find the value of $\lim_{x=0} \{f(x)\}$ if

(i) $f(x) = 1$ for $x > 0$, (ii) $f(x) = \dfrac{1-\cos x}{x^2}$.

For (ii), use the fact that

$$\cos x = 1-\tfrac{1}{2}x^2+\tfrac{1}{24}x^4$$

for small values of x.

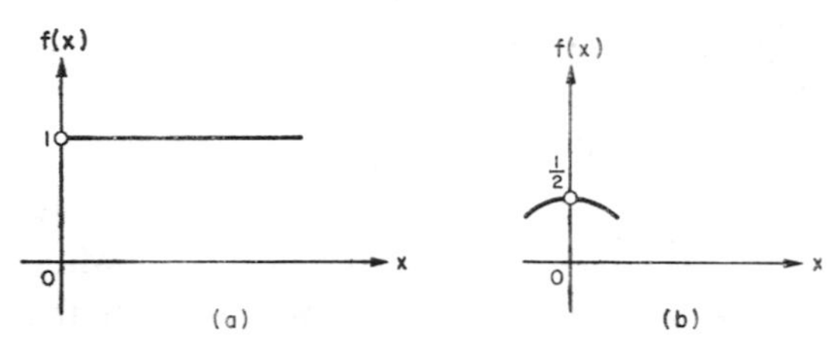

FIG. 2.61.

(i) The graph of $f(x) = 1$ for $x > 0$ is shown in Fig. 2.61a. We can clearly make it continuous at $x = 0$ by setting $f(0)$ equal to 1, and thus $\lim_{x\to 0+} \{f(x)\} = 1$.

(ii) If $\cos x = 1-\frac{1}{2}x^2+\frac{1}{24}x^4$ for small values of x, $f(x)$ can be represented by the expression

$$\frac{1-(1-\frac{1}{2}x^2+\frac{1}{24}x^4)}{x^2} \quad \text{or} \quad (\tfrac{1}{2}-\tfrac{1}{24}x^2).$$

From this expression we can see that as $x\to 0+$, the value of $f(x) \to \frac{1}{2}$, the $\frac{1}{24}x^2$ term disappearing when $x = 0$ (Fig. 2.61b). Since this is the case whether $x \to 0$ from above or below,

$$\lim_{x\to 0+} \{f(x)\} = \lim_{x\to 0-} \{f(x)\} = \tfrac{1}{2},$$

or, more concisely,

$$\lim_{x\to 0} \{f(x)\} = \tfrac{1}{2}.$$

Example 2.11

A function $g(x)$ is defined by the equations $g(x) = x$ if $x > 0$, $g(x) = 2$ if $x < 0$.

Sketch the graph of the function, and state what you can about the values of:

(i) $\lim_{x \to 0+} \{g(x)\}$, (ii) $\lim_{x \to 0-} \{g(x)\}$, (iii) $\lim_{x=0} \{g(x)\}$

A sketch of the graph is given in Fig. 2.62: the right-hand sloping line approaches the origin but does not actually contain it (since $g(0)$ is undefined). Similarly the left-hand line of zero gradient approaches the value 2, but it does not contain the point (0, 2).

(i) As x approaches 0 from above, $g(x)$ decreases to zero in direct proportion. Since for any small range neighbourhood N about the value 0 we can find a domain neighbourhood with lower boundary at $x = 0$ whose image lies entirely within N, we can write that $\lim_{x \to 0+} \{g(x)\} = 0$.

(ii) As x approaches 0 from below, the point mapping x with $g(x)$ moves along the horizontal line $g(x) = 2$ towards the $g(x)$-axis. Thus the value of $g(x)$ maintains and approaches the value 2 on the range axis, and for any small neighbourhood N about $g(x) = 2$, a domain neighbourhood with upper boundary $x = 0$ whose image lies entirely within N can be found. We can therefore write that $\lim_{x \to 0-} \{g(x)\} = 2$.

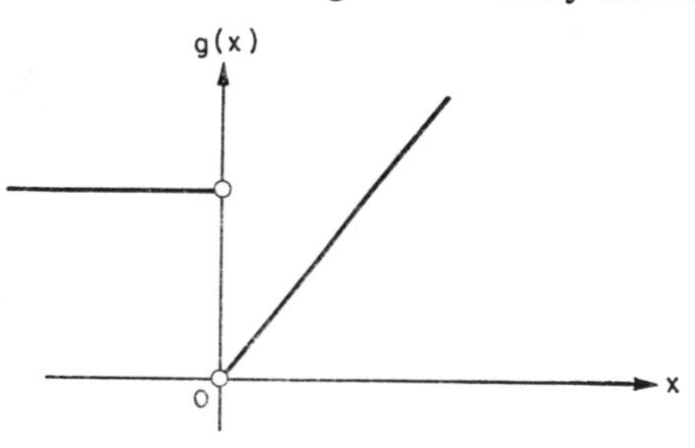

FIG. 2.62.

(iii) Since $\lim_{x \to 0+} \{g(x)\} \neq \lim_{x \to 0-} \{g(x)\}$, we cannot define a value for $g(0)$ so that the graph is continuous. This is also obvious from Fig. 2.62. Thus the static limit of $g(x)$ at $x = 0$ does not exist.

Example 2.12

Find the values of the limits:

(a) $\lim_{x \to 1+} \left\{\frac{x-2}{x^2+x-2}\right\}$; (c) $\lim_{x \to \infty} \left\{\frac{4x-3}{2x+7}\right\}$;

(b) $\lim_{x \to 1-} \left\{\frac{x-2}{x^2+x-2}\right\}$; (d) $\lim_{x \to 1} \left\{\frac{x-1}{x^2+x-2}\right\}$.

(a) Since $\frac{x-2}{x^2+x-2} = \frac{x-2}{(x-1)(x+2)}$,

we can see that as x approaches 1 from above, the numerator approaches -1, and the denominator zero (from above). Thus the fraction as a whole increases without bound, or, symbolically,

$$\lim_{x \to 1+} \left\{\frac{x-2}{x^2+x-2}\right\} = \infty.$$

(b) If $x \to 1$ from below, the denominator approaches zero from below, for if $x < 1$, $(x-1) < 0$. Thus the fraction as a whole decreases without bound; or

$$\lim_{x \to 1-} \left\{\frac{x-2}{x^2+x-2}\right\} = -\infty$$

(c) We can divide the fraction $(4x-3)/(2x+7)$ throughout by x provided $x \neq 0$, and thus consider the equivalent value of

$$\lim_{x \to \infty} \left\{\frac{4-3/x}{2+7/x}\right\}.$$

As x increases without bound, $3/x$ and $7/x$ approach zero, and so the fraction as a whole approaches the ratio 4/2 or 2. Thus,

$$\lim_{x \to \infty} \left\{\frac{4x-3}{2x+7}\right\} = 2.$$

Alternatively, this can be seen by considering that the $4x$ term in the numerator makes the -3 insignificant in comparison with x, and similarly the $2x$ becomes the predominant term in the denominator. The ratio thus tends to take on the value of $4x/2x$, or 2 as already found.

(d) Rewriting the fraction in the form

$$\frac{x-1}{(x-1)(x+2)},$$

we can divide it throughout by $(x-1)$ provided that $x \neq 1$. Now the value of

$$\frac{1}{x+2} \rightarrow \frac{1}{3}$$

both as $x \rightarrow 1+$ and as $x \rightarrow 1-$. We can therefore write that

$$\lim_{x\to 1+}\left\{\frac{x-1}{x^2-x-2}\right\} = \lim_{x\to 1-}\left\{\frac{x-1}{x^2-x-2}\right\}$$

$$= \lim_{x=1}\left\{\frac{x-1}{x^2-x-2}\right\}$$

$$= \frac{1}{3}$$

EXERCISE 2f

1. Discuss the values of:

(i) $\lim_{x\to 1+}\left(\frac{x}{1-x}\right)$, (iv) $\lim_{x\to 1-}\left\{\frac{3x^2-x-2}{x-1}\right\}$,

(ii) $\lim_{x\to -1+}\left\{\frac{x^2+1}{x+1}\right\}$, (v) $\lim_{h\to 0+}\left\{\frac{(a+h)^2-a^2}{h}\right\}$.

(iii) $\lim_{x\to 1+}\left\{\frac{x^2-1}{x-1}\right\}$,

2. Write down the values of:

(i) $\lim_{x\to\infty}\left(1+\frac{1}{x}\right)$, (iii) $\lim_{x\to\infty}\left(\frac{3x+1}{x^2+2}\right)$,

(ii) $\lim_{x\to\infty}\left(\frac{5x+3}{x+2}\right)$, (iv) $\lim_{x\to\infty}\left(\frac{3-2x}{x^2+1}\right)$.

3. Draw the graph of $\tan x/x$ for $0°1' \leqslant x \leqslant 1°0'$. Hence estimate the value of $\lim_{x\to 0+} (\tan x/x)$.

4. The repeating decimal $0{\cdot}33\dot{3}$ can be written as $(\frac{3}{10}+\frac{3}{10^2}+\frac{3}{10^3}+\frac{3}{10^4}+\ldots)$. Write down an expression for the sum to n terms of this geometrical progression, and deduce the limit of the sum as n increases without bound.

5. 0·142857 142857 142857... (an infinitely recurring decimal) can be written as

$$\frac{142857}{10^6}+\frac{142857}{10^{12}}+\frac{142857}{10^{18}}+\ldots$$

Write down an expression for the sum to n terms of this G.P., and deduce the value of its limit as n increases without bound.

6. The value of s_n is defined by the equation $s_n = 1+\frac{1}{2}+(\frac{1}{2})^2+\ldots+(\frac{1}{2})^n$.

Thus to take two examples, $s_1 = 1+\frac{1}{2}$, $s_3 = 1+\frac{1}{2}+(\frac{1}{2})^2+(\frac{1}{2})^3$.

To what does the value of s_n tend as n increases without bound? Draw a sketch graph of the values of s_n against n, marking in the limit of s_n as $n \rightarrow \infty$.

***7.** Express the following statements symbolically:

(i) The strength of light from a bulb at a given point depends on the distance of that point from the bulb. The light due to the bulb becomes negligible at large distances from the bulb.

(ii) The quantity of energy needed to change the temperature of any body by a given amount becomes excessively large as the temperature of the body approaches the value of $-273°$ C. The nearer the temperature is to this value, the larger the amount of heat required.

8. A function $f(n)$ is defined to have a value $\frac{3}{10}+\frac{3}{10^2}+\frac{3}{10^3}\ldots+\frac{3}{10^n}$ (i.e. $f(n)$ is $0{\cdot}33\dot{3}$ to n places of decimals).

Write down the value of $\lim_{n\to\infty}\{f(n)\}$.

9. Write down the limits of the following functions at $x = 0$:

(i) $\frac{\sin 2x}{x}$ $\left(\text{use the relationship } \frac{\sin 2x}{x} = \frac{2\sin 2x}{2x}\right)$

(ii) $\frac{\sin 3x}{\sin x}$,

(iii) $\frac{\sin 4x - \sin 3x}{x}$,

(iv) $\frac{1-\cos x}{x^2}$ (use the relationship $1-\cos x = 2\sin^2(\frac{1}{2}x)$).

10. What is the value of

$$\lim_{h=0}\left\{\left[\frac{1}{h}\right]\left[\frac{1}{x+h}-\frac{1}{x}\right]\right\}?$$

†11. Sketch the graph of $|1/x|$ in the region $-2 \leqslant x \leqslant 2$. What happens as x approaches zero from (a) above, (b) below?

If $y = |1/x|$ for all values of x except $x = 0$, can the functional relationship between x and y be defined to make its graph continuous?

†12. Achilles can run at 20 km/h, and the tortoise with whom he is having a race can run at 2 km/h. Achilles gives the tortoise a start of 2 km, and they have a race. During the time that Achilles takes to run the first kilometre (3 min), the tortoise will have gone a further tenth of a kilometre. Achilles will therefore need more time to cover this further distance, and while he is doing so, the tortoise will go still further. And for each bit of time that Achilles needs to get to the point where the tortoise was, the latter will travel a bit further away from the start. Achilles will therefore never catch up with the tortoise, i.e. Achilles will get nearer and nearer the tortoise but never quite catch it up. The limit of Achilles' journey is therefore the tortoise. Discuss. (Zeno's paradox.)

†**13.** Sketch the graph of $y = \sin(1/x)$.

†**14.** Sketch the graphs of

(i) $\dfrac{\cos x}{x}$, (ii) $\dfrac{x}{\cos x}$.

†**15.** State the positions of any discontinuities in the graphs of

(i) $y = |x|$, (ii) $y = \dfrac{1}{x^2-9}$, (iii) $y = \dfrac{x-1}{x+1}$.

†**16.** Evaluate the following limits:

(i) $\lim\limits_{h=0} \left\{\dfrac{\sin(x+h)-\sin x}{h}\right\}$,

(ii) $\lim\limits_{h=0} \left\{\dfrac{\cos(x+h)-\cos x}{h}\right\}$

(use the appropriate sum and difference formulae).

†**17.** What is the value of $\lim\limits_{h=0} \left\{\dfrac{(x+h)^n-x^n}{h}\right\}$? (Use the binomial theorem to expand $(x+h)^n$ in increasing powers of h.)

†**18.** A body B revolves with a constant speed v in a circle centre C and radius r (see Fig. 2.63). If the velocity v of the body remained constant in size and direction, it would travel to P in a time t, where $BP = vt$. But the body is travelling in a circle, and instead of arriving at P it arrives at Q, even though its speed is constant. As a result it can be said to have "fallen" towards the centre of the circle through a distance PQ or s. Show that $2s(r+s/2) = v^2t^2$.

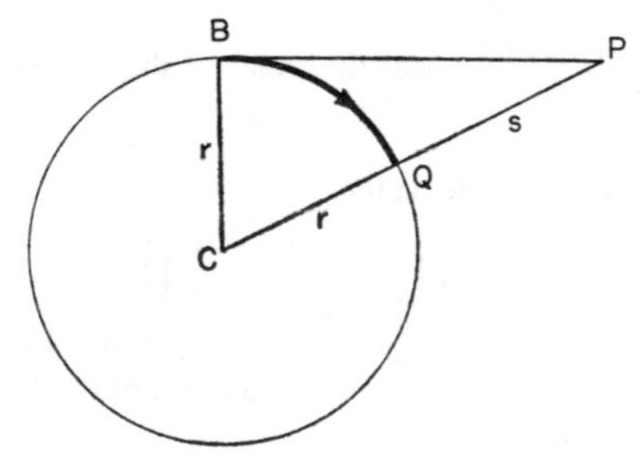

Fig. 2.63.

For a body falling from rest with acceleration a, the distance covered in a time t is $\frac{1}{2}at^2$. Replace s by this expression, and by considering the behaviour of the equation you obtain as $t \to 0$, show that the instantaneous acceleration of the body B is v^2/r towards the centre of the circle.

†**19.** Prove that

$$\lim_{x=a} \{f(x)+g(x)\} = \lim_{x=a} \{f(x)\} + \lim_{x=a} \{g(x)\}.$$

†**20.** Sketch the graphs of:

(i) $y = 2^{-1/x}$, (ii) $y = 2^{1/x}$,

and discuss the behaviour of the curves near $x = 0$. Is it possible to find a value for y at $x = 0$ to make either curve continuous at $x = 0$?

†**21.** Discuss the behaviour of the function $f(x)$ defined by the equation

$$f(x) = \lim_{n\to\infty} \frac{x^n}{1+x^n}.$$

Consider in particular the ranges $-1 < x < 1$, $x > 1$, and $x < -1$, and also the value of $f(x)$ at $x = 1$ and $x = -1$. Sketch the graph of the function, and state where discontinuities occur.

†**22.** Sketch the graph of the electric field strength inside and outside a hollow spherical charged conductor. Explain what happens to the lines of electric induction at the discontinuity.

†**23.** Explain the difference between an open interval $a < x < b$ and a closed interval $a \leq x \leq b$.

Explain also the statements that

(i) a function $f(x)$ is continuous in an open interval $a < x < b$ if it is continuous at every point in that interval;

(ii) a function $f(x)$ is continuous in a closed interval $a \leq x \leq b$ if it is continuous at every point between a and b, and

$$\lim_{x\to a+} \{f(x)\} = f(a) \quad\text{and}\quad \lim_{x\to b-} \{f(x)\} = f(b).$$

Draw a sketch of the graph of a function which is continuous within a closed interval, but for which

$$\lim_{x\to a-} f(x) \neq f(a) \quad\text{and}\quad \lim_{x\to b+} f(x) \neq f(b).$$

†**24.** A circle of unit radius has its centre O. A radius is drawn from the centre of the circle to an external point X, and two radii OP and OQ, both on the same side of OX, are drawn so that $P\hat{Q}X = x$ and $Q\hat{O}P = h$. The feet of the perpendiculars from Q and P to OX are A and B, respectively, and the foot of the perpendicular from P to QA is C. Show that

$$\frac{\sin(x+h)-\sin x}{h} = \frac{QC}{\text{arc } PQ}.$$

Deduce that

$$\lim_{h\to 0+} \left\{\frac{\sin(x+h)-\sin x}{h}\right\} = \lim_{h\to 0+} \left\{\frac{QC}{\text{chord } PQ}\right\} = \lim_{h\to 0+} \{\cos(x+\tfrac{1}{2}h)\} = \cos x.$$

†**25.** By rationalising the numerator, show that $\lim\limits_{x\to 0+} \dfrac{\sqrt{(x+1)}-1}{x} = \dfrac{1}{2}$.

†**26.** Write down the values of

$$\lim_{x\to a-} \phi(x), \quad \lim_{x\to a+} \phi(x),$$

if the graph of $\phi(x)$ is that of Fig. 2.64. What is the value of $\phi(a)$?

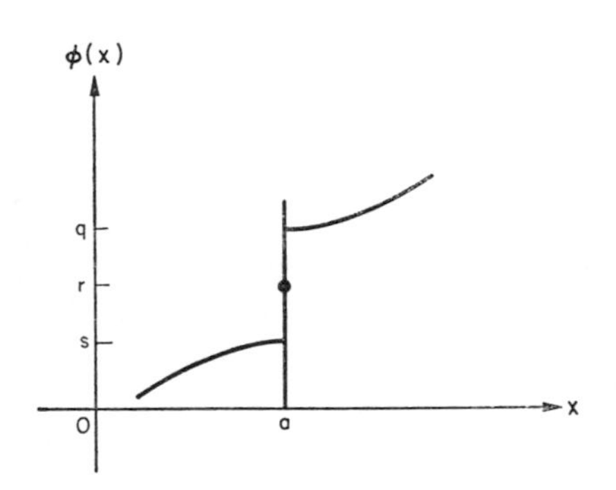

FIG. 2.64.

†**27.** Show that the area of an n-sided regular polygon inscribed in a circle of radius r is

$$\tfrac{1}{2}nr^2 \sin\left(\frac{2\pi}{n}\right).$$

Given that $(\sin x)/x \to 1$ as $x \to 0$, find the value of

$$\lim_{n\to\infty}\left\{\tfrac{1}{2}nr^2 \sin\left(\frac{2\pi}{n}\right)\right\},$$

and explain why this is the area of a circle.

†**28.** Sketch the graphs of $y = 1/(x-1)$, $y = 1/(x-2)$ and $y = 1/(x-1)(x-2)$.

†**29.** Draw, on a single piece of graph paper, the graphs of $y=1$, $y=(\sin x)/x$, and $y=\cos x$ (where x should be measured in radians). An interval of $0 \leqslant x \leqslant 1{\cdot}5$ will be sufficient. Does your graph display the inequality $\cos x < (\sin x)/x < 1$ for $x > 0$? By considering the value of x for which $\cos x = 0{\cdot}99$, find the range of x over which $(\sin x)/x$ is within $0{\cdot}01$ of 1.

†**30.** Sketch the graph of $f(x) = [1-x]$, where $[x]$ is the greatest integer not greater than x or, colloquially, the "whole number" part of x. In what interval is $f(x) = 0$? Does

$$\lim_{x\to 0+}\{f(x)\} = f(0)?$$

Is $f(x)$ continuous at $x = 0$?

2.7. NUMERICAL APPROXIMATION*

A theorem of Weierstrass states that every function that is continuous in an interval (a, b) can be represented in that interval to any desired degree of accuracy by a polynomial. Put into a more specific form, this theorem tells us that, if we choose the coefficients $c_0, c_1, c_2, \ldots$ c_n carefully enough, we can replace any function $f(x)$ which is continuous in an interval by an expression of the form $c_0+c_1x+c_2x^2+\ldots+c_nx^n$, and the polynomial approximation will give us as accurate a substitute for the function $f(x)$ in the specified interval as we wish, provided that we make n large enough.

We can determine the optimum values for $c_0, c_1, c_2, \ldots, c_n$ if we are given $(n+1)$ values of $f(x)$ for known values of x (say $x_0, x_1, x_2, \ldots$ x_n): all that is then necessary is to choose the coefficients c_i so that

$$\begin{aligned} f(x_0) &= c_0+c_1x_0+c_2x_0^2+c_3x_0^3+\ldots+c_nx_0^n, \\ f(x_1) &= c_0+c_1x_1+c_2x_1^2+c_3x_1^3+\ldots+c_nx_1^n, \\ f(x_2) &= c_0+c_1x_2+c_2x_2^2+c_3x_2^3+\ldots+c_nx_2^n, \end{aligned}$$

and so on. (In graphical terms this is equivalent to making the graph of $p(x)=c_0+c_1x+c_2x^2+\ldots+c_nx^n$ go through $(n+1)$ points which also lie on the graph of $f(x)$: and, clearly, the

*Some readers may prefer to omit this at a first reading.

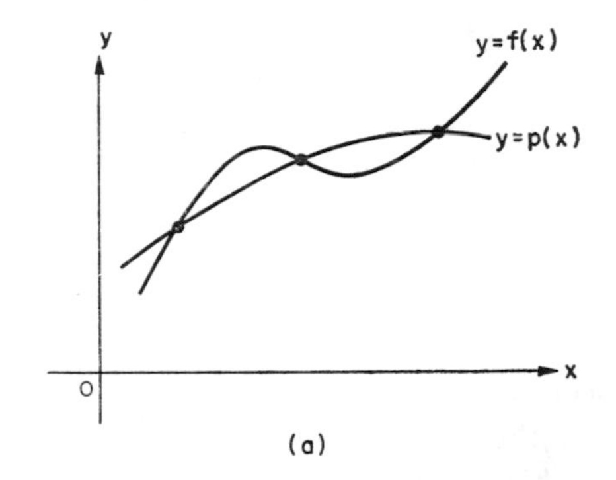

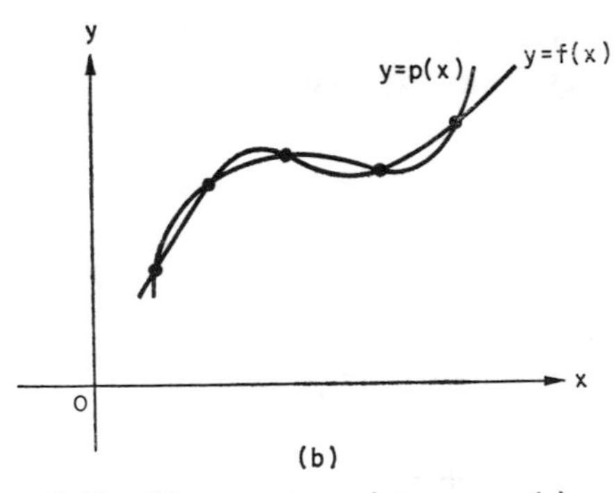

FIG. 2.65. The more points $y = p(x)$ and $y=f(x)$ have in common, the smaller will be the differences between the curves at any intermediate point.

larger the value of n, the smaller will be the gap between the two graphs at any intermediate position (Fig. 2.65).)

Knowing the form of the substitute polynomial $p(x)$, we can then use it to calculate the approximate values of $f(x)$ at values of x which are intermediate to the x_i at which $f(x)$ is specified, and which therefore lie in the range $x_0 < x < x_n$. The method thus enables us to obtain an *interpolated* value for $f(x_i+\alpha)$, where $x_i < (x_i + \alpha) < x_{i+1}$, and we can make this interpolation as accurate as we please by choosing n sufficiently large (always providing that the table of values of $f(x)$ with which we are supplied is sufficiently large too).

In practice, such a procedure is rather laborious, and it pays us to obtain a general equation expressing the polynomial coefficients c_i in terms of the given values of the function $f(x_i)$. In practice, too, it is convenient if the given values $f(x_i)$ are available for evenly spaced values of x. Under these conditions, we can replace $x_0, x_1, x_2, \ldots, x_n$ by $a, a+h, a+2h, a+3h, \ldots, a+nh$, where a is the smallest value of x for which the corresponding value of $f(x)$ is available, and where, if we are thinking of an interval, $a \leqslant x \leqslant b$, h is the equal interval between the values of x at which $f(x)$ is specified, and therefore equal to $(b-a)/n$. (Under these conditions we are in effect fitting as high an order polynomial curve to the given points as our table of values will allow.) The resulting expression is best written in terms of differences between successive values of $f(x)$, and simple algebraic manipulation shows that the most accurate interpolation for the value of $f(a+\theta h)$, where $0 < \theta < 1$, is

$$f(a+\theta h) \simeq f(a)+\theta\delta f(a)+\frac{\theta(\theta-1)}{2!}\delta^2 f(a) + \frac{\theta(\theta-1)(\theta-2)}{3!}\delta^3 f(a)+\ldots + \frac{\theta(\theta-1)\ldots(\theta-n+1)}{n!}\delta^n f(a),$$

and where

$$\delta f(a) = f(a+h)-f(a),$$
$$\delta^2 f(a) = \delta f(a+h)-\delta f(a),$$
$$\delta^n f(a) = \delta^{n-1} f(a+h)-\delta^n f(a).$$

This equation for the interpolated value of $f(a+\theta h)$ is known as the *Gregory–Newton* formula, and apart from the obvious use just discussed and developed in the examples below, it will be of some interest at a number of points throughout this book. It is important to note that its use depends on having a table of values of $f(x)$ at equal intervals of x, and that the value of x at which the interpolation is desired lies at or near the lower value end of the table. For interpolation under other conditions, some other formula must be used.

Example 2.13

Find by interpolation the approximate value of $10^{0.872}$, given the following table of values.

x	0·87	0·88	0·89	0·90	0·91	0·92	0·93	0·94
10^x	7·413	7·585	7·762	7·943	8·128	8·317	8·511	8·710

We first construct a table showing the required values of $\delta f(a)$, $\delta f(a+h)$, $\delta f(a+2h)$, ..., $\delta^2 f(a)$, $\delta^2 f(a+h)$, ..., $\delta^3 f(a)$, etc. In this case, $a = 0{\cdot}87$, and since the interval in x at which $f(x)$ is tabulated is 0·01, we shall choose h as 0·01. Proceeding from definition, we then obtain

x	$10^x = f$	δf	$\delta^2 f$	$\delta^3 f$
0·87	7·413	0·172	0·005	−0·001
0·88	7·585	0·177	0·004	0·000
0·89	7·762	0·181	0·004	0·000
0·90	7·943	0·185	0·004	0·001
0·91	8·128	0·189	0·005	0·000
0·92	8·317	0·194	0·005	
0·93	8·511	0·199		
0·94	8·710			

As we have seen before (questions 21, 22, p. 32), the differences rapidly decrease to zero. In this case we can say that the second-order differences are constant, and there is therefore no need to develop the Gregory–Newton formula beyond the third term. (Or, alternatively, there is no point in fitting a polynomial of degree greater than 3 to our tabulated values.)

We are now in a position to estimate the value of $10^{0\cdot872}$: with the given table of values, we are concerned with making the interpolation in the interval $0\cdot87 < x < 0\cdot88$, and with $a = 0\cdot87$ and $h = 0\cdot01$, we must set θ equal to 0·2. We then need only add the first three terms of the Gregory–Newton formula together, and proceed as follows:

$$f(0\cdot87) = 7\cdot413$$

$$\theta\,\delta f(0\cdot87) = 0\cdot2\times0\cdot172 = 0\cdot0344$$

$$\frac{\theta(\theta-1)}{2!}\delta^2 f(0\cdot87) = \frac{(0\cdot2)(-0\cdot8)}{2!}(0\cdot005) = 0\cdot0004$$

$$\therefore\ f(0\cdot872) = 7\cdot448,$$

and the required answer is therefore that $10^{0\cdot872} \simeq 7\cdot448$.

Exercise 2g

1. Explain the meaning of the terms *polynomial*, *polynomial of degree n*.

2. If a function $f(x)$ is tabulated at $x = a$, $x = a+h$, $x = a+2h$, $x = a+3h$, etc., and $\delta f(a) = f(a+h)-f(a)$, $\delta^2 f(a) = \delta f(a+h)-\delta f(a)$, etc., show that $\delta^2 f(a) = f(a+2h)-2f(a+h)+f(a)$, and that $\delta^3 f(a) = f(a+3h)-3f(a+2h)+3f(a+h)-f(a)$.

3. Obtain as accurate an interpolated value for $\sin^{-1}(0\cdot2211)$ as the following table of values allows:

x	0·22	0·23	0·24	0·25	0·26	0·27
$\sin^{-1}x$	0·2218	0·2321	0·2424	0·2527	0·2631	0·2734

4. What is meant by "linear" interpolation? Derive the expression $f(a+\theta h) \simeq f(a)+\theta\delta f(a)$ from the Gregory–Newton formula. Under what conditions do you consider it useful to use this approximation?

5. If $f(x) = x^3+ax^2+bx+c$, what are the general expressions for $\delta f(x)$, $\delta^2 f(x)$, $\delta^3 f(x)$ and $\delta^4 f(x)$? (Take $\delta f(x) = f(x+1)-f(x)$, $\delta^2 f(x) = \delta f(x+1)-\delta f(x)$, etc.)

6. Thinking of the Gregory–Newton formula for $f(a+\theta h)$ as a function of θ, a and h being constants, explain why you would expect $\delta^{n+1} f(a)$ to be zero for a function whose proper expression is a polynomial of degree n (cf. questions 21 and 22, p. 32).

7. If $f(x)$ is a polynomial of degree n in x, deduce that $\delta f(x)$ is a polynomial of degree $(n-1)$ in x. What can be said about the degree of $\delta^2 f(x)$, $\delta^n f(x)$, and $\delta^{n+1} f(x)$?

8. If $f(x) = a\theta(x)+b\phi(x)+c\psi(x)$, show that
$$\delta f(x) = a\delta\theta(x)+b\delta\phi(x)+c\delta\psi(x).$$

9. The following table shows values of a so-called Bessel function $J_0(x)$ in the range $0\cdot4 \leq x \leq 1\cdot0$:

x	0·4	0·5	0·6	0·7	0·8	0·9	1·0
$J_0(x)$	0·9604	0·9385	0·9120	0·8812	0·8463	0·8075	0·7652

Find by interpolation approximate values for $J_0(0\cdot55)$, $J_0(0\cdot45)$ and $J_0(0\cdot35)$. Why must care be exercised in applying the Gregory–Newton formula to this last case?

SUMMARY

A *scalar* quantity is one which can be regarded as having magnitude only, and a *vector* quantity is one which can be regarded as having two components representable by line magnitude and direction. Scalars can be compounded algebraically; vectors must be compounded with a triangle law.

Displacement and velocity are vector quantities; *velocity* is defined as the change in displacement occurring in an interval of time divided by the length of the interval. When the displacement occurring in a problem occurs along a straight line, it is convenient to represent distances to one side of a fixed reference point by positive displacements, and distances in the opposite direction by negative displacements.

A unity scale graph is one which employs unit length along both axes to represent unit value of the variable associated with the axes: a standard scale graph employs a common but non-unit length along its axes to represent unit values of the variables.

The tangent of the angle which the line joining two points on a unity or a standard scale displacement–time graph makes with the time axis is equal to the average velocity of the body whose motion is represented by the graph for the interval between the instants represented by the two points. The area between the unity scale velocity–time graph, the time axis and two ordinates is numerically equal to the distance travelled by the body whose motion is represented by the graph during the interval between the instants represented by the two ordinates.

It is possible to describe geometrical figures by algebraic equations; two perpendicular axes are constructed, the vertical, range or dependent variable axis carrying possible values for a variable y, and the horizontal axis carrying the values of an independent or domain variable x. The position of any point in the graphical plane can be described by the values of x and y opposite which the point is situated, and these numbers are known as the *coordinates* of the point. The x-coordinate is sometimes called the *abscissa*, and the y-coordinate the *ordinate*. A concise way of expressing the location of a point is to write the coordinates as a number pair inside a bracket, and when this is done, the convention is to give the value of the x-coordinate as the first number of the pair. This system of coordinates is referred to as the *Cartesian* system.

Algebraic equations relating y to x describe the locus of a point which can move in the Cartesian graphical plane subject to the law expressed by the equation. The coordinates of points lying on the locus are said to satisfy the equation; points lying above the locus have ordinate values greater than those given by the equation, and points lying below the locus have ordinate values less than those given by the equation.

An increase in the value of the variable can be regarded as a positive change, and on a graph, this results in an upward or rightward movement of the point representing the value of the variable; a decrease can be regarded as a negative change, and the graphical movement is now leftward or downward.

A functional relationship or a graph is continuous at a particular domain value if, for any small neighbourhood drawn about the associated range value, a domain neighbourhood can be found such that its entire image lies within the small range neighbourhood, no matter how small this latter neighbourhood. A function is said to be continuous in an interval if it is continuous at every value within that interval.

If a function is undefined at any point, and a value can be given it for that point so that the function becomes continuous at the point, this value is said to be the *limit* of the function at the point. The equation

$$\lim_{x=a} \{f(x)\} = b$$

implies that $f(x)$ can be made continuous at $x = a$ by defining $f(a)$ to equal b.

A quantity which increases without bound is said to become infinite: the equation

$$\lim_{x \to a+} \{f(x)\} = \infty$$

implies that the image $f(x)$ of x lies within any neighbourhood bounded on the lower side if x is greater than a and sufficiently close to it. The equation

$$\lim_{x \to \infty} \{f(x)\} = a$$

implies that the image $f(x)$ of x can be made to lie within any small neighbourhood about the range value a if x is chosen sufficiently large.

The Gregory–Newton formula for the interpolated value of $f(a+\theta h)$ is

$$f(a+\theta h) \simeq f(a)+\theta\delta f(a)+\frac{\theta(\theta-1)}{2!}\delta^2 f(a)\ldots$$

Miscellaneous Exercise 2

1. A body moves so that its distance x from a fixed point X varies with the time t according to the equation $x = 3t - t^2$. Draw a graph of the motion from $t = 0$ to $t = 5$, and interpret its shape.

2. The height h of a body above ground level varies with time t according to the equation $h = 8t - 5t^2$, where h is in metres and t is in seconds. Draw a graph of the first 3 seconds of motion, and from your graph deduce the maximum height the body reaches, and the value of t at which it occurs.

3. The displacement s of a body from a fixed point is given by the equation $s = t(t-2)(t-3)$, where t is the time after the start of the motion. Draw a graph of the motion for values of t between 0 and 5, and interpret its shape.

4. A stone is thrown upward from the top of a cliff with a velocity of 20 m/s. After t seconds, its height h relative to the cliff top is given by the equation $h = 20t - 5t^2$. Find its position at the end of the first, third and fifth seconds, plot a graph of its motion, and deduce its greatest height.

5. A boat pointing directly across a stream is rowed with a velocity of 16 km/h relative to the water, the speed of the water being 12 km/h. Calculate the position of the boat relative to its starting point after

(i) 6 minutes, (ii) 12 minutes,

and write down its velocity relative to the bank. If the journey from one bank of the stream to the other takes 15 minutes, what is the displacement of the boat when it reaches the other bank relative to its initial position?

6. Write down the relationship which exists between (i) the tangent of the angle θ made by a chord joining two points on a displacement–time graph with the time axis, and (ii) the average velocity v of the body whose motion the graph represents between the two instants represented by the two points, if

(a) the graph is a unity scale graph,
(b) the scale used on the time axis is twice that on the displacement axis,
(c) the scale used on the displacement axis is five times that used on the time axis,
(d) the graph is a standard scale graph.

7. A curve is defined by the rule that for all points lying on it, the ordinate is equal to 7 more than twice the square of the abscissa. What are the coordinates of the points whose abscissae are

(a) 1, (b) 2, (c) -2, (d) 0,

and what are the abscissae of the points whose ordinates are

(e) 25, (f) 39?

8. A circle centre the origin has three of the following five points lying on it. Find which they are, and deduce the radius of the circle: $A(3, 4)$; $B(-1, 5)$; $C(0, -5)$; $D(3, 3)$; $E(-4, 3)$.

9. Show that the triangle whose vertices are $P(3, 2)$, $Q(1, 3)$ and $R(-1, -1)$ is right angled. At which vertex is the right angle?

10. Determine which of the following points lie on the line $y = 2x-5$: $A(2, -1)$; $B(0, -5)$; $C(1, 3)$; $D(1, -6)$; $E(-2, -9)$.

11. A curve has an equation of the form $ax^2+bx+c = y$. If the curve goes through the points (0, 4), (2, 0) and $(-1, 3)$, find the values of a, b and c.

12. Find the equations of the lines passing through the following pairs of points:

(i) $(2, -5)$ and $(4, 1)$, (ii) $(3, 5)$ and $(-1, -3)$.

13. Find the equation of the line joining (a, b) to (c, d).

14. Find the coordinates of the intersection of the line $x+2y-8 = 0$ with the curve $y^2 = x$.

15. Write down the x and y intercepts of the lines and curves:

(i) $y = 2x-7$, (iv) $y = 3x^2$,
(ii) $2y = x+5$, (v) $y = x^2-4$,
(iii) $x+y+1 = 0$, (vi) $x^2+y-1 = 0$.

16. Sketch the graphs of the lines $y = 7-2x$ and $y = 4+x$. Shade in the area occupied by the points which satisfy the inequalities $y < 7-2x$ and $y < 4+x$ simultaneously.

17. Draw a sketch of the lines $2x+3y = 7$ and $x+5y = 4$, and indicate in which area lie the points which satisfy the inequalities $2x+3y > 7$ and $x+5y > 4$ simultaneously.

18. A variable point P moves so that its distance from the origin is five times its distance from the point (12, 0). Prove that the locus of P is a circle $x^2+y^2-25x+150 = 0$, and find the centre and radius of this circle. Find also the equation of the tangent of this circle at the point (11, 2). (OC)

19. A point (x, y) moves so that its distance from $(a, 0)$ is equal to its distance from the line $x = -a$. What is the relationship between x and y?

20. Sketch the graphs of the functions $y = 2\sqrt{x}$ and $y = -2\sqrt{x}$. What equation expresses a relation between y and x and incorporates both of these functions? Explain why this equation does not represent a function.

21. Factorise the equation $x^2-y^2+6y-9 = 0$, and sketch the graphs of the two functions it incorporates.

22. Sketch the graph of $y = (x^2+8)/(x^2-4)$.

23. Sketch the graph of the function $f(x)$ defined by the equations $f(x) = 1+x$ if x is rational, and $f(x) = 0$ if x is irrational.

24. A function $g(x)$ is defined by the equations $g(x) = 1$ if $x \geqslant 0$, $g(x) = 2$ if $-2 < x < 0$, $g(x) = 3$ if $x < -2$. Sketch a graph of the function, and write down its domain and range.

25. Write down the values of the following limits:

(i) $\lim_{x\to\infty}\left\{\frac{1+x}{1-x}\right\}$,

(ii) $\lim_{x\to\infty}\left\{\frac{2}{3+7x}\right\}$,

(iii) $\lim_{x\to\infty}\left\{\frac{2x}{3-7x}\right\}$,

(iv) $\lim_{x\to\infty}\left\{\frac{x^2-9}{2x^2+x-4}\right\}$,

(v) $\lim_{x\to a}\left\{\frac{x^2-ax+a^2}{x+b}\right\}$,

(vi) $\lim_{x\to 2}\left\{\frac{x^2+4x+3}{x^2-4x-3}\right\}$,

(vii) $\lim_{x\to 3}\left\{\frac{x^2+4x+3}{x^2-4x-3}\right\}$.

26. Sketch the graphs of the following functions:

(i) $\sin(x-30°)$,

(ii) $\sin 3x$,

(iii) $x\cos x$,

(iv) $\tan 2x$,

(v) $\sqrt{x}$.

27. The lengths of the sides of a triangle are denoted by a, b and c. What is the functional relation between the area of the triangle, two of the three sides and the angle A? Express the relationship with (i) b, (ii) A as the dependent variable (or subject).

†**28.** A particle moves so that its displacement is given by the equation $s = 2^{-t}\sin t$. Draw a sketch graph of the displacement–time curve for the particle, and describe the motion.

CHAPTER 3

Some Applications of Graphs

3.1. RATES OF CHANGE

The definition of speed as the distance covered in unit time is sufficiently commonplace to pass it by without appreciating its full significance. It is but one example of standardisation—an essential process in any science. To boast of an investment which yields an income of £800 a year gives hardly any measure of its return at all: we need to know how much the investment yields in a year for every pound invested. To report that Great Britain is overcrowded because there are 60 million people living in it does not give adequate grounds for using the word "overcrowded": what is of greater importance is the average number of people living in one square mile.

Reducing to unity the value of one of the variables in situations involving two is often necessary, and this is also the case in problems where we are interested in the change in the value of one variable caused by a change in the other. We cannot easily compare the properties of expansion of two metals if we are told that a bar of the first expands by 0·00130 cm when the temperature rises by 64°C while a bar of equal length of the second expands by 0·00035 cm when its temperature rises by 17°C. But if we standardise the results, comparison is easier. The first bar expands by 0·0000203 cm and the second by 0·0000206 cm for every 1°C rise in temperature, and this statement enables the slightly greater rate of expansion of the second to be readily appreciated.

Standardisation of the change in one variable with respect to another is achieved by determining the ratio of the change in the dependent variable to that in the independent variable producing it. A change of 0·0013 cm in the bar length caused by a 64°C change in temperature gives an "average" expansion of (0·0013/64), or 0·0000203 cm, for a 1°C rise in temperature. Similarly, a change of δy in a variable y caused by a change of δx in the (independent) variable x can be standardised as a change of $(\delta y/\delta x)$ in y for every unit change in x. This ratio is sensibly called the "rate of change of y with respect to x", and the word "rate" need not necessarily be reserved to mean time rate. Time changes may be the most useful or commonplace, but many problems arise in which time is not the independent variable.

Example 3.1

When the voltage across an electric fire increases by 20 volts, the current increases by $\frac{1}{3}$ ampere. What is the rate of change of the current with respect to the voltage?

For a change of one volt, the current increases by ($\frac{1}{3}$/20) or $\frac{1}{60}$ ampere, and the rate of change of current with respect to voltage is thus $\frac{1}{60}$ ampere per volt.

Example 3.2

In a given interval of time, a car factory produced 400 more cars than usual, and as a result the average cost of each car fell by £18. What is the rate of change of production cost per car with respect to the number of cars produced?

The ratio of the dependent variable change to the independent variable change in this case is 18/400 or 0·045, and since the fraction is composed of a number of pounds divided by a number of cars, the unit is which the fraction is measured is £ per car. Thus, the rate of change of production cost per car with respect to the number of cars produced is £ 0·045 per car.

The value of the rate of change between two dependent variables can be related to the shape of the graphical map of their relation. If two variables y and x are related in the manner described by the graph in Fig. 3.1,

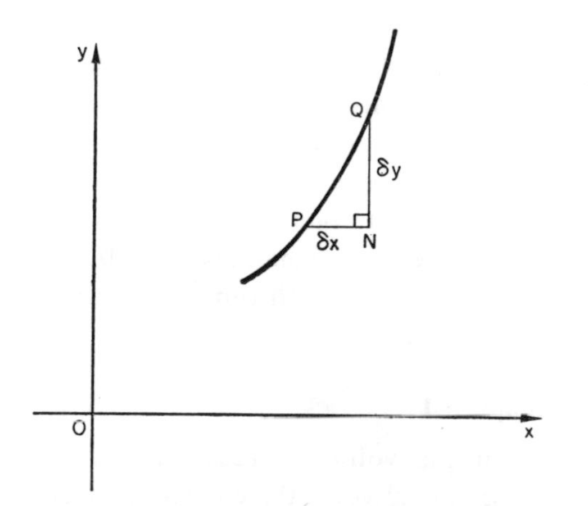

FIG. 3.1. A change of δx in x leads to a change of δy in y; the rate of change of y with respect to x is $\delta y/\delta x$ or QN/PN.

a small change δx in the value of x will produce an associated change δy in the value of y. Representing the changes in x and y by PN and QN, and completing the right-angled triangle PNQ, we can write that the rate of change of y with respect to x ($= \delta y/\delta x$) is equal to the ratio QN/PN (assuming that the graph is drawn with unity of standard scaling). Now if the rate of increase of y with respect to x is large, the change δy in y resulting from or associated with a small change δx in x will be comparatively large. Thus, QN will be large for small values of PN, and the graph will be steep. Similarly, flatter curves indicate lower values of rates of change, and downward-sloping curves represent negative rates of change (when an increase in the independent variable leads to a decrease in the dependent variable) (Figs. 3.2 and 3.3).

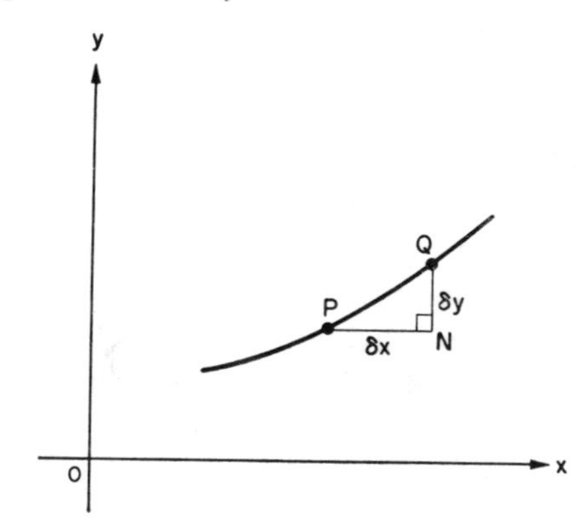

FIG. 3.2. Flatter curves indicate lower rates of change if the graph is drawn with standard scaling.

Another reason for the convention of assigning the vertical and horizontal axes to the dependent and independent variables respectively is now apparent. Interchanging the axes (Fig. 3.4) gives a misleading impression if x is the independent variable, for the rate of change of y with respect to x is large, whereas the graph appears flat—and, in fact, its steepness is a measure of the ratio $\delta x/\delta y$ or the rate of change of x with respect to y. And

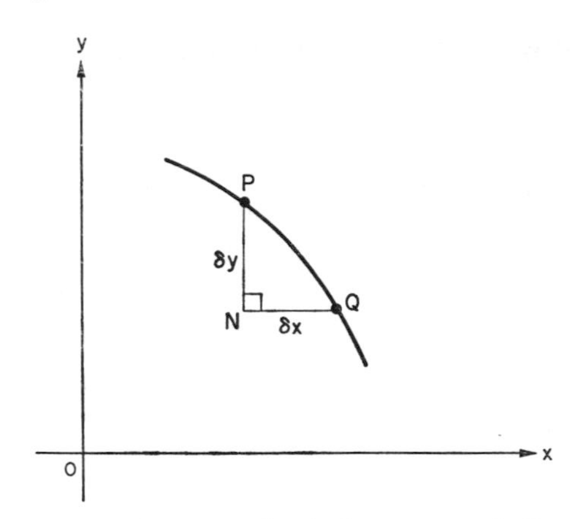

FIG. 3.3. Downward-pointing curves indicate negative rates of change (when, for example, an increase in x leads to a decrease in y).

since we are rarely concerned with a change in the independent variable resulting (if it can) from the change in the dependent variable, the rate of change of the independent variable with respect to the dependent variable is of little or no real meaning or use. Similarly, the importance of the scales used on the axes is emphasised by considering rates of change: distension of the independent variable axis leads to an apparent flattening of the curve, whereas the rate of change of the dependent variable with respect to the independent variable still has the same value (Fig. 3.5). It will be recalled that numerical significance can be

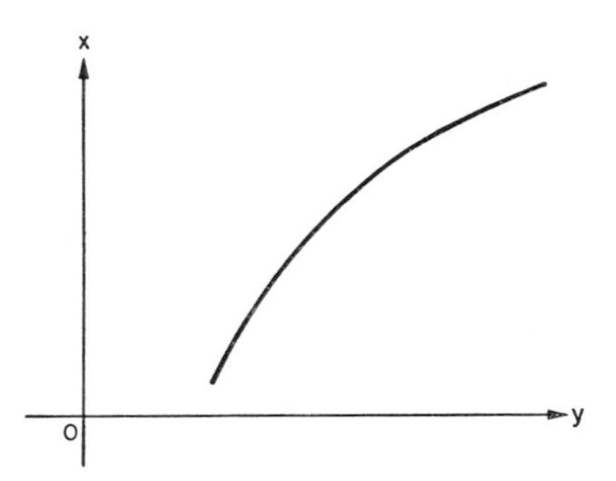

FIG. 3.4. Interchanging the x- and the y-axes gives a misleading impression of the relationship between a dependent and an independent variable: this graph represents the same information as the graph of Fig. 3.1.

attached to the geometrical appearance of a graph only when standard or unity scaling is employed.

A comment must be added concerning the units in which rates of change are expressed. As we have seen, a rate of change is a ratio; and it is common usage to refer to the unit of the ratio as "units of y per unit of x"—e.g. a ratio of new pence over centimetres is measured in new pence per centimetre (or new pence/centimetre as it is usually written). But some confusion might arise if one of the variable is already a ratio: for example, the specific heat capacity of a substance can be

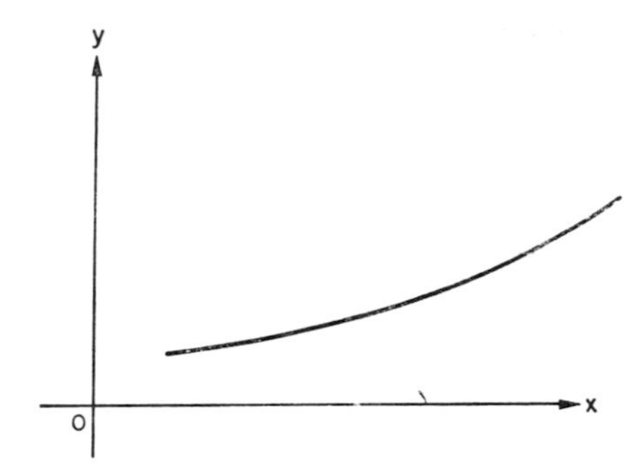

FIG. 3.5. And distension of the independent variable axis leads to the appearance of a flatter curve and therefore a lower rate of change.

defined as the number of joules required to raise the temperature of 1 kg of it by 1°C. We can easily determine that a given body of mass 10 kg requires 15 joules to raise its temperature by 15°C: and from this we deduce that 1 kg of the substance of which the body is made will require 1·5 joules to raise its temperature 15°C, i.e. it needs 1·5 J/kg for 15°C. The specific heat capacity is thus 1·5 J/kg divided by 15°C, but it is ambiguous to write the answer as 0·1 J/kg/°C. This could mean that (J/kg) is the unit of the dependent variable and °C that of the independent variable, or joules the unit of the dependent variable and (kg/°C) that of the independent variable. As a result, the use of negative powers is becoming increasingly common, and a unit such as J/kg is often written J kg^{-1}. The ratio 1·5 J kg^{-1}/15°C is thus unambiguously quoted as 0·1 J kg^{-1} °C^{-1}, and we can now see that joules/(kg/°C) are in fact

$$\frac{\text{J}}{\text{kg/°C}} = \frac{\text{J °C}}{\text{kg}} \quad \text{or} \quad \text{J °C kg}^{-1}.$$

3.2. RATE OF CHANGE AS AN ENLARGEMENT FACTOR

A graphical way of looking at the rate of change can be established by recalling that any graph maps all the members of the independent variable included in the domain onto

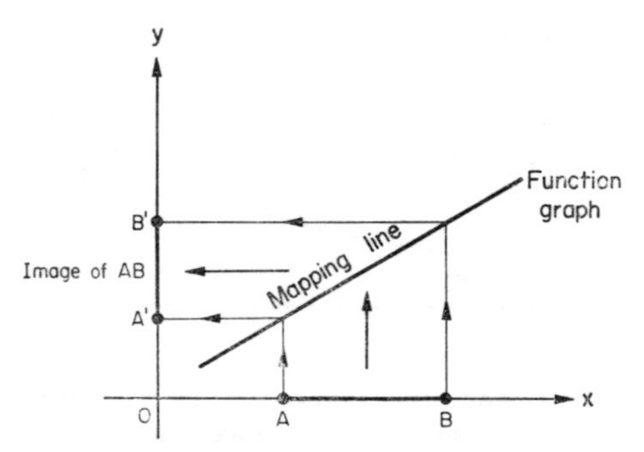

FIG. 3.6. The mapping represented by the function graph produces the image $A'B'$ from AB; the enlargement factor is $A'B'/AB$.

the associated members of the dependent variable. If we consider a group of adjacent x-axis points, representing a continuous spread of independent variable values (the group AB in Fig. 3.6), we can quite easily find the group of y-axis points which represent the corresponding spread of dependent variable values ($A'B'$). Now the line segment $A'B'$ is—as it were—an "image" of the independent variable line segment AB produced by the function mapping, and we can quite sensibly think of the ratio $A'B'/AB$ as the enlargement factor associated with the particular function whose mapping is relevant to this case.

If we draw the picture in a slightly different manner (Fig. 3.7), we can see that the ratio $A'B'/AB$ also represents the rate of change of the dependent variable with respect to the independent variable. Thus the graphical enlargement factor has a value identical to the rate of change.

This pictorial image of the rate of change is also valid from the point of view of physical units. If, for example, a car travels 40 kilometres in half an hour, the graph of its motion will map a time axis segment of $\frac{1}{2}$ hour onto a distance axis segment of 40 kilometres. Working in axial units, the enlargement factor corresponding to the mapping will be 40 km/$\frac{1}{2}$ hour or 80 km/hour. This is indeed equal to the

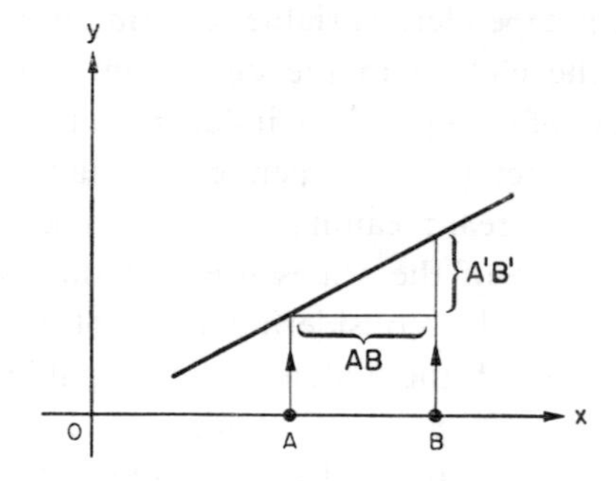

FIG. 3.7. The ratio $A'B'/AB$ also represents the rate of change of y with respect to x.

average rate of change of the distance with respect to the time over the interval in question, and both are, in turn, equal to the average speed in that interval.

EXERCISE 3a

1. The population of Great Britain increased from 11,944,000 in 1801 to 22,259,000 in 1851. What was the average rate of increase per year? How does this compare with the average rate of increase between 1851 and 1901, and 1901 and 1951, if the population in 1901 and 1951 was 38,257,000 and 50,225,000 respectively?

2. The numbers of people found guilty of indictable offences in England and Wales in 1960 and 1961 were 203,775 and 182,217 respectively. What rate of change do these figures suggest? If the population of England and Wales was 45,561,000 in 1960 and 46,072,000 in 1961, find the number of people found guilty of indictable offences per 1,000 population in the two years, and calculate the rate of change of this number with respect to time. Estimate the number of people found guilty for every thousand of the population in (i) 1962, (ii) 1963, and comment on the accuracy of your results.

3. Two variables x and y are related by the equation $y = 2^x$. What is the average rate of change of y with respect to x between (i) $x = 0$ and $x = 1$, (ii) $x = 1$ and $x = 2$, (iii) $x = 2$ and $x = 3$, (iv) $x = a$ and $x = b$?

4. Calculate how many times faster x^2 changes than x in the intervals between: (i) 0 and 2, (ii) 1 and 3, (iii) 0 and a, (iv) 1 and a.

5. A traveller's cheque for £10 can be exchanged for 338 Czech crowns. What rate of exchange does this represent? The official rate is 34 crowns to the £; what is the rate of deduction for encashment?

6. A firm sells ten typewriters normally costing £28·75 each for £273·125 to a customer. What percentage rate of discount does this represent?

7. A car increases its speed from 10 m/s to 15 m/s in 4 seconds. What rate of change of velocity does this represent? Give the units of your answer.

8. The speed of a car increases from 60 km/h to 90 km/h in 6 seconds. What rate of change of velocity does this represent? Give the units of your answer.

***9.** In the equation $\eta = F/av$, F is measured in newtons, a in m and v in m/s. What are the units of η?

***10.** The resistance of a piece of wire rises from 7·8 Ω at 20°C to 8·4 Ω at 100°C. What is the average rate of change of resistance with respect to temperature between 20°C and 100°C?

***11.** The length of a bar increases from 81·2 cm at 9°C to 81·8 cm at 100°C and 82·3 cm at 200°C. What is the average rate of change of length with temperature between

(i) 9°C and 100°C, (ii) 100°C and 200°C,

and what is the average rate of change of the coefficient of expansion with respect to temperature between 9°C and 200°C? Give the units of your answer.

12. A body thrown up from the earth's surface has an equation of motion $s = 20t - 5t^2$, where s is the height (in metres) of the body above the earth's surface t seconds after it has been thrown. Plot a graph of the motion and state the range of t over which the equation is valid.

What is the average rate of change of s with respect to t between

(i) $t = 0$ and $t = 2$,
(ii) $t = 3$ and $t = 5$,
(iii) $t = 6$ and $t = 8$?

Explain the meaning of your answers in physical terms.

13. The graphs of (i) $y = 3x$, (ii) $y = 5x$, and (iii) $y = 7x$ are drawn with unity scaling. By considering the y-axis image of the x-axis segment representing $1 \leqslant x \leqslant 3$, find the enlargement factors of the mapping for each of the three cases.

14. A function of x is defined to have the value $(ax+b)$ at x, a and b being constants. What relation, if any, is there between the value of the unity scale mapping and (i) a, (ii) b?

What meaning can be attached to a negative value of a?

15. Draw a sketch graph of $y = x^2$, and discuss what happens to the value of the unity scale enlargement factor as x varies between a large negative value and a large positive value.

***16.** The resistance of a piece of wire varies with the temperature as shown in the following table:

Resistance (ohms)	100	104	110	116	121	126
Temperature (°C)	0	20	40	60	80	100

Plot a graph of these figures, and deduce from it the temperature coefficient of resistance of the specimen of wire.

***17.** A column of air is trapped in a capillary tube by a thin thread of mercury, and the length of the trapped column varies with temperature as follows:

Length (cm)	20·7	21·0	21·4	22·0	22·4	23·2	23·9	24·8	26·3
Temperature (°C)	20·0	25·1	30·1	38·3	45·1	54·9	65·1	79·8	99·0

Plot a graph of these figures, and determine the coefficient of linear expansion of air between 20°C and 100°C, assuming the tube to be uniform in cross-section.

At what temperature would the length of the air column be zero?

18. (i) A meteorological balloon is released from a point 500 metres distant horizontally from an observer. If the balloon ascends vertically and its angle of elevation, as measured by the observer, increases from 30° to 60° during an interval of 1 minute, calculate its average speed of ascent in m/s.

(ii) A man walks down a straight path which is on a plane hillside and which makes an angle of 60° with the line of greatest slope. After walking for 300 m, the man finds that he has descended through a vertical distance of 30 m. Find the angle of slope of the hillside. (OC)

3.3. INSTANTANEOUS RATES OF CHANGE

Consider now the motion of a pebble dropped from the top of a cliff 45 m high (Fig. 3.8). If we record the position of the stone at half second intervals, the motion of the pebble can be summarised by the following table of results:

t (seconds after release)	0	$\frac{1}{2}$	1	$1\frac{1}{2}$	2	$2\frac{1}{2}$	3
s (position in metres below the top of the cliff)	0	1·25	5	11·25	20	31·25	45

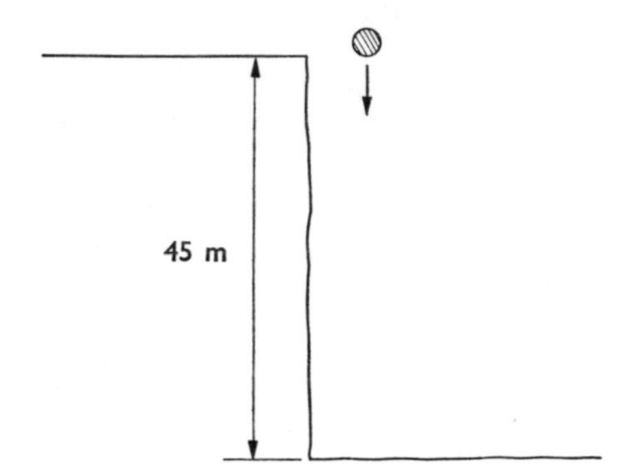

FIG. 3.8. A pebble is dropped from a cliff 45 m high.

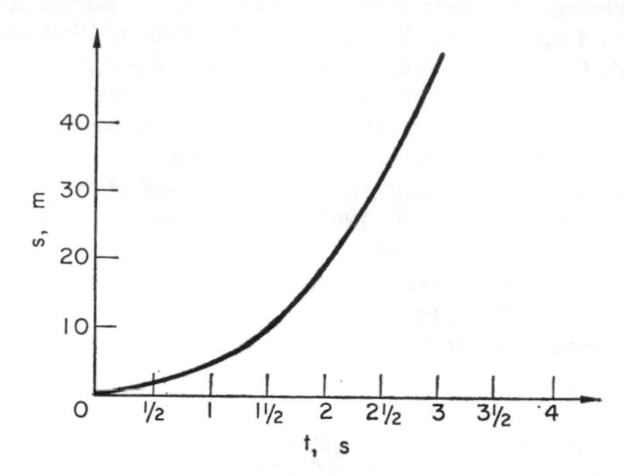

FIG. 3.9. The graph of the pebble's motion.

As the pebble increases its speed under gravity, the distances it travels in each $\frac{1}{2}$ second interval increase, and the graph of its motion displays the expected shape shown in Fig. 3.9. The steepness increases progressively—and this is simply an expression of the fact that the rate of change of displacement with time (i.e. the velocity) is increasing throughout the motion.

We can calculate the average velocity during any second from the table above. Taking, for example, the 2nd second (i.e. the motion between $t = 1$ and $t = 2$), the average velocity in this interval can be evaluated as $(20-5)/(2-1)$, or 15 m/s. Similarly, the average velocity for the interval from $t=1$ to $t=1\frac{1}{2}$ can be found as $(11{\cdot}25-5)/(1\frac{1}{2}-1)$, or 12·5 m/s. The lower value is to be expected, of course, for the mid-point of the latter interval occurs at an earlier time and when the speed is lower. It might now be asked whether it is possible to determine the average velocity for an even shorter interval.

Having plotted carefully the curve of Fig. 3.9, we can in fact "fit" an equation to describe it: the curve appears quadratic in shape, and indeed the equation $s = 5t^2$ expresses the relation between all the pairs of values in the table exactly. From this equation we can find the value of s at $t = 1\frac{1}{5}$; this can be calculated as 7·2 (m). Thus the average velocity during the interval from $t = 1$ to $t = 1\frac{1}{5}$ is $(7{\cdot}2-5)/(1\frac{1}{5}-1)$, or 11 m/s.

Once again this figure is lower than the value obtained for the previous intervals, and clearly the shorter we make the interval over which the average velocity is taken, the lower will be our result. We can investigate whether or not there is a limit to this average velocity as the interval is made smaller as follows.

Let us consider an interval of time from $t = 1$ to $t = 1+\delta t$. The value of s when $t = (1+\delta t)$ is $5(1+\delta t)^2$, and thus the average velocity during the interval is

$$\frac{5(1+\delta t)^2-5}{(1+\delta t)-1}.$$

Multiplying out the numerator, we obtain

$$\frac{5\{1+2\delta t+(\delta t)^2\}-5}{(1+\delta t)-1},$$

and simplifying, we obtain

$$\frac{10\delta t+5(\delta t)^2}{\delta t} \quad \text{or} \quad \{10+5(\delta t)\} \text{ m/s}$$

as the final expression for the average velocity between $t = 1$ and $t = (1+\delta t)$. This result agrees with those already obtained, for if we put δt equal to 1, $\frac{1}{2}$ and $\frac{1}{5}$ in turn, we obtain 15, $12\frac{1}{2}$ and 11 m/s, respectively, as before. But we can now also see precisely how the value of the average velocity over an interval varies with the size of the interval, and further deduce that as the size of the interval is reduced (i.e. as $\delta t \to 0+$), the limiting value of 10 m/s will be approached. Thus in an abbreviated form,

lim (average velocity as the time interval $\to$ 0)

$$= \lim_{\delta t \to 0+} (10+5\delta t)$$
$$= 10 \text{ m/s}.$$

It seems sensible to call this value of 10 m/s the velocity at the instant $t = 1$, and by this we mean that if the velocity of the pebble suddenly stopped increasing at that instant, it would cover 10 m in the next second. But before we define what is usually referred to as the instantaneous velocity, we must allow ourselves the thought that we could have considered how the average velocity for an interval *ending* at $t = 1$ varies with the size of the time interval. This backward looking approach is equally relevant, and we can easily deduce that the average velocity between two such times is still given by the expression $\{10+5(\delta t)\}$ m/s, except that δt is now negative or less than zero. Repeating the argument, we then find that as $\delta t \to 0-$, the value of $\{10+5(\delta t)\}$ $\to$ 10 m/s.

So the limiting values of v as $\delta t \to 0+$ and $\delta t \to 0-$ are both the same: and in this case we can dispense with the idea of the dynamic limit, writing that

$$\lim_{\delta t = 0} \left(\frac{\delta s}{\delta t}\right) = 10.$$

It is in fact this static limit which is used to define the quantity called *instantaneous velocity*, and although it might appear pedantic to demand that the values of

$$\lim_{\delta t \to 0+} \left(\frac{\delta s}{\delta t}\right) \quad \text{and} \quad \lim_{\delta t \to 0-} \left(\frac{\delta s}{\delta t}\right)$$

are equal before we speak of an instantaneous velocity, it is a condition worth making. If

$$\lim_{\delta s \to 0+} \left(\frac{\delta s}{\delta t}\right) \neq \lim_{\delta t \to 0-} \left(\frac{\delta s}{\delta t}\right),$$

the velocity of the body in question changes abruptly at the instant we are considering. And if immediately before an instant the body is travelling with a velocity of, say, 16 m/s, and immediately after the instant it is travelling with a velocity of 80 m/s, its velocity is discontinuous and there is no point in speaking of any velocity at the instant of change.

These ideas are extended to rates of change in general. We have already seen how to define the rate of change between two variables y and x over an interval—namely as the value of the ratio $\delta y/\delta x$—and we can now define the *instantaneous rate of change* between the two variables as the limiting value of the ratio $(\delta y/\delta x)$ at $\delta x = 0$ (again implying that if this limit does not exist, there is no point in thinking of an instantaneous rate of change). We shall see in the next chapter that it is often quite a simple matter to obtain an expression for an instantaneous rate of change if the equation relating the two variables involved is known.

Example 3.3

Calculate the average rate of change of y with respect to x for the intervals (i) $x = 0$ to $x = 5$, (ii) $x = -1$ to $x = 1$, (iii) $x = 2$ to

$x = 2+a$, given that y and x are related by the equation $y = x^2$.

By considering the limiting value of (iii) at $a = 0$, deduce the instantaneous rate of change of y with respect to x at $x = 2$.

(i) When $x = 0$, $y = 5$, $y = 25$. Thus $\delta y = 25$ when $\delta x = 5$, and therefore the rate of change of y with respect to x over the interval from $x = 0$ to $x = 5$ is $\delta y/\delta x = 25/5 = 5$.

(ii) Since $y = 1$ both when $x = -1$ and $x = 1$, $\delta y = 0$ for this interval. Thus the average rate of change over the interval is 0.

(iii) When $x = 2$, $y = 4$; when $x = 2+a$, $y = (2+a)^2 = 4+4a+a^2$. Thus $\delta y = 4a+a^2$, and therefore the average rate of change over the interval is equal to

$$\left(\frac{4a+a^2}{a}\right) \quad \text{or} \quad (4+a).$$

If $a \to 0$ from either side in (iii), the value of $(4+a)$ approaches the common limit 4. Thus the instantaneous rate of change of y with respect to x at $x = 2$ is 4.

EXERCISE 3b

1. Find the average velocity of the pebble dropped from the cliff top described on p. 78 during the intervals from

(i) $t = 2$ to $t = 3$,
(ii) $t = 2$ to $t = 2\frac{1}{2}$,
(iii) $t = 2$ to $t = 2\frac{1}{4}$.

What is the average velocity between $t = 2$ and $t = 2+\delta t$? Hence find the average velocity between $t = 2$ and $t = 2{\cdot}1$, and the instantaneous velocity at $t = 2$.

2. Find the average velocity of the pebble dropped from the cliff top described on p. 78 during the intervals

(i) $t = \frac{1}{2}$ to $t = \frac{1}{2}+\delta t$,
(ii) $t = 1\frac{1}{2}$ to $t = 1\frac{1}{2}+\delta t$,
and (iii) $t = 2\frac{1}{2}$ to $t = 2\frac{1}{2}+\delta t$.

Calculate the instantaneous velocities at $t = \frac{1}{2}$, $1\frac{1}{2}$ and $2\frac{1}{2}$ seconds from your results.

3. The displacement of a body falling in air is given by the equation $s = 30t+5t^2$ where s is in metres and t in seconds. Calculate the average rate of change in displacement with respect to time during the intervals between

(i) $t = 0$ and $t = 2$, (iii) $t = 0$ and $t = \frac{1}{2}$,
(ii) $t = 0$ and $t = 1$, (iv) $t = 0$ and $t = 0+\delta t$.

Hence find the instantaneous rate of change when $t = 0$. What physical meaning has the rate of change of displacement with respect to time?

4. Find the average velocity of a body whose motion is described in question 3 during the intervals between

(i) $t = 1$ and $t = 2$, (iii) $t = 1$ and $t = 1\frac{1}{4}$,
(ii) $t = 1$ and $t = 1\frac{1}{2}$, (iv) $t = 1$ and $t = 1+\delta t$.

Write down the instantaneous velocity at $t = 1$.

5. Referring again to the body whose motion is described in question 3, find the average velocity between $t = 2$ and $t = 2+\delta t$. What is the instantaneous velocity at $t = 2$?

6. The height s (in metres) of a body thrown up from the earth's surface t seconds after projection is given by the equation $s = 40t-5t^2$. Find the average rate of change of s with respect to t between

(i) $t = 1$ and $t = 1+\delta t$, (ii) $t = 0$ and $t = 0+\delta t$.

Hence write down the instantaneous velocities at $t = 1$ and $t = 0$, and explain the meaning of the results.

7. Repeat question 6 for the intervals

(i) $t = 2$ to $t = 2+\delta t$, (ii) $t = 3$ to $t = 3+\delta t$.

8. A body is dropped from a cliff top and after t seconds has fallen through s metres, where the relation between s and t is described by the following table:

t (seconds)	0	0·2	0·4	0·6	0·8	1·0
s (m)	0	0·2	0·8	1·8	3·2	5·0

Write down an equation relating s to t and find the instantaneous velocity when $t = 0{\cdot}6$.

9. Referring to the motion described in question 8, find an expression for the instantaneous velocity at a time t by considering the average velocity in the interval from t to $t+\delta t$. How long after the body is dropped from the cliff top will the velocity be

(i) 6 m/s, (ii) 15 m/s, (iii) 100 km/h?

10. A bullet is fired vertically from the surface of the earth and its equation of motion is found to be $h = 70t-5t^2$, where h is the height of the bullet in metres above the earth's surface t seconds after the firing. What is the instantaneous velocity when

(i) $t = 2$, (ii) $t = 0$?

From your results find the velocity of firing in km/h.

11. Referring again to the motion of the bullet described in question 10, find the instantaneous velocity of the bullet when (i) $t = 1$ and (ii) $t=13$. Explain your results, and give a sketch graph of the equation $h = 70t-5t^2$.

12. Referring once more to the motion described by the equation $h = 70t-5t^2$, find the instantaneous velocity at an indefinite time t (see question 9). When is the velocity of the bullet

(i) 38 m/s, (ii) 200 km/h?

***13.** An electron in an electric field moves in a straight line and according to the equation $s = 10t^2$, where s is the distance travelled in metres by the electron t milliseconds after it starts from rest. Find its instantaneous velocity when

(i) $t = 1$, (ii) $t = 2$, (iii) $t = 3$ milliseconds.

***14.** An electron moves in an electric field so that its displacement (s) along a straight line varies with the time t as shown in the following table:

t (second)	0	0·1	0·2	0·3	0.4	0·5
s (m)	0	0·4	1·6	3·6	6·4	10·0

Write an equation for the relationship between s and t, and find the instantaneous velocity when $t = 1$, assuming the motion to continue according to your equation. How far will the electron have travelled by then?

15. Find the instantaneous rate of change of y with respect to x for the function $y = x^2+8$ when (i) $x = 2\frac{1}{2}$, (ii) $x = 0$.

16. Find the instantaneous rate of change of y with respect to x for the function $y = 8-x^2$ when (i) $x = 10$, (ii) $x = 4$.

17. Find the instantaneous rate of change of p with respect to q for the function $p = 3q^2-q$ when (i) $q = 1$, (ii) $q = 0$.

18. Find the instantaneous rate of change of a with respect to b if $a = 7-6b-b^2$ when (i) $b = 4$, (ii) $b = 0$.

19. Draw a graph of $y = (x-1)(x-2)$. What is the average rate of change of y with respect to x over the intervals

(a) $x = 3$ to $x = 4$,
(b) $x = 0$ to $x = 1$,
(c) $x = -1$ to $x = 1$,
(d) $x = 1$ to $x = 1+\delta x$?

What is the instantaneous rate of change of y with respect to x when $x = 1$?

20. Two variables p and q are related by the equation $p = 9q+7$. Draw sketch graphs of (i) p against q, (ii) q against p. What is the rate of change of (a) p with respect to q, (b) q with respect to p?

21. The statement that the "average rate of change over an interval is greater than the instantaneous rate of change at the lower end of the interval if the curve is upward bending" can be made with reference to Fig. 3.1. What relative magnitude does the average rate of change bear to the instantaneous rate of change at the upper end of an interval for such a curve? What do the relations become if the curve is downward bending?

†**22.** £100 is invested at 5% p.a. compound interest. What is the amount invested at the end of each of the first 5 years? What is the average rate of simple interest over the 5 years, and the instantaneous rate of interest at the end of the 5th year?

***23.** The length of a bar of iron is given by the equation

$$l = 100+0{\cdot}0015t+0{\cdot}00003t^2,$$

where $l =$ the length of the bar in cm, and $t =$ the temperature in °C.

Draw a sketch graph of the length for $0 \leqslant t \leqslant 1000$, and calculate the

(i) average rate of change of length with temperature for the ranges

(a) $0 \leqslant t \leqslant 100$,
(b) $0 \leqslant t \leqslant 500$;

(ii) the instantaneous rate of change of length with respect to temperature at

(c) $t = 0$°C,
(d) $t = 100$°C,
(e) $t = 1000$°C.

What relationship do these values bear to the coefficient of linear expansion of iron?

†**24.** Find an expression for the instantaneous rate of change of y with respect to x at an indefinite value of x if $y = a+bx+cx^2$.

3.4. GRADIENTS

The slope of a road is a measure of its steepness: and the steepness is an indication of the vertical height through which the road rises or falls for a given distance along it. A slope of 1 in 10 indicates that if a car travels 10 m along the road, its height above sea-level will have changed by 1 m. The slope defined in this way is clearly the sine of the angle the road makes with the horizontal (Fig. 3.10), and an alternative way of thinking of the steepness of the road is to regard it as the change in height for every foot travelled along the road. Mathematically, however, the slope or gradient of a line on a graph is more usefully defined in

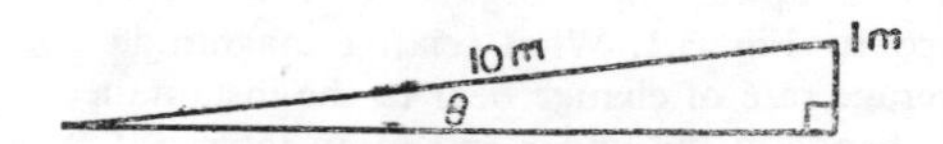

FIG. 3.10. The slope of the road is commonly quoted as the sine of the angle the road makes with the horizontal.

terms of the rate of change between the variables mapped together. If, for example, the line maps x with y, its gradient is defined as the ratio of any change in the value of the dependent variable y to the change in the value of the independent variable x which produces it. Thus, in this case (Fig. 3.11), the tangent of

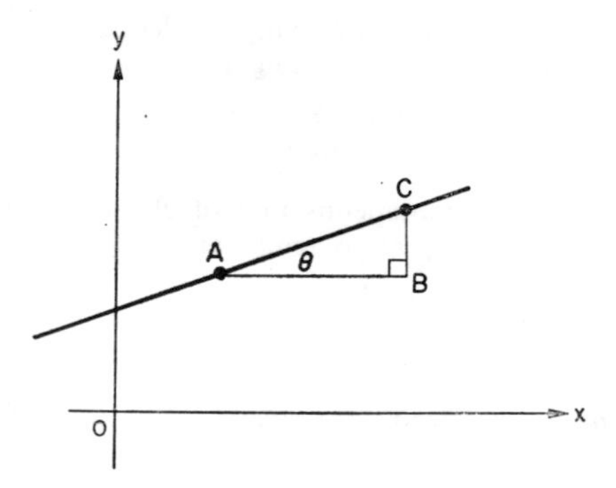

FIG. 3.11. The gradient of the line on a standard scale graph is understood to be the tangent of the angle between the line and the direction of the x-axis.

the angle that the line makes with the direction of the x-axis is the trigonometrical function which gives the gradient, and the value of the gradient is equal to the rate of change of y with respect to x. This definition is also convenient if the coordinates of two points on the line (e.g. A and C in Fig. 3.11) are known, for it is easier to calculate the difference in the x coordinates (AB) and the difference in the y coordinates (BC) than the length along the line (AC) which will involve a square root. In the case of the road slopes, of course, it is easier to measure AC and BC than AB (which will be into the side of the hill) and BC. Thus, we can summarise our definition and interpretation of the gradient of the line joining two points (x, y) and $(x+\delta x, y+\delta y)$ on a standard scale graph by the equation

$$\begin{aligned}\text{gradient } \delta y/\delta x &= \text{rate of change of } y \text{ with respect to } x \text{ in the interval } x \text{ to } x+\delta x \\ &= \tan\theta \quad \text{(Fig. 3.11).}\end{aligned}$$

(If the coordinates of the points joined by the line are given in the form (x_1, y_1), (x_2, y_2), then δy can be replaced by y_2-y_1 and δx by x_2-x_1 (Section 1.7). The gradient of the line is then more usefully expressed by the ratio $(y_2-y_1)/(x_2-x_1)$.)

If a line slopes downwards from left to right, an increase or positive change in x will lead to a decrease or negative change in y. The ratio

$$\frac{\text{change in } y}{\text{change in } x}$$

will therefore have a negative sign and the gradient of such a line will then be negative (Fig. 3.12). It will be noticed that in both cases, the larger the value of θ, the larger the gradient: to get a rough idea of the size and the

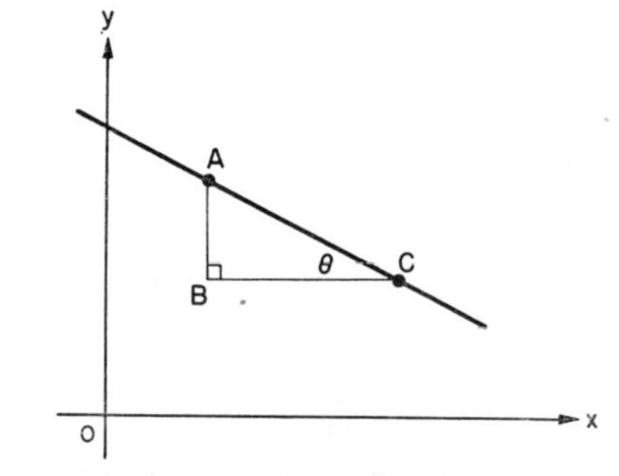

FIG. 3.12. Downward-pointing lines have negative gradients.

sign of the gradient, imagine that the line represents a road seen in cross-section in the side of a hill. If it is steep, the gradient is large; and looked at as a road proceeding from left to right, uphill is positive (gaining height), and downhill is negative (losing height). The gradient of a horizontal line is zero (since for any change in the value of x there is no change in

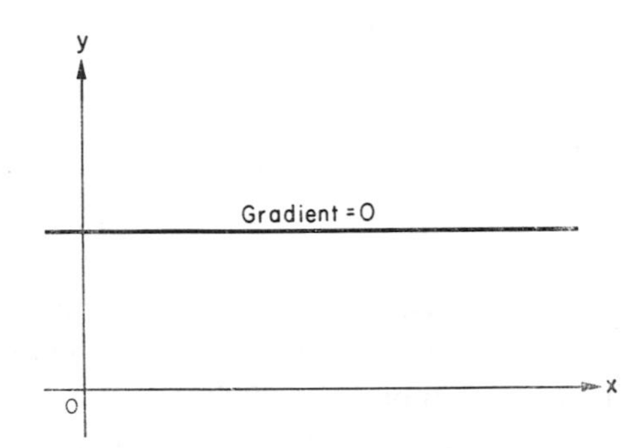

FIG. 3.13. A horizontal line has zero gradient.

the value of y, and therefore the numerator of the ratio defining the gradient is zero) (Fig. 3.13), but it is rather meaningless to speak of the gradient of a vertical line. A line which rotates from the horizontal towards the vertical has, however, a gradient which increases without bound, and which increases extremely rapidly as the line approaches the vertical. Its gradient can therefore be said to become infinite, and the phrase that the gradient of a line parallel to the y axis is infinite is often (if rather loosely) made (Fig. 3.14).

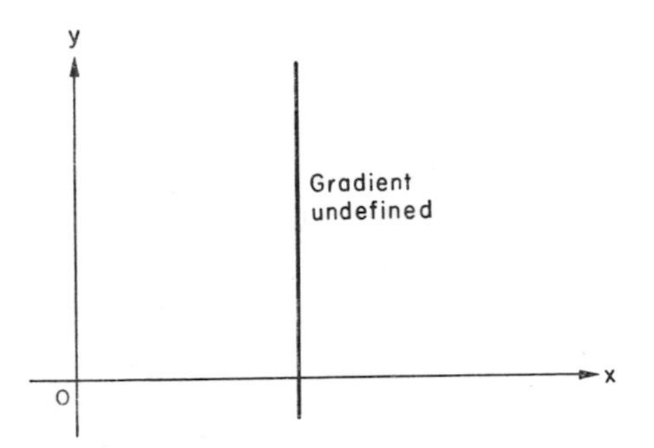

FIG. 3.14. The gradient of a line parallel to the y-axis is undefined.

Recalling the comments of Section 2.4, it can now be seen that the gradient of the line $y = mx$ is m, for the line passes through the origin (see Fig. 3.15), and the ratio of the change in y to the change in x for any movement of any point on it must equal m. Similarly, $y = mx+c$ is a parallel line displaced by c along the y-axis (Fig. 3.16), and its gradient is also m.

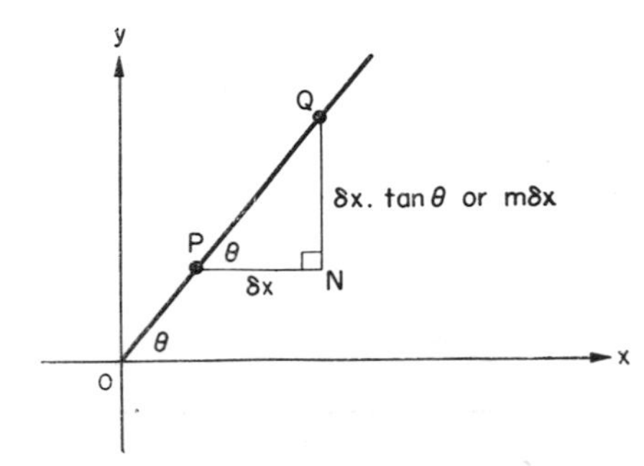

FIG. 3.15. The gradient of the line $y = mx$ is m...

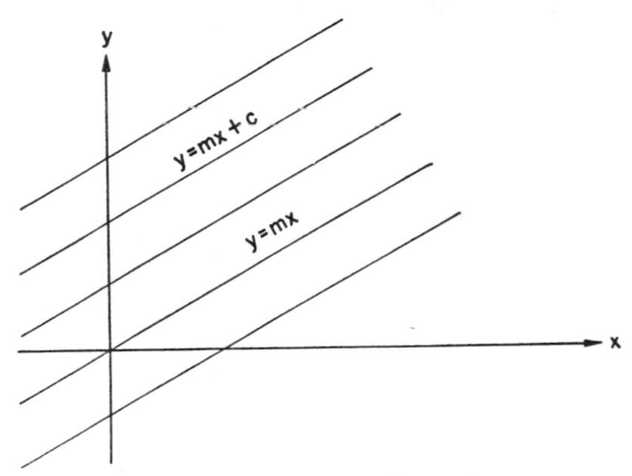

FIG. 3.16. ... and so is the gradient of $y=mx+c$.

The gradient of $ax+by+c = 0$ is best obtained by rewriting the equation in the form

$$y = -\frac{a}{b}x - \frac{c}{a},$$

when, by comparison with $y = mx+c$, it can be seen that the gradient is $-a/b$. The gradient of any line is given by the coefficient of x when the coefficient of y is unity, the x and the y terms being on the opposite sides of the equation.

The concept of the gradient of a curve presents a new problem. Thought of as the cross-section of a road running from left to right, it seems sensible to say that the gradient of the curve in Fig. 3.17 increases going from left to right, for moving towards right of P, the road become steeper. But we have not yet coined a precise definition for the gradient of a curve and we must digress for a moment to do so.

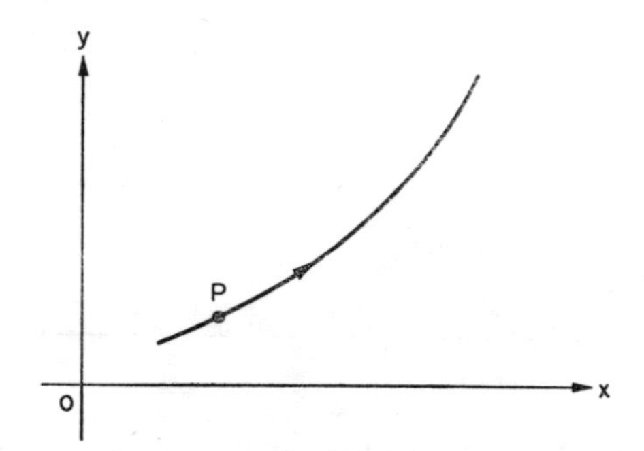

FIG. 3.17. Intuitively we feel that the gradient of an upward bending curve increases with x.

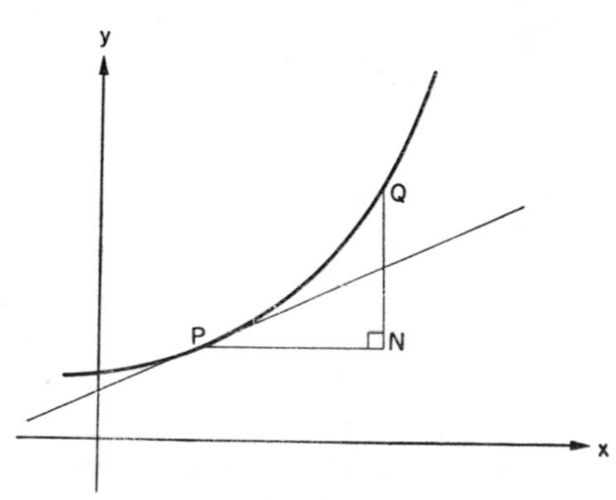

FIG. 3.18.

If we mark two points P and Q on a curve (Fig. 3.18), we can evaluate the gradient of the chord PQ by determining the value of the ratio QN/PN. Now if Q slides down the curve towards P through q', q'', q''', etc. (Fig. 3.19a), it is clear that the value of QN/PN and therefore the gradient of PQ will decrease. Similarly if Q approaches P from below (Fig. 3.19b), the gradient of PQ will increase.

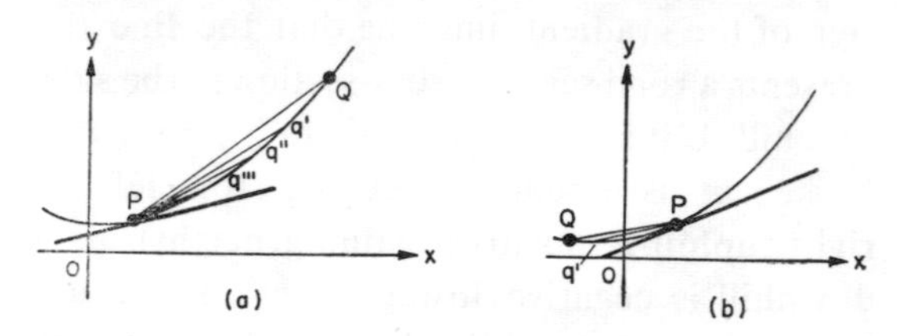

FIG. 3.19. (a) As Q approaches P, the gradient of QP approaches that of the tangent at P; (b) and this is also true if Q approaches P from below.

Representing these results graphically leads us to a graph with a discontinuity (Fig. 3.20), for when Q has the same abscissa as P, there is no longer a chord PQ. To make the gradient graph continuous, we must insert a gradient value equal to the common limit of the chord gradients as Q approaches P from below and from above, and we can define this common limit value as the gradient of the curve at P.

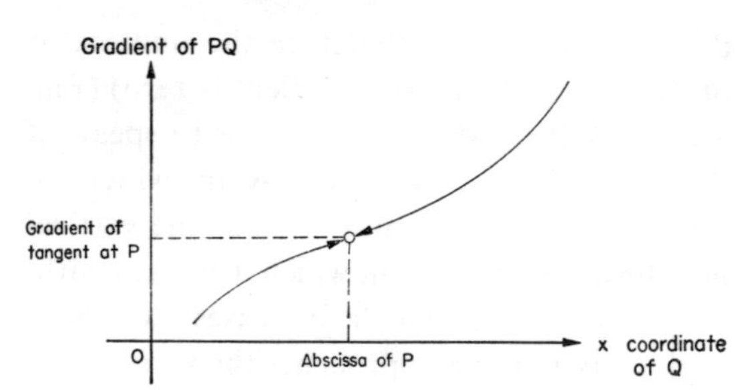

FIG. 3.20. We can make the graph continuous by putting the gradient of PQ equal to that of the tangent at P when P and Q are coincident.

We can develop this a stage further. The "tangent" to a curve at any point P is conveniently defined as the line which passes through P and occupies the limiting position of the secant PQ as Q approaches P from either side. It is clear that under these circumstances, the gradient of a curve at any point is the same as the gradient of the tangent at that point (Fig. 3.21). Although we shall rarely need to use such a method, we can none the less obtain some estimate of the gradient of a given curve at any point by constructing the tangent to the

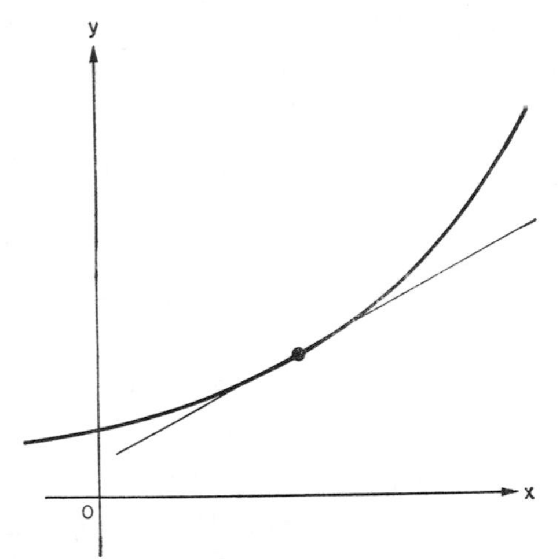

FIG. 3.21. The gradient of a curve at a point is equal to the gradient of the tangent there.

curve at the point in question, and measuring the tangent's gradient directly.

Adopting these definitions completes the association between the gradient of a curve or line and the rate of change of the dependent variable with respect to the independent variable. If we denote the length QN of Fig. 3.18 by δy, and that of PN by δx, we can rewrite our definition of the tangent to a curve at any point P as the line which passes through the point P and whose gradient has the value

$$\lim_{\delta x=0}\left(\frac{\delta y}{\delta x}\right).$$

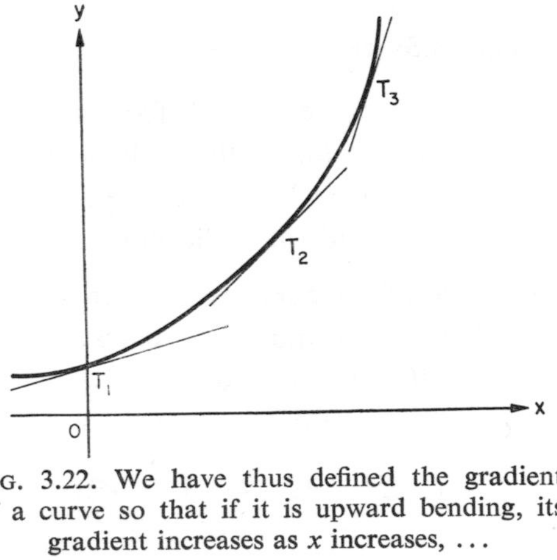

FIG. 3.22. We have thus defined the gradient of a curve so that if it is upward bending, its gradient increases as x increases, ...

Now the expression $\delta y/\delta x$ gives the "*incremental*" *rate of change** between the variables over the interval δx, and we have already defined the limiting value of this ratio at $\delta x = 0$ to be equal to the instantaneous rate of change at the point in question. Thus, the instantaneous rate of change has a value equal to that of the gradient of the tangent to the curve at the point representing the mapping of the instant under consideration.

We can finally see that the gradient of a curve does not have a constant value. If a series of tangents is constructed at a succession

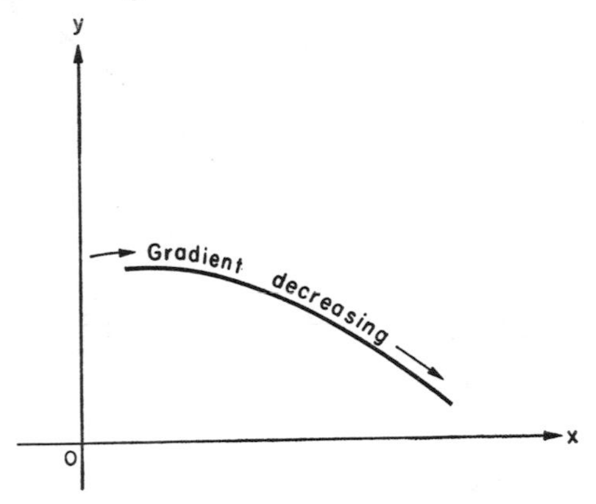

FIG. 3.23. ... and if the curve is downward bending, the gradient decreases as x increases.

of points on a curve such as that shown in Fig. 3.22, we can see that the gradient of any particular tangent is different from that of its immediate neighbours. The instinctive suggestion made earlier that the gradient of the curve in Fig. 3.17 increases progressively with the abscissa thus receives confirmation from the definitions we have chosen to adopt. We need only add that the choice we have made demands in a similar fashion that the gradient of a downward bending curve such as that shown in Fig. 3.23 progressively decreases (i.e. becomes more negative) as the value of the abscissa increases.

* The word increment means a small increase, and the incremental rate of change refers to the value of $\delta y/\delta x$ for a small δx beginning at P.

Example 3.4

Calculate the gradients of the lines joining the points PQ, QT, RS and PS, where the coordinates of P, Q, R, S and T are (1, 3), (7, 8), (−1, 4), (−3, −1) and (0, −5), respectively. If U is a 6th point with abscissa 2 and ordinate −1, what will the gradient of SU be?

The points can be plotted on graph paper (see Fig. 3.24), and to find the gradient of PQ, the triangle PAQ can be completed. PQ is the hypotenuse, and the perpendicular sides PA and AQ are drawn parallel to the x- and y-axes, respectively. The gradient of PQ is the value of the ratio AQ/PA, and if QA produced meets the x-axis at N, QN = the y-coordinate of Q (i.e. 8). But AN = the y-coordinate of P (i.e. 3), and so

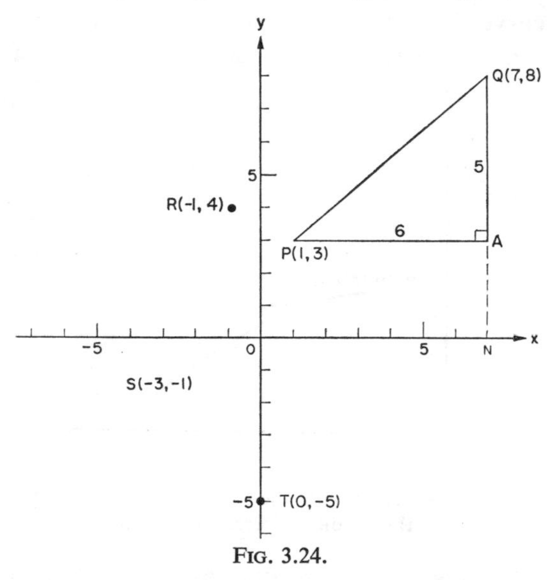

FIG. 3.24.

$$\begin{aligned} QA &= QN - AN, \\ &= 8-3, \\ &= 5. \end{aligned}$$

Similarly, PA = the difference between the x-coordinates, which is $(7-1)$ or 6. Thus the gradient of $PQ = 5/6$.

Similarly, the gradient of QT is given by the ratio of the difference between the y-coordinates of Q and P to the corresponding difference in the x-coordinates (note that the subtraction must be carried out in the same order).

$$\text{Thus gradient of } QT = (8-(-5))/(7-0) = 13/7.$$

Similarly,

$$\text{gradient of } RS = (4-(-1))/(-1-(-3)) = 2\tfrac{1}{2},$$

$$\text{and gradient of } PS = (3-(-1))/(1-(-3)) = 1.$$

Since the y-coordinate of U is the same as that of S, the gradient of SU will be 0/5, i.e. 0.

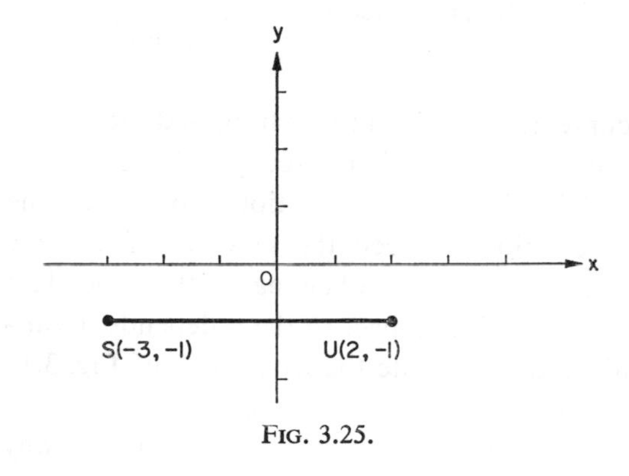

FIG. 3.25.

The line SU is therefore flat, as can be seen by considering a figure with the points S and U plotted on it (Fig. 3.25).

Example 3.5

Sketch the curve $y = x^2$ for values of x from −3 to 3. Mark the points A and B whose abscissae are 3 and −2, respectively, on it. What is the gradient of the chord AB?

A sketch of the curve is given in Fig. 3.26, and the points A and B are marked in. The triangle ABL is completed, and it can be seen that the gradient of AB is equal to the ratio

$$AL/LB = (9-4)/(3-(-2)) = 1.$$

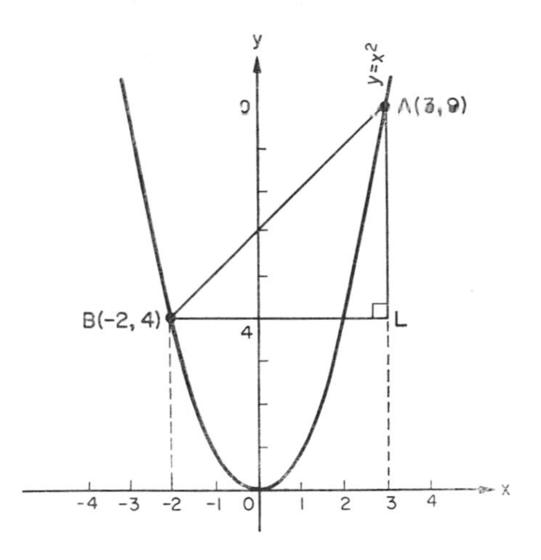

FIG. 3.26.

Example 3.6

Calculate the average speed of the body whose motion is represented in the graph of Fig. 3.27 between the times represented by the points P and Q.

We can calculate the speed of the body whose motion is represented in Fig. 3.27 from the numerical values associated with P and Q. By simple geometric alignment, P can be found to be the point (1, 90) and Q the point

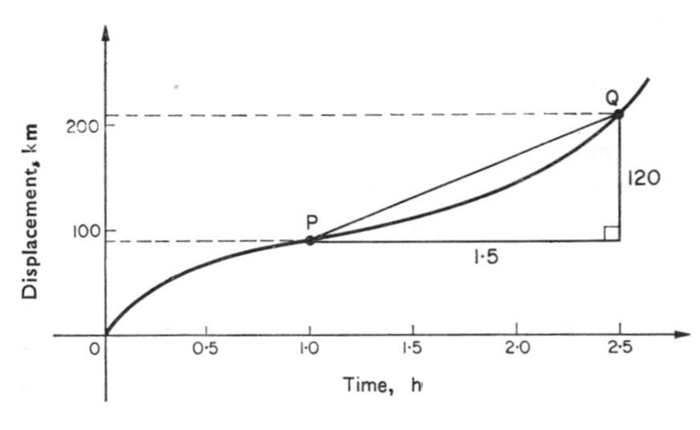

FIG. 3.27.

(2·5, 210), so we can write that the average speed between P and Q is

$$\frac{210-90}{2\cdot5-1} = \frac{120}{1\cdot5}, \quad \text{or} \quad 80 \text{ km/h}.$$

EXERCISE 3c

1. What is the gradient of a line joining the point $A(1, 3)$ to the point B if the coordinates of B are:

(i) $(-1, 2)$, (ii) $(-2, -2)$, (iii) $(4, 4)$, (iv) $(5, -3)$, (v) $(-7, 3)$?

2. A curve passes through the points $A(4, 2)$, $B(8, -1)$, $C(0, 6)$. What are the gradients of the chords AB, AC and BC?

3. A curve passes through the points $P(-1, -3)$, $Q(-2, -8)$, and $R(2, -5)$. What are the gradients of the chords PQ, QR and PR?

4. Draw a graph of the line $4x-2y = 1$, using the same scale for both axes. Take any two points on the line, and by determining the difference between their x-coordinates and their y-coordinates, find the gradient of the line. Measure the angle that the line makes with the x-axis, and find the value of its tangent.

Redraw the graph using half the y-axis scale for the x-axis (e.g. if 1 cm represents 1 on the y-axis, 1 cm represents 2 on the x-axis). Repeat the calculations. What do you notice about the value of the tangent?

5. Sketch the following lines on a piece of graph paper:

$$y = 3x+5, \quad y = 3x+1,$$
$$y = 3x+1, \quad y = 2x-1,$$

all for the range $-5 \leqslant x \leqslant +5$. What are the gradients of the lines? What will be the gradients of the lines $y = x$, $y = -4x+6$, $y = 7$ and $2x+3y+5=0$?

6. Write down the gradients of the lines:

(i) $y = 5x$, (ii) $y = -2x$, (iii) $27 = x$, (iv) $y = 3x-5$, (v) $2y = 5x+1$.

7. Write down the gradients of the lines:

(i) $3y = x$, (ii) $y+x = 0$, (iii) $y+3x = 4$, (iv) $2y = 3x-1$, (v) $x+2y = 7$.

8. Write down the gradients of the lines:

(i) $x-y = 9$, (ii) $2x+3y = 4$, (iii) $5x-y+8 = 0$, (iv) $3y-2 = 0$, (v) $2x+9y-17 = 0$.

9. An engine in a shunting yard has a displacement s relative to the signal box which varies with time t as follows:

t (min)	0	$\frac{1}{4}$	$\frac{1}{2}$	1	$1\frac{1}{4}$	$1\frac{1}{2}$	$1\frac{3}{4}$
s (hm)	1·0	1·8	2·6	3·0	3·0	2·1	1·2

t (min)	2	$2\frac{1}{4}$	$2\frac{1}{2}$	$2\frac{3}{4}$	3	$3\frac{1}{4}$	$3\frac{1}{2}$
s (hm)	0·3	−0·6	−1·5	−1·5	−1·0	−0·5	0

Draw a graph of the motion, and calculate the engine's average speed in hm/min during the three parts of the shunting operation.

10. Draw the graph of $y = x^2$ for the range $-1 \leqslant x \leqslant 1$. Determine the gradient at $x = -\frac{3}{4}$, $x = -\frac{1}{4}$ and $x = \frac{1}{2}$ by drawing tangents to the curve at these points. Calculate the gradients of the chords joining the points whose abscissae are

(i) 0 and 1, (ii) $\frac{1}{4}$ and $\frac{3}{4}$, (iii) $-\frac{1}{2}$ and $+\frac{1}{2}$.

***11.** The relationship between the distance x of an object from the first principal focus of a lens, the distance y of the image of the object from the second principal focus, and the focal length f of the lens is $xy = f^2$ (Newton's expression). A set of values of x and y were found to be as follows:

x	30·1	19·8	15·0	9·9
y	7·1	10·9	14·3	21·7

Plot a linear graph of the results, and find the value of f.

***12.** The length L of a simple pendulum was varied, and the times T for one swing at the various lengths determined. The results were as follows:

L(cm)	31·6	41·0	50·7	60·1
T (s)	1·10	1·26	1·40	1·53

Given that the relationship between T and L is $T = 2\pi\sqrt{(L/g)}$, where g is the acceleration of gravity, plot a linear graph of the results and deduce from it the value of g.

***13.** A set of readings to verify the relation $T^2 = a/h + bh$ were taken and recorded as follows:

h	2·0	4·0	6·0	8·0	10·0
T	3·4	3·2	4·0	4·6	5·4

Plot a linear graph representing these figures, and find the values of a and b.

***14.** The relationship between the volume v of a fixed mass of gas and its pressure p when undergoing a change in which no heat can enter or leave the system is $pv^\gamma = \text{constant}$. Pairs of values of p and v were found as follows by experiment:

p	1·0	1·5	2·0	2·5	3·0
v	2·28	1·69	1·39	1·18	1·04

Plot a linear graph of these observations, and deduce the value of γ.

***15.** The graph shown in Fig. 3.28 shows how the potential varies along the length of a discharge tube.

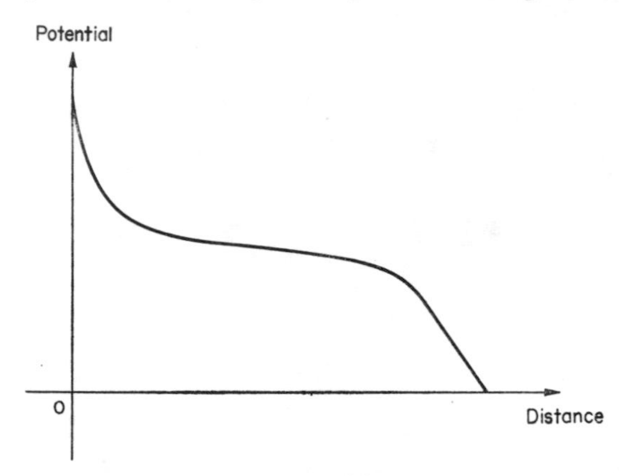

FIG. 3.28.

Given that the electric field V at any point is equal to the potential gradient at that point, with sign reversed, draw a sketch graph showing how V varies along the length of the tube.

***16.** A calorimeter is half filled with warm water, and a cooling curve plotted from the following readings which show how the temperature (θ) varies with time (t):

θ (°C)	50·1	47·6	45·3	43·2	41·2	39·4
t (min)	0	1	2	3	4	5

The calorimeter was then emptied and half refilled with a liquid, and the experiment repeated, giving the following results:

θ (°C)	49·8	46·2	43·0	40·2	37·7	35·5
t (min)	0	1	2	3	4	5

Plot both cooling curves, and find the rate of fall of temperature when $\theta = 45$°C in both cases. Given that the rate of loss of heat in both cases at 45°C is the same, and the thermal capacity of the calorimeter and its contents in the first instance was 227 J/°C, calculate the thermal capacity of the calorimeter and its contents in the second case.

***17.** The temperature (θ) at various points along a bar which is lagged and heated at one end varies with the distance from the heated end (L) as follows:

L (cm)	0	5	10	15	20
θ (°C)	100	89·7	79·4	69·1	58·8

Plot a graph of these results, and explain why it demonstrates that the temperature gradient is constant along the bar. What is the equation relating θ to L?

†**18.** Prove that the tangent of the angle that the line $y = mx+c$ makes with the x-axis is equal to m, provided the graph is drawn with unity or standard scaling.

†**19.** Convert the following slopes into

(i) % steepness,
(ii) angles with the horizontal:

(a) Rhine in Holland, 1 in 10,000;
(b) Radstädter–Tauern Pass (Austria), 1 in 5;
(c) Grossglockner Pass (Austria), 1 in 8;
(d) Porlock Hill, Somerset, 1 in 3.

†**20.** Draw a large-scale graph of $y = \frac{1}{2}x^2$ for $0 \leqslant x \leqslant 1$. Mark a point P with abscissa 0·2, and points q_1, q_2, q_3 and so on at $x = 1$, 0·9, 0·8, ... down to 0·4. Calculate the gradients of Pq_1, Pq_2 Pq_3, ..., Pq_7, and tabulate these results against the difference between the abscissae of the points P and q. Represent this table graphically, and estimate as accurately as you can the value of $\lim_{q \to P}$ (gradient of Pq).

3.5. GRADIENT FUNCTIONS

As we have seen already, the gradient of a curve is not constant; and for any particular curve the value of the gradient must vary in a regular way if the curve itself is regular. For example, if a graph of $y = x^2$ is drawn, it can be seen that the gradient to the left of the y-axis is negative, while to the right it is positive (Fig. 3.29). Furthermore, the gradient at P is fairly large and negative; it decreases in magnitude (i.e. changes to a more positive value) as x increases going to the right, reaching the value 0 at $x = 0$. It then becomes positive and increases more and more as x increases. For many regular curves the relation between the gradient at a point and the x-coordinate of that point can be expressed in an algebraic form, and the function which gives the gradient at any point is known as the gradient function. Using the symbol $g(x)$ for the gradient function, we can write the equation "gradient $= g(x)$", and then, substituting a value of x into $g(x)$, the gradient at the point whose x-coordinate is equal to that value can be determined. This, of course, is much quicker and much more accurate than measuring the gradient from a plotted graph: and since many problems involve knowing the gradient of a curve for their solution, knowledge of the gradient function for any particular curve in question is a valuable time-saver, as well as giving an exact answer as opposed to the approximate answer one is bound to get from drawing a graph and tangent.

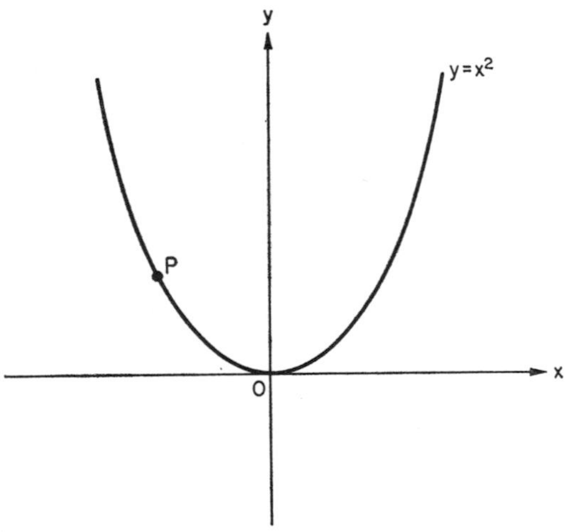

FIG. 3.29. The gradient of $y = x^2$ at P increases as P moves to the right.

The derivation of the gradient function of the equation of the curve—i.e. the steps which enable $g(x)$ to be obtained from the function $f(x)$ which appears on the right-hand side of the equation $y = f(x)$ describing the curve—is one of the starting points of calculus, the so-called differential calculus. The technique enabling this derivation to be made will be discussed in the next chapter.

Example 3.7

Sketch the curve $y = 4x^3$. Describe how its gradient changes as x increases from a large negative value to a large positive value.

Given that the gradient function is $12x^2$, write down the gradient of the curve at $x = -2$, $x = 2$, the point (1, 4) and the point $(-3, -108)$.

By substitution into the equation of the curve, the value of y is found to be 0 when $x = 0$. The curve therefore passes through the origin. As x increases from 0 towards $+1$, y increases to $+4$, but more slowly, since the cube of a number less than 1 is smaller than the number itself. When $x = 1$, $y = 4$, and as x increases beyond this, y increases rapidly.

For negative values of x, y is also negative; the curve is similar in shape. The sketch thus appears as in Fig. 3.30(a)).

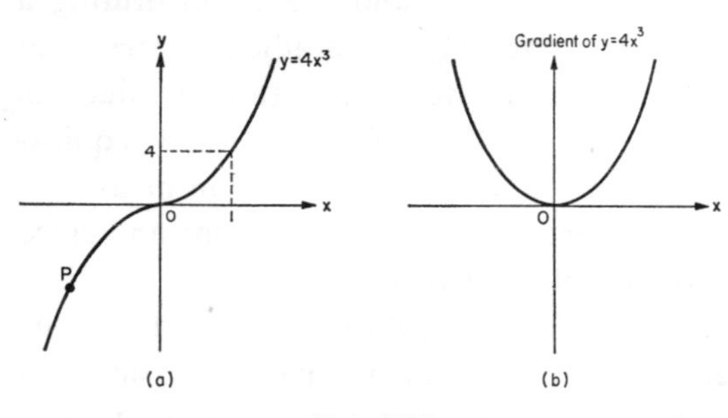

FIG. 3.30.

The gradient of the curve can be seen to be large and positive when x is large and negative (e.g. at P), and as x increases, the gradient decreases (i.e. it becomes less "steep" going from left to right). At the origin the gradient is 0, and as x increases beyond this, the gradient increases without bound. The variation is represented by the sketch graph in Fig. 3.30b.

The gradient at any point on the curve can be found by substituting the value of the abscissa into the gradient function. Thus, at $x = -2$, the gradient is equal to $12(-2)^2$, or 48. At $x = 2$, gradient $= 12(2)^2 = 48$, also. At (1, 4), gradient $= 12(1)^2 = 12$, since the x coordinate of the point in question is 1. At $(-3, -108)$, gradient $= 12(-3)^2 = 108$.

The gradients at the given points are therefore 48, 48, 12 and 108.

EXERCISE 3d

1. Draw the graph of $y = \frac{1}{2}x^2$ for $-10 \leqslant x \leqslant 10$. Tabulate the values of the gradient at the points whose abscissae are $-8, -6, -4, -2, 0, 2, 4, 6$ and 8, against the corresponding values of x. Do you notice any connection?

2. The gradient function of a certain graph is $x-3$. What is the gradient at the point whose abscissa is: (i) 4; (ii) $\frac{1}{2}-$; (iii) 6; (iv) 0?

3. The gradient function of a given curve is $2x$. What is the gradient of the curve at the following points: (1, 1); (3, 9); $(-2, 4)$?

4. What are the gradient functions of the lines $y = 7x+5$ and $y = 4x+3$?

†**5.** The gradient function of a given curve is x. What is the gradient of the curve at the points whose abscissae are 0, -3 and $+5$, and at the points (2, 2) and $(-4, 8)$? Given that the curve passes through these last two points, suggest a suitable equation to represent it.

†**6.** Draw a sketch graph of $y = |x|$, and sketch the graph of its gradient function. What can be said about the gradient of $y = |x|$ at $x = 0$?

SUMMARY

If a change of δx in a domain variable x can be associated with a change δy in the related range variable y, the rate of change of y with respect to x over the interval δx is the value of the ratio $\delta y/\delta x$.

The instantaneous rate of change of y with respect to x at a particular value of x is the value of

$$\lim_{\delta x=0} \left(\frac{\delta y}{\delta x}\right),$$

assuming this latter limit exists, and where any interval δx is measured from the particular value of x in question.

If we consider the displacement s from a fixed point of a body whose motion is related to the time t, the instantaneous velocity of the body is defined to be the value of

$$\lim_{\delta t=0} \left(\frac{\delta s}{\delta t}\right),$$

if this limit exists at the particular value of t in question.

The gradient of a line on a standard scale graph is defined to take the value of the tangent of the angle the line makes with the positive direction of the x-axis. If the line represents the (linear) relation between two variables x and y, the gradient can be equated with the value of $\delta y/\delta x$, where δy and δx are corresponding changes in the values of y and x. Upward sloping lines therefore have positive gradients (and represent positive rates of change); downward sloping lines have negative gradients (and represent negative rates of change). The gradient of the line $y = mx+c$ is m.

The gradient of a curve at any point is the gradient of the tangent to the curve at that point. The function which describes the variation in the gradient of the curve with the abscissa is called the gradient function.

Miscellaneous Exercise 3

1. The number of divorce petitions filed in England and Wales rose from 9970 in 1938 to 27,070 in 1960. Assuming the population of England and Wales to be 48 million in 1938 and 52 million in 1960 calculate

(i) the number of divorce petitions filed per 1000 of population in each of the years 1938 and 1960;

(ii) the average rate of change in the number of petitions filed per 1000 population between 1938 and 1960.

2. In 1939 the rate of exchange between France and England was 176·10 francs to the £: in 1963 it was 13·71 new francs to the £ (1 new franc = 100 (old) francs). What is the average rate of change in the exchange rate per annum?

3. An aeroplane accelerates from rest to 60 m/s in 5·4 seconds. What is its average acceleration? If the acceleration is twice this value at the beginning, and falls steadily to zero after 5·4 seconds, what is the rate of change of the acceleration?

4. In the equation $y = a\sqrt{(b/c)}$, a is measured in cm, b in cm and c in cm/s^2. What are the units of y?

***5.** If ϱ denotes the resistivity of a sample of wire, the total resistance R of a piece of wire of length L and cross-sectional area A is given by the equation $R = \varrho L/A$. What are the units of ϱ on the mks system if R is measured in ohms? If R changes from 1·2 ohms to 1·5 ohms when the temperature increases from 20·5°C to 130·1°C, what is the average rate of change of resistivity with temperature, given that L and A remain constant at 100 cm and 0·003 cm^2 respectively?

6. A trolley is accelerating along a bench top. A paper tape attached to it runs through a vibrator, and the vibrator makes a mark on the paper tape every $\frac{1}{50}$th second. The distances of a number of vibrator marks from the beginning (i.e. the first mark) of the tape are given in the table below. Plot a distance/time graph of the trolley's motion, and by drawing a tangent to the curve at $t = 2$, estimate the trolley's speed at this instant.

Vibrator mark	0	25	50	75	100	125	150
Distance from first mark (cm)	0	2·2	8·7	19·9	35·4	55·3	79·6

Closer inspection of the marks around the 100th show the following spacing:

Vibrator mark	97	98	99	100	101	102	103
Distance (cm)	33·2	34·0	34·7	35·4	36·1	36·8	37·6

Determine the average speed for the three intervals on either side of the 100th mark, and hence estimate the instantaneous speed at $t = 2$.

7. Explain why the gradient of a velocity/time curve represents on acceleration. Discuss the meaning of a negative acceleration.

The graph of Fig. 3.31 shows the variation in the velocity of a car. Sketch its acceleration/time curve, and describe the motion.

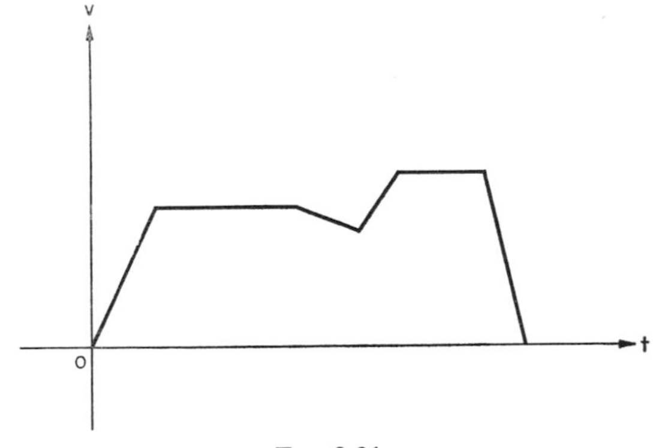

Fig. 3.31.

***8.** The resistance of a piece of wire varies with temperature as follows:

Resistance R (ohms)	6·4	10·0	13·4	16·6	19·6	22·4
Temperature T (°C)	−50	0	50	100	150	200

Draw a graph showing the variation in resistance with temperature, and find the average temperature coefficient of resistance for the ranges:

(i) −50°C to 0°C, (iii) 0°C to 200°C,
(ii) 0°C to 100°C, (iv) 100°C to 200°C.

What is the instantaneous temperature coefficient of resistance at 20°C?

***9.** The length of a bar of metal varies with the temperature as follows:

Length (cm)	128	130·5	135·5	143·5	154	167
Temperature (°C)	−100	0	100	200	300	400

Draw a graph representing these figures, and find the average coefficient of linear expansion for the range (i) −100°C to 0°C, (ii) 0°C to 100°C, (iii) 100°C to 200°C.

What is the instantaneous coefficient of linear expansion at 300°C?

10. Write down the gradients of the lines joining the points

(i) (0, −7), (1, 10), (iv) (−11, −9), (2, 5),
(ii) (−11, −9), (2, 5), (v) (−3, −1), (−7, −8).
(iii) (1, 3), (5, 3),

11. Two variables x and y are related by an equation of the form $y = ax^n$. Plot a logarithmic graph of the following pairs of values of x and y, and use it to find the values of a and n.

y	1·22	1·45	1·90	2·45	2·88	3·34
x	150	210	360	600	830	1120

†**12.** Sketch a graph whose gradient (i) increases, (ii) decreases progressively as the value of the domain variable increases.

†**13.** The graph of the function f has a "kink" in it at P (Fig. 3.32). If Q is any point on the curve, and

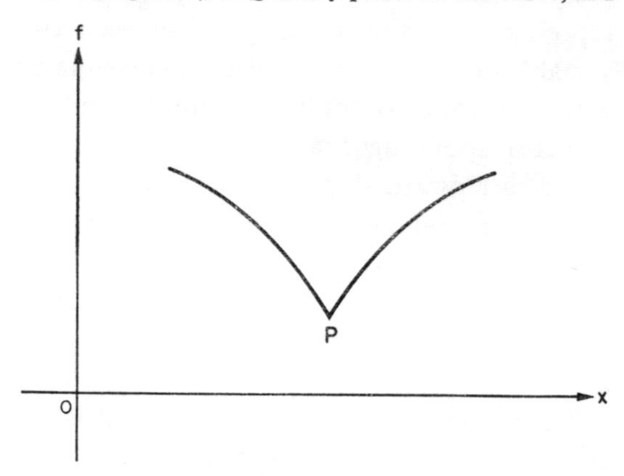

FIG. 3.32.

R is the foot of the perpendicular from P on to the ordinate at Q, explain why

$$\lim_{Q\to P+}\left(\frac{QR}{PR}\right) \neq \lim_{Q\to P-}\left(\frac{QR}{PR}\right)$$

†***14.** The "chord" resistance of an electrical component is defined as the value of the voltage/current ratio when it is in its operative state. The "incremental" resistance is defined as the value of the ratio (change in voltage)/(change in current) when a small change is made in its operating situation. Estimate the chord and incremental resistance of the diode whose characteristics are summarised in the following table when the applied voltage is (i) 200, (ii) 100.

Applied voltage (V)	0	50	100	150	100	250
Current (mA)	0·0	3·0	8·0	15·0	22·0	28·0

If the voltage takes the value 200·2 V, what will the current be?

PART 2

THE DIFFERENTIAL CALCULUS

CHAPTER 4

Introduction to Differentiation

4.1. DERIVING THE GRADIENT FUNCTION

We have seen that the gradient of a curve at any point (x, y) is defined as the limiting value of the ratio $\delta y/\delta x$ at $\delta x = 0$. If we can obtain an expression for the value of this limit at the general point (x, y), we shall have obtained the function of x which describes the variation in the gradient of the curve.

Consider first the curve $y = x^2$. As we have seen, its gradient is variable, the variation being expressed by the gradient function. If we take two points P and Q having abscissae x and $x+\delta x$, respectively (see Fig. 4.1, which is greatly magnified in the $PQRST$ region), then we can write that

$$PT = x^2,$$

$$QS = (x+\delta x)^2.$$

Now,

$$\begin{aligned} QR &= \delta y \\ &= (QS-RS) \\ &= (QS-PT), \text{ since } PRST \text{ is a rectangle,} \end{aligned}$$

and thus

$$\begin{aligned} \delta y &= (x+\delta x)^2 - x^2 \\ &= x^2+2x\,\delta x+(\delta x)^2 - x^2 \\ &= 2x\,\delta x+(\delta x)^2. \end{aligned}$$

Now the gradient of the chord $PQ = QR/RP = \delta y/\delta x$, and since $\delta y = 2x\,\delta x+(\delta x)^2$, we can rewrite the gradient in the form*

$$\begin{aligned} \text{gradient of chord } PQ &= \frac{2x\,\delta x+(\delta x)^2}{\delta x} \\ &= 2x+(\delta x). \end{aligned}$$

$\therefore$ gradient of curve at P

$$\begin{aligned} &= \lim_{Q=P} (\text{gradient of chord } PQ) \\ &= \lim_{\delta x=0} \left(\frac{\delta y}{\delta x}\right) \\ &= \lim_{\delta x=0} (2x+\delta x) \\ &= 2x. \end{aligned}$$

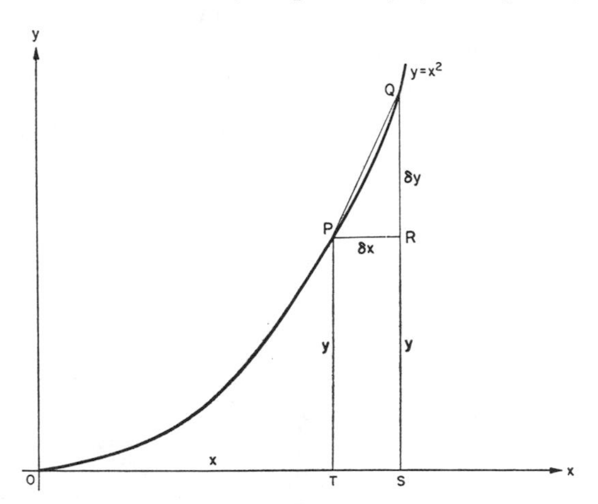

FIG. 4.1. Derivation of the gradient function of $y = x^2$. The point P is fixed but indefinite, and $PT = OT^2$.

* Provided that $\delta x \neq 0$, which by the very meaning of our argument it will not.

This is a function of x, and is in fact the gradient function: the gradient of $y = x^2$ at any point is equal to twice the x-coordinate of that point.

The derivation of the gradient function of a straight line $y = mx+c$, where m and c are constant, is rather trivial. If x changes by 1, y will change by m; if x changes by 2, y will change by $2m$, and the ratio

$$\frac{\text{change in } y}{\text{change in } x}$$

will be equal to m whatever the positions of P and Q. (The gradient of a straight line is, of course, everywhere constant, and therefore does not vary with x.) But a proof in the above manner would be as follows.

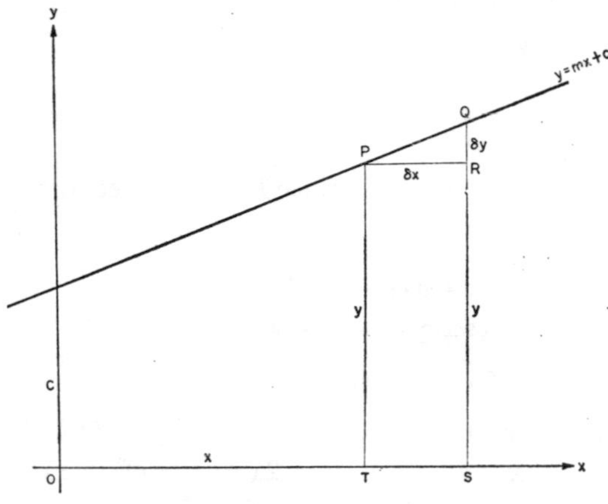

FIG. 4.2. Gradient of a straight line. $PT = m(OT)+c$.

Referring to Fig. 4.2, and noting that the ordinate of any point on the line is equal to the abscissa multiplied by m, plus a constant c, we have

$$PT = mx+c;$$

and $$QS = m(x+\delta x)+c,$$

since the abscissa of Q is $(x+\delta x)$.

Thus,

$$\begin{aligned} y = QR &= QS-PT \\ &= [m(x+\delta x)+c]-[mx+c] \\ &= m\,\delta x. \end{aligned}$$

$$\begin{aligned} \therefore \text{ Gradient at } P &= \lim_{\delta x=0}\left(\frac{\delta y}{\delta x}\right) \\ &= \lim_{\delta x=0}\left(\frac{m\,\delta x}{\delta x}\right) \\ &= \lim_{\delta x=0}(m) \\ &= m, \end{aligned}$$

since m is not affected in any way by the value of δx.

The procedure followed in these two proofs is typical of the reasoning behind the derivation of any gradient function; it will be analysed more closely in Section 4.7, and then applied to a variety of functions, but it would be desirable for the reader to ensure that he can follow quite clearly the steps in the argument at this stage.

Example 4.1

Derive from first principles the gradient function of the curve $y = 3x^2+x$. Hence find the gradient at the point (2, 14).

Consider a portion of the curve as shown in Fig. 4.3. (Even if the curve has not this particular shape, the reasoning will still be valid.) Take two points P and Q on the curve whose abscissae are x and $(x+\delta x)$, respectively (and Q will therefore be to the right of P if δx is positive). The ordinate of P (which

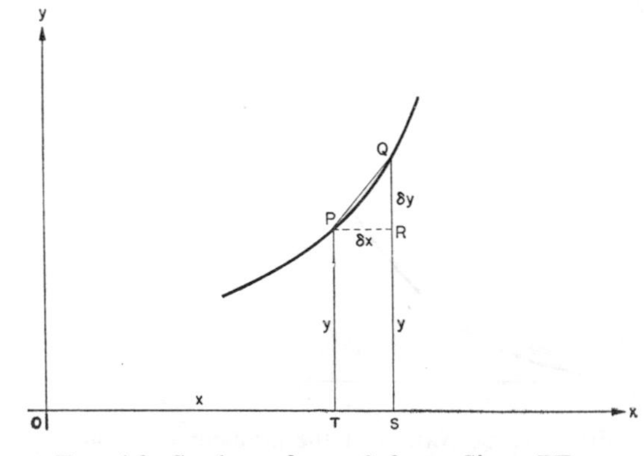

FIG. 4.3. Section of $y = 3x^2+x$. Since $PT = y$ and $OT = x$, $PT = 3(OT)^2+OT$.

is PT) equals $(3x^2+x)$ from the equation of the curve, and the ordinate of Q is $[3(x+\delta x)^2+(x+\delta x)]$: i.e. $QS = (3x^2+6x\delta x+3(\delta x)^2+x+\delta x)$.

But gradient of chord $PQ = \dfrac{QR}{PR}$

$$= \frac{QS-PT}{PR}$$

$$= \frac{\delta y}{\delta x}$$

$$= \frac{[3x^2+6x\,\delta x+3(\delta x)^2+x+\delta x]-(3x^2+x)}{\delta x}$$

$$= \frac{6x\,\delta x+3(\delta x)^2+\delta x}{\delta x}$$

$$= 6x+1+3(\delta x).$$

Now the gradient of the curve at P

$$= \lim_{\delta x=0}\left(\frac{\delta y}{\delta x}\right)$$

$$= \lim_{\delta x=0}(6x+1+3\,\delta x), \text{ in this case,}$$

$$= 6x+1,$$

and this is the required gradient function.

The value of x at the point (2, 14) is 2; thus the gradient at this point is the value of $(6x+1)$ with x equal to 2. This gives 13 as the required value. *Answer:* The gradient function is $(6x+1)$; the gradient at (2, 14) is 13.

Exercise 4a

1. What are the ordinates of the points on the curve $y = 2x^2$ whose abscissae are $\frac{1}{2}$, 1 and 2? What are the gradients of the chords joining pairs of these points?

2. Calculate the gradient of the chord joining the points $x = 1$ and $x = 2$ on the curve $y = -x^4$.

3. Derive an expression for the gradient of the chord joining two points whose abscissae are x and $(x+\delta x)$ and which lie on the curve $y = x^2-x$. What is the limiting value of this as $\delta x \to 0+$?

4. Calculate the gradient of the chord of $y = 3x^2$ between the points with abscissae x and $(x+\delta x)$. Hence calculate the gradients of the chords between $x = 1$ and $x = 1{\cdot}1$; $x = 1$ and $x = 1{\cdot}01$; $x = 1$ and $x = 1{\cdot}001$; and $x = 1$ and $x = 1{\cdot}0001$.

5. Derive from first principles the gradient function of the curve $y = 2x^2+1$. Hence, find the gradient at $x = 3$, $x = -1$ and $x = \frac{1}{2}$.

6. By considering a point $P(1, 1)$ on the curve $y = x^3$, and taking another point Q which moves so that its abscissa has the values 1·1, 1·01, and 1·001 successively, derive the gradient of the chord PQ for these different positions of Q. Deduce the value of the gradient of the curve at P.

7. If $P(x, y)$ and $Q(x+\delta x, y+\delta y)$ lie on the curve $y = 3x^3$, show that $\delta y/\delta x = 9x^2+9x(\delta x)+3(\delta x)^2$. Hence write down the value of

$$\lim_{\delta x=0}\left(\frac{\delta y}{\delta x}\right).$$

8. Find the gradient of $y = x^3$ at $x = \frac{1}{3}$ and $x = \frac{1}{4}$.

†9. Sketch the graph of the gradient of the function $y = |x|$ for values of x on either side of the origin. Explain why

$$\lim_{\delta x=0}\left(\frac{\delta y}{\delta x}\right)$$

does not exist for $y = |x|$ at $x = 0$, and state whether

(i) y exists at $x = 0$,
(ii) the gradient of y is continuous at $x = 0$,
(iii) the derivative of y exists at $x = 0$ (see § 4.2),
(iv) the derivative of y is continuous at $x = 0$.

Write an expression for the derivative of $y = |x|$ at any point (x, y). (You may answer with two expressions for different ranges of x if you wish.)

†10. Find the gradient of the chord joining (x, y) to $(x+\delta x, y+\delta y)$ on the curve $y = x^3$ as a function of x and δx. What is the gradient of the curve at (x, y)?

†11. Calculate the gradient of the chord of the curve $y = \sin x$ between the points whose abscissae are 30° and 60° ($\pi/6$ and $\pi/3$ radians, respectively).

†12. Explain why a graph with a kink in it can be continuous at the kink but its derivative there is non-existent and therefore discontinuous (see Fig. 4.4).

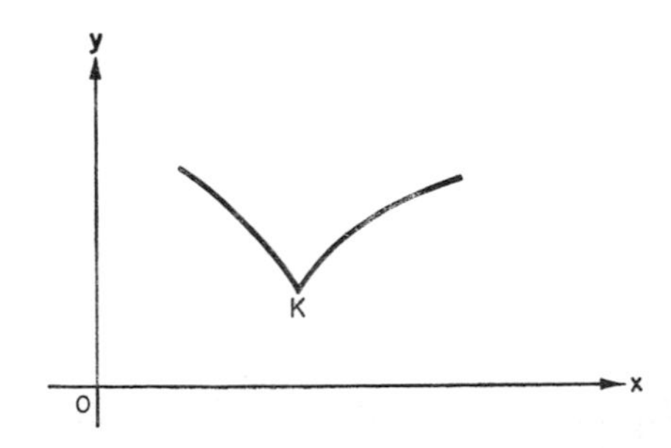

FIG. 4.4. The curve is continuous at K, but does not possess a derivative there.

†**13.** Explain how a function $\delta y/\delta x$ can have a "limiting value" at $\delta x = 0$ although δy and δx are zero.

†**14.** Explain why (i) the gradient of the line $y = a$, where a is a constant, is zero everywhere; (ii) the gradient of the line $x = a$ is "infinite".

†**15.** Sketch the graph of $y = x-[x]$ where $[x]$ is the largest integer less than or equal to x. Give a sketch graph showing the variation in its gradient.

4.2. DERIVATIVES

For the reason that the gradient function "gradient $= 2x$" can be derived from the equation $y = x^2$ in the manner shown in the last section, $2x$ is often referred to as the *derivative* of x^2. A synonymous expression sometimes used is *differential coefficient*. Thus m can be said to be the derivative or differential coefficient of $(mx+c)$.

In much of the text of the next three chapters we shall be concerned with obtaining the derivatives of a variety of functions. We can usefully prepare the way by summarising the technique employed in obtaining the derivatives of the elementary functions x^2 and $mx+c$ in the last section.

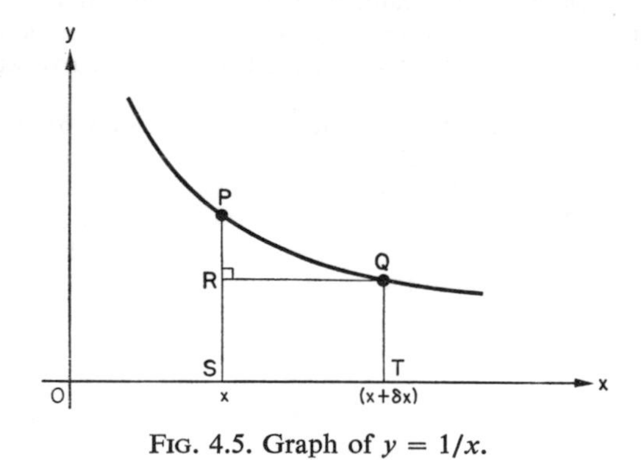

FIG. 4.5. Graph of $y = 1/x$.

If a function $f(x)$ has a graph something like that shown in Fig. 4.5, then we can say that

(i) the values of the function at any position x and the closely neighbouring position $(x+\delta x)$ can be represented by two points P and Q on the graph of the function;

(ii) the difference between the ordinates at Q and P (the length PR) can be written as an expression for δy, where δy is the change in y corresponding to the change δx in the value of x. (If the curve slopes downwards, δy will be negative (since $QT < PS$), and this is consistent with the interpretation of δy and negative gradients.)

Analytically, $QR = QT-PS$; or in function notation,

$$\delta y = f(x+\delta x)-f(x).$$

(iii) If the ordinates of P and Q are PS and QT, and if the form of the function $f(x)$ is known, we can evaluate PS as $f(x)$ and QT as $f(x+\delta x)$.

(iv) If we then manipulate this expression for δy to give a term involving δx as a factor, collecting the remaining terms in powers of δx if the powers are low, or in one single bracket of terms in δx to the second or greater power,

(v) and obtain an expression for the ratio $\delta y/\delta x$ by dividing the expression by δx (giving the gradient of the chord PQ),

(vi) and find the limiting value of this ratio at $\delta x = 0$ (i.e. $Q = P$), we shall have obtained an expression for the required derivative.

These steps can be summarised in function notation—and the summarising equation

$$\text{derivative} = \lim_{\delta x=0} \left\{ \frac{f(x+\delta x)-f(x)}{\delta x} \right\}$$

can be taken as a definition of the derivative of $f(x)$ with respect to x.

We can now use this definition as the starting point for deriving an expression for any derivative, and there will be no need for recourse to any graphical work. As we shall see, the only problems to be surmounted involve manipulating the expression for

$$\frac{f(x+\delta x)-f(x)}{\delta x}$$

into a form which enables the value of its limit at $\delta x = 0$ to be found.

Example 4.2

Find the derivatives of (i) x^4, (ii) $1/x$.

(i) If we associate the function $f(x)$ with x^4, we can write that

$$f(x+\delta x) = (x+\delta x)^4$$

and

$$f(x) = x^4.$$

Thus

$$\frac{f(x+\delta x)-f(x)}{\delta x} = \frac{(x+\delta x)^4-x^4}{\delta x}.$$

A convenient manipulation to employ at this stage is to expand $(x+\delta x)^4$ by the binomial theorem, giving

$$\frac{f(x+\delta x)-f(x)}{\delta x}$$

$$= \frac{x^4+4x^3(\delta x)+6x^2(\delta x)^2+4x(\delta x)^3+(\delta x)^4-x^4}{\delta x}$$

$$= 4x^3+6x^2(\delta x)+4x(\delta x)^2.$$

Thus $\lim_{\delta x=0}\left\{\frac{f(x+\delta x)-f(x)}{\delta x}\right\}$

$$= \lim_{\delta x=0}\{4x^2+6x^2(\delta x)+4x(\delta x)^2\}$$

$$= 4x^3.$$

(ii) Associating $f(x)$ with $1/x$ in this case, we obtain

$$\frac{f(x+\delta x)-f(x)}{\delta x} = \frac{\frac{1}{x+\delta x}-\frac{1}{x}}{\delta x} = \frac{\frac{x-(x+\delta x)}{x(x+\delta x)}}{\delta x}$$

$$= -\frac{\delta x}{\delta x.x(x+\delta x)} = -\frac{1}{x(x+\delta x)}.$$

Thus $\lim_{\delta x=0}\left\{\frac{f(x+\delta x)-f(x)}{\delta x}\right\}$

$$= \lim_{\delta x=0}\left\{-\frac{1}{x(x+\delta x)}\right\}$$

$$= -\frac{1}{x^2}.$$

EXERCISE 4b

Find from first principles the derivatives of

1. $5x^2$. **2.** $2x^4$. **3.** (x^2-8x). **4.** $1/x^2$.

5. If $f(x) \equiv \sqrt{x}$, write down an expression for $\{f(x+\delta x)-f(x)\}$.

Express this in the form

$$\frac{Ax+B\,\delta x}{\sqrt{(x+\delta x)}+\sqrt{x}}$$

by multiplying your expression by

$$\frac{\sqrt{(x+\delta x)}+\sqrt{x}}{\sqrt{(x+\delta x)}+\sqrt{x}},$$

and hence find the derivative of $\sqrt{x}$.

4.3. THE DERIVATIVE OF x^n WITH n POSITIVE INTEGRAL

We can now turn to the problem of finding a general expression for the derivative of x^n. To do so requires that we place some restriction on the value of n, and in this first situation we will demand that n is a positive integer. We then proceed as follows.

Let $f(x) \equiv x^n$; then

$$\frac{f(x+\delta x)-f(x)}{\delta x} = \frac{(x+\delta x)^n-x^n}{\delta x}.$$

Now we can expand $(x+\delta x)^n$ by the binomial theorem, obtaining

$$(x+\delta x)^n = x^n+nx^{n-1}\,\delta x+\frac{n(n-1)x^{n-2}(\delta x)^2}{2!}+$$

terms in the cube and higher powers of δx to $(\delta x)^n$.

Thus

$$\frac{(x+\delta x)^n-x^n}{\delta x} = nx^{n-1}+\frac{n(n-1)x^{n-2}}{2!}(\delta x)+$$

terms in the square and higher powers of δx to $(\delta x)^{n-1}$.

Thus

$$\lim_{\delta x=0}\left\{\frac{f(x+\delta x)-f(x)}{\delta x}\right\}$$

$$= \lim_{\delta x=0}\{nx^{n-1}+\text{terms in }\delta x\text{ and higher powers}\}$$

$$= nx^{n-1}. \tag{4.1}$$

This is the required derivative.

4.4. DIFFERENTIATION

The shorter mental process of obtaining $2x$ as the gradient function of x^2 without actually deriving it is referred to as *differentiation.* To discover the nature of this process, consider the following special cases of the expression (4.1):

The function		derivative	
	x^2;	$2x$	$(n = 2)$
	x^3;	$3x^2$	$(n = 3)$
	x^4;	$4x^3$	$(n = 4)$
	x^5;	$5x^4$	$(n = 5)$

In these examples it can be observed that two changes are made to the function in obtaining the derivative:

(i) The power of x in the original function is placed in front of the x as a multiplying factor (e.g. in the last example, the power of x in the function itself is 5; this 5 appears as a factor in front of the x in the derivative $5x^4$).

(ii) The power of x in the derivative is always one less than that in the function being considered (e.g. in the same example, the power of x is reduced from 5 in the original function to 4 in the derivative).

To generalise back to the original expression (4.1), we can say that in forming the derivative of x^n two steps are necessary:

(i) copy the power (n) into the place immediately in front of the term as a multiplying factor: $x^n \rightarrow nx^n$;

(ii) reduce the power of x by one:

$$nx^n \rightarrow nx^{n-1}.$$

With a very small amount of practice these two alterations can be made in one step, and the derivative of x^n written down immediately as nx^{n-1}.

In conclusion, we will look at five special cases and corollaries more closely:

(i) With $n = 2$, the derivative of x^2 is $2x$. The power of one is omitted in the derivative for brevity, in the same way as it is in writing x^5 instead of $1x^5$. In this case n equals 2, and the $(n-1)$, which is 1, is understood without actually being written.

(ii) With $n = 1$, the derivative of x is 1. This is again no exception to the rule; x stands for x^1, so the derivative is therefore $1x^0$. Now x^0 is defined to be 1, and so the derivative $1x^0$ is simply 1.

(iii) The derivative of, say, $7x^2$ is seven times the derivative of x. This is apparent from graphical considerations; on the curve $y = x^2$, y increases by 3 if x increases from 1 to 2. In the case of $y = 7x^2$, y increases by 21, i.e. seven times as much. This will be the case for any range of x, and so the latter curve is seven times as steep. The derivative of $7x^2$ is therefore $14x$. Analytically, the derivative of ax^n is nax^{n-1}.

(iv) The derivative of any constant or number is zero. This can be reasoned in two or three ways:

(a) The graph of, say, $y = 7$ is a straight line parallel to the x-axis. The gradient is therefore zero. Or,

(b) $y = 7$ can be written as $y = 7x^0$. The derivative is therefore $7.0x^{-1}$, and this is zero (again showing this to be no exception to the general rule for x). Or,

(c) if $f(x) = 7, \lim \{f(x+\delta x)-f(x)/\delta x\} = 0$.

(v) The derivative of a sum of terms is equal to the sum of the derivatives. This follows from the proofs from first principles (cf. Example 4.1). Thus the derivative of x^3+3x^2+7 is $3x^2+6x$ $(+0)$.

Example 4.3

Differentiate the following functions with respect to x:

(i) x^9, (iii) $5x^2+7$,
(ii) $7x^5$, (iv) $4x^4+5x+6$.

(i) By applying the rules just given, or by regarding the function as the particular case of the derivative of x^n with $n = 9$, the required derivative is $9x^8$.

(ii) The derivative of x^5 is $5x^4$ by the same rules, and thus the derivative of $7x^5$ is $35x^4$.

(iii) The derivative of a sum of terms is the sum of the derivatives, and the answer is therefore $10x+0$, i.e. $10x$, since the derivative of a constant such as 7 is 0.

(iv) The derivative of $4x^4+5x+6$ is similarly $16x^3+5$.

The answers are therefore:

(i) $9x^8$, (iii) $10x$,
(ii) $35x^4$, (iv) $16x^3+5$.

4.5. THE DERIVATIVE AS THE GRADIENT FUNCTION

It will be remembered that the gradient function for any curve $y = f(x)$ can be found by evaluating the expression

$$\lim_{\delta x=0}\left(\frac{\delta y}{\delta x}\right).$$

Since $\delta y = f(x+\delta x)-f(x)$, and the derivative of $f(x)$ has been defined as

$$\lim_{\delta x=0}\left\{\frac{f(x+\delta x)-f(x)}{\delta x}\right\},$$

we can remind ourselves that the gradient function for $y = f(x)$ is the same function as the derivative of $f(x)$.

The process of differentiation is an operation which has to be frequently carried out, and, as usual, a shorthand notation is useful. If we pay particular attention to the equation $y = f(x)$, we can imagine that the function $f(x)$ is replaced by the variable y, and instead of speaking of the derivative of $f(x)$, we can refer to the derivative of y. The function which is the derivative of y (with respect to x) is represented by $\mathrm{d}y/\mathrm{d}x$, and this is usually read for short as "dy by dx". An alternative layout, $\dfrac{\mathrm{d}}{\mathrm{d}x}(y)$, read as "d by d$x$ of y", is often convenient if y is complicated.

The phrase "with respect to x" is necessary, as it indicates which variable (if there is more than one) has to be altered, or if there is any doubt about which letter is the variable, then the instruction to differentiate "with respect to x" indicates quite clearly that x can vary. Thus, if $y = a+bt$, a and b would, in the absence of any other instructions, be taken to be constants; the expression $\mathrm{d}y/\mathrm{d}t$ indicates clearly that t is to be regarded as the variable for the purpose of differentiation, and the other letters are to be treated as constants. This is not quite the full story, but it will suffice for the moment. Notice also that the letter d which appears twice in the quantity $\mathrm{d}y/\mathrm{d}x$ is not an algebraic quantity and cannot, for example, be cancelled or squared (cf. the delta quantities). Since the group $\mathrm{d}/\mathrm{d}x$ implies an instruction to carry out the operation of differentiation with respect to x on whatever follows it is sometimes termed an *operator*: the function y is "operated on" or altered in an altogether non-algebraic way.

Recalling the procedure adopted in Section 4.1 for obtaining the derivative, it can be seen that

$$\frac{\mathrm{d}y}{\mathrm{d}x} = \lim_{\delta x=0}\left(\frac{\delta y}{\delta x}\right)$$

since the value of the derivative at any point with abscissa x is equal to the limiting value of the gradient of the chords through that (indefinite) point as $\delta x \to 0$. That the value of the derivative at any point is the gradient at that point then follows simply, and we can note in passing that as far as the commonly used symbols are concerned, the δ's change to d's as the limit of zero is approached.

EXERCISE 4c

Write down the derivatives of:

1. x^8. **4.** $6x$. **7.** -3.

2. $5x^8$. **5.** 8. **8.** $-2x$.

3. x. **6.** $\frac{1}{4}x$. **9.** $3x+5$.

10. $2x^2-7$.

Differentiate the following functions with respect to x:

11. $\frac{1}{2}x^7+3$. **14.** x^{20}. **17.** $-x^4$.

12. $(7x-5)$. **15.** $7x^2+5x$. **18.** $10x^5$.

13. $\frac{1}{2}x^2$. **16.** $x^{30}-6x^{15}-1$.

Find the derivatives with respect to x of the following functions:

19. $x^3+x^2+7x-20$.

20. $(x+1)^2$. **23.** (p^2-x^2). **27.** $ax^{11}-bx^7$.

21. $x(x-3)$. **24.** $x^{13}-x^7$. **28.** $a^2+b^2+c^2$.

22. p^2x^2. **25.** $x^{13}.x^7$. **29.** a^2b^2.

26. $x^8.x^{18}$.

Find the differential coefficients of:

30. $5(x^2+2)$. **34.** $x^3-\frac{1}{4}x^2$.

31. $(3x)^2$. **35.** x^{2n-1}.

32. $\frac{1}{4}(x^2-x-4)$. **36.** $2x^{+3n}$.

33. x^2-ax+b.

37. Find the gradient function of the curve $y = x^2-17x+1$.

38. What can you say about the gradient of $y = 3x^2-6x+1$ as x changes from -3 to $+3$?

39. Find the gradient of the curve whose equation is $y = x^3-7x+1$ at the points

(i) $(1, -5)$, (ii) $(-1, 7)$, (iii) $(0, 1)$.

40. At what point is the gradient of the curve $y = x^2-2x-3$ zero?

41. Find dy/dt for the functions:

(i) $y = 7-3t$, (iii) $y = 5t^{+2}$,

(ii) $y = t^3+17t-10$, (iv) $y = at^{30}-(bt^{30})^2$.

42. Write down expressions for the functions given by:

(i) $d(3+p^2-2p^3)/dp$, (iv) $d(2\pi r)/dr$,

(ii) $d(2y^2-y+7)/dy$, (v) $d(4\pi r^2)/dr$,

(iii) $d(8t^5-27t^{10})/dt$, (vi) $d(\frac{4}{3}\pi r^3)/dr$.

43. Show that when $x=2$ the gradient of the curve $y = -1+3x-\frac{1}{4}x^2$ is double that when $x = 4$. Find also the abscissa of the point on the curve at which the gradient is -1. (OC)

44. (a) For what values of x is the gradient of (i) $y = x^2-3+7$, (ii) $y = 4x^2+5x-1$, positive?

(b) For what values of x is the gradient of $y = x^3-5x^2+3x+18$ negative?

45. Find the coordinates of the points on the curve $y = x^3-2x^2+4$ where the gradient is -1.

†**46.** Obtain the equation of the tangent to the curve $y = x^2(3-x)$ at the point whose abscissa is (i) 2, (ii) -2, (iii) 0.

†**47.** Prove that the gradient of $y = 5x+2$ is 5 at all points.

†**48.** Prove that the gradient function of $y = t^3-7t$ is $3t^2-7$.

†**49.** Sketch the curve $y = x^3-3x^2-9x+22$ in the region $-4 \leqslant x \leqslant 4$, indicating where the gradient is zero.

†**50.** Prove from first principles that the gradient function of x^3 is $3x^2$.

†**51.** Find the points of intersection of the curve $y = x^3-4x^2-2x$ and the x-axis, and derive the equations of the tangents at these points.

†**52.** What is the value of $\lim_{\delta q=0} \{\delta p/(p^2\delta q)\}$ at $q = 2$ if $p = q^2-2$?

†**53.** Find, in degrees and minutes, the angles which the tangents to $y = 5+x-x^2$ at (0, 5), (1, 5), (2, 3) and (3, -1) make with the x-axis.

4.6. THE DERIVATIVE OF x^n WITH n NEGATIVE INTEGRAL

The derivation of nx^{n-1} as the derivative of x^n in Section 4.3 is valid only if n is positive and integral. If n is negative, the expansion of $(x+\delta x)^n$ by the binomial, and the subsequent reasoning, require some adaptation. We have to recast our argument in the manner shown opposite.

We might first reflect a little longer on the situation in which we are interesting ourselves. We have already seen that the derivative of $1/x$ is $-1/x^2$ (Example 4.2), and if we rewrite the result in a form using negative powers, we can state that

$$\text{the derivative of} \quad x^{-1} \quad \text{is} \quad -x^{-2}.$$

Example 4 of Exercise 4b asked for the derivative of $1/x^2$, and carrying out the usual operations correctly leads to the result $-2/x^3$. Rephrasing this in negative powers leads to the statement that

$$\text{the derivative of } x^{-2} \quad \text{is} \quad -2x^{-3}.$$

Now in both of these cases, the derivative bears the same relation to the original function as does nx^{n-1} to x^n, for in the first example, setting n equal to -1 leads us from the function x^{-1} to the derivative $-x^{-2}$. Similarly, putting n equal to -2 in the second example leads us from x^{-2} to $-2x^{-3}$. The rule that the derivative of x^n is nx^{n-1} seems to apply to negative integers as well as positive ones: the proof goes as follows.

Instead of deriving an expression for the derivative of x^n with n negative integral, we will start with the function $1/x^n$, where n is integral. Proceeding as usual,

$$\frac{f(x+\delta x)-f(x)}{\delta x}$$

$$= \frac{1}{(x+\delta x)^n} - \frac{1}{x^n} \text{ if } f(x) \equiv \frac{1}{x^n}$$

$$= \frac{x^n-(x+\delta x)^n}{x^n(x+\delta x)^n}$$

$$= \frac{x^n - \left(x^n+nx^{n-1}\delta x+\dfrac{n(n-1)}{2!}(\delta x)^2+\ldots+(\delta x)^n\right)}{x^n(x+\delta x)^n}$$

$$= \frac{-nx^{n-1}\delta x - \dfrac{n(n-1)}{2!}\,x^{n-2}(\delta x)^2 - \ldots - (\delta x)^n}{x^n(x+\delta x)^n}$$

$$= -\frac{n\delta x}{x(x+\delta x)^n} - \frac{n(n-1)(\delta x)^2}{2!x(x^2+\delta x)^n} - \ldots - \frac{(\delta x)^n}{x^n(x+\delta x)^n}.$$

Thus

$$\lim_{\delta x=0}\left\{\frac{f(x+\delta x)-f(x)}{\delta x}\right\}$$

$$= \lim_{\delta x=0}\left\{-\frac{n\delta x}{x(x+\delta x)^n} - \frac{n(n-1)(\delta x)}{2!x^2(x+\delta x)^n} - \ldots - \frac{(\delta x)^{n-1}}{x^n(x+\delta x)^n}\right\}$$

$$= -\frac{n}{x^{n+1}}, \quad \text{or} \quad -nx^{-n-1}.$$

It can thus be seen that the general expression nx^{n-1} for the derivative of x^n with n positive integral carries over into any situation where n is negative integral: in both cases the derivative is obtained by multiplying x^n by its power and reducing this power by one.

We have already considered how this general conclusion applies to $1/x$ and $1/x^2$. Its further use enables us to differentiate such functions as $1/x^5$, where, for example, by setting $n = -5$ we obtain the derivative as $-5x^{-6}$.

Three small points should be noted. The first concerns the reduction of the negative power of x^{-n} by one. In the examples with n positive we become accustomed to reducing such functions as x^4 to x^3, and the final power of 3 is numerically smaller than the initial 4. With n negative we must be careful to reduce x^{-4} to x^{-5}, for it is -5 which is smaller than -4 by one, and not -3.

We must secondly guard against writing $3x^{-4}$ as

$$\frac{1}{3x^4} \quad \text{instead of} \quad \frac{3}{x^4}.$$

The 3 should not appear in the denominator, for the negative power -4 applies only to the x. Similarly,

$$\frac{1}{3x^4} \quad \text{is} \quad \frac{1}{3}x^{-4}, \quad \text{and not} \quad (3x)^{-4},$$

and if we wish to invert a product as a whole, brackets should be used. Thus $(2x)^{-1}$ can be correctly rewritten as

$$\frac{1}{2x}.$$

The final point demanding attention involves fractions such as

$$\frac{1}{x^3+x},$$

where the denominator has more than one term.

The derivative of

$$\frac{1}{x^3+x}$$

cannot be written down at first sight with the techniques we have discussed so far, and any temptation to recast the original function in the incorrect forms

$$\left(\frac{1}{x^3}+\frac{1}{x}\right) \quad \text{or} \quad (x^{-3}+x^{-1})$$

must be resisted.

4.7. SIMPLE PRODUCTS AND QUOTIENTS

A function such as $x(x^2-2)$ is best dealt with by multiplying out the brackets, giving x^3-2x, whence its derivative is obtained as $3x^2-2$; similarly,

$$\frac{x^3-x^2+7}{x} = x^2-x+7x^{-1},$$

and its derivative is $(2x-1-7x^{-2})$. If the powers are low, therefore, rewrite the function as a set of terms, and differentiate term by term.

Example 4.4

Differentiate with respect to x:

(i) $1/x^6$, (ii) $(x-1)(x^2+2x+5)$, (iii) $\frac{1}{4x^2}$, (iv) $\frac{2}{x^3}$.

(i) $1/x^6$ can be written as x^{-6}, and either by simple differentiation (see p. 100), or by using the general expression with n equal to -6, we obtain the derivative as $-6x^{-7}$ or $-6/x^7$.

(ii) Multiplying out the product, we get x^3+x^2+3x-5. Differentiating this term by term, the answer $3x^2+2x+3$ is obtained.

(iii) $\frac{1}{4x^2}$ can be written as $\frac{1}{4}x^{-2}$, and the derivative is therefore $\frac{1}{4}(-2x^{-3})$, $\frac{-2}{4x^3}$, or, in its simplest form, $-1/2x^3$.

(iv) $\frac{2}{x^3} = 2x^{-3}$, and the derivative is $2(-3x^{-4})$ or $-6/x^4$.

Exercise 4d

Write down the derivatives with respect to x of:

1. $\frac{2}{x}$.
2. $\frac{1}{2x}$.
3. $\frac{3}{2x}$.
4. $\frac{-1}{x}$.
5. $\frac{1}{x^2}$.
6. $\frac{1}{x^3}$.
7. $\frac{1}{x^4}$.
8. $\frac{-7}{x^7}$.
9. $6x^{-4}$.
10. $-\frac{1}{2}x^{-8}$.
11. $\frac{1}{3x^6}$.
12. $5x^{-2}$.
13. $\frac{1}{5x^2}$.
14. $(5x)^{-2}$.
15. $\frac{-15}{x^3}$.
16. $8x^{-3}$.
17. $\left(\frac{1}{x^2}\right)^3$.
18. $(3x^{-2})^2$.
19. $\frac{x^{-2}}{x^4}$.
20. $\frac{3x+7}{x^8}$.

Differentiate with respect to x the following functions:

21. $x(x^2-1)$.
22. $x^3(2x+7)$.
23. $\frac{x+1}{x}$.
24. $\frac{x^2-1}{x-1}$.
25. $(x-1)^2$.
26. $\frac{2x-3}{x^4}$.
27. $\frac{x^{-1}+x}{3x}$.
28. $\frac{(3x+1)^2}{x^5}$.
29. $\frac{(x-8)(x+1)}{x}$.
30. $x^2(x^2+2)$.
31. $x(x-3)(x+1)$.
32. $x^{-8}(x^2+6)$.
33. $\frac{x^{-3}+x}{7x^2}$.
34. $\frac{(x^2-2)^2}{x}$.

†35. Prove from first principles that the derivative of $\frac{1}{x^4}$ is $-\frac{4}{x^5}$.

†36. Write down an expression for

(i) $\frac{d}{dt}\left\{\frac{1+t^3}{1+t}\right\}$,

(ii) $\frac{d}{dy}\left\{\frac{1-y^3}{1-y}\right\}$,

(iii) $\frac{d}{dz}\left\{\frac{1-z^4}{1-z}\right\}$.

†37. The wear on a car tyre is given by an expression of the form $0{\cdot}00001s^2$, where s is the speed of the car in km/h, and where the wear is expressed in cm of tread per 1000 km. What is the wear at an average speed of (i) 60 km/h, (ii) 80 km/h?

Find the rate of change of wear with speed at (iii) 60 km/h, and (iv) 100 km/h.

†38. Prove from first principles that $d(ax^{-n})/dx = -anx^{-n-1}$.

†39. Two variables x and y are related by the equation $y = x^{\frac{1}{n}}$, where n is a positive integer. Explain the meaning of the expression $x^{\frac{1}{n}}$, and rewrite the equation in the form $x = f(y)$.

If x changes by δx, y changes by δy. Derive an expression for $\delta y/\delta x$ in terms of y, n and δy (starting from the equation $x = f(y)$), and hence find an expression for dy/dx in terms of x and n only.

Is this expression the same as nx^{n-1} with n replaced by $1/n$?

†40. Two variables x and y are related by the equation $y^q = x^p$, where p and q are positive integers. By considering the behaviour of this equation when x and y are replaced by $(x+\delta x)$ and $(y+\delta y)$ respectively, find an expression for $\delta y/\delta x$ and hence dy/dx.

†41. At what points on the graph of $y = (x+4)(x-2)^2$ is the tangent parallel to the x-axis? Sketch the graph.

†42. At what points on the graph of $y = x^3-2x$ does the tangent make an angle of (i) 45°, (ii) 135°, (iii) 225° with the positive direction of the x-axis?

4.8. DIFFERENTIATING MORE THAN ONCE

It is often necessary to differentiate a function twice or more in succession; the notation for the function obtained by differentiating twice in succession wrt (with respect to) x is d^2y/dx^2, or $d^2(y)/dx^2$. The resulting function is called the *second-order derivative*, in order to distinguish it from the *first-order derivative* dy/dx. Similarly, the third-order derivative is obtained by differentiating the original function three times in succession, the symbolic representation being d^3y/dx^3. Notice the position of the 2's and the 3's; between the d and the y in the numerator, and after the x or the other variable in the denominator. For this reason the term is read as "d two y by dx squared". Notice also the difference between d^2y/dx^2 and $(dy/dx)^2$; the former is obtained by differentiating the original function twice with respect to x, whilst the latter is obtained by differentiating once, and then squaring. These two quantities are not equal, as can be seen by taking any particular function—e.g. if $y = x^3$, then $dy/dx = 3x^2$, while $d^2y/dx^2 = 6x$. On the other hand, $(dy/dx)^2$ is equal to $9x^4$.

The second-order derivative can be given a graphical interpretation. The expression for d^2y/dx^2 can be obtained by differentiating the expression for dy/dx with respect to x, and since this latter (first-order) derivative is an expression for the rate of change of y with x—i.e. the gradient of the y-x graph—the second-order derivative is the rate of change of the gradient of the graph. Thus a rapidly curving graph has a high value of d^2y/dx^2; and a gently bending one has a low value of d^2y/dx^2 (Fig. 4.6).

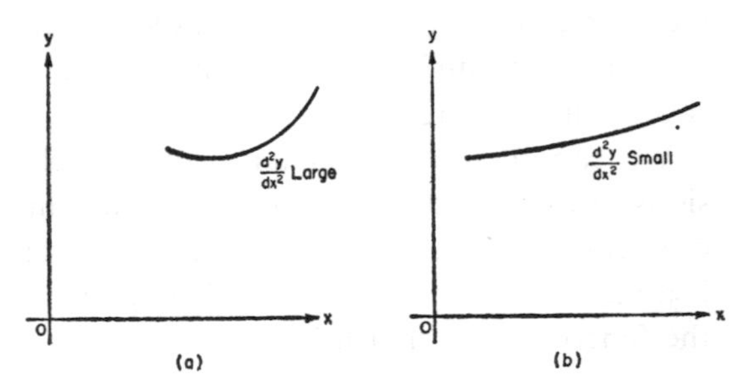

FIG. 4.6. (a) The value of d^2y/dx^2 at any point on a graph is a measure of the rate of change of the gradient there. A rapidly curving graph has a large value for its second-order derivative. (b) A "flat" curve has a small value of d^2y/dx^2.

It is also useful to associate the sign of the second-order differential coefficient with the "concavity" of a graph of the function. We can see in Fig. 4.7a that the gradient of the curve increases from a negative value at A through zero at B to a positive value at C: the gradient of the curve thus increases with x, and $\mathrm{d}(\mathrm{d}y/\mathrm{d}x)/\mathrm{d}x$ or $\mathrm{d}^2y/\mathrm{d}x^2$ will thus be positive. Using ordinary language, we could say that the concavity of the curve is "upwards", and this is directly associated with the positive value of $\mathrm{d}^2y/\mathrm{d}x^2$ throughout the interval from A to C.

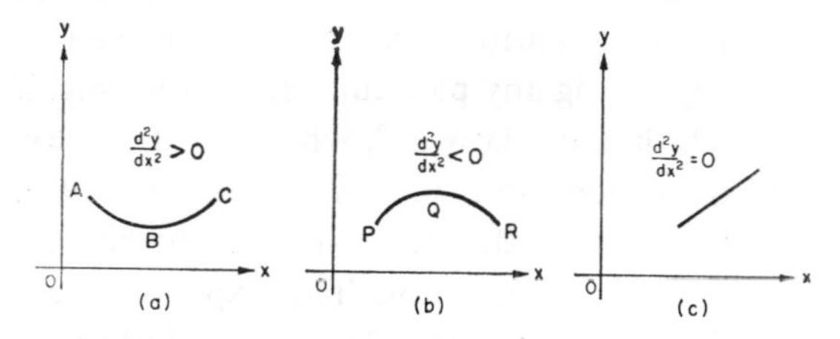

FIG. 4.7. The sign of the second-order derivative indicates the concavity.

The opposite situation in Fig. 4.7b depicts a downwards concavity, and since the gradient of the curve decreases from a positive value at P through zero at Q to a negative value at R, the second-order differential coefficient $\mathrm{d}^2y/\mathrm{d}x^2$ will be negative throughout the interval PQR.

The straight line exhibits no change in gradient or concavity: as might be expected, the second-order differential coefficient of a straight line is zero.

We can therefore summarise our conclusions concerning the value of the second-order differential coefficient, and say that its value is a measure of the curvature of the graph of the function, and that it is

(i) positive for an upwards concavity,

(ii) negative for a downwards concavity,

(iii) zero for a straight line.

Example 4.5

What can be said about the values of $\mathrm{d}y/\mathrm{d}x$ and $\mathrm{d}^2y/\mathrm{d}x^2$ at the points P and Q on the curves shown in Fig. 4.8?

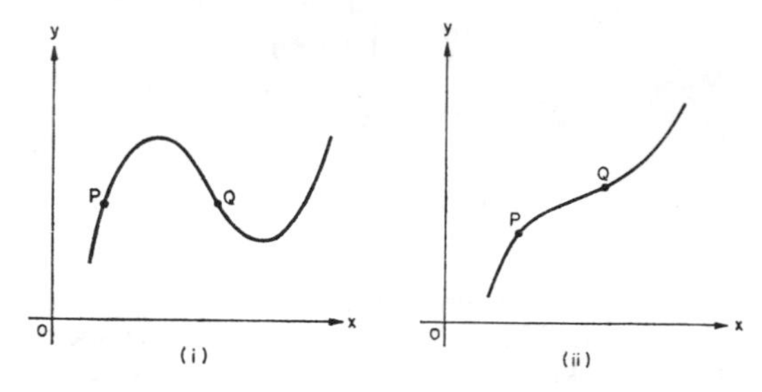

FIG. 4.8.

(i) At P the gradient is positive: thus $(\mathrm{d}y/\mathrm{d}x)_P > 0$. It is, however, decreasing —or the concavity is downwards— and thus $(\mathrm{d}^2y/\mathrm{d}x^2)_P < 0$ (i.e. $\mathrm{d}^2y/\mathrm{d}x^2$ is negative).

At Q the gradient is negative, and increasing towards a zero value with the concavity upwards as a result. Thus, $(\mathrm{d}y/\mathrm{d}x)_Q < 0$ and $(\mathrm{d}^2y/\mathrm{d}x^2)_Q > 0$.

(ii) At P the gradient of the curve is positive: thus, $(\mathrm{d}y/\mathrm{d}x)_P > 0$. Passing through P from left to right on the curve, the gradient decreases: thus $\mathrm{d}(\mathrm{d}y/\mathrm{d}x)/\mathrm{d}x$ or $(\mathrm{d}^2y/\mathrm{d}x^2)_P < 0$, and the concavity is downwards.

At Q, the gradient of the curve is still positive, but less than at P: thus $(\mathrm{d}y/\mathrm{d}x)_Q > 0$ and $(\mathrm{d}y/\mathrm{d}x)_Q < (\mathrm{d}y/\mathrm{d}x)_P$.

The curve is momentarily "flat" (i.e. the gradient is not changing and the curve is linear) at Q_1, and thus the rate of change of the gradient is zero. Thus $(\mathrm{d}^2y/\mathrm{d}x^2)_Q = 0$, and the concavity is neither upwards or downwards.

4.9. OTHER NOTATIONS

It has been explained that $f(x)$ stands for "a function of x"; if it is required to represent its derivative in symbolic form, this can be done by writing $f'(x)$, the second-order derivative being $f''(x)$, and so on, the number of dashes representing the number of times that $f(x)$ has been differentiated. Thus, if $y = f(x)$, then $dy/dx = f'(x)$ and $d^2y/dx^2 = f''(x)$. The quantity $f'(5)$ will therefore represent the first-order derivative of $f(x)$ with the value 5 substituted for x.

Another common symbolism writes dy/dx as y'. The symbol y' is clearly an analogue of $f'(x)$, and the fact that y is a function of x is taken for granted. The symbol y'' represents d^2y/dx^2, and in all cases the independent variable is understood to be x, and no other letter.

Many problems arising in physical situations lead to a demand for differentiating with respect to time. We have already seen, for example, that the time rate of change of displacement is the velocity, and that the time rate of change of velocity is the acceleration. As a result of this common occurrence, a special notation has been coined for time derivatives, and it is commonly recognised that $\dot{x}$ represents the first-order time derivative of x (i.e. dx/dt, where t is the time), that $\ddot{x}$ represents the second-order time derivative (d^2x/dt^2), and so on. Similarly, $\dot{y}$ represents the derivative dy/dt.

Exercise 4e

Write down the first- and second-order derivatives with respect to x of the following functions.

1. $2/x^7$.
2. $2/3x^3$.
3. x^{-4}.
4. $3/8x^4$.
5. $(3x)^{-2}$.
6. $5x^{-12}$.
7. $-7x^{-2}$.
8. $(x-1)/x^2$.
9. $(2x+7)/x$.
10. $(p^2-x^2)(p^2+x^2)$.
11. $x(x^2-5)$.
12. $(2x-3)(x+7)$.
13. $(ax+b)^2$.
14. $(3x-5)^3$.
15. $(x^3+1)(x+1)$.
16. $(x^{-1}-x)/3x^2$.
17. $x^{-3}(x-1)$.
18. $x(x^{-4}-3)$.
19. $x^2(1-x)^2$.
20. $(x+2)^2/x$.
21. $(1-x)(1+x)$.
22. x^{2p+1} (where p is integral).

23. Find the gradient and the rate of change of the gradient of the curve $y = 7x-(4/x)+1$ at the point (4, 28).

24. Find dy/dx, d^2y/dx^2 and $(dy/dx)^2$ for the curve $y = x(x-1)$.

25. A function $f(x)$ has the form $(x^2-5x+4)/x$. Evaluate $f(1)$, $f'(1)$, $f''(1)$ and $f'''(1)$.

26. If $y = 8x^{-2}$, write down expressions for y', y'' and $(y')^2$?

27. What are the first- and second-order differential coefficients of x^{-9}, $7x^{-4}$, $(2x)^{-3}$, $-8/x^4$ and $(x^2-1)^2$?

28. Write expressions for $\dot{x}$, $\ddot{x}$ and $(\dot{x})^2$ if

(i) $x = 3(t^2)^3$, (ii) $x = t^2-5t+7$.

29. If $x = 3t^2-5$ and $y = (t^2-1)^2$, what are the functions $\dot{x}$, $\dot{y}$, $\ddot{x}$, $\ddot{y}$ and $\dot{y}/\dot{x}$?

30. Sketch the graph of the functions $f(x)$ for which

(i) $f'(x) > 0$ for all $x < a$, $f'(x) < 0$ for all $x > a$, and $f'(a) = 0$;

(ii) $f'(x) < 0$ for all $x < a$, $f'(a) = 0$, $f'(x) > 0$ for $a < x < b$, $f'(b) = 0$, and $f'(x) < 0$ for all $x > b$;

(iii) $f'(x) < 0$ for all $x < a$, $f'(x) = 0$, and $f'(x) < 0$ for all $x > a$.

Write down the abscissae of the points at which the gradient of $y = f(x)$ is zero.

†**31.** Sketch the graphs of the functions $f(x)$ for which

(i) $f'(x) > 0$ for all $x \neq a$, and $f'(a) = 0$;

(ii) $f'(x) < 0$ and $f''(x) > 0$ for $x < a$, $f'(a) = 0$, $f'(x) > 0$ for all $x > a$, $f''(x) > 0$ for $a < x < b$, $f''(b) = 0$, and $f''(x) < 0$ for $x < b$;

(iii) $f'(x) < 0$ and $f''(x) < 0$ for $x < a$, $f'(x) < 0$ and $f''(x) > 0$ for $x > a$.

†**32.** The graphs shown in Fig. 4.9 are graphs of five functions $y = f(x)$. For what ranges of x in each case is

(i) $f(x) < 0$,
(ii) $f'(x) < 0$,
(iii) $f''(x) < 0$,
(iv) $f''(x) = 0$,
(v) $f'(x) = 0$?

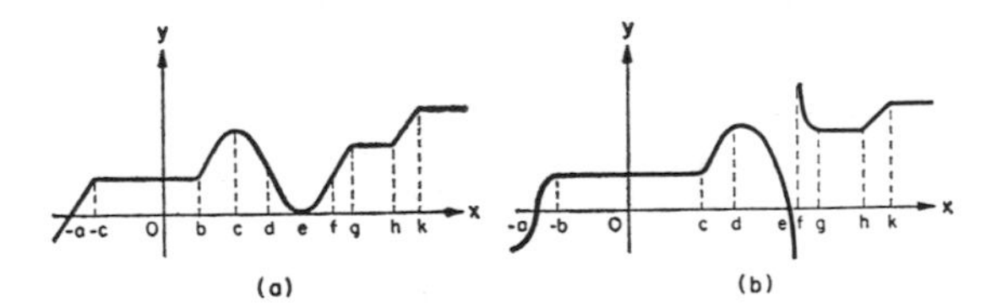

Fig. 4.9.

Sketch the corresponding graphs of $f'(x)$ and $f''(x)$.

†**33.** Show that the curve $y = x^3-2x^3+x$ touches the x-axis at (1, 0), but does not cross it at this point.

†34. It is known that the gradient function of a curve is $2x-7$. If the curve goes through the point (0, 5), find its equation.

†35. Find the equation of the tangent to the curve $y = 4/x$ at the point (2, 2). Prove that the area of the triangle formed by this tangent and the axes is 8. Prove that the area of the triangle formed by the axes and the tangent to the curve at $(a, 4/a)$ is also equal to 8 whatever the value of a. (OC)

†36. State whether the gradients of the curves in Fig. 4.10 at the points P are greater or less than zero

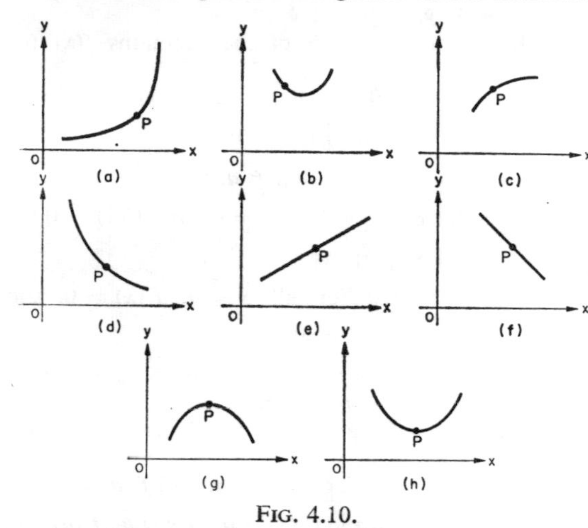

FIG. 4.10.

and also whether they are increasing or decreasing. What can be said about the value of d^2y/dx^2 in each of the graphs? In which curves is the concavity (i) upwards, (ii) zero?

†37. Write down an expression for $d^r/dx^r(x^n)$, given that n and r are integral. Are there any values of r for which the expression is not correct?

†38. By differentiating with respect to x, show that the concavity of the curve $y = ax^n$ (where a and n are constants) depends on the product $an(n-1)x^{n-2}$. Hence show that:

(i) the concavity of a straight line is zero;

(ii) the concavity of an even powered function (i.e. n is even) is in the same sense for all values of x, being upwards if $a > 0$ and downwards if $a < 0$.

†39. If

$$\theta(x) = 1+\frac{x}{1}+\frac{x^2}{2\cdot 1}+\frac{x^3}{3.2.1}+\frac{x^4}{4.3.2.1}+\dots+\frac{x^r}{r(r-1)\dots 3.2.1}+\dots,$$

show that

$$\frac{d\{\theta(x)\}}{dx} = \theta(x).$$

You may assume that it is legitimate in this case to differentiate the infinite series term by term.

†40. If $\theta(x)$ is defined as in question 39, show that

$$\frac{d\{\theta(x^2)\}}{dx} = 2x\{\theta(x^2)\}.$$

†41. If

$$\phi(y) = 1-\frac{y^2}{2!}+\frac{y^4}{4!}-\frac{y^6}{6!}+\dots(-1)^r\frac{y^{2r}}{(2r)!}+\dots,$$

show that $$\frac{d^2}{dy^2}\{\phi(y)\} = -\phi(y).$$

Make the assumption given in question 39.

†42. The function 2^x can be represented by the series

$$1+ax+\frac{a^2x^2}{2!}+\frac{a^3x^3}{3!}\dots+\frac{a^rx^r}{r!}+\dots$$

where a is a constant. Show that $d(2^x)/dx = a.2^x$, and write down an expression for $d^n(2^x)/dx^n$. Make the assumption given in question 39.

†43. Write down the range of x for which the gradients of

(i) $2x^3-6x^2+6x+11$, (ii) $3x^4-12x^3+3$,

are (a) greater than 0, (b) less than 0.

For what range of x is the concavity of each (c) upwards, (d) downwards?

4.10. THE DERIVATIVES OF $\sin x$ AND $\cos x$

If we identify $f(x)$ with $\sin x$, our original definition of the derivative of $f(x)$ leads us to the evaluation of

$$\lim_{\delta x=0}\left\{\frac{\sin(x+\delta x)-\sin x}{\delta x}\right\}.$$

The most suitable manipulation to apply to this expression uses the so-called difference formula

$$\sin A-\sin B=2\sin\{(A-B)/2\}\cos\{(A+B)/2\},$$

and identifying $(x+\delta x)$ with A and x with B leads to the equation

$$\frac{d}{dx}(\sin x) = \lim_{\delta x=0}\left\{\frac{2\sin(\delta x/2)\cos\{(2x+\delta x)/2\}}{\delta x}\right\}$$

$$= \lim_{\delta x=0}\left\{\frac{\sin(\delta x/2)}{\delta x/2}\cos\left(x+\frac{\delta x}{2}\right)\right\}.$$

Now $\lim_{\theta=0}(\sin\theta/\theta) = 1$ (see Appendix), and we therefore conclude that

$$\frac{d}{dx}(\sin x) = \lim_{\delta x=0}\left\{1 \cos\left(x+\frac{\delta x}{2}\right)\right\}$$

and

$$\boxed{\frac{d}{dx}(\sin x) = \cos x.}$$

Similar reasoning can be used to show that

$$\boxed{\frac{d}{dx}(\cos x) = -\sin x,}$$

and these two important results should be committed to memory.

4.11. THE DERIVATIVE OF x^n WITH n RATIONAL

We have obtained the function nx^{n-1} as the derivative of x^n when n is integral in Sections 4.3 and 4.6. We now investigate the derivative of x^n when n is rational. An expression for such a derivative is necessary whenever a function like $\sqrt[3]{x}$ or $\sqrt[4]{x}$ has to be differentiated, and although such situations are less common than those with n integral, there are never the less many occasions when rational powers do occur.

If we have two variables x and y related by the equation $y = x^{p/q}$ where p and q are integral, we can say that we have an expression for the derivative of $x^{p/q}$ if we know the form of dy/dx. This can be obtained by evaluating

$$\lim_{\delta x=0}\left(\frac{\delta y}{\delta x}\right),$$

and we can therefore proceed as follows.

$$\text{If} \quad y = x^{p/q}, \quad y^q = x^p.$$
$$\text{Thus} \quad (y+\delta y)^q = (x+\delta x)^p,$$

and expanding both sides of the latter by the binomial for positive integral indices,

$$y^q + qy^{q-1}\delta y + \frac{q(q-1)}{2!}y^{q-2}(\delta y)^2 + \ldots + (\delta y)^q$$

$$= x^p + px^{p-1}\delta x + \frac{p(p-1)}{2!}x^{p-2}(\delta x)^2 + \ldots + (\delta x)^p.$$

Now $y^q = x^p$, and so, cancelling and rearranging,

$$\delta y\left\{qy^{q-1} + \frac{q(q-1)}{2!}y^{q-2}(\delta y) + \ldots + (\delta y)^{q-1}\right\}$$
$$= \delta x\left\{px^{p-1} + \frac{p(p-1)}{2!}x^{p-2}(\delta x) + \ldots + (\delta x)^{p-1}\right\}.$$

Thus

$$\frac{\delta y}{\delta x} = \frac{px^{p-1} + \frac{p(p-1)}{2!}x^{p-2}(\delta x) + \ldots + (\delta x)^{p-1}}{qy^{q-1} + \frac{q(q-1)}{2!}y^{q-2}(\delta y) + \ldots + (\delta y)^{q-1}}.$$

We need only demand that $\delta y \to 0$ as $\delta x \to 0$ to arrive at the simple conclusion that

$$\frac{dy}{dx} = \lim_{\delta x=0}\left(\frac{\delta y}{\delta x}\right) = \frac{px^{p-1}}{qy^{q-1}}.$$

If we wish to have the derivative in terms of x only (the usual form), we must replace the y^{q-1} in the denominator by $(x^{p/q})^{q-1}$ or $x^{(p-p/q)}$. Then

$$\frac{dy}{dx} = \frac{px^{p-1}}{qx^{(p-p/q)}} = \frac{p}{q}x^{p-1-p+p/q} = \frac{p}{q}x^{(p/q-1)}.$$

This can be seen to be the expression given by nx^{n-1} if n is set equal to p/q, and so once again we can use the function nx^{n-1} as the derivative of x^n.

Example 4.6

Write down the derivatives of:

(i) $x^{2\frac{1}{2}}$, (ii) $3x^{-\frac{1}{3}}$.

Regarding $x^{2\frac{1}{2}}$ as x^n with $n=2\frac{1}{2}$, the derivative of this function can be written as $2\frac{1}{2}x^{1\frac{1}{2}}$.

The derivative of $3x^{-\frac{1}{3}}$ can be similarly obtained by substituting $-\frac{1}{3}$ for n in the general expression. This gives $3\left(-\frac{1}{3}x^{-\frac{4}{3}}\right)$, or $-x^{-\frac{4}{3}}$.

EXERCISE 4f

Find from first principles the derivatives of the following functions:

1. $1/x^3$. **2.** x^3. **3.** $\cos x$. **4.** $\sin 2x$. **5.** $\sqrt{x}$. **6.** x^{-4}.

7. Sketch the graphs of $\sin x$ and $\cos x$ for the region $0 \leqslant x \leqslant 2\pi$, and explain how they illustrate the fact that the derivative of $\sin x$ is $\cos x$.

8. For what values of x is the gradient of the curve $y = \cos x$ negative?

9. Sketch the graphs of $y = x^n$ for (a) $n < 1$, (b) $0 > n > 1$, (c) $0 < n < 1$, (d) $n > 1$.

Mark on each graph two points P and Q whose abscissae are x and $(x+\delta x)$ respectively ($\delta x > 0$). If the ordinates of P and Q are y and $(y+\delta y)$ respectively, state in which cases δy is negative.

Write down the derivatives with respect to x for the following functions:

10. $x^{\frac{7}{4}}$.
11. $2x^{\frac{5}{3}}$.
12. $-3x^{\frac{3}{2}}$.
13. $x^{-\frac{1}{2}}$.
14. $x^{-\frac{3}{2}}$.
15. $2x^{-\frac{3}{2}}$.
16. $-3x^{-\frac{1}{2}}$.
17. $x^{-\frac{1}{6}}$.
18. $x^{-0.1}$.
19. $8x^{0.1}$.
20. $3x^{\frac{1}{2}}-7x^{-\frac{1}{2}}$.
21. $x^{\frac{1}{2}}(x+1)$.
22. $2x-7/x^{\frac{3}{2}}$.
23. $\left(x^{\frac{1}{2}}-x^{-\frac{1}{2}}\right)^3$.
24. $\sin x$.
25. $-3\sin x$.
26. $\frac{1}{4}\cos x$.
27. $x^2-\sin x$.
28. $5(\sin x-x)+\sqrt{x}$.

29. Show that $d^2y/dx^2 = -y$ if $y = \sin x$.

30. If $y = \cos x$, write expressions for y', y'' and y'''.

31. Write expressions for $d(\sin x+\cos x)/dx$ and $d^2(\sin x+\cos x)/dx^2$.

32. If $F(x) \equiv 3\cos x$, what are the values of $F(0)$, $F'(0)$, $F''(\pi/2)$ and $F'''(\pi/3)$?

†**33.** Find from first principles the derivative of $\tan x$.

†**34.** Find from first principles the derivative of $-\cos 5x$.

†**35.** Find the derivatives of (i) $\sec x$, (ii) $\cot x$, (iii) $\sin^2 x$.

†**36.** Show that $\sin(x+\pi/2) = \cos x$, $\sin(x+\pi) = -\sin x$ and $\sin(x+3\pi/2) = -\cos x$.

If $f(x) = \sin x$, explain why $f'(x) = \sin(x+\pi/2)$, $f''(x) = \sin(x+\pi)$ and $f'''(x) = \sin(x+3\pi/2)$. Hence write down an expression for $f^r(x)$.

†**37.** Prove from first principles that the derivative of $x^{\frac{3}{4}}$ is $\frac{3}{4}x^{-\frac{1}{4}}$.

†**38.** If $g(x, y)$ represents the function $x^2+xy+2y^2$, derive expressions for

$$\text{(i)}\ \lim_{\delta x=0}\left\{\frac{g(x+\delta x, y)-g(x, y)}{\delta x}\right\},$$

$$\text{(ii)}\ \lim_{\delta y=0}\left\{\frac{g(x, y+\delta y)-g(x, y)}{\delta y}\right\}.$$

†**39.** If $f(x) = 2^x$, show that $f'(x) = 2^x f'(0)$, where

$$f'(0) = \lim_{\delta x=0}\left\{\frac{2^{\delta x}-1}{\delta x}\right\}.$$

†**40.** State whether either or both of the following statements are true, and explain:

(i) if a function has a derivative at $x = a$, it must be continuous at $x = a$;
(ii) if a function is continuous at $x = a$, it must have a derivative at $x = a$.

†**41.** Explain why $\sqrt{x}$ does not possess a derivative at $x = 0$. Relate your explanation to a sketch of the graph of $y = \sqrt{x}$.

4.12. NUMERICAL DIFFERENTIATION

We can obtain some estimate of the value of the derivative of a function $f(x)$ even though the algebraic form of $f(x)$ is not available or expressible or differentiable. To do this we require a table of values of $f(x)$, and if the value of x at which the value of $f'(x)$ is required is at the lower end of the table, and the tabulated values of $f(x)$ are given for equally spaced values of x, we need only return to the Gregory–Newton formula to acquire the result.

Treating the Gregory–Newton formula (Section 2.7)

$$f(a+\theta h) = f(a)+\theta\,\delta f(a)+\frac{\theta(\theta-1)}{2!}\delta^2 f(a)+\frac{\theta(\theta-1)(\theta-2)}{3!}\delta^3 f(a)+\dots$$

as a function of θ, we can differentiate it term by term with respect to θ, and put $\theta = 0$ to obtain

$$hf'(a) = \delta f(a)-\tfrac{1}{2}\delta^2 f(a)+\tfrac{1}{3}\delta^3 f(a)-\tfrac{1}{4}\delta^4 f(a)+\dots,$$

or

$$f'(a) = \left[\frac{\mathrm{d}}{\mathrm{d}x}\{f(x)\}\right]_{x=a} = \frac{1}{h}\{\delta f(a)-\tfrac{1}{2}\delta^2 f(a)+\tfrac{1}{3}\delta^3 f(a)-\tfrac{1}{4}\delta^4 f(a)+\dots\}.$$

The value of $f'(a)$ is then readily obtainable from the difference values $\delta f(a)$, $\delta^2 f(a)$, etc.

Example 4.7

Find the value of $\mathrm{d}[10^x]/\mathrm{d}x$ at $x = 0{\cdot}87$, using the table of values of 10^x given on p. 68.

From the difference table constructed for Example 2.13 on p. 68, we have that $\delta f(0{\cdot}87) = 0{\cdot}172$, $\delta^2 f(0{\cdot}87) = 0{\cdot}005$, and $\delta^3 f(0{\cdot}87) = -0{\cdot}001$. Since $h = 0{\cdot}01$, we obtain

$$f'(0{\cdot}87) = \{0{\cdot}172-\tfrac{1}{2}(0{\cdot}005)+\tfrac{1}{3}(-0{\cdot}001)\}/h = 100(0{\cdot}1692) = \underline{16{\cdot}92}.$$

Thus $\mathrm{d}[10^x]/\mathrm{d}x_{x=0.87} = \underline{16{\cdot}92}$.

EXERCISE 4g

1. Derive an expression for $f''(a)$ from the Gregory–Newton formula. Use it to obtain an approximate value of $\mathrm{d}^2[10^x]/\mathrm{d}x^2$ at $x = 0{\cdot}87$.

2. Construct a table of values of $\log_{10} x$ for $5 \leqslant x \leqslant 20$, tabulating at intervals of 1. Use your table to find an approximate value of $\mathrm{d}\{\log_{10} x\}/\mathrm{d}x$ at (i) $x = 5$, (ii) $x = 10$, (iii) $x = 15$. What do you notice about the results?

Given that $\log_e x = 2{\cdot}303 \log_{10} x$, find the approximate of $\mathrm{d}(\log_e x)/\mathrm{d}x$ at the same values of x.

3. From the table of values on p. 69, find an approximate value of $\mathrm{d}\{J_0(x)\}/\mathrm{d}x$ at $x = 0{\cdot}4$.

4. The following table shows a set of values of e^x for $0 \leqslant x \leqslant 0{\cdot}6$. Find an approximate value of $\mathrm{d}(e^x)/\mathrm{d}x$ at $x = 0$, $x = 0{\cdot}1$, and $x = 0{\cdot}2$.

x	0	0·1	0·2	0·3	0·4	0·5	0·6
e	1·0000	1·1052	1·2214	1·3499	1·4918	1·6487	1·8221

5. The displacement of a body varies with time as follows:

Time (s)	0	1	2	3	4	5
Displacement (m)	0	5	15	30	50	75

Find its velocity at $t = 0$ by numerical differentiation, and check your result by finding an equation to describe the motion and differentiating it.

6. The velocity of a car of mass 1000 kg varies with time as follows:

Time (s)	0	2	4	6	8	10	12
Velocity (m/s)	0	1	3	6	10	15	21

Find an approximate value of the acceleration at (i) $t = 0$, (ii) $t = 2$, (iii) $t = 4$ and (iv) $t = 6$, and calculate the power at which the engine is working at these times. Assume that frictional forces and wind resistance are zero.

SUMMARY

The first-order derivative or differential coefficient of y with respect to x is denoted by the symbols $\mathrm{d}y/\mathrm{d}x$, y' or $f'(x)$, where $y = f(x)$, and is defined by the equation

$$\frac{\mathrm{d}y}{\mathrm{d}x} = \lim_{\delta x=0}\left(\frac{\delta y}{\delta x}\right) = \lim_{\delta x=0}\frac{f(x+\delta x)-f(x)}{\delta x}.$$

The following results can be derived using this definition:

$$\frac{\mathrm{d}}{\mathrm{d}x}(x^n) = nx^{n-1} \quad \text{for all } n,$$

$$\frac{\mathrm{d}}{\mathrm{d}x}(\sin x) = \cos x,$$

$$\frac{\mathrm{d}}{\mathrm{d}x}(\cos x) = -\sin x.$$

Simple products and quotients are best dealt with by arranging them in single terms and differentiating the resulting expressions term by term.

The second-order derivative of y with respect to x can be defined as the derivative of the differential coefficient of y with respect to x and is denoted by d^2y/dx^2, y'' or $f''(x)$. Its value at any point is a measure of the concavity or curvature of the graph of the function at that point.

Differentiation is the name given to the process of finding the derivative of a function: the second and higher order derivatives can be obtained by differentiating more than once.

In cases where the function to be differentiated is available only in numerical form, an estimate of the value of $f'(x)$ at $x = a$ can be made from the expression

$$\{\delta f(a) - \tfrac{1}{2}\delta^2 f(a) + \tfrac{1}{3}\delta^3 f(a) \ldots\}/h,$$

where h is the interval of tabulation.

MISCELLANEOUS EXERCISE 4

Differentiate the following with respect to x:

1. x^{27}. **9.** $x^{\frac{1}{5}}$. **17.** $1/x^3$.

2. x^{-19}. **10.** $x^{\frac{3}{4}}$. **18.** $6/x^{2\frac{1}{4}}$.

3. ax^{30}. **11.** $4x^{\frac{1}{8}}$. **19.** $x(5-x^2)$.

4. x^{-30}. **12.** $7x^{-\frac{1}{5}}$. **20.** $3x^2(2x^2-5)$.

5. $18x^3$. **13.** $\sqrt[3]{(x^4)}$. **21.** $(x-7)(9-x+x^2)$.

6. $9x^{15}$. **14.** $11x^{-1\frac{1}{2}}$. **22.** $(x^2-x)/x$.

7. $6x^{-2}$. **15.** $\sqrt[4]{(x^3)}$. **23.** $(3-x)/\sqrt[4]{x}$.

8. $3x^{-8}$. **16.** $1/x^{\frac{2}{3}}$. **24.** $(3-x)^2/\sqrt{x}$.

25. Differentiate with respect to t:

(i) at^2+7, (ii) $(a-t)(b-t)$.

26. Differentiate with respect to p:

(i) $(7-3p^2)/p$, (ii) $(p^2)^2$.

27. If $f(y) = y^2+6y-17$, evaluate $f(2)$, $f'(4)$ and $f''(-1)$.

28. If $f(x) = (x^2-a)^2$, what must a be if $f'(1) = 5$?

29. Over what range is the gradient of $y = x^3-3x^2-9x+1$ positive?

30. Over what range is the gradient of $y = x^4-18x^2+5$ (a) positive, (b) negative?

31. Over what range of values of x is the concavity of x^3-7x^2+3x+1 (a) upwards, (b) downwards?

32. Sketch the graphs of:

(a) $y = \frac{1}{5}(x+x^3)$ for $-3 \leqslant x \leqslant 3$,

(b) $y = \frac{1}{8}(4x-x^3)$ for $-4 \leqslant x \leqslant 4$.

33. For what values of x is the concavity of $x^4+2x^3-12x^2+15x-7$ (a) upwards, (b) downwards?

34. Sketch the graph of $y = \frac{1}{4}x^4$ in the region $-2 \leqslant x \leqslant 2$. Explain why d^2y/dx^2 is zero at $x = 0$, but there is no inflexion there.

35. Prove that the gradient of the graph of (x^2-4) is greater than 0 for all values of x greater than 0. For what values of x is the gradient less than 0? What can you say about the concavity of the graph of (x^2-4)?

36. Differentiate (i) $(x^2-3)^2$, (ii) $(x^3-2)/x$. (OC)

37. Differentiate $x+x^{\frac{1}{2}}+x^{-\frac{1}{2}}+x^{-1}$ with respect to x. (OC)

38. Differentiate $2x^2+5+4x^{-2}$, and $\sqrt{x}$, each with respect to x. Find the value of each result when $x = 4$. (OC)

†**39.** Discuss the continuity of (i) the graph, (ii) the first-order derivative, (iii) the second-order derivative, of a step function (see question 15, p. 59).

†**40.** Draw a sketch graph of a function $f(x)$ for which the values of

$$\lim_{\delta x \to 0+} \left\{\frac{f(a+\delta x)-f(a)}{\delta x}\right\} \quad \text{and} \quad \lim_{\delta x \to 0-} \left\{\frac{f(a+\delta x)-f(a)}{\delta x}\right\}$$

are not equal.

Discuss whether (i) $f(x)$ is continuous at $x = a$, (ii) $f(x)$ has a derivative at $x = a$.

CHAPTER 5

Applications of Differentiation

5.1. VELOCITY AS THE DERIVATIVE OF DISPLACEMENT

It was explained in Section 3.3 that the velocity of a body at any instant is given by the limiting value of the average velocity during an interval described about the instant as the interval size is reduced to zero.* Referring to the displacement–time graph of Fig. 5.1, we can

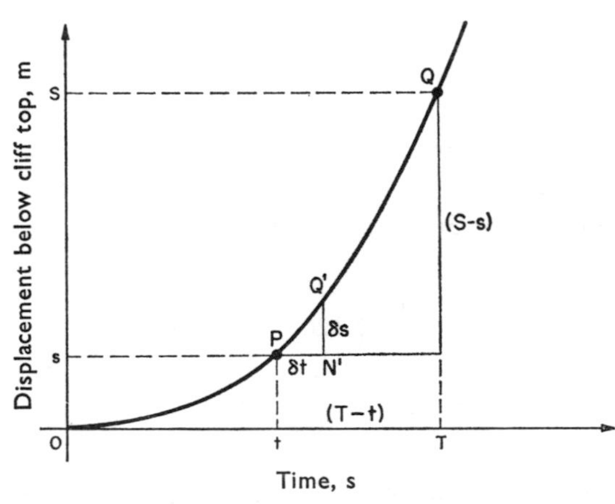

FIG. 5.1. The instantaneous velocity of P is the limiting value of $\delta s/\delta t$ as $\delta t \to 0$.

write that the average velocity of the body whose motion it represents during the interval between times t and T as $(S-s)/(T-t)$, where s and S are the values of the displacement at the beginning and end of the interval, respectively. For a small interval between times t and $t+\delta t$, this average velocity can be written as $\delta s/\delta t$, and thus the value of the instantaneous velocity is given by the value of

$$\lim_{\delta t=0} \left(\frac{\delta s}{\delta t}\right),$$

i.e. the value of the derivative ds/dt. If, therefore, the relationship between the displacement s and the time t is known, an expression for the velocity of the body concerned at any instant of the motion can be obtained by differentiation.

Taking, for example, the motion described by the graph of Fig. 5.1 to be that of a pebble dropped from a cliff top as in Section 3.3 and Fig. 3.9, we find that the height s feet through which the pebble has fallen t seconds after its release is given by the relation $s = 5t^2$. Differentiating with respect to t, we obtain the expression for the instantaneous velocity —namely, $v=10t$. Or, for a small bob oscillating on the end of a spring with the displacement equation $s = 10 \sin t$, the instantaneous velocity v will vary with time according to the relationship $v = 10 \cos t$ (Fig. 5.2).

The units in which the velocity is expressed are governed in the first instance by the units of s and t. If, for example, in the equation relating s to t, s is quoted in metres and t in seconds,

* More correctly, the common limit (if it exists) for reducing intervals on either side.

the values of ratios of the form $(S-s)/(T-t)$, $\delta s/\delta t$ or ds/dt derived from the equation will all be in m/s; or if s is in metres and t is in minutes, the values of the ratios will be expressed in metres per minute. If the velocity is required in km/h a simple conversion will have to be carried out (e.g. multiply by 3/50 to convert m/min into km/h, or by 3·6 to convert m/s into km/h).

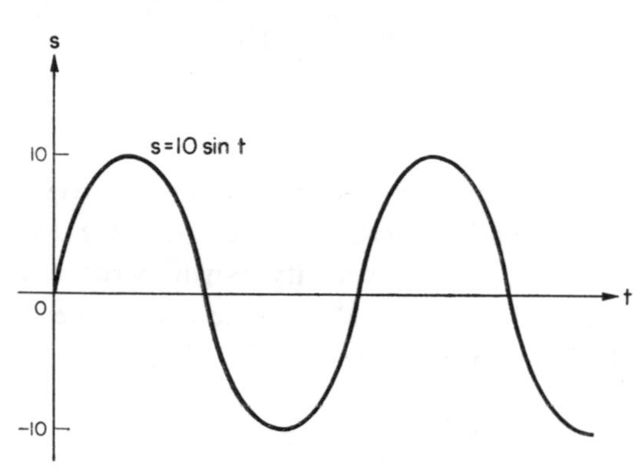

FIG. 5.2. The displacement–time graph of an oscillating particle.

The value of the derivative at any moment has been shown to equal the gradient of the tangent at the point representing that moment on a standard scale or a unity scale graph (Section 2.2): if the equation of motion is not available or even expressible in algebraic form, the velocity at any instant can be calculated by constructing the tangent on a standard or a unity scale graph of the motion.

Example 5.1

Find the velocity of a body whose equation of motion is $s = 5t\sqrt{t}-3t$ when (i) $t = 0$, (ii) $t = 4$. At what time is the velocity zero?

Rewriting the equation of motion in the form $s = 5t^{\frac{3}{2}}-3t$, we obtain by differentiation the equation $v = \frac{15}{2}t^{\frac{1}{2}}-3$.

Thus, when $t = 0$, $v = -3$; and when $t = 4$, $v = \frac{15}{2}\sqrt{4}-3 = 12$.

If $v = 0$, the time t must be such that $\frac{15}{2}t^{\frac{1}{2}}-3 = 0$.

Thus $$t^{\frac{1}{2}} = \frac{6}{15} = \frac{2}{5},$$
and $$t = \frac{4}{25}.$$

Answer: (i) -3, (ii) 12; $v = 0$ when $t = \frac{4}{25}$.

Example 5.2

An aeroplane takes off, and t seconds later is s km away from the airport where

$$s = \frac{t^2}{240}-\frac{t^3}{108\,000}.$$

Derive an expression for its velocity t seconds after take-off, and convert this to give the answer in km/h. What is the velocity of the aeroplane when (i) $t = 10$, (ii) $t = 20$, (iii) $t = 60$ seconds?

By differentiation, $v = t/120-t^2/36\,000$; and with t expressed in seconds, this equation gives an expression for the velocity in km/s. A speed of x km/s is equivalent to $60x$ km/min, or $3600x$ km/h, and so

$$v = 3600\left\{\frac{t}{120}-\frac{t^2}{36\,000}\right\}$$
$$= \left(30t-\frac{t^2}{10}\right) \text{ km/h},$$

where the value of t is to be given in seconds.

Thus, when

$t = 10$, $\quad v = 300-\frac{100}{10} = 290$ km/h,

$t = 20$, $\quad v = 600-\frac{400}{10} = 560$ km/h,

$t = 60$, $\quad v = 1800-\frac{3600}{10} = 1540$ km/h.

Example 5.3

The graph of Fig. 5.3 represents the motion of a particle. Estimate the velocity $12\frac{1}{2}$ seconds after the start of the motion.

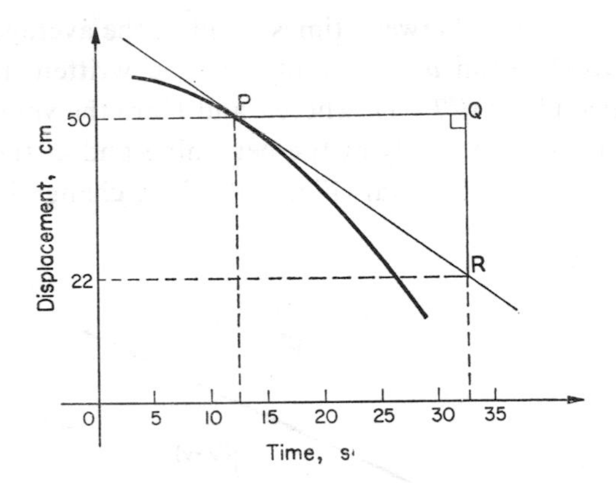

FIG. 5.3.

The position of the particle $12\frac{1}{2}$ seconds after the start is represented by the point P on the curve. To obtain the velocity, the tangent PR is constructed at P, and its gradient is given by the ratio QR/PQ. The ordinate at Q is 50, and that at R 22, so that $QR = -28$. The gradient of the tangent—i.e. the velocity—is therefore equal to $-28/20$ or $-1{\cdot}4$ cm/s.

Note that since the displacement is decreasing, the velocity is negative (cf. Section 2.2), and also that the gradient of the tangent can be evaluated by constructing any right-angled triangle with hypotenuse along the tangent PR. As pointed out before, it is worth choosing the length of PQ to be a simple number, since this then gives a whole number denominator in the fraction from which the gradient is calculated.

Answer: The velocity $12\frac{1}{2}$ seconds after the start of the motion is $-1{\cdot}4$ cm/s.

EXERCISE 5a

1. Find an expression for the velocity of a body whose displacement s varies with the time t according to the equation:

(i) $s = 4t^3-7t$, (ii) $s = 9-2t^2$, (iii) $s = 2\cos t$.

2. Repeat question 1 for the equations:
(i) $s = 5\sqrt{t}+3$, (ii) $s = a-bt$, (iii) $s = 2t^p$.

3. Write down an expression for the velocity of a body whose equation of motion is:

(i) $s = 6t^3-8t$, (ii) $s = 4t^{\frac{5}{2}}$, (iii) $s = 3\sin t-\cos t$.

4. The displacement s of a body varies with time t according to the equation $s = 8+t^{\frac{5}{2}}$, where s is measured in metres and t in seconds. Derive an expression for its velocity v in (i) m/s, (ii) km/h.

5. Calculate the velocity of a body which moves according to the equation $s = \frac{55}{9}(2+t)^2$, where s is the displacement in metres and t the time in seconds when (i) $t = 0$, (ii) $t = \frac{1}{2}$, (iii) $t = 2$, giving your answer in km/h.

6. Write down the velocity of a body whose displacement s at time t is given by the equation $s = 0{\cdot}8\sin t$, where s is in cm and t is in seconds, when (i) $t = 0$, (ii) $t = \frac{1}{2}\pi$, (iii) $t = \frac{1}{2}$, (iv) $t = 1$.

7. Derive an expression for the velocity of a body whose displacement s from a fixed point varies with time t according to the equation $s = 0{\cdot}04\cos t$, where t is quoted in milliseconds, and s in cm. Use your expression to find the velocity in cm/s when (i) $t = 0$, (ii) $t = \frac{1}{2}$, (iii) $t = 1$, (iv) $t = 1{\cdot}571$, (v) $t = 3{\cdot}142$, (vi) $t = 4{\cdot}713$.

What are the values of s at the last three instants of time?

8. A car starts from rest so that its distance s from its starting point varies with time t as follows:

s (km)	0	0·6	1·4	2·4	3·4	4·4	5·0	6·2	7·6	9·4	11·6
t (min)	0	1	2	3	4	5	6	7	8	9	10

Draw a graph to represent the motion, and estimate the velocity when (i) $t = 1{\cdot}5$, (ii) $t = 4$, (iii) $t = 7{\cdot}5$, (iv) $t = 10$.

9. (a) Find, from first principles, the differential coefficient with respect to x of x^2-3x+2.

(b) The distance s moved in a straight line by a particle in time t is given by $s = at^2+bt+c$, where a, b and c are constants. If V is the velocity of the particle at time t, show that $V^2-b^2 = 4a(s-c)$.

(NUJMB)

5.2. ACCELERATION

To introduce the ideas of acceleration, we will use a slightly different example of motion.

A stone is projected with a certain velocity vertically from the earth's surface, and its height is found to vary with time as follows:

Time after projection (s)	0	1	2	3	4	5	6
Height above ground (m)	0	25	40	45	40	25	0

The graph of these results is given in Fig. 5.4, and it is comparatively simple to find that the equation relating the height s to the

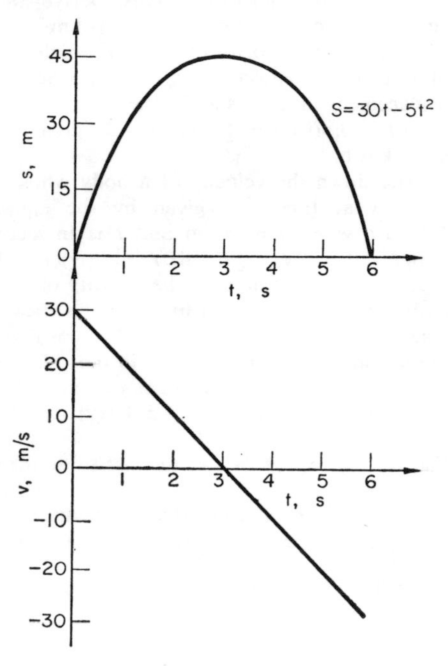

FIG. 5.4. The displacement–time and velocity–time graphs of a stone projected from the earth's surface.

time t has the form $s = 30t-5t^2$. We can then deduce that the instantaneous velocity varies with time according to the equation $v = 30-10t$, and tabulating the velocity at the end of each second, we obtain:

Second	1st	2nd	3rd	4th	5th	6th
Velocity	20	10	0	−10	−20	−30

These results are also represented in Fig. 5.4, and both from the graph and the table it is apparent that the velocity decreases by 10 m/s every second. This change in velocity every second is called the *acceleration*; defined over an interval between times t and T, the average acceleration a of a body can be written as $a = (V-v)/(T-t)$, where v and V are the velocities of the body at the beginning and at the end of the interval (Fig. 5.5). If the change in

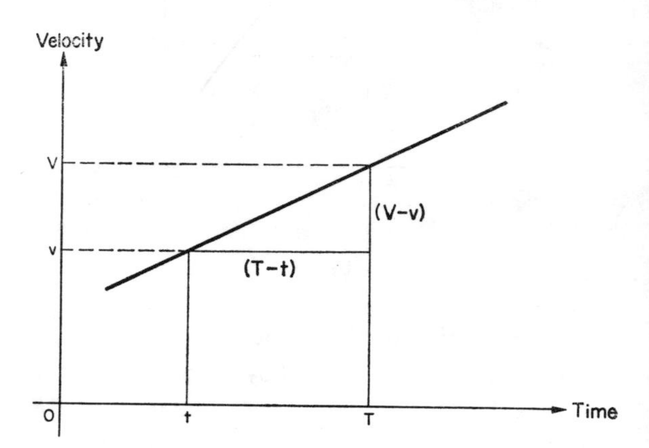

FIG. 5.5. The acceleration of a body is the change in velocity per unit time interval: here the acceleration is uniform and equal to $(V-v)/(T-t)$.

velocity $(V-v)$ is measured in m/s, and the interval $(T-t)$ in seconds, the value of the acceleration will be obtained from this fraction in m/s. In the case we have just considered the acceleration is constant and equal to -10 m/s²: like velocity, acceleration is a vector and must therefore be given a sign or direction where appropriate.

If we consider a general case of motion with constant acceleration a, we can see that in a time t, the velocity will increase by $(a\times t)$. And if the motion commences with a velocity u, the velocity v after time t is given by the equation

$$v = u+at.$$

The variation in the velocity can then be represented graphically as shown in Fig. 5.6.

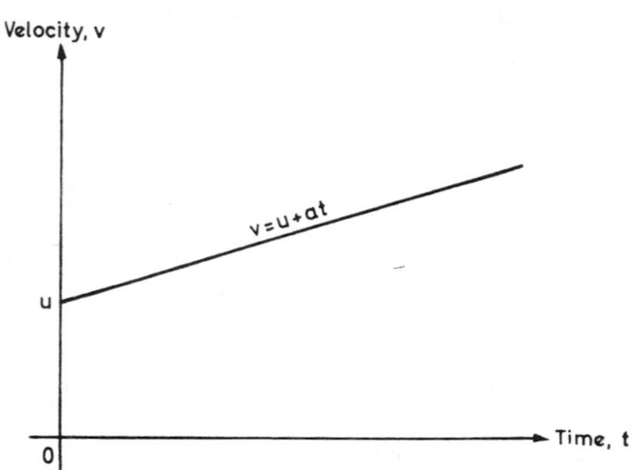

FIG. 5.6. The velocity of a uniformly accelerated body increases linearly with time.

5.3. INSTANTANEOUS ACCELERATION

The average acceleration during an interval for a motion whose velocity–time graph is not linear is still the ratio $(V-v)/(T-t)$ (Fig. 5.7):

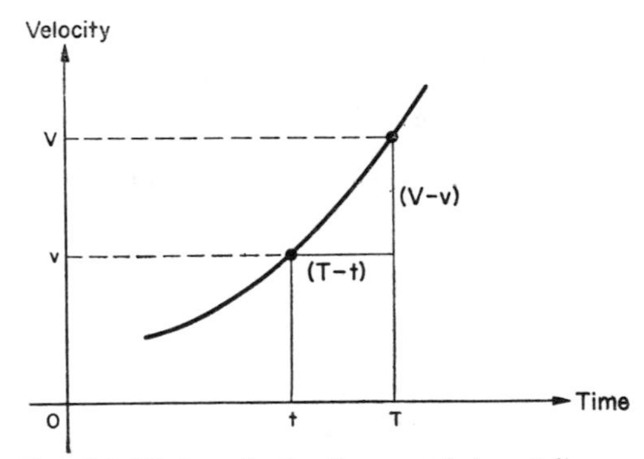

FIG. 5.7. If the velocity–time graph is not linear, the average acceleration during the interval from t to T is $(V-v)/(T-t)$ ...

as such, the numerical value of the average acceleration in the interval from t to T is given by the gradient of the chord joining the points representing the velocities at the ends of the interval.

If the interval is short, the ratio $(V-v)/(T-t)$ can be written as $\delta v/\delta t$ (Fig. 5.8): and the limiting value of this ratio at $\delta t = 0$ gives *the instantaneous acceleration*. As before, the numerical value of the acceleration at any instant t is given by the gradient of the tangent to the v/t curve at the point representing the velocity at that instant (Fig. 5.9), and if the equation

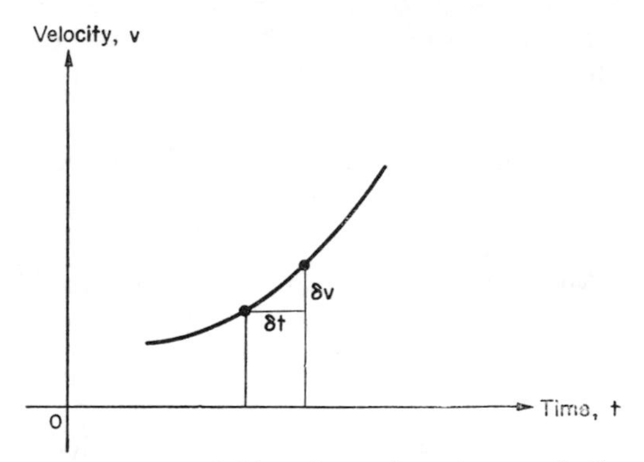

FIG. 5.8. ...and this ratio can be written as $\delta v/\delta t$ in preparation for defining the instantaneous acceleration as $\lim_{\delta t=0}\left(\frac{\delta v}{\delta t}\right)$.

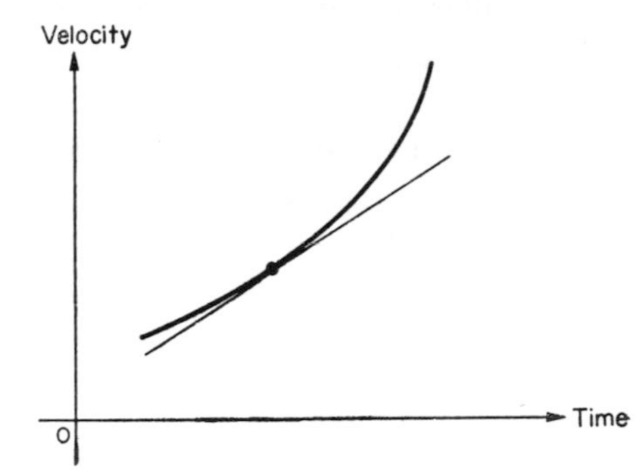

FIG. 5.9. The instantaneous acceleration of a body is numerically equal to the gradient of the tangent at the point whose abscissa represents the instant.

relating the displacement or the velocity to the time is known, the value of the acceleration at any instant can be calculated by evaluating the derivative of the equation. Symbolically, we can write that

$$a = \lim_{\delta t=0}\left(\frac{\delta v}{\delta t}\right) = \frac{dv}{dt} = \frac{d^2s}{dt^2}.$$

Example 5.4

The displacement s of a particle from a fixed point t seconds after motion begins is given by the equation $s = 12t-\frac{1}{10}t^2$, where s is measured in cm. What are the displacement, velocity and acceleration at the start, and one second and three seconds after the start?

The values of the displacement can be obtained by substituting 0, 1 and 3 directly into the displacement–time equation. Thus the displacement at the start is 0 cm; after 1 second, $s = 12-\frac{1}{10} = 11{\cdot}9$ cm; and after 3 seconds, $s = 36-\frac{9}{10} = 35{\cdot}1$ cm.

An expression for the velocity is obtained by differentiating the equation with respect to t, whence $v = 12-\frac{1}{5}t$. When $t = 0$, the velocity is 12; when $t = 1$, $v = 12-\frac{1}{5} = 11{\cdot}8$; and when $t = 3$, $v = 12-\frac{3}{5} = 11{\cdot}4$. The velocities at the three times are thus 12, 11·9 and 11·4 cm/s, respectively.

The expression for the acceleration f is obtained by differentiating the expression for the velocity, whence $a = -\frac{1}{5}$ is obtained. The acceleration is thus independent of the time, and has the constant value of $-0{\cdot}2$ cm/s², the minus sign indicating a retardation (i.e. velocity decreasing with time).

The summarised answers are therefore:

Time after start	0	1	3	s
Displacement	0	11·9	35·1	cm
Velocity	12	11·8	11·4	cm/s
Acceleration	−0·2	−0·2	−0·2	cm/s²

Exercise 5b

1. The velocity of a body increases from 134 cm/s to 200 cm/s in 4 seconds. What is its average acceleration?

2. A car starts from rest and acquires a speed of 20 m/s in 8 seconds. What is its average acceleration?

3. The velocity of a body varies with time as shown in the following table: determine its average acceleration for each of the intervals 0–5, 5–10, 10–15 and 15–20 seconds.

Time (s)	Velocity (cm/s)
0	0
5	80
10	200
15	360
20	420

4. Calculate the average acceleration of a car in (i) km/h², (ii) m/s², if it acquires a speed of 120 km/h in 10 seconds starting from rest.

5. The velocity of a body falling freely under gravity varies with time according to the table:

v	10·14	11·12	12·10	13·08	14·06	m/s
t	0	0·1	0·2	0·3	0·4	s

Show that the acceleration is uniform, and deduce its value.

6. A body falling freely under gravity accelerates from a speed of 204 cm/s to 2656 cm/s in $2\frac{1}{2}$ seconds. What is the numerical value of the acceleration due to gravity?

7. Write down expressions for the acceleration of a body whose velocity varies with time according to the equation

(i) $v = 47t-4$, (iii) $v = 2\cos t$,
(ii) $v = 4\sqrt{t}$, (iv) $v = 3(2-t)^2$.

8. The velocity of a body travelling at a speed of 9 m/s due north changes by 12 m/s in a direction due East. If the change takes place in 2 seconds, what is the acceleration, and what is its final velocity?

9. A body experiences an acceleration of 5 m/s² in a downward direction for 6 seconds. By what does its velocity change? Calculate its final velocity given that its velocity before it experienced the acceleration was

(i) 20 m/s downwards,
(ii) 10 m/s upwards,
(iii) 30 m/s horizontally,
(iv) 25 m/s at an upward angle of $\tan^{-1}(\frac{3}{4})$ with the horizontal,
(v) 30 m/s at a downward angle of 45° with the horizontal.

(You may obtain your answer by drawing if you wish.)

10. A body falling freely under gravity drops through a distance s metres in t seconds where $s = 4{\cdot}9t^2$. What is its velocity after t seconds, and what is the acceleration due to gravity? How does your working show this to be constant?

11. A stone is thrown vertically downwards from the top of a cliff and the equation relating its distance s (metres) below the cliff top and the time t (seconds) after it was thrown is $s = 10t+5t^2$. Find an expression for its velocity, and determine how fast the stone was travelling when (i) $t = 0$, (ii) $t = 1$, (iii) $t = 2$.

What is its acceleration?

12. A bullet is fired vertically upwards and the relationship giving the height it reaches after t seconds is $h = 300t-5t^2$.

Find an expression for its velocity, and determine

(i) its speed at the moment of firing,
(ii) the value of t when it is at its greatest height,
(iii) the greatest height it reaches.

13. A train leaves a station and moves so that its distance from the station t minutes later is given by the equation $s = \frac{1}{4}t^2 - \frac{1}{90}t^3$, where s is measured in kilometres. What is the train's velocity in km/h (a) 1 minute, (b) 2 minutes, after it has left the station? Derive an expression for its acceleration in km/h^2, and state when this is zero. What is the velocity when the acceleration is zero?

***14.** An electron moves from a negative electrode to a positive electrode so that its distance s cm from the negative electrode is given by the equation $s = 5 \times 10^{-9} t^2$, where t is the time in nanoseconds (1 nanosecond $= 10^{-9}$ s). What is its velocity in cm/s after (a) 10 nanoseconds, (b) 1 microsecond, (c) 1 millisecond, and what is its acceleration in cm/s^2?

15. The displacement s of a body is related to the time t by the equation of the form $s = 5 \sin t$. Derive an expression for the velocity of the body and write down its value when

(i) $t = 0$,
(ii) $t = 0{\cdot}5$,
(iii) $t = 1$,
(iv) $t = \frac{1}{2}\pi$,
(v) $t = \pi$,
(vi) $t = 5$.

Derive an expression for the acceleration, and sketch a displacement–time, velocity–time and acceleration–time graph of the motion.

16. A mass on the end of a spring bobs up and down so that its displacement from the central position is given by the equation $s = 3 \cos t$. Find the values of the displacement, velocity and acceleration when t is equal to (i) 0, (ii) 1, (iii) $\frac{1}{2}\pi$, (iv) $\frac{3}{2}\pi$, and explain the meaning of your results.

17. A body is dropped from an aeroplane, and its distance beneath the aeroplane varies with the time after its release according to the values shown in the following table:

t (s)	0	1	2	3	4	5	6	7	8	9	10
s (m)	0	5	21	46	81	126	180	238	298	358	418

Plot a graph representing the motion of the body, and use it to estimate the instantaneous velocity at $t = 3$, 6 and 9. Comment on your results.

†18. The velocity of a body varies with the time according to the equation $v = 6t$. Draw a sketch graph of this variation, give the value of the acceleration, and write down an expression for the distance covered in the first t seconds of the motion.

†19. The height h m of a rocket above its launching pad t seconds after firing is given by an expression of the form $h = at^3 + bt^2 + ct + d$. Derive expressions for the velocity, acceleration and the rate of change of acceleration with respect to the time in terms of t. If the rocket starts from rest, what can be said about the values of b and c?

Given that the rate of change of acceleration with respect to time must not be greater than $240g$ per second at any instant, and that the acceleration should not be greater than $30g$, where g denotes the acceleration due to gravity, write down an expression for the velocity during the first $\frac{1}{8}$th of a second in terms of g, assuming that the rocket reaches its maximum acceleration in the shortest possible time. What does the expression become for $t > \frac{1}{8}$?

Deduce also the expression for the height of the rocket above its launching pad during the first $\frac{1}{8}$th second, and write down the corresponding expression for time values greater than the time taken to reach maximum acceleration.

20. The point P moves in a straight line so that after t seconds its distance x m from a fixed point O in the line is given by the equation $x = \sin t + 2 \cos t$. Its velocity is then v m/s and its acceleration a m/s^2.

(i) Show that $v^2 = 5 - x^2$ and that $a = -x$.
(ii) Find the greatest distance of P from O.
(iii) Find the velocity of P after 2 seconds, taking one radian as 57°18′. (OC)

21. A particle is moving in a straight line, and its distance x m from a fixed point O in the line at time t seconds is given by

$$s = 2 + 3 \sin t + \cos t.$$

Find the value t for which the particle first comes to rest, and find also the acceleration and distance from O at this instant. (OC)

22. A particle moves along the x-axis, its position when t seconds have elapsed being given by $x = 27 - 36t + 12t^2 - t^3$. Show that:

(i) for the first 2 seconds the particle moves in a negative direction:
(ii) for the next 4 seconds it moves in a positive direction.

Find also the accelerations at the instants when the particle has zero velocity. (OC)

†23. A dart is thrown with a speed of 10 m/s in a horizontal direction. It strikes a board 5 m away with a velocity at an angle of $\tan^{-1}(\frac{1}{2})$ with the horizontal. Assuming that the horizontal velocity remains constant, find the value of the acceleration due to gravity. (Assume the board to be vertical and a horizontal distance of 5 m from the point of projection.)

†24. The constant acceleration a experienced by a body makes an acute angle θ with the direction of the initial velocity u. Show that the magnitude of the final velocity v is given by the equation

$$v^2 = u^2 + a^2t^2 + 2uat \cos \theta,$$

where t is the time for which the acceleration acts.

†**25.** Consider a body rotating in a circle of radius r with an angular velocity ω. If at any moment its velocity is v, at a short time t later its velocity will be v at an angle of ωt with its initial velocity (Fig.5.10).

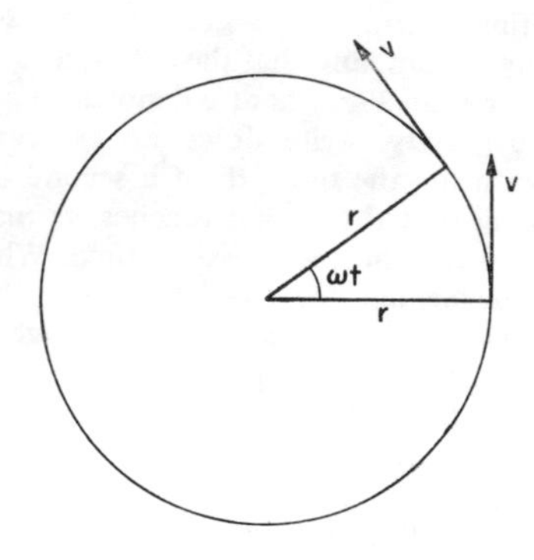

FIG. 5.10.

Show that the magnitude of the change in the velocity of the body is given by $2r\omega \sin(\frac{1}{2}\omega t)$, and hence write down an expression for the average acceleration during the interval t. Deduce the value of the instantaneous acceleration.

5.4. RATES OF CHANGE

It will be recalled that the instantaneous rate of change of a variable y with respect to a variable x over an interval δx about x is defined as the limiting value of the ratio $\delta y/\delta x$ at $x = 0$, where δy is the change in y associated with the change δx in x. We can now understand that the value of the rate of change of y with respect to x is given by the value of the derivative dy/dx.

To deduce the value of an actual change in y, knowing the size of an associated change in x and the instantaneous rate of change, needs either more information or the assumption that the value of the rate of change of y with respect to x is constant while x is changing. This assumption is valid only if the relation between y and x is linear in form, and if this is so, we can say that the value of the rate of change of y with respect to x tells us how many times greater or smaller is a change in y than the associated (causative) change in x (cf. the enlargement factor of Section 3.2). Thus, if $dy/dx = 4{\cdot}7$, and an increase of $0{\cdot}1$ is made in the value of x, an increase of $(4{\cdot}7\times0{\cdot}1)$ in the value of y will result. Or if x changes by $0{\cdot}3$, y will change by $(4{\cdot}7\times0{\cdot}3)$; or if x is decreased by $0{\cdot}07$, the corresponding decrease in y will be $(4{\cdot}7\times0{\cdot}07)$. A negative value of the derivative or rate of change of y with respect to x indicates that an increase in x is associated with a decrease in y and vice versa.

The situation for other forms of relationship is discussed further in the next section.

Example 5.5

Two variables p and q are related by the equation $p = 5q^2+1/q$. Derive an expression for the instantaneous rate of change of p with respect to q, and find its value when (i) $q = 1$, (ii) $q = 2$.

The instantaneous rate of change of p with respect to q is given by the derivative,

and $$\frac{dp}{dq} = 10q-\frac{1}{q^2}.$$

When $q = 1$, $$\frac{dp}{dq} = 10-\frac{1}{1} = 9,$$

and when $q = 2$, $$\frac{dp}{dq} = 20-\frac{1}{4} = 19{\cdot}75.$$

Answer: The instantaneous rate of change of p with respect to q is $10q-1/q^2$: the values when $q = 1$ and $q = 2$ are 9 and 19·75 respectively.

EXERCISE 5c

Find the rates of change of y with respect to x for the following functions at the values indicated:

1. $y = (8-x)^2$; $x = 2$.

2. $y^2 = 8x$; $x = 2$.

3. $y = \sin x$; $x = 0{\cdot}5$.

4. $y = 4\cos x-5$; $x = -0{\cdot}5$.

5. $y = \sqrt{-1/\sqrt{x}}$; $x = 4$.

6. $y = (7x+1)^2$; $x = -0{\cdot}1$.

7. Write an expression for the rate of change of

(i) p with respect to q,
(ii) q with respect to p,

if the equation relating p and q is $p^2q = 2$.

8. The depth x of liquid in a vessel being filled with water varies with the time t according to the equation $x = 10t-5t^2+\frac{5}{3}t^3$, where x is in centimetres and t is in minutes. Find the rate of change of depth with respect to time when (i) $t = 0$, (ii) $t = \frac{1}{2}$, (iii) $t = 2$, giving your answer in cm/s.

9. What is the significance of a negative value of a rate of change of one variable with respect to another?

10. Derive expressions for the rate of increase of:

(i) the area of a circle with respect to the radius,
(ii) the circumference of a circle with respect to the radius,
(iii) the volume of a sphere with respect to the radius.

***11.** The period t of a simple pendulum of length L is given by the equation

$$t = 2\pi\sqrt{\frac{L}{g}},$$

where g is the acceleration due to gravity. Write down an expression for the rate of change of t with respect to (i) L, (ii) g.

Explain the meaning of the sign of your answers.

12. Derive expressions for the rate of change of

(i) the area of a circle with respect to its circumference,
(ii) the volume of a sphere with respect to its surface area.

13. The radius of a circle is increasing at a constant rate of p cm/s. Find the rate of increase of the area at the instant when the circumference is q cm long. (L, pt. Quest.)

***14.** If a body moves through a distance δx when acted on by a force F, the change δE in its potential energy is approximately $-F\,\delta x$. Explain why this is so, deduce the relation $F = -\mathrm{d}E/\mathrm{d}x$, and hence show that the force experienced by the body in any direction is equal but opposite to the rate of change of its potential energy with respect to that direction.

***15.** If the density of the atmosphere at a particular height h is ϱ, and the acceleration due to gravity g, show that the change in the atmospheric pressure δp, moving from the level h to a higher level $(h+\delta h)$, is given by the equation $\delta p = -g\varrho\delta h$. Hence show that the rate of change of atmospheric pressure with height is $-g\varrho$.

***16.** The number of nuclear disintegrations occurring in 1 second in a radioactive substance varies with time according to the relation

$$N = 10^6\left(1-\frac{t}{1!}+\frac{t^2}{2!}-\frac{t^3}{3!}+\frac{t^4}{4!}-\frac{t^5}{5!}\cdots\right).$$

Write down an expression for the rate of change for the number of disintegrations per second with respect to time, quoting the units of your answer, and show that $\mathrm{d}N/\mathrm{d}t = -N$. Interpret your result physically.

***17.** The electrical resistance R of a piece of wire varies with the temperature t according to the relationship

$$R = R_0(1+\alpha t+\beta t^2),$$

where R_0 = resistance at 0°C, and α and β are constants. What is the temperature coefficient of resistance (i.e. the rate of change of resistance of 1 ohm with respect to temperature) at (i) 0°C, (ii) 100°C, (iii) 500°C, (iv) 1000°C, if $\alpha = 4\times10^{-3}$, $\beta = 2\times10^{-3}$?

†**18.** Show that the rate of change of a variable x with respect to another variable y is equal to the reciprocal of the rate of change of y with respect to x. What assumptions do you have to make?

***19.** The potential energy of body when x cm away from a fixed point P is given by (i) k/x, (ii) k, (iii) kx, k being a constant. Find the force experienced by the body in each of the three cases, paying careful attention to the direction in which it acts (see question 14).

***20.** The energy of a parallel plate capacitor is given by the expressions $\frac{1}{2}CV^2$ or $Q^2/2C$, where C is the capacitance of the parallel plates, V the potential difference between them, and Q the charge on them. If $C = A/4\pi x$, where A is a constant and x the distance between the plates, find an expression for the force between the plates when Q is constant. (V is a variable under these conditions.)

If the capacitor is connected to a battery, Q varies but V remains constant. Find the force between the plates under these conditions, given that the energy of the battery is $(T-Q)V$, where T is a constant.

Are the forces ones of attraction or repulsion?

***21.** The current I flowing "across" a capacitor of value C is equal to the time rate of change of the charge Q on its plates. If the charge on the plates of the capacitor is C times the applied voltage V, find an expression for the current in the circuit when an alternating voltage $E_0 \cos 2\pi ft$ is applied to the capacitor. (f is the frequency of alternation of the applied voltage.) Discuss the meaning of the statements:

(i) the current "leads" the voltage;
(ii) the impedance of a capacitor is inversely proportional to the frequency of the applied voltage;
(iii) a capacitor "passes" a high-frequency voltage but "blocks" a low-frequency one.

5.5. SMALL INCREMENTS

The distinction between the average rate of change of one variable with respect to another over an interval and the instantaneous rate of change at one particular value has been made in Section 3.3. The problem of calculating the small change produced in the dependent variable by a small change in the independent variable frequently arises, and if we know the (non-linear) algebraic form of the relation between the two variables, it is often quickest to use a method which gives only an approximate answer. (If the relation between the variables is linear we can proceed as suggested on p. 120 and obtain an exact answer.) If the change in the independent variable is small, the difference between the approximate answer and the accurate answer will often be negligible, and the economy in effort more than makes up for the loss in accuracy in the answer. Again, in problems involving physical variables, any numbers quoted to more than 3 or 4 figures are often meaningless in that they suggest an accuracy greater than that with which most of the variables can be measured.

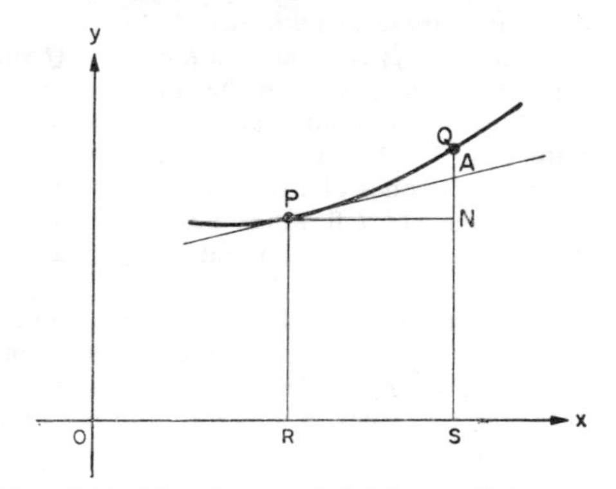

FIG. 5.11. The change of RS in x will lead to a change in QN in y. If PA is the tangent at P, and if RS is small, $QN \simeq AN$.

Combining Figs. 3.1 and 3.18 in Fig. 5.11, we can see that if a small change RS is made in the independent variable, a change of NQ will be produced in the dependent variable. Now if the tangent at P is constructed, it will cut the ordinate SQ at A as shown; and if RS is very small, AQ will in general be small and less than NQ. We can suppose, in fact, that NA is very nearly equal to NQ. But

$$NA = PN.\tan A\hat{P}N,$$

and the tangent of the angle APN is the gradient of the curve at P. Since the gradient of the curve is given by the value of $\mathrm{d}y/\mathrm{d}x$ at that point, we can see that

$$NA = PN.\frac{\mathrm{d}y}{\mathrm{d}x}. \qquad 5.1)$$

Denoting the length RS or PN by δx, and NQ by δy, we can make the approximation that $NA \simeq \delta y$, and the relation

$$\delta y \simeq \frac{\mathrm{d}y}{\mathrm{d}x}.\delta x$$

is obtained.*

To see how this can be used in a numerical problem, consider what happens when a metal disc is heated through a small temperature change and the radius of the disc changes from 2 cm to 2·0001 cm. Since the area of a circular disc given by the relationship $A = \pi r^2$, where A is the area of the disc and r is its radius, we can see that the area will, as a result of the expansion, change by a small amount. Using the approximation

$$\delta A \simeq \frac{\mathrm{d}A}{\mathrm{d}r}\cdot\delta r,$$

we can, knowing the value of δr, obtain an approximate value for δA by deriving an expression for $\mathrm{d}A/\mathrm{d}r$.

Now $\qquad \mathrm{d}A/\mathrm{d}r = \pi.2r,$

so $\qquad \delta A \simeq 2\pi r.\delta r.$

Thus if $r = 2$ cm (little error is introduced by setting the variable equal to the more convenient number whether this occurs before or after the change; and, in any case, in this problem the tangent PA was drawn from the

* It will be realised that this approximation expresses the closeness of value between an incremental rate of change (see p. 85) and the derivative.

earlier point P on the curve of Fig. 5.11, and the value of the abscissa at P is 2), and $\delta r = 0{\cdot}0001$ cm,

$$\delta A \simeq 2\pi . 2(0{\cdot}0001)$$
$$= 0{\cdot}0004\pi \text{ cm}^2.$$

Another application of eqn. (5.1) arises in finding the approximate value of the root α of an equation $f(x) = 0$. Suppose there is a value x_1 for which $f(x_1)$ is very small. By writing

$$\frac{f(\alpha)-f(x_1)}{\alpha - x_1} \simeq f'(x_1),$$

and using the fact that $f(\alpha) = 0$, we obtain the equation

$$-f(x_1) \simeq (\alpha - x_1)f'(x_1),$$

i.e.

$$-\frac{f(x_1)}{f'(x_1)} \simeq \alpha - x_1,$$

and

$$\alpha \simeq x_1 - \frac{f(x_1)}{f'(x_1)}.$$

This approximation to α will in general be better than x_1, and we can calculate a value for a second approximation x_2 using the equation

$$x_2 = x_1 - \frac{f(x_1)}{f'(x_1)}.$$

Repeating the procedure, we can then make an even better estimation of α, using the relation

$$x_3 = x_2 - \frac{f(x_2)}{f'(x_2)},$$

and by continuing as long as patience and perseverance last, we can obtain a solution to the equation $f(x) = 0$ to any desired degree of accuracy.*

* To obtain a solution correct to, say, three significant figures, the guesses x_n are quoted and used with three figures, and the process represented by the equation

$$x_{n+1} = x_n - \frac{f(x_n)}{f'(x_n)}$$

is repeated for successive values of n until the three figures obtained for x_{n+1} are the same as those of x_n. This final common figure can then be taken as the required root correct to three significant figures.

It should be realised that this method can, under certain circumstances, break down. To improve on the first approximation by using the relation $x_2 = x_1 - f(x_1)/f'(x_1)$ demands that the derivative of $f(x)$ does not change too rapidly in the neighbourhood of x_1, that $f'(x)$ does not equal zero at x_1 or between x_1 and the desired root α, and that $f(x)$ has no inflexion between x_1 and α. In graphical terms, x_2 can be seen to be the intersection of the tangent at x_1 with the x-axis (Fig. 5.12), and if, for example, the

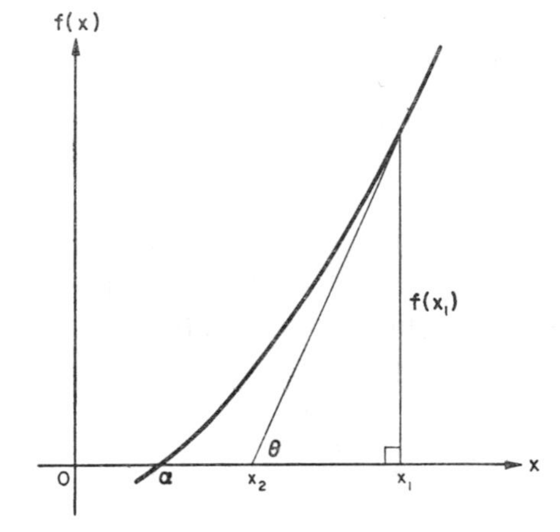

FIG. 5.12.

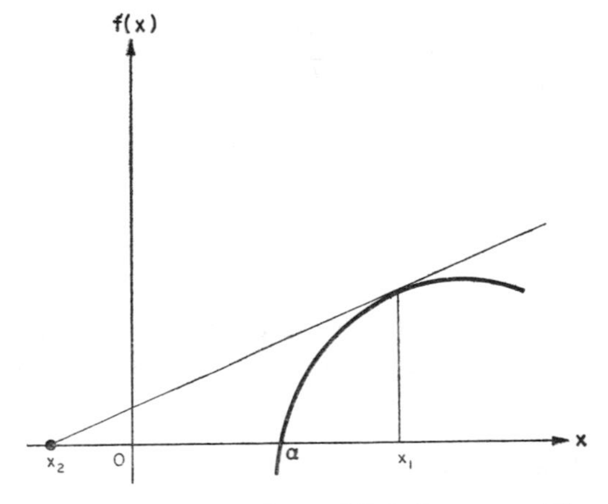

FIG. 5.13. If the gradient of $f(x)$ changes rapidly in the neighbourhood of x_1, the value obtained for x_2 may be further from α than x_1.

gradient of $f(x)$ at x_1 is markedly different from the gradient in the vicinity of the root α, this intersection may well be further away from α than x_1 (Fig. 5.13). To improve on the initial guess in this case requires a better choice for the starting-point, and provided such a choice is made, it can be shown that the sequence $x_1, x_2, x_3, \ldots, x_n \ldots$ converges towards the required root α as n increases without bound. The method is often referred to as the Newton–Raphson method.

Example 5.6

The equation $x^3-3x-1 = 0$ has a root near 2: find its value to one place of decimals.

Setting $f(x) = x^3-3x-1$, we can write that $f'(x) = 3x^2-3$ or $3(x^2-1)$. Then using $x_1 = 2$ as a first approximation, we obtain our second approximation as

$$x_2 = 2-\frac{(8-6-1)}{3(4-1)} = 1{\cdot}89.$$

Then

$$x_3 = 1{\cdot}89-\frac{(6{\cdot}75-5{\cdot}67-1)}{3(3{\cdot}57-1)}$$

$$= 1{\cdot}89-\frac{0{\cdot}08}{7{\cdot}71}$$

$$= 1{\cdot}88,$$

and

$$x_4 = 1{\cdot}88-\frac{(6{\cdot}65-5{\cdot}64-1)}{3(3{\cdot}54-1)}$$

$$= 1{\cdot}88.$$

Since the last two iterations give the same answer, we can write that the root is 1·9.

EXERCISE 5d

1. The radius of a circle expands on heating from 8 to 8·01 cm. What is the approximate increase in area?

2. The radius of an iron sphere expands from 2 to 2·007 cm on heating. What is the approximate increase in volume?

3. Find approximate expressions for δy in terms of x and δx if: (i) $y = \pi x^2$, (ii) $y = \sin x$, (iii) $y = 1/x$, (iv) $y = 2x(x-1)$.

4. Variables p and q are related by the following equations: find the value of δp for each relationship if $\delta q = 0{\cdot}05$ and $q = 1$. (i) $p = (q+1)^2$, (ii) $p = 1/q^2$, (iii) $p = \sqrt{q}$, (iv) $p = 2 \cos q$.

5. The population P of a country increases with time so that T years after the population is P_0, the value of P is given by the equation $P = P_0(1+0{\cdot}05T+ +0{\cdot}0025T^2)$. Find an approximate value for the increase in population in the 10th year after it is 50 million.

6. The radius of a circle increases from r to $(r+\delta r)$. Show that the increase in the circumference C is given by the exact equation $\delta C=(\mathrm{d}C/\mathrm{d}r)\delta r$, and state under what general condition the equation $\delta y=(\mathrm{d}y/\mathrm{d}x)\delta x$ would be correct.

7. The displacement of a body moving under gravity is given by the equation $s = 78t-5t^2$. Find approximately the distance travelled between $t = 1$ and $t = 1{\cdot}1$.

8. Determine the percentage inaccuracy caused by using the equation $\delta y = (\mathrm{d}y/\mathrm{d}x)\delta x$ if $y = x^2$, $x = 2$ and (i) $\delta x = 0{\cdot}5$, (ii) $\delta x = 0{\cdot}1$, (iii) $\delta x = 0{\cdot}01$.

***9.** Show that the rate of change of the period t of a simple pendulum with respect to its length L is $2\pi^2/gt$. Hence, find an approximate value for the change in the period of a 1-second pendulum if its length increases by 1 mm ($g = 9{\cdot}80$ m/s²). (The period of a 1-second pendulum is 2 seconds.)

***10.** Show that the change δt in the period t of a simple pendulum of length L caused by a change of δg in the acceleration due to gravity is given by the equation

$$\frac{\delta t}{t} \doteq -\frac{\delta g}{2g}.$$

Hence find the percentage change in t if g changes by (i) 1%, (ii) 0·01%.

11. Two variables x and y are related by the equation $y = ax^n$, where a and n are constants. Find an approximate expression for the change δy in y when x changes by δx, and hence show that

$$\frac{\delta y}{y} \doteq n\frac{\delta x}{x}.$$

If x changes by P%, by how many per cent does y change? Will this change in y be in the same direction as that in x?

12. A function $\psi(x)$ can be represented by the expression $(1+\frac{1}{2}x-\frac{1}{8}x^2)$ if x is small. If x changes from 0·005 to 0·008, find how much the value of $\psi(x)$ changes.

13. Find approximate values for the square roots of 25·1, 49·2 and 226 without using square root tables.

14. Find an approximate value for $\sqrt[3]{125{\cdot}8}$.

15. The equation $x^3-5x^2-16x+20 = 0$ has a root near -3. Find its value correct to one place of decimals.

16. The equation $x^3-6x^2-13x+18 = 0$ has a root near -2. Find its approximate value using the Newton–Raphson method.

17. Find an approximate root of the equation $x^4-4x-1 = 0$ using the Newton–Raphson method, and starting with $x_1 = 0$.

18. "The use of the equation $x_2 = x_1-f(x_1)/f'(x_1)$ for improving on a guess for the solution of $f(x) = 0$ is equivalent to drawing the tangent to $y = f(x)$ at x_1, and setting x_2 equal to the value of x where the tangent cuts the x-axis." Explain with reference to Fig. 5.12 why

$$\text{(i)}\quad \cot\theta = \frac{1}{f'(x_1)},$$

$$\text{(ii)}\quad x_1-x_2 = \frac{f(x)_1}{f'(x_1)},$$

and deduce the validity of the above statement for a function whose graph is like that shown.

If an initial guess of 1 is taken for a root of the equation $2x^3-6x+1 = 0$, use of the equation

$$x_2 = x_1-\frac{f(x_1)}{f'(x_1)}$$

is not feasible. Explain why this is so, and explain graphically what feature about the point whose abscissa is 1 makes the method impracticable. Make another guess for x_1, and find an approximate value for a root of the equation $2x^3-6x+1 = 0$ near $x = 0$.

19. Explain why the value of $f(a+1)-f(a)$ is a good approximation to $f'(a)$ if a is large.

The square root of 985 is 31·385; that of 986 is 31·401. By letting $f(x) = \sqrt{x}$, and setting x equal to 985, deduce that an approximate value of $1/(2\sqrt{x})$ for $x = 985$ is 0·016.

20. Solve the equation $x = \cos x$ using the Newton–Raphson method, obtaining your answer correct to three significant figures.

21. Explain the meaning of the equation $f(a+h)-f(a) \simeq hf'(a)$. Given that $\sqrt{63{\cdot}2} = 7{\cdot}950$, and that $\sqrt{63{\cdot}3} = 7{\cdot}956$, deduce that an approximate value of $1/(2\sqrt{x})$ for $x = 62{\cdot}2$ is 0·06.

22. A spherical soap bubble has a radius r and a volume V. If r is subject to a slight variation, show that the percentage increase in r is approximately one-third of the percentage increase in V. If r increases from 1 cm to 1·03 cm, find, to two significant figures, the increase in V. (NUJMB)

23. Draw two separate graphs to show the approximate positions of x_1, x_2 and x_3, given that

$$x_1 = 3,\quad x_2 = x_1-\frac{f(x_1)}{f'(x_1)},\quad x_3 = x_2-\frac{f(x_2)}{f'(x_2)},$$

and $\qquad f(x) \equiv x^2-3x-4.$

On another graph, show the approximate positions of x_2 and x_3', where x_2 is defined in the same manner as before, but

$$x_3' = x_2-\frac{f(x_2)}{f'(x_1)}.$$

Discuss the difference in the equations for x_3. Which do you think is the more (i) rapidly convergent, (ii) convenient?

24. A sphere is heated slightly and its volume increases by $p\%$. Show that the approximate increase in its radius is $\frac{1}{3}p\%$.

By what approximate percentage does its surface area increase?

***25.** The time of oscillation of a simple pendulum of length l is $2\pi\sqrt{(l/g)}$, where g is the acceleration due to gravity. If l changes by $p\%$, show that the time of oscillation changes by approximately $\frac{1}{2}p\%$.

†26. Explain, using a graph, why the Newton–Raphson method can break down if an initial guess of x_1 is made for the root α of $f(x) = 0$, and

(i) $f'(x_1) = 0$,

(ii) $f''(x) = 0$ between x_1 and α, and $f'(x_1) = f'(x_2)$.

5.6. GEOMETRICAL APPLICATIONS

In the realms of geometry, differentiation has a ready application in finding the equations of tangents and normals to curves whose equations are known. That the equation of a line of gradient m and y-intercept c is $y = mx+c$ should by now be well known, and if we are finding the equation of the tangent touching a particular curve at (x_1, y_1), we can also write that $y_1 = mx_1+c$, since the equation of the tangent must be satisfied at (x_1, y_1). Subtracting these two equations, we obtain the useful form

$$y-y_1 = m(x-x_1)$$

for the equation of a line of gradient m passing through the point (x_1, y_1).

The gradient function of the curve in question can be found by differentiation, and the value of m then obtained by substitution of x_1 or y_1, or both, into the gradient function expression (Fig. 5.14).

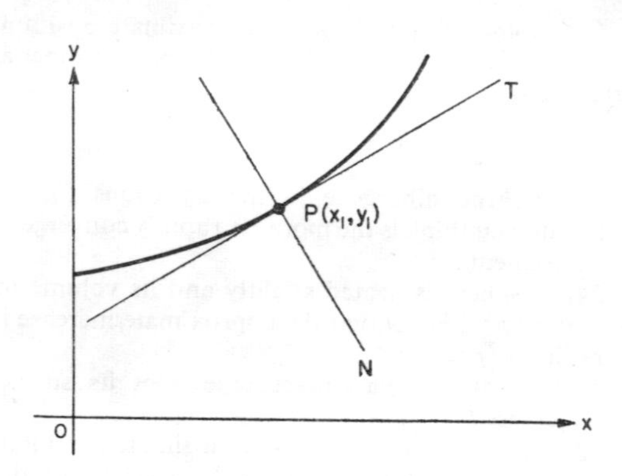

FIG. 5.14. The equation of the tangent PT of gradient m passing through (x_1, y_1) is $y-y_1 = m(x-x_1)$; the equation of the normal PN is $y-y_1 = -(1/m)(x-x_1)$.

The equation of the normal can easily be obtained as well. The normal is perpendicular to the curve at the point in question, and thus the value of m in this case is the reciprocal of the gradient of the curve at (x_1, y_1) with a negative sign, since for two perpendicular lines, $m_1 = -1/m_2$.

Example 5.7

Find the equations of the tangent and normal to the curve $y = 2x^2-3x+7$ at the point (2, 9).

Differentiating the curve equation, we obtain the gradient function

$$\frac{dy}{dx} = 4x-3.$$

Substituting the value $x = 2$ into this equation, we obtain the gradient of the curve at (2, 9) as $\{4(2)-3\}$ or 5. The equation of the tangent is thus

$$y-9 = 5(x-2)$$

or

$$5x-y-1 = 0.$$

The gradient of the normal will be $-1/5$: and thus its equation is

$$y-9 = -\frac{1}{5}(x-2),$$

or

$$x+5y-47 = 0.$$

It will be noticed that the coefficients of x and y in the equation of the normal are the same as those in the equation of the tangent, but interchanged and with their relative signs reversed. This is always the case, for the lines $ax+by+c = 0$ and $bx-ay+d = 0$ are perpendicular whatever the values of a and b. Knowing this, we can devise an alternative and possibly quicker method for finding the equation of the normal to a curve at any point once the equation of the tangent at that point is known.

Thus taking the example given above, we can find the equation of the tangent in the usual way. Then knowing it to be $5x-y-1 = 0$ at (2, 9), we can immediately say that the equation of the normal at that point will have form $x+5y+d = 0$, where d is a constant yet to be determined. The value of d is most easily found by using the fact that the equation must be satisfied by the coordinates (2, 9). Thus,

$$2+5.9+d = 0,$$

whence

$$d = -47.$$

The equation of the normal is therefore $x+5y-47 = 0$.

Example 5.8

Find the equations of the tangent and the normal to the curve $y = x^3-6x^2+11x-6$ at (1, 0), and find the angle between the curve and the tangent at the point where they meet again.

The gradient function of the curve is $3x^2-12x+11$, and thus the gradient at (1,0) is 2. The tangent therefore has an equation of the form $2x-y+c = 0$, and the normal $x+2y+d = 0$. Both these lines pass through the point (1, 0), and therefore $c = -2$ and $d = -1$. The required equations are thus

$$2x - y-2 = 0 \text{ (tangent)},$$

and

$$x+2y-1 = 0 \text{ (normal)}.$$

The tangent meets the curve again at the point whose coordinates satisfy the equations $2x-y-2 = 0$ and $y=x^3-6x^2+11x-6$ simultaneously. Substituting for y in the second equation, we obtain

$$x^3-6x^2+9x-4=0$$

as the equation for the abscissae of the intersections. Knowing one solution already ($x = 1$, the point of contact), we can factorise the equation in the form $(x-1)(ax^2+bx+c)=0$. Evaluating a, b, and c mentally gives

$$(x-1)(x^2-5x+4) = 0,$$

whence

$$(x-1)(x-4)(x-1) = 0$$

and

$$x = 1 \quad \text{or} \quad 4.$$

Thus the tangent and curve meet again at the point whose abscissa is 4, and whose ordinate (from the tangent equation $2x-y-2 = 0$) is therefore 6.

The gradient of $y = x^3-6x^2+11x-6$ at (4, 6) is, from the gradient function, 11. Now the angle between lines of gradient m_1 and m_2 is $\tan^{-1}\{(m_1-m_2)/(1+m_1m_2)\}$, and substituting for m_1 and m_2 gives the required answer $\tan^{-1}(9/23)$.

EXERCISE 5e

Find the equations of the tangents and normals to the following curves at the points indicated:

1. $y = 3x^2-4x+1$; (1,0)

2. $y = 7-\sqrt{x}$; (4, 5)

3. $y = 3x-1/2$; $x = 1$.

4. $y = \sin x$; $x = 7\pi/6$.

5. $y = 7(3-x)^6$; $x = -2$.

6. $xy^2 = 8$; $y = 1$.

7. $3x^2+4y-9 = 0$; $x = -3$.

8. Find the equation of the normal to the curve $100y = x^4$ at the point (5, 25/4). (OC)

9. Show that the equation of the tangent to the parabola $y^2=4ax$ at the point $(at^2, 2at)$ is $x-ty+at^2=0$. Show also that the normal to the parabola at the same point has the equation $y+tx = 2at+at^3$.

10. Find the equation of the tangent to the curve $y = 36/x$ at the point (4, 9). Prove also that the point (4, 9) is the mid-point of that part of this tangent cut off by the axes.

If $(6c, 6/c)$ are the coordinates of the point on this curve at which the slope of the tangent is -4, find a positive value for c, and find also the equation of this tangent. (OC)

11. The coordinates of a point P on the curve $y=x^2$ are (3, 9). Find the equation of the tangent at P.

If this tangent intersects the y-axis at T, prove that the x-axis bisects PT.

Prove that the same result follows for the tangent at a point (c, c^2), for any value of c. (OC)

12. Prove that the two curves $y = x^2/2$ and $y=4x^{\frac{1}{2}}$ meet at the point (4,8). Find the equations of their tangents at this point; and find also the angle between these tangents, correct to the nearest minute. (OC)

13. The curve whose equation is $y = x^3+ax^2+b$ passes through the point (1, 2) and has a gradient 1 at that point. Find the values of a and b and the points where the gradient is 8. Find also the equation of the tangent to the curve at the point where $x = 2$, and find where it cuts the tangent to the curve at the point (1, 2). (SU)

14. An expression of the second degree is denoted by $f(x)$. If $f(1) = 7$, $f(2) = 23$, $f(3) = 17$, find the gradient of the graph of $f(x)$ if $x = 2$. (OC)

15. A curve whose equation is $y = x^3+ax^2+bx+1$ touches the x-axis where $x = 1$. Find the values of a and b; and verify that the curve passes through the point (2, 3).

If the curve crosses the x-axis at A and the y-axis at B, find the equations of the tangents to the curve at A and B. (SU)

†**16.** Find the equation of the tangent to the curve at the point where $x = 3$.

Find the coordinates of the point at which this tangent meets the curve again and prove that it is a normal to the curve at this point. (OC)

5.7. TURNING POINTS AND STATIONARY VALUES

5.7.1. *Minima.* Another important application of differentiation arises in problems involving functions whose values pass through a maximum or minimum. To take an example, we might consider the design of a cylindrical can which is to hold a given volume of soup. Clearly a long, thin can is inconvenient to handle and uses too much metal: and the same criticisms apply to a very wide flat one. Somewhere in between lies a set of dimensions which, for the given volume of soup, uses a minimum quantity of metal.

Expressed more mathematically, the amount of metal used to make the can varies with the diameter of the can. We can guess that the graph showing the variation in the amount of metal used will look like that of Fig. 5.15. By constructing an expression for the relationship between the surface area of the can and its diameter, the graph can be plotted accurately, and the minimum point found. The diameter for which the volume of metal is a minimum is then the best to employ. But this procedure is a long one, and the final answer will be limited by the accuracy with which the graph can be constructed and drawn.

The characteristic of the point on the graph we are interested in is that the gradient there is zero. By finding the equation of the variation depicted by the curve in Fig. 5.15, we can easily find the abscissa of the point for which the gradient is zero, and without drawing the graph we shall then have found the required diameter exactly.

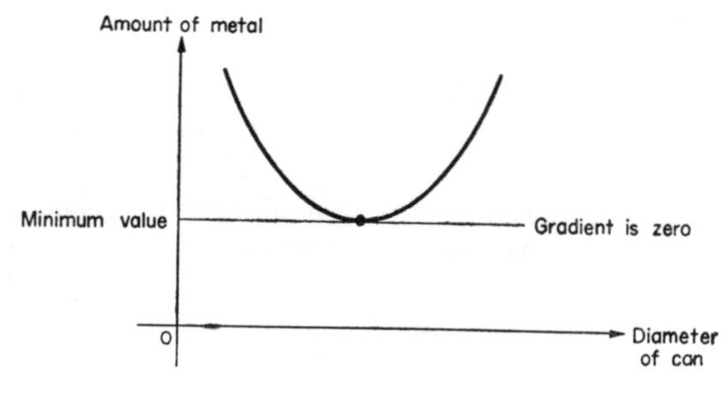

FIG. 5.15. The gradient of a graph at the minimum point is zero.

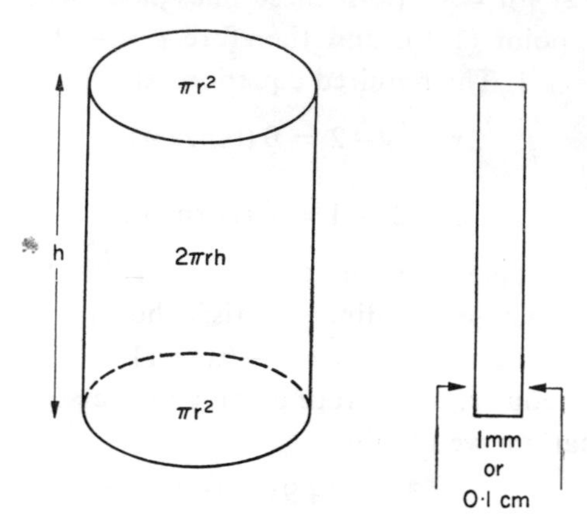

FIG. 5.16.

Thus, specifically, if the radius of the can is r cm, the height h cm and the thickness of the metal 1 mm (Fig. 5.16), the volume of metal V used in making the can is equal to

$$(2\pi r^2 + 2\pi rh)(0{\cdot}1) \text{ cm}^3 .$$

This equation for V involves two variables h and r, and to express V as a function of r only means that h must be replaced. The variables h and r are related since the amount of soup to go into the can has been decided: if this amount is, for example, 200 cm³, the equation relating h to r can be obtained by equating 200 and the volume of the can. Thus

$$200 = \pi r^2 h,$$

and therefore

$$h = 200/\pi r^2.$$

We can then write V as a function of r only:

$$V = \left(2\pi r^2 + \frac{2\pi r \,.\, 200}{\pi r^2}\right)(0{\cdot}1)$$

$$V = 0{\cdot}2\pi r^2 + \frac{40}{r}.$$

This is the equation of the curve in Fig. 5.15, and the gradient function is obtained by differentiation:

$$\frac{dV}{dr} = 0{\cdot}4\pi r - \frac{40}{r^2}.$$

Now this gives a zero gradient when

$$0 = 0{\cdot}4\pi r - \frac{40}{r^2},$$

i.e. $$0{\cdot}4\pi r^3 = 40$$

or $$r = \sqrt[3]{\frac{40}{0{\cdot}4\pi}} = \sqrt[3]{\frac{100}{\pi}}\ \text{cm}.$$

The radius giving the least volume of metal can thus be calculated immediately.

Example 5.9

The expression x^2+4x-1 has a minimum value. Find the value of x for which this occurs, and find the minimum value.

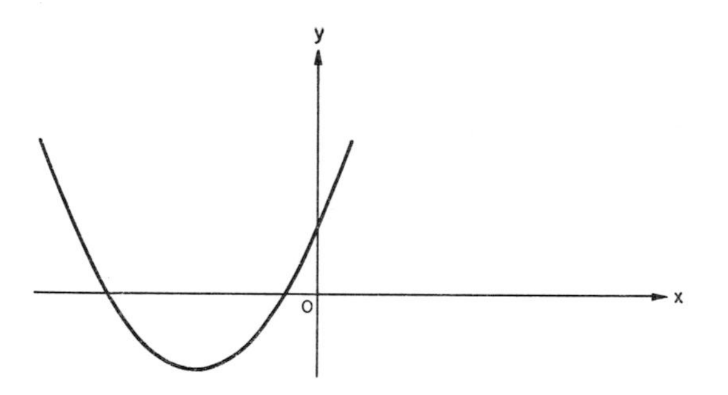

FIG. 5.17.

Introducing the letter y for the expression gives the equation $y = x^2+4x-1$ for the curve shown in Fig. 5.17. The gradient function dy/dx is $2x+4$, and this has a zero value when $2x+4 = 0$, or $x = -2$.

Thus the expression x^2+4x-1 has a minimum value when or $x = -2$.

Substituting into the original expression, the minimum value is obtained as

$$\{(-2)^2+4(-2)-1\} \text{ or } -5.$$

EXERCISE 5f

1. Calculate the diameter of a cylindrical container which will hold the following amounts of liquid and require the least amount of metal in its manufacture:

(i) contents 400 cm^3, thickness of metal 1 mm,
(ii) contents 500 cm^3, thickness of metal t cm,
(iii) contents V cm^3, thickness of metal p cm.

2. The following expressions have minimum values; find the values of x for which they occur:

(a) x^2+6x-7, (c) x^2-4x+5,
(b) x^2+3x+1, (d) x^2-8x.

Can you infer the value of x for which the expression (x^2+ax+b) has a minimum value?

3. Find the value of p for which the following expressions have a minimum value; and evaluate these minimum values:

(a) $2p^2+3p-5$, (c) $5p^2-2p+17$,
(b) $4p^2-8p+1$, (d) p^4-7p^2+2.

4. Find the position of the minimum in the graphs of

(i) $y = x+1/x$, for $x > 0$,
(ii) $y = x^2-3x+2$,

and sketch their forms.

†**5.** What can you say about the value of d^2y/dx^2 at a minimum point?

6. A closed rectangular box of volume 72 dm^3 is to be constructed with its length double its breadth. Find the least possible surface area of the box. (C)

7. Establish the identity $x^2+2x+5 \equiv (x+1)^2+4$, and hence deduce algebraically that the minimum value of x^2+2x+5 is 4.

8. Write down the least values of x^2+4x+1, x^2+8x+8 and x^2-6x-4, using only algebraic methods.

9. The cost S new pence and the time t minutes of manufacture of a certain article are connected by the formula

$$S = \frac{16}{t^3}+\frac{3t^2}{4}.$$

Find (i) the rate of change of cost when $t = 4$,
(ii) the minimum cost. (NUJMB)

†**10.** The focus S of the parabola $y^2=4ax$ is $(a, 0)$. Show that the distance from the focus to any point $P(x, y)$ on the parabola is $(x+a)$.

Sketch a graph showing how the length of SP varies with the x coordinate of P. Is this graph continuous at $x = 0$? Does it possess a derivative at $x = 0$?

The value of SP goes through a maximum or minimum at $x = 0$. Is its turning point at $x = 0$ a maximum or a minimum, and why does $d(SP)/dx$ not equal 0 at this point?

5.7.2. *Maxima*. A curve such as that shown in Fig. 5.18 is said to go through a maximum value at the point P. One characteristic of

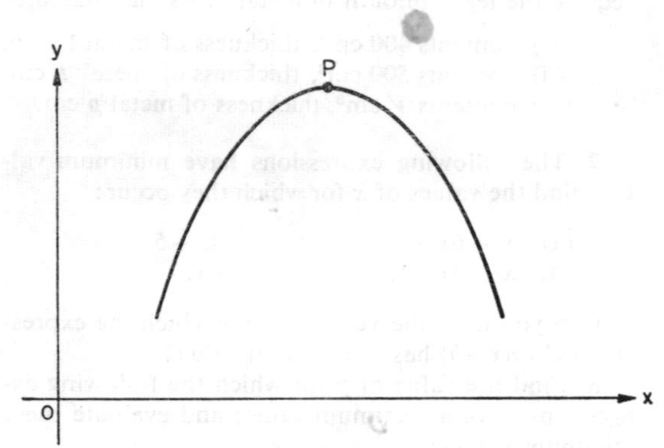

FIG. 5.18. P represents the maximum value.

this point is that the gradient is zero, since a tangent to the curve constructed at this point is parallel to the x-axis. Thus this characteristic of a maximum value is the same as that for a minimum value discussed above. If we know whether we are involved with a maximum or a minimum value problem, the fact that the gradient at both maximum and minimum points is zero leads to no confusion,

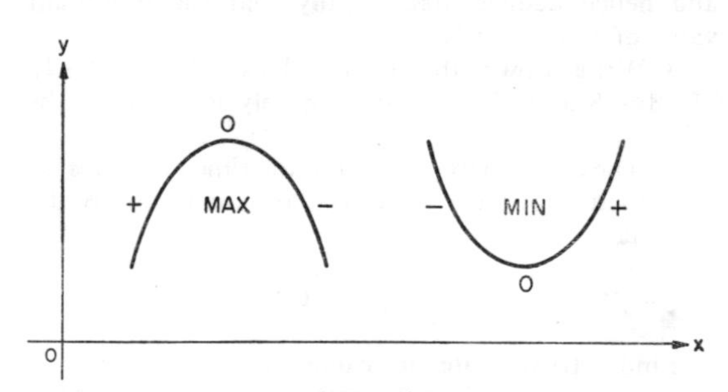

FIG. 5.19. The gradient of a curve changes sign at a maximum or a minimum point.

but if we have to decide analytically whether the point at which the gradient is zero—the so-called turning point—is a maximum or minimum, we need some further theory.

There are in fact two procedures of general application. Both follow from the observation that the gradient of any curve changes sign about a turning point (Fig. 5.19). Thus about a minimum point, the gradient of the curve changes from a negative value through zero (at the minimum point) to a positive value as x increases, and, similarly, about a maximum point, the gradient changes from positive to negative as the curve passes through the maximum point. Evaluation of the sign of the gradient of the curve about the turning point can thus be used to decide which of the two situations exists.

Example 5.10

The curve $y = x^3-3x^2-9x+7$ has two turning points. Find their abscissae, and determine whether they are maximum or minimum values.

The gradient of the curve is given by $dy/dx = 3x^2-6x-9$, and when this is equal to zero, the curve has a turning point. Thus the turning points occur when

$$3x^2-6x-9 = 0,$$

or $$3(3-x)(x+1) = 0,$$

i.e. when $x = 3$ or -1.

We now rewrite the gradient function in its factorised form (which is more convenient for the next stage of the calculation): i.e. we rewrite $(3x^2-6x-9)$ as $3(x-3)(x+1)$. We can then see that:

(i) when x is just less than 3, the gradient is negative, since of the three factors 3, $(x-3)$ and $(x+1)$, the first and last are positive, and the middle one negative;
(ii) when $x = 3$, the gradient is zero as found already;
(iii) when x is just greater than 3, the gradient is positive since all three factors are positive.

Thus diagrammatically

Value of x:	$x < 3$	$x = 3$	$x > 3$
Gradient:	\	—	/

Being careful to arrange the values of x in increasing magnitude, the fact that the turning point at $x = 3$ is a *minimum* becomes apparent from the lower line sketch of the gradient variation.

Similarly, when x is just less than -1 (for example, equal to $-1 \cdot 1$), the gradient is positive, since the first factor is positive, and the last two negative;

when $x = -1$, the gradient is zero,

when x is just greater than -1 (for example, $x = -0.9$), the gradient is negative, since the first and third factors of the gradient function are positive, and the middle factor is negative. Thus the summary in tabular form is

Value of x:	$x < -1$	$x = -1$	$x > -1$
Gradient:	/	—	\

and the turning point at $x = -1$ is a *maximum*.

Example 5.11

Investigate the nature of the turning points in the graph of the function $x^{\frac{1}{2}} + x^{-\frac{1}{2}}$.

The gradient function is given by the derivative $\frac{1}{2}x^{-\frac{1}{2}} - \frac{1}{2}x^{-\frac{3}{2}}$, and this has a zero value when $(x-1)/2x\sqrt{x} = 0$, i.e. $x = 1$.

Writing the gradient function in the form $(x-1)/2x\sqrt{x}$, the gradient can be seen to be < 0 and > 0 when $x < 1$ and $x > 1$, respectively. Thus:

x value	$x < 1$	$x = 1$	$x > 1$
Gradient:	\	—	/

and thus the turning point at $x = 1$ is a minimum.

An alternative method which avoids determining the sign of the gradient either side of $x = 1$ is applicable here. This method makes use of the curve sketching approach. If x is small, $1/\sqrt{x}$ will be large, and the value of $1/\sqrt{x}$ will increase as x decreases. If x is very large, $\sqrt{x}$ will also be very large, and thus the function increases to large values as $x \to 0$ as well as when x increases to large values. An intermediate turning point must therefore be a minimum.

Exercise 5g

1. Show that the turning points in the curves

(i) $y = -3x^2$, and
(ii) $y = -ax^2$ (where $a > 0$),

are maxima.

2. Show that the turning point of the curve $y = ax^2$ is a maximum if $a < 0$, and a minimum if $a > 0$.

3. Describe the position and nature of the turning points of the curves

(i) $y = 2x^2 + 4x - 1$,
(ii) $y = 3 - 2x - x^2$.

4. Show that the nature of the turning-point of the curve $y = ax^2 + 2bx + c$ depends only on the sign of a.

5. Investigate the turning-points of the functions

(i) $x^3 - 3x^2$, (ii) $x^3 - 9x^2 + 24x$,
(iii) $x^3 - 5x^2 + 3x - 18$.

†**6.** The probability density of a statistic x, normally distributed with mean $\bar{x}$ and standard deviation 1, is $e^{-(x-\bar{x})^2/2}$. Given that

$$\frac{d}{dx}(e^{\phi}) = \phi' e^{\phi},$$

where ϕ is any continuous function of x, show that the maximum probability density occurs at $x = \bar{x}$. Given also that $y^n e^{-y^2}$ converges to zero as $y \to \infty$, sketch the graph of $e^{-(x-\bar{x})^2/2}$.

7. Using a scale for x of 2 cm for $\pi/3$ radians and a scale for y of 2 cm for $\frac{1}{2}$ unit, draw the graph of $y = \sin x$ in the range $0 \leqslant x \leqslant 2\pi$.

By drawing suitable straight lines on the graph, determine the maximum and minimum values of $(2 \sin x - 1)/2x$ for $0 \leqslant x \leqslant 2\pi$.

8. A piece of wire, 150 cm long, is cut into two parts; one part is bent to form a square, the other part to form a circle. Find the lengths of the two parts when the sum of the two areas formed is least. (Take π as 3·142.) (SU)

5.7.3. *Use of the second derivative.* The second method of determining the nature of the turning point is to find the value of the second derivative at the point in question. It was explained in Section 4.8 that the second-order derivative of the equation for a curve gives the expression for the rate of change of the gradient of the curve. Now if the gradient is increasing as x increases, the value of the rate of change of the gradient of the curve with respect to x will in general be positive. This is the case for a minimum point, for the gradient changes from negative through zero to positive going through the point (see Fig. 5.19).

Similarly, the decreasing gradient of the curve going through a maximum point leads to a negative value of the rate of change of the gradient, and the value of the second-order derivative at the maximum point will in general be less than zero. Summarising these results, we have that, in general,

at a MAXIMUM	$dy/dx = 0$	$d^2y/dx^2 < 0$
at a MINIMUM	dy/dx* $= 0$	$d^2y/dx^2 > 0$

It occasionally happens that the value of the second-order derivative at a point is zero; this does not give conclusive identification to the nature of the behaviour of the curve at the point. In these cases it is better to examine the change in sign of dy/dx as x passes through the point considered. This method is also preferable, of course, when determination of the second-order derivative is algebraically inconvenient, as often happens, for example, when square roots are involved.

Mention must finally be made of the exact nature of these so-called "turning points". The sense in which the terms maximum and minimum have been used has implied only a "relative" maximum or minimum: that is to say, the value of the function at the turning point concerned is greater than or less than the value at any points close to it. Nevertheless, the value of the function may easily exceed that value at a point far removed from the turning point—as it does at the point Q with the maximum at P in Fig. 5.20. The term "absolute"

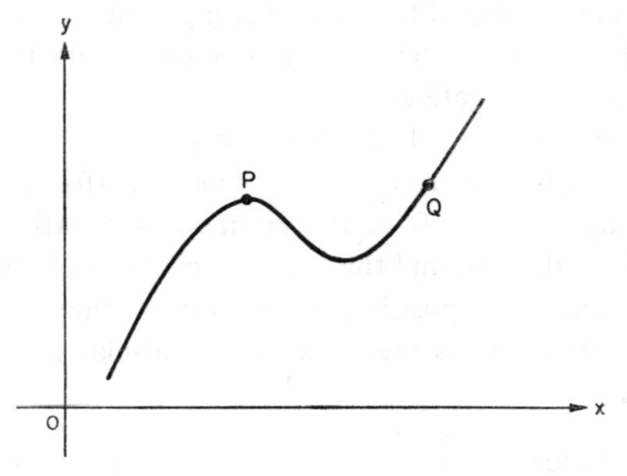

FIG. 5.20.

maximum is used to indicate a value which is exceeded nowhere else (e.g. at P in Fig. 5.18), and similarly the value at an absolute minimum is the lowest value the function ever attains (Fig. 5.21). Often, however, the distinction is not important and the words maximum and minimum values are used loosely to mean either sort.

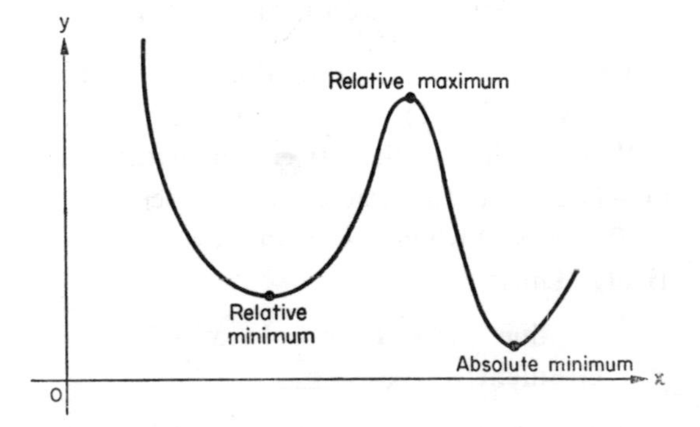

FIG. 5.21. A relative maximum or minimum is only a local maximum or minimum.

* This assumes that the derivative exists at the turning point (see Exercise 5h, questions 35, 36). If dy/dx does not exist, a turning point can still be present.

Example 5.12

Describe the turning points of the curve given by the equation $y = x^4-2x^2+1$.

Differentiating the equation with respect to x,

$$dy/dx = 4x^3-4x.$$

This equals zero when $4x(x^2-1) = 0$,

i.e. $x = 0, \ 1 \ \text{or} \ -1.$

The second-order derivative d^2y/dx^2 is obtained by differentiating again,* giving the equation

$$d^2y/dx^2 = 12x^2-4,$$

and substituting the values 0, 1 and -1 for x gives the values of the second order derivatives at the successive turning points:

(i) at $x = 0$, $d^2y/dx^2 = -4$. This is < 0, and thus there is a *maximum* at $x = 0$;
(ii) at $x = 1$, $d^2y/dx^2 = 12-4 = 8$. This is > 0, and the turning-point at $x = 1$ is a *minimum*;
(iii) at $x = -1$, $d^2y/dx^2 = 12-4 = 8$. The turning-point at $x = -1$ is also a *minimum*.

The values of y at these turning points are obtained by substituting the values 0, 1 and -1 for x in the original equation. This gives:

for the maximum at $x = 0$, $y = 1$,
for the minimum at $x = 1$, $y = 0$,
for the minimum at $x = -1$, $y = 0$.

Noting that the curve is a positive quartic curve, it will be appreciated that the value of y is large and positive for large and positive values of x, and thus the maximum at $x = 0$ is a relative maximum. The minima at $x = 1$ and $x = -1$ are absolute minima, and a sketch of the function is given in Fig. 5.22.

* If the general shape of the quartic curve is known, it will be unnecessary to investigate the turning points analytically. The succession for an "upright" quartic must be a minimum, maximum and minimum with increasing x, and the results can be obtained this way.

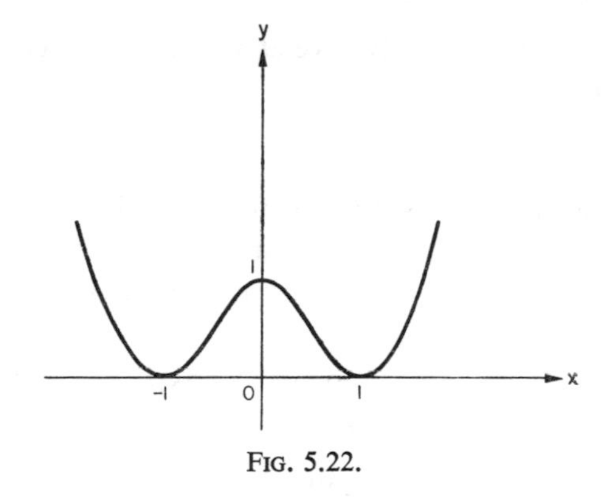

Fig. 5.22.

Example 5.13

A fence of length 100 m is to be used to form three sides of a rectangular enclosure, the fourth side being a wall. Find the maximum area which can be enclosed by the fence.

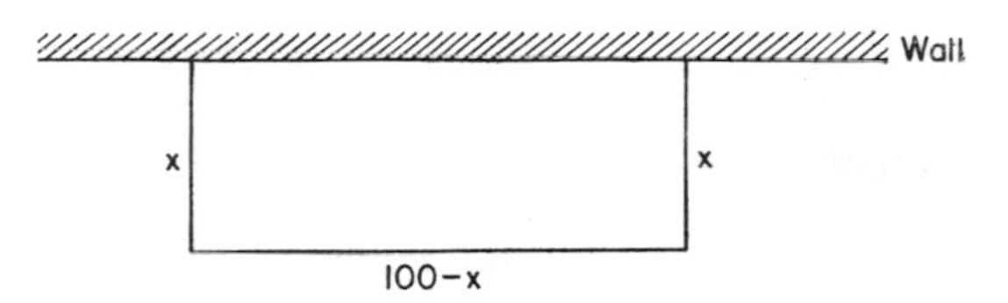

Fig. 5.23.

Letting one of the equal sides be of length x m (see Fig. 5.23), the side parallel to the wall must be of length $(100-2x)$ m. The area of the enclosure thus equals $x(100-2x)$ m², and denoting this by A, the equation

$$A = 100x-2x^2$$

can be written. Differentiating with respect to x,

$$dA/dx = 100-4x,$$

and this rate of change of A with respect to x must equal zero for a maximum value. Thus

$$100-4x = 0;$$

and $x = 25$ (m).

To check that this is the maximum rather than the minimum value, we can evaluate the second order derivative at $x = 25$.* Differentiating again,

$$d^2A/dx^2 = -4,$$

and this is less than 0 for all values of x. Thus the turning point at $x = 25$ is a maximum, and the value of the area enclosed by the fence is equal to

$$A = 100.25-2.(25)^2$$
$$= 2500-1250$$
$$= 1250 \text{ m}^2.$$

The maximum area which can thus be enclosed is 1250 m², and it might finally be noted that it is often necessary to substitute the value of the independent variable at which the turning point occurs into the original equation to obtain the numerical answer to the problem.

Example 5.14

Investigate the turning points of the curve given by the equation

$$y = x^4-4x^3+6x^2-4x+1.$$

Differentiating to obtain the gradient function of the curve, we obtain

$$dy/dx = 4x^3-12x^2+12x-4 = 4(x-1)^3.$$

The gradient is therefore zero at $x = 1$ only, and at this point the second order derivative

$$d^2y/dx^2 = 12x^2-24x+12 = 12(x-1)^2$$
$$= 0.$$

The nature of the turning point at $x = 1$ must therefore be investigated by the curve sketching method, and considering the sign of the gradient at the points to the left of $x = 1$ and to the right of $x = 1$, we obtain:

value of x:	<1	1	>1
sign of $x-1$:	$-$	0	$+$
sign of $(x-1)^3$:	$-$	0	$+$
gradient:	\	—	/

The turning point at $x = 1$ is therefore a minimum.

The value of y at the turning point is given by the original equation, from which it can be seen that $y_{min} = 1-4+6-4+1 = 0$.

* In examples such as this it is not really necessary to check the nature of the turning point mathematically. Clearly the area enclosed by a long fence close to the wall, and that included in a long thin enclosure perpendicular to the wall, are both small. Areas tending towards the square shape must be larger, and the turning point will be a maximum.

Exercise 5h

Describe the nature of the turning points of the following functions:

1. $x^3-6x^2-15x+7$. **4.** x^6+6x.
2. $2x^3-3x^2-12x-1$. **5.** $x^3-12x-4$.
3. x^4-32x.

6. Sketch the graph of $y = x^4-18x^2+9$, giving in full an analysis of the nature of the turning points.

7. Discuss the nature of the turning points on the graph of $y = x^4-2a^2x^2$, and explain what happens as $a \to 0$.

8. Sketch the following curves:

(i) $y = x^4+4x^3+6x^2+4x+1$,
(ii) $y = x+4/x$,
(iii) $y = x^2+3/x^2$,

and state whether the turning points exhibited are relative or absolute.

9. A fence of length 50 m is to enclose a rectangular area, using a wall to form one side of the rectangle. Find the maximum area which can be enclosed.

10. A fence of length 100 m is used to form a rectangular enclosure. Find the maximum area which can be enclosed.

11. A rectangular enclosure of 400 m² is to be made using

(i) a fence and a wall to form one side of the enclosure,
(ii) a fence.

What is the least length of fencing that will be necessary?

12. A closed box with a square base is to be made to hold 100 cm³ of liquid. What must the side of the base be so that as little metal as possible is used to make the box?

13. A box with a square base and open at one end is to have a volume of 1000 cm^3. What must the dimensions be if as little metal as possible is to be used in its construction?

14. What is the maximum volume of a can built from 200 cm^2 of metal which can be supplied in any shape or shapes if the shape of the can is

(i) rectangular on a square base,
(ii) cylindrical, the can being closed at both ends?

15. A piece of cardboard 12×10 dm is to be made into a rectangular, open container after squares have been cut from its four corners, and the cardboard folded along the dotted lines as shown in Fig. 5.24. What is the maximum volume of the box which can be made in this manner?

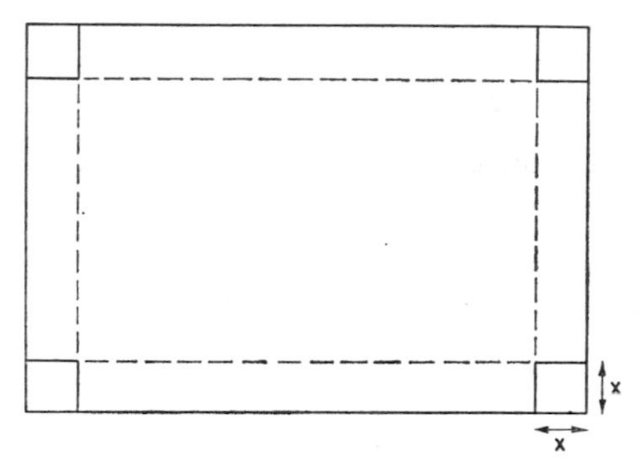

FIG. 5.24.

16. The displacement of a particle from a fixed point t seconds after motion commences is given by the equation $s = 108t-4t^3$. What are the maximum and minimum values of the displacement? How far does the particle travel before it arrives at the original point again?

17. The distance of a particle travelling up an inclined plane above the point where the plane meets the horizontal is given by the equation $s = 30t-8t^2$, where t is the time elapsing after the particle has passed the junction of the plane and the horizontal. Find the maximum distance the particle travels up the plane.

Explain the meaning of ds/dt and d^2s/dt^2 in terms of the motion of the particle.

18. The displacement of a particle is given by the equation $s = 10 \sin t$. What is the maximum value of the displacement s, and what is the maximum value of the velocity (v) of the particle? When do these maximum values occur, and where does the maximum velocity occur?

19. The curve shown in Fig. 5.25 (a), (b) and (c) are described as positively J-shaped, negatively J-shaped and U-shaped, respectively. What conditions must dy/dx and d^2y/dx^2 satisfy throughout the range of the curve for these shapes to be exhibited?

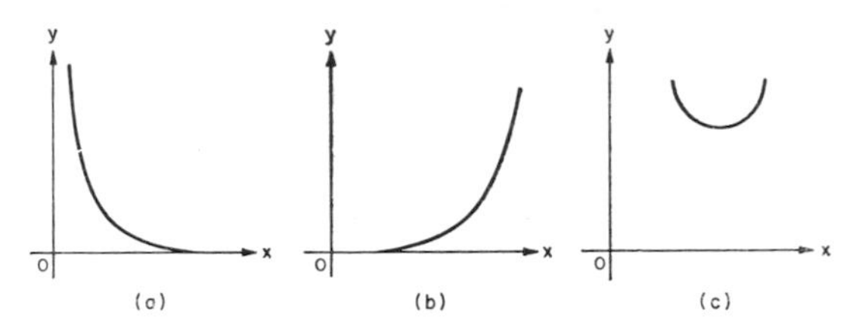

FIG. 5.25.

20. The point O is the centre of the base of a right circular cone C_1 of given volume V cubic units and given base radius r units. A second right circular cone C_2 is drawn inside C_1 with its vertex at O and with the edge of its base lying on the curved surface of C_1. Taking the base radius of C_2 as tr units, where t can vary from 0 to 1, find an expression for the volume of C_2 in terms of t and V only. Hence show that the maximum volume of C_2 is $4V/27$ cubic units. (The volume of a right circular cone of height h units and base radius r units is $\frac{1}{3}\pi r^2 h$ cubic units.) (NUJMB)

21. If the radius of the base of an open cylindrical can is r, its total surface area S and its volume V, show that $V = \frac{1}{2}r(S-\pi r^2)$.

Hence, or otherwise, show that the maximum volume, for a given surface area, of an open cylindrical can occurs when the height of the can is equal to the radius of the base. Show also that in this case the volume of the can is $1\frac{1}{2}$ times the volume of the hemisphere which will just fit inside the can. (SU)

22. Find the greatest values of $1-2x-x^2$, $4-x-x^2$ and $5+2x-x^2$ by an algebraic method.

23. Deduce that the minimum value of $(x+1/x)$ occurs at $x = 1$. Write down this minimum value, and verify that it is a minimum turning point by differentiation.

†**24.** Discuss the nature of the function $\left(1-x^{\frac{2}{3}}\right)$ at $x = 0$.

†**25.** As a point P moves round the ellipse $x^2/a^2+y^2/b^2 = 1$, the focus $(-ae, 0)$ of the ellipse being S, the length r of SP varies according to the equation $r = a+ex$, where $b^2 = a^2(1-e^2)$, $e < 1$ and x is the abscissa of P. Draw a graph showing how the value of r varies with x, and explain why the length of SP is a maximum when $x = a$, and a minimum when $x = -a$. Why is it that dr/dx is not zero for these values of x?

†**26.** Discuss the nature of and sketch the graph of $y = x^{\frac{5}{2}}$ where $x = 0$.

†**27.** The coefficient of friction between a particle travelling up an inclined plane and the plane is 0·1. If the angle of inclination of the plane to the horizontal is 30°, and the particle starts moving up the plane with a velocity of 10 m/s (Fig. 5.26), what is the maximum distance the particle travels up the plane?

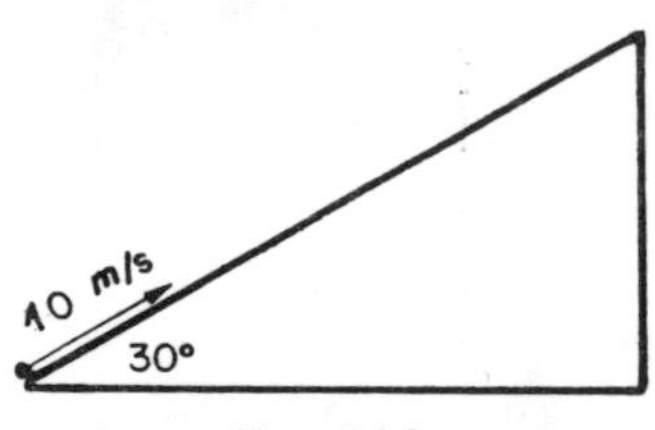

FIG. 5.26.

†28. If the equation $xf''(x)+[f'(x)]^2 = x^2+1$ is satisfied by a function f of the variable x, what can be said about the nature of a turning point occurring at $x = a$, where (i) $a < 0$, (ii) $a > 0$?

†29. Show that the area enclosed by a rectangular enclosure of perimeter p is greatest if the rectangle is a square. Show that the area then enclosed is less than that enclosed by a circle of the same perimeter.

†30. Discuss whether a cylindrical can whose surface area is S cm² holds more or less than a hollow sphere of the same surface area.

†31. A motor-car, X, uniformly accelerated at 2 metres per second per second, passes a point P on a straight road at 6 m/s; 2 seconds later a motor-car, Y, uniformly accelerated at $1\frac{1}{2}$ metres per second per second, and moving in the same direction as X, passes P at 15 m/s. Prove that Y overtakes X at a point Q on the road 6 seconds after X passes P, and that X overtakes Y at point R 12 seconds later still.

Prove also that $QR = 360$ m and that the maximum distance separating X and Y between Q and R is 9 m. (OC part question)

†32. Describe the nature of the turning points at X and N in the graph shown in Fig. 5.27, and discuss whether or not the derivatives $f'(x)$ and $f''(x)$ exist at these points.

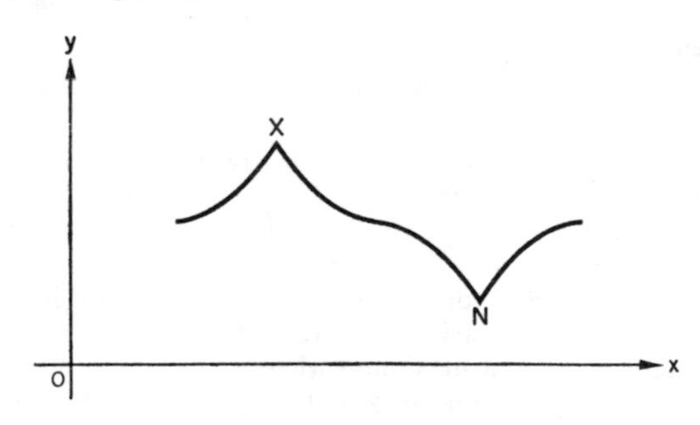

FIG. 5.27.

†33. What is the greatest area of a rectangle which can be fitted in the area between the x- and the y-axes and the curve $y = 2-2\sqrt{x}$ (see Fig. 5.28)?

†34. Two variables x and y are known to be related linearly, and a number of pairs of corresponding values (x_i, y_i) are experimentally determined and recorded (the symbol x_i stands for any one of the

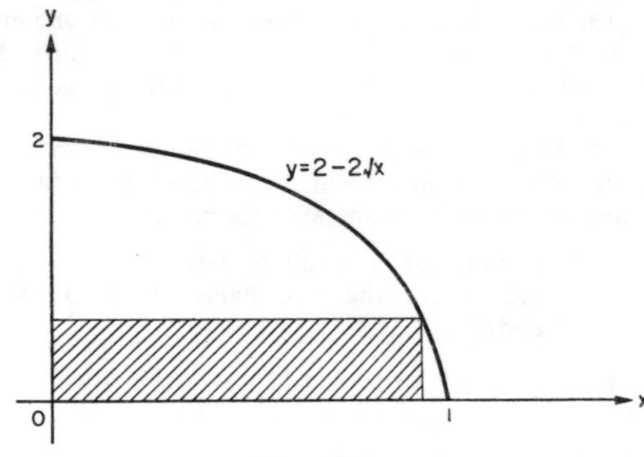

FIG. 5.28.

observations x_1, x_2, x_3, etc.). If the line expressing the relation between x and y has an equation of the form $y = mx$, we can write that for many observations (x_i, y_i), $y_i = mx_i$, while for many others, (x_i, y_i) will not lie on $y = mx$ due to experimental error. We can use the value of (y_i-mx_i) as a measure of how far (x_i, y_i) lies from the line: or, better, we can use the square of this error $(y_i-mx_i)^2$. The line of best fit through a set of points (x_i, y_i) is defined as the line for which the sum of the square deviations

$$S = \sum_{i=1}^{n} (y_i-mx_i)^2$$

of the points from the line is a minimum. Expand the bracket into separate terms such as

$$\sum_{i=1}^{n} y_i^2 \quad\text{and}\quad -2m\sum_{i=1}^{n} x_i y_i$$

(treating m as a constant which can be taken outside the summation sign), and then differentiate S with respect to m to find the value of m for which S is a minimum. Hence, write down the equation of the line of "best fit" (i.e. that with the gradient m for which S is a minimum) through the experimentally determined points in terms of

$$\sum_{i=1}^{n} x_i^2,\quad \sum_{i=1}^{n} x_i y_i \quad\text{and}\quad \sum_{i=1}^{n} y_i^2.$$

†35. Sketch the graph of $y = |x|$.
Discuss whether or not:

(i) $|x|$ is continuous at $x = 0$,
(ii) $|x|$ takes a minimum value at $x = 0$,
(iii) the derivative of $|x|$ exists at $x = 0$,
(iv) the derivative of $|x|$ is zero at $x = 0$.

Is it possible to think of the second-order derivative of $|x|$ at $x = 0$?

†36. Sketch the graph of $y = x^{\frac{2}{3}}$. Find an expression for the gradient function dy/dx, and sketch its graph.

Discuss whether or not $y = x^{\frac{2}{3}}$ goes through a minimum value at $x = 0$.

*37. The gravitational force on a body of mass m at a distance R from the centre of the earth is GMm/R^2, where G is a constant, and M is the mass of the earth. Show that the rate of change of the gravitational attraction experienced by a body on or above the earth's surface with respect to height is greatest at the earth's surface.

†38. The probability of one success in n trials is $np(1-p)^{n-1}$, where p is the probability of success in one trial. What must p equal if the probability of one success in (i) two, (ii) three, (iii) four trials is to be as high as possible?

5.7.4. *Inflexions.* A further situation can arise when the gradient of the curve is zero, and a simple example is given by the graph of $y = x^3$ (Fig. 5.29(i)). The point at which the

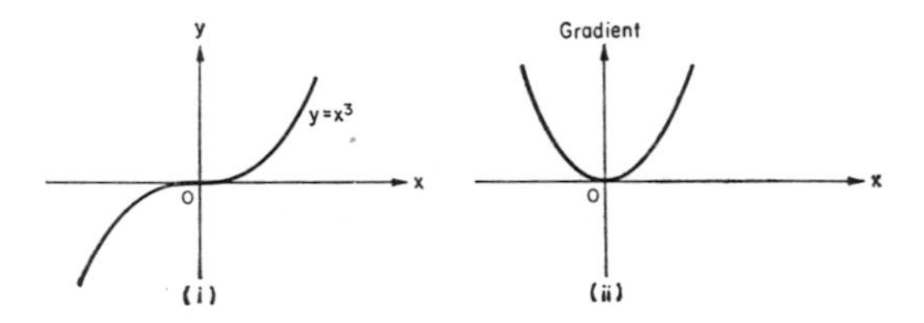

FIG. 5.29. The curve $y = x^3$ shows an inflexion at the origin.

gradient is zero is neither a maximum nor a minimum point, and the gradient has the same sign on either side. Such a region is called an *inflexion*, the characteristic of an inflexion being that the concavity of the curve at the inflexion changes sense (see Section 4.8). This means that it is the gradient of the curve rather than the curve itself which goes through a maximum or a minimum value at the point (Fig. 5.29(ii)), and, as we have seen before, this leads to the consequence that $d^2y/dx^2 = 0$ for the particular value of x in question.

Notice that a curve can exhibit an inflexion even if the gradient is not zero (Fig. 5.30); and that although the value of the second-order derivative is zero at an inflexion, it does not necessarily follow that an inflexion occurs whenever d^2y/dx^2 is zero. For example, the

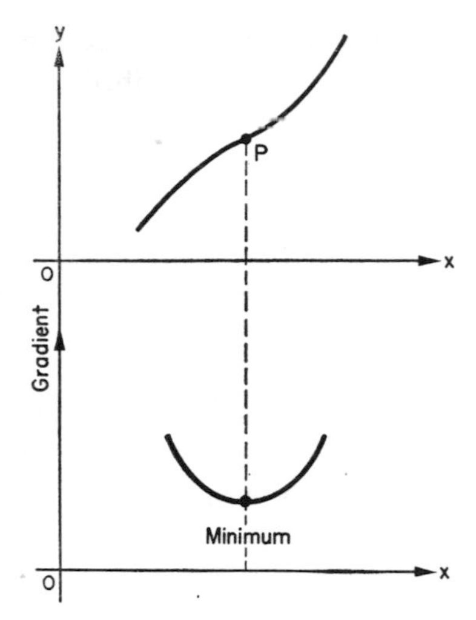

FIG. 5.30. The characteristic of an inflexion is that the concavity of the curve changes at the point of inflexion, and this implies that the gradient of the curve goes through a maximum or a minimum.

curve $y = x^4$ shows a minimum at $x = 0$ although $d^2y/dx^2 = 0$ when $x = 0$ (Fig. 5.31). The safest way of verifying whether or not an inflexion is present is to use the curve sketching method described on p. 131.

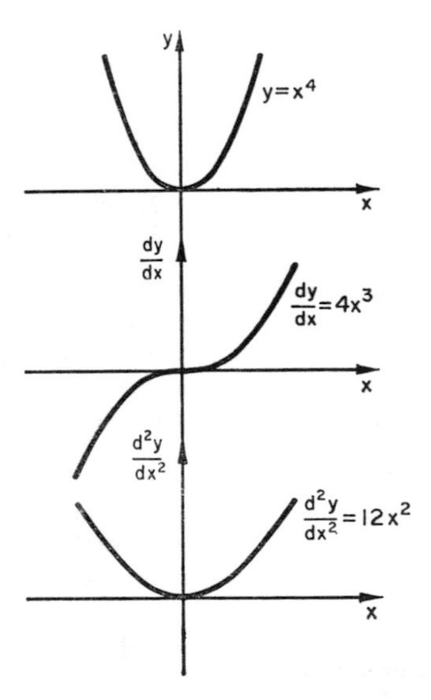

FIG. 5.31. The value of d^2y/dx^2 must be zero at an inflexion but an inflexion does not always occur when the second-order derivative is zero.

The values of functions at which the three types of behaviour just described occur (i.e. maxima, minima and horizontal inflexions), are referred to collectively as *stationary values*: at these points, the value of the functions do not change as x changes over a small range. The turning points are points at which maxima or minima occur, and we can conveniently state the conditions which are usually satisfied for these behaviours to be exhibited using the symbol $f(x)$ to represent the expression whose graph is being considered. Thus,

for a MAXIMUM at $x=a$	$f'(a) = 0$	$f''(a) < 0$
for a MINIMUM at $x=a$	$f'(a) = 0$	$f''(a) > 0$
for a HORIZONTAL INFLEXION	$f'(a) = 0$	$f''(a) = 0$,

assuming that in each case $f'(a)$ exists, and where the last condition ($f''(a) = 0$) is a necessary but not sufficient condition for a horizontal inflexion.

Exercise 5i

Give a description of the stationary values which occur in the following functions:

1. x^3-3x^2+3x-1.
2. x^3+3x^2+3x+1.
3. x^4-7.
4. x^3-7.
5. x^5-7.
6. $x^3-27x+5$.
7. x^5-5x+7.
8. $x^3-6x^2+12x-8$.

9. Describe the turning points which occur in the graph of $y^2+\{(4+x^2)/x\}$, and sketch the graph.

10. Draw a sketch graph of $y = x^3(x-8)$.

11. Find the maximum and minimum values of the expression $(1+3x-x^3)/4$ and distinguish between them.

Draw a rough sketch of the curve $y = (1+3x-x^2)/4$ between $x = -3$ and $x = +3$, and calculate the maximum positive gradient on this part of the curve. (OC)

12. A function $f(x)$ has a maximum value of 4 at $x = 1$. The limiting values of $f(x)$ as $x \to \infty$ and $x \to -\infty$ are both 1. The function $f(x)$ is continuous. Sketch the graph of $f(x)$ if $f(x)$ is an (i) even, (ii) odd function, and show also the graph of $f'(x)$. What is the smallest number of inflexions $f(x)$ can have? (An "even" function is one for which $f(-x) = f(x)$ and an "odd" function is one for which $f(-x) = -f(x)$.)

13. A farmer erects a fence along three sides of a rectangle in order to make a sheep-fold; the fourth side of the rectangle is provided by a hedge already in existence. Find the maximum area of the enclosure if the total length of the fence is to be 80 metres. (OC)

14. Given that the function $y = ax^3+bx^2$, where a and b are constants, has a stationary value of -4 when $x = 2$, find the values of a and b. (OC)

15. A motor-car A travelling along a straight road, with uniform velocity u, passes at A a car B travelling, with uniform acceleration f, in the same direction, the velocity of B at X being u_1 ($< u$); B overtakes A at Y, thereafter travelling with uniform velocity. Find

(i) the maximum distance between the cars between X and Y, and
(ii) the uniform acceleration to be given to A at Y if A overtakes B at Z, where $YZ = XY$. (OC)

16. Show that the function ax^2+bx+c, where a, b and c are constants, can be expressed in the form

$$a\left(x+\frac{b}{2a}\right)^2+\left(c-\frac{b^2}{4a}\right).$$

Hence find:

(i) the maximum value of the function if $a < 0$;
(ii) the minimum value of the function if $a > 0$.

17. Find the maximum and minimum values of the function

$$y = \tfrac{1}{3}(4x^3-3x-1).$$

Sketch the graph of the function and find the greatest and least values of y for values of x in the range $-1{\cdot}5$ to $1{\cdot}5$ inclusive. (NUJMB)

18. Find dy/dx for the two curves $y = x-4/x$ and $y = 4x^3/3-1/x^2$, and prove that the curves have the same gradient when $x = \frac{1}{2}$. Find the range of values of x for which the gradient of the curve $y = x-(4/x)$ is greater than 5. Determine whether the curve has any turning points. (C)

19. A given volume of metal is melted down and made into a single solid body which consists of a circular cylinder of radius r and height h surmounted by a right circular cone of radius r and height r. Prove that the total surface area of the body is least when $h = \sqrt{2}r$. (SU)

20. A function $f(x)$ is defined by the equations $f(x) = -1$ for $-\pi \leqslant x \leqslant \pi$, and $f(x) = \cos x$ for $x < -\pi$ and $x > \pi$. Sketch a graph of the function, and state whether it is continuous or not at $x = -\pi$ and $x = +\pi$. Find the positions of the turning-points in the range $-2\pi \leqslant x \leqslant 2\pi$.

21. Express the function $3 \sin x+4 \cos x$ in the form $R \sin (x+\alpha)$, and deduce the maximum value of $(3 \sin x+4 \cos x)$ from this expression, stating also at what values of x it occurs.

22. Find the maximum and minimum values of $a \sin x+b \cos x$ by a trigonometric method.

23. A shop sells a certain article at \$100 each, offering a discount of $60n$ cents per article where n is the number of articles purchased. Find the number of articles in the single sale yielding the greatest income if there are only (i) 150, (ii) 75 articles available.

***24.** A d.c. motor has an armature of resistance r ohms, and the current i flowing in it when an e.m.f. of E volts is applied is $(E-V)/r$ amps, where V is the back e.m.f. developed in the rotating armature. The power developed by the motor is Vi watts. Show that this is a maximum when $V = \frac{1}{2}E$ and that its value is then $E^2/4r$. State what requirements this leads to for an efficient and powerful motor, and discuss whether there is theoretically any limit to the power obtainable from a motor on a given supply voltage.

***25.** Explain why the current flowing in a circuit consuming W watts at a potential difference of E volts is W/E amps. Hence show that the voltage loss in a transmission cable of resistance r is Wr/E, and deduce an expression for the power P available at the supply or output end in terms of the input voltage E, the input power W and r. Show that for a given transmission cable and voltage, the maximum power available at the output end occurs when $W = E^2/2r$. Find the maximum value of P, and deduce at what non-zero value of W the output power is zero. Explain why it is possible to have no power available when there is a non-zero input.

Explain also how your expression for P shows that the higher the input voltage, the less the power loss in the cable.

26. In the rectangle $ABCD$ the side AB is x cm and the side BC is $3x$ cm; $PQRS$ is a square. The sum of the perimeters of the rectangle $ABCD$ and the square $PQRS$ is 8 cm. Show that the sum, A cm², of the rectangle and the square is $(4-8x+7x^2)$ cm².

Calculate the value of x for which A is a minimum and sketch the graph of A against x for possible values of x. (NUJMB)

27. The demand for the product of a firm is measured by how many thousand items they sell a week. If this number is x (thousand) and the price of each item is p (thousand) pounds, x and p are related by the equation $x+\frac{3}{8}p = 1$. Write down an expression for the value of the sales, £S million per week, and show that the marginal income dS/dx is equal to $\frac{8}{3}(1-2x)$. For what value of p is the marginal income zero? Explain the significance of your answer.

28. The marginal product of a firm is defined as the change in total production divided by the change in the size of the labour force necessarily employed by the firm to produce the extra output. Explain why the marginal product is given by the gradient of the tangent to the unity scale product/labour curve, and explain why the average product per unit of labour for any particular value is given by the gradient of the chord joining the origin to the point representing the situation on the graph. Explain further why, when the average product is increasing, the marginal product is greater than the average product, and when the average product is decreasing, the marginal product is less than the average product. Is it true to say that the marginal product equals the average product when the average product is at a maximum?

29. The relationship between the demand x articles per day and the price p of a certain manufactured article is $x+ap = 1$, where a is a constant. Write down an expression for the revenue R from the sale of the x articles, expressing R as a function of a and x only. If the cost of manufacture varies according to the relation $C = ax^2$, find the value of x for which the profit P $(= R-C)$ is a maximum. Show that this maximum occurs when the marginal revenue dR/dx and the marginal cost dC/dx are equal, and explain.

30. Prove that the area of the quadrilateral $ABCD$ is $\frac{1}{2}AC.BD \sin\theta$, where θ is the angle between the diagonals AC and BD. If the sum of the lengths of the diagonals is 8 cm, find the greatest possible area of the quadrilateral. (L)

31. Show that the gradient of $y = x^4-8x^2$ is positive when x is greater than 2 and determine whether it is positive for any other values of x.

Find the maximum and minimum values of x^4-8x^2 and sketch the graph of this function. (L)

32. State, without proof, conditions for the value $f(a)$ of the function $f(x)$ to be (a) a maximum, (b) a minimum.

Find the stationary values of the function

$$f(x) = 1-\frac{9}{x^2}+\frac{18}{x^4}$$

and determine their nature.

Sketch the curve $y = f(x)$. (L)

†33. A curve described by an equation of the form $y=ax^n$, where n is integral and greater than 1, has a stationary value at the origin, since the first derivative $y' = nax^{n-1}$ is zero if $x = 0$. By putting n equal to 2, 3, 4, 5, and 6 in turn, determine the nature of the stationary value if $a > 0$, and relate this to the order and value of the first non-zero derivative. (For example, if $n = 3$, the stationary value can be found to be an inflexion. The first derivative $y'= 3ax^2$, and if $x = 0$ this derivative has a zero value. Similarly, the second derivative $y'' = 6ax$, and the third derivative $y''' = 6a$. If $a > 0$ this latter derivative is positive. Thus the order of the first non-zero derivative is 3, and for $n = 3$, the turning point at the origin is an inflexion.)

†34. Describe the nature and position of the inflexion in the graph of $y = x^3-6x^2+4x+8$.

†35. By considering the general shape of the graph of the cubic function ax^3+bx^2+cx+d, state whether the stationary value at $x = p$ is a maximum or minimum, given that the other stationary value occurs at $x = q$, where $q > p$, and that $a > 0$. What happens if $a < 0$?

†36. Given that $a > 0$, find the relative maximum value of ax^3+bx^2+cx+d.

SUMMARY

The time derivative of the displacement of a body gives its velocity:

$$v = \frac{ds}{dt}.$$

The time derivative of the velocity gives the acceleration:

$$a = \frac{dv}{dt} = \frac{d^2s}{dt^2}.$$

The instantaneous rate of change of one variable with respect to another is given by the derivative. A small change δy in a variable y produced by a small change δx in the related variable x can be calculated approximately by the relation

$$\delta y \doteq \frac{dy}{dx}\delta x.$$

For larger values of δx, it is necessary to find the values of y_1 and y_2 corresponding to x_1 and x_2, from the equation relating x and y.

The equation of the tangent at (x_1, y_1) to a curve can be found by substituting into the equation

$$y - y_1 = m(x - x_1),$$

where m takes the value of the derivative at (x_1, y_1). The equation of the normal at the same point is

$$y - y_1 = -\frac{1}{m}(x - x_1).$$

If a stationary value of $f(x)$ occurs at $x = a$, the value of the first-order derivative $f'(a)$—if it exists—must be zero. In general,

if the stationary value is a minimum, $f''(a) > 0$, and
if the stationary value is a maximum, $f''(a) < 0$.

(Often the nature of the stationary point will be obvious from other considerations and need not be checked mathematically.)

If $f''(a) = 0$, the nature of the stationary value should be investigated by evaluating the sign of the gradient on either side of $x = a$.

An inflexion is the point at which the concavity of the curve changes: the second-order derivative $f''(a)$—if it exists—must be zero, but if $f''(a) = 0$, an inflexion need not necessarily be present.

MISCELLANEOUS EXERCISE 5

1. A particle is moving in a straight line and its distance (s cm) from a fixed point in a line after t seconds is given by the equation $s = 12t - 15t^2 + 4t^3$. Find

(i) the velocity and acceleration of the particle after 3 seconds;
(ii) the distance travelled between the two times when the velocity is instantaneously zero. (OC)

2. (i) Differentiate the expression $3x^{\frac{1}{2}} + 4 - 5x^{-\frac{1}{2}}$ with respect to x.
(ii) Find a maximum and a minimum value of the expression $x^3 - 3x$, and give a rough sketch of the curve $y = x^3 - 3x$. (OC)

3. Find the equation of the tangent to the curve $y = x^2 - 4x + 5$ at the point (5, 10), and also the equation of the line through (5, 10) which is perpendicular to the tangent.

Find the coordinates of the point on the curve at which the tangent will be parallel to the line $y = 2x$. (OC)

4. A fast train from Taunton to Paddington is s km from Taunton, t hours after it has left the station, where $s = 12t^2(15 - 4t)$. Find the distance between Taunton and Paddington, and the time taken by the train to cover it. Find also the maximum speed of the train.

5. A square sheet of metal of side 12 dm has four equal square portions removed at the corners, and their sides are then turned up to form an open rectangular box. Prove that, when the box has maximum volume, its depth is 2 dm. (OC)

6. A particle moves along the axis of x and at the end of time t its position is given by $x = t^3 - 9t^2 + 24t$. Show that as t increases from zero,

(i) the particle moves in a positive sense until it reaches the point $x = 20$;
(ii) it then moves backwards until it reaches the point $x = 16$;
(iii) thereafter it moves in a positive sense.

Find the velocity of the particle on the second occasion it is at the point $x = 20$. (OC)

7. The bottom of a rectangular tin box is a square of side $x/3$ dm and the box is open at the top. It is designed to hold 4 dm³ of liquid. Express in terms of x the total area of the bottom and four sides of the box. Find the value of x for which this area is a minimum.

8. Prove that the function $(2-x)(1+x+x^2)$ has a maximum value at $x = 1$ and a minimum value at $x = -\frac{1}{3}$. (OC)

9. A body is projected in a viscous medium and moves s cm in t seconds before coming to rest. The relationship between s and t during the motion is $s = 4t-\frac{1}{2}t^3$. Find

(i) the distance the body travels,
(ii) its average speed while in motion,
(iii) its initial speed,
(iv) its speed at half time.

10. The gradient of the graph of a function $y = f(x)$ is such that dy/dx increases steadily from -1 at $(-1, -3)$ to 2 at $(4, 17)$. It is zero at the point whose abscissa is $-\frac{1}{2}$. Sketch the graph of the function, and deduce the sign of d^2y/dx^2. Does the point $x = -\frac{1}{2}$ represent a maximum or a minimum value?

11. From the equations $z = x^3+y^3-15x$ and $x-y = 3$, find an equation connecting z and x. Find then the maximum and minimum value of z. (OC)

12. (i) Find dy/dx when $y = (1+3x)^2$. Deduce the equation of the tangent at (2,49) to the curve whose equation is $y = (1+3x)^2$.
(ii) The formula $v = 8\sqrt{x}$ gives the velocity, v m/s, of a particle when it has travelled a distance x m from rest. Prove that, if δv and δx denote corresponding small increases in v and x,

$$\delta v \simeq \frac{32\delta x}{v}.$$

Hence find the approximate increase in the velocity of the particle when x increases from 36 to 37. (OC)

13. (i) Differentiate $(x+2)(x^2+3)$ and $(2x-3/x)^2$ with respect to x.
(ii) Find the coordinates of points on the curve

$$y = 2x^3+3x^2-12x+6$$

at which x has a stationary value and state in each case whether the value of y is a maximum or a minimum. (OC)

14. Sketch the portion of the graph of $y = f(x)$ for which $f'(x)$ and $f''(x)$ are

(i) both positive,
(ii) both negative,
(iii) alternatively positive and negative.

15. The displacement of a particle from a fixed point varies with time according to the equation $s = t(t-2)(2t-5)$. Describe the motion and sketch graphs showing the variation in its displacement, velocity and acceleration.

16. Find the values of x for which

(i) $\dfrac{dy}{dx}$, (ii) $\dfrac{d^2y}{dx^2}$ are zero,

given that $y = (x-1)(x+1)(x+2)$.

17. Find the equation of the tangent to the curve $y = 4/x$ at the point (2, 2); and prove that the area of the triangle formed by the axes and this tangent is equal to 8.

Prove that the area of the triangle formed by the axes and the tangent at the point $(a, 4/a)$ is also equal to 8 for all values of a. (OC)

18. (i) Differentiate with respect to x the expression $4x^5+x-5x^4$.
(ii) Find the maximum value, and also the minimum value, of the expression $x^3-3x^2-9x+27$. (OC)

19. Prove that the point (5, 6) lies on the curve $y = 2x^2-7x-9$, and find the equation of the tangent of the curve at this point.

Find also the equation of the tangent which is parallel to the line $y = x$. (OC)

20. (i) Differentiate (a) $(1-x)(2-3x+x^2)$; (b) $6x^3-4x/x^2$.
(ii) Find the range of values of x between which the function $x^3-3x^2-9x+11$ decreases in value as x increases. (OC)

21. A circular piece of metal of radius $2x$ cm has a rectangular hole x cm by 1 cm punched from it. For what value of x is the remaining area a maximum?

22. The breaking stress at a point x m from one end of a certain rod of total length 1 m is given by the expression $kx(1-x)^2$. Given that k is a constant, find where the rod is most likely to break.

23. A rectangular box without a lid is made of cardboard of negligible thickness. The sides of the base are $2x$ cm and $3x$ cm, and the height is y cm. If the total area of the cardboard is 200 cm², prove that $y = 20/x-3x/5$.

24. (i) Find dy/dx if (a) $y = (x^2-3)^2$, (b) $x^2y-x^3+1 = 0$.
(ii) Find the equation of the tangent at the point (2, 2) on the curve $y = x^3-3x$, and the coordinates of the point at which the tangent meets the curve again. (OC)

25. The straight line $y = a^2$ meets the curve $y = x^2$ at the points P, Q. The coordinates of a point C are (0, 12). Given that a^2 is less than 12, find an expression in terms of a for the area of the triangle CPQ, and find the value of a when this area is a maximum. (OC)

26. While a train is travelling from its start at A to its next stop at B, its distance x km from A is given by $x = 180t^2-90t^3$, where t hours is the time taken.

Find in terms of t its velocity and its acceleration, after time t.

Hence find

(i) the time taken by the journey from A to B,
(ii) the distance AB,
(iii) the greatest speed attained. (OC)

27. If y is a cubic function of x of the form $y = ax^3+bx^2+c$, complete the table below:

x	0	-1	-2
y	-4		
$\frac{dy}{dx}$		-3	
$\frac{d^2y}{dx^2}$			-6

Describe the nature of the points on the graph of this function at which (i) $x = 0$, (ii) $x = -1$, (ii) $x = -2$. (OC)

28. A cylinder of radius x is inscribed in a cone of height h and base radius a, the two figures having a common axis of symmetry. Find an expression for the cylinder in terms of x, a and h.

Prove that the cylinder of greatest volume that can be so inscribed has a volume 4/9 of the volume of the cone. (OC)

29. A wheel spins so that after t seconds each spoke makes an angle $\theta°$ with its initial position where $\theta = t(1-t)(2-t)$. What is the angular velocity in degrees per second when (i) $t = 0$, (ii) $t = 1$, (iii) $t = 3$?

Describe the motion, and find the angular retardation when (iv) $t = 0$.

30. The volume V of a doughnut of external radius R and internal radius r is given by the equation $2\pi^2Rr^2$ (Fig. 5.32). Derive an expression for the approximate value of the increase δV in the volume of the doughnut when the internal radius is increased by δr. What can you deduce from this?

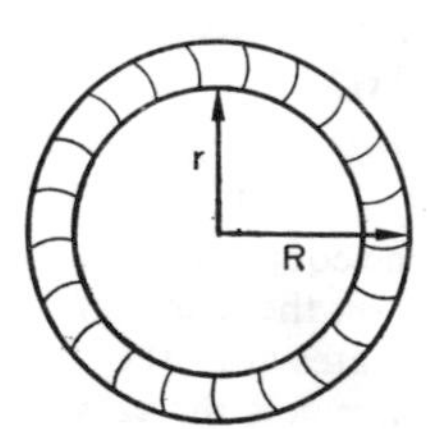

FIG. 5.32.

31. If $y = x^4$, calculate the approximate change in y if x changes from 2 to 2·05.

32. If $p = q-1/q$, calculate the approximate change in p if q changes from 5 to 5·007.

33. Use the relation $y = \sqrt{x}$ to calculate the approximate square root of 4·01, putting x equal to 4.

34. Calculate the approximate square roots of 1·09, 1·04, 1·0001, 4·1, 25·3.

35. Use the equation $y = \sin x$ to find without recourse to tables, approximate values for sin 30°5′ and sin 30°8′.

36. The time of swing t seconds of a pendulum of length l cm is given by the equation $t = 2\pi\sqrt{(l/980)}$. If $l = 24{\cdot}82$ cm, find the percentage change in the period if the length increases by 0·01 cm due to expansion. What gain or loss would this cause in a day?

37. Calculate the approximate increase in the volume of a sphere which increases in radius from 2 cm to 2·1 cm.

38. Find the approximate fourth roots of 15·9998 and 10000·1.

39. The coefficient of linear expansion of a certain metal is 3.10^{-5} per °C. What will be the approximate increase in area of a circular disc of the metal of radius 5·0 cm at 0°C if it is heated to 25°C?

40. A function $f(x)$ exists for which $f'(x) = af(x)$. If, for a small change in x, the percentage change in the value of $f(x)$ is three times as large as that in x, find a.

41. With the same axes and the same scales draw the graphs of the functions x^2-1 and $2-1/x$, plotting points on the graphs for values of x from $-2\frac{1}{2}$ to $+2\frac{1}{2}$ at $\frac{1}{2}$-unit intervals and for the value 0·25 of x.

Prove that the abscissae of the points of intersection of the two graphs are the roots of the equation $x^3-3x+1 = 0$, and find from your graphs the values of these roots, correct to 0·05. Prove that, of the tangents to the second of the two graphs, one (and only one) passes through the origin. Find the angle which this tangent makes with the x-axis. (OC)

42. Write down what you can about the values of $f'(x)$ and $f''(x)$ at the points marked in Fig. 5.31.

†**43.** The sum of the squares of the deviations of a set of n observations x_i about an arbitrary reference a is defined to be the value of the expression

$$\sum_{i=1}^{n}(x_i-a)^2.$$

Show that this expression is a minimum if a is the mean of the n observations.

CHAPTER 6

Techniques of Differentiation

6.1. FUNCTIONS OF A FUNCTION

As suggested in Section 4.4, the derivative of a function such as $(5-3x^2)^7$ can be obtained by expanding the bracket into a series of terms in increasing powers of x. But with a power as high as 7, such a procedure is both tedious and clumsy, and we will discuss a better method which can be devised by employing a new variable, z, defined by the equation $z = (5-3x^2)$.

Denoting the original function $(5-3x^2)^7$ by y, we can write

$$y = z^7 \tag{6.1}$$

and

$$z = 5-3x^2 . \tag{6.2}$$

From these we can easily obtain expressions for dy/dz and dz/dx, and we shall see later that these are sufficient to derive the desired expression for dy/dx. For the moment, however, it is more appropriate to consider some relations which exist between δx, δy and δz.

It will be recalled that δx represents a small increase in the value of x, and δy is associated with a corresponding change in y. From equation (6.2) it can be seen that a change in x will involve a change in z; this change will be denoted by δz.

Providing that $\delta z \neq 0$,* we can rewrite the ratio $\delta y/\delta x$ in the form $(\delta y/\delta x)(\delta z/\delta x)$. It should be pointed out that it is not axiomatic to write a similar equation with dy, dx or dz; the single expressions dy, dx and dz have not been defined in their own right and are therefore meaningless. Now

$$\frac{dy}{dx} = \lim_{\delta x=0}\left(\frac{\delta y}{\delta x}\right),$$

and so

$$\frac{dy}{dx} = \lim_{\delta x=0}\left(\frac{\delta y}{\delta z}\cdot\frac{\delta z}{\delta x}\right).$$

A theorem of limits tells us that the limit of a product is equal to the product of the separate limits, and thus this equation can be rearranged in the form

$$\frac{dy}{dx} = \lim_{\delta x=0}\left(\frac{\delta y}{\delta z}\right)\cdot\lim_{\delta x=0}\left(\frac{\delta z}{\delta x}\right).$$

By equation (6.2), if δx is zero, δz will be zero also. Replacing the limit of the first factor at $\delta x = 0$ by the limit at $\delta z = 0$, we obtain

$$\frac{dy}{dx} = \lim_{\delta z=0}\left(\frac{\delta y}{\delta z}\right)\cdot\lim_{\delta x=0}\left(\frac{\delta z}{\delta x}\right).$$

But the limits are the derivatives, and therefore finally

$$\frac{dy}{dx} = \frac{dy}{dz}\cdot\frac{dz}{dx}. \tag{6.3}$$

Referring to equations (6.1) and (6.2), dy/dz can be readily expressed as $7z^6$ and

* If $\delta z = 0$ when $\delta x \neq 0$, we can deduce that $dz/dx = 0$. And since y is a function of z only (eqn. (6.1)), δy will equal 0 if no change occurs in z. Thus $dy/dx = 0$ also, and equation (6.3), $dy/dx = (dy/dz)(dz/dx)$, is satisfied unless the value of $\delta y/\delta z$ increases without bound as $\delta z \to 0$.

dz/dx as $-6x$. The gradient of the curve $y = (5-3x^2)^7$ is thus given by the expression $7z^6.(-6x)$, and, substituting for z in terms of x, the gradient of the curve can be written as

$$-42x(5-3x^2)^6.$$

It will be seen that this is a much neater and quicker process than straightforward differentiation after expansion, and also one which gives the factors of the derivative. Its only restriction is that z must be such that δz is zero when δx is zero: or, expressed in an alternative manner, the derivative of z must exist at the value of x at which the derivative dy/dx is required. In the vast majority of cases where we need expressions for derivatives this will be the case: if difficulty is encountered, it should be foreseeable in the nature of the function for which z is substituted.

The problem of differentiating an expression such as $(5-3x^2)^7$ is referred to as differentiation of a "function of a function" because it is, in fact, a compound function. The "inner" function $(5-3x^2)$ is raised (by the "outer" function) to a power of 7: or denoting $(5-3x^2)$ by $g(x)$, and z^7 by $f(z)$, we can sensibly represent $(5-3x^2)^7$ by the symbol $[f\{g(x)\}]$.

Using this notation, we can summarise the procedure leading to equation (6.3) in the following steps:

(i) if the derivative of $y = f\{g(x)\}$ is required, make the substitution $z = g(x)$ (the variable y can now be written as $f(z)$, i.e. in terms of z only);

(ii) derive expressions for dy/dz (from $y = f(z)$) and dz/dx (from $z = g(x)$), and multiply them together to give the expression for dy/dx.

In simpler language, if we require the derivative of a function raised to a high power, introduce a new variable z to replace the "inner" function (the expression in the bracket of higher power). From the equation defining z, write down the derivative dz/dx, and from the equation expressing y as a simple power of z (e.g. $y = z^7$ above), find the expression for dy/dz. The product

$$\frac{dy}{dz}\cdot\frac{dz}{dx}$$

then gives the derivative dy/dx.

It may aid the memory to suppose that the equivalence of dy/dx and

$$\frac{dy}{dz}\cdot\frac{dz}{dx}$$

is trivial, imagining that the dz's appearing in the numerator and the denominator cancel,

i.e. $$\frac{dy}{dx} = \frac{dy}{\cancel{dz}}\cdot\frac{\cancel{dz}}{dx}.$$

As pointed out earlier, one cannot in fact express the mathematics behind this equivalence in this sort of language, but it is a manner of thinking which will enable equivalent expressions such as

$$\frac{da}{db} = \frac{da}{dx}\cdot\frac{dx}{db}$$

or

$$\frac{dy}{dt} = \frac{dy}{dx}\cdot\frac{dx}{dt}$$

to be quickly written down.

Example 6.1

Differentiate $(9+4x^2)^5$ with respect to x.

Let $y = (9+4x^2)^5$, and $z = 9+4x^2$. Then the required derivative dy/dx can be found by evaluating the product $(dy/dz)(dz/dx)$.

Since

$$y = z^5 \quad \text{and} \quad z = 9+4x^2,$$

$$\frac{dy}{dz} = 5z^4 \quad \text{and} \quad \frac{dz}{dx} = 8x.$$

Thus $$\frac{dy}{dx} = 5z^4.8x$$
$$= 40x(9+4x^2)^4.$$

Example 6.2

If $y = (1+2\sqrt{x})^8$, derive an expression for dy/dx.

Let $z = (1+2\sqrt{x})$; and then $y = z^8$.
Differentiating, we get

$$\frac{dz}{dx} = 2.\tfrac{1}{2}x^{-\frac{1}{2}} = \frac{1}{\sqrt{x}},$$

and
$$\frac{dy}{dz} = 8z^7.$$

Thus
$$\frac{dy}{dx} = \frac{dy}{dz}.\frac{dz}{dx} = 8z^7.\frac{1}{\sqrt{x}}$$

and substituting for z,

$$\frac{dy}{dx} = \frac{8(1+2\sqrt{x})^7}{\sqrt{x}}.$$

EXERCISE 6a

Differentiate the following functions with respect to x:

1. $(2x+3)^7$. **5.** $(1+x^2)^{\frac{3}{2}}$.
2. $(5-x)^{10}$. **6.** $(1-x+2x^2)^{\frac{5}{4}}$.
3. $1/(4+3x)^4$. **7.** $(11-3x^2)^{-3}$.
4. $\sqrt{(2-x)}$. **8.** $1/(8+7x)^3$.

9. Differentiate with respect to t:

(i) $(1-t^2)^{-2}$, (iii) $(8t-1^2)^8$.
(ii) $1/\sqrt{(7+4t)}$,

10. Write down an expression for the derivative $d\{x+\sqrt{(1+x^2)}\}/dx$.

11. Two variables, x and y, are related by the equation $y = \sin x$.

What is the equation relating dy/dt and dx/dt?

Write down the value of dy/dt when $x = \frac{1}{4}\pi$ and $dx/dt = 0{\cdot}1$; and that of dx/dt when $x = \frac{1}{3}\pi$ and $dy/dt = 2$.

12. Write down the value of dy/dt, given that $dx/dt = 0{\cdot}1$ and $x = 1$, where the two variables x and y are related by the following equations:

(i) $y = x^2$, (iv) $y = \cos x$,
(ii) $y = (2x-3)^6$, (v) $y = 1/x^2$.
(iii) $y = \sqrt{(2-x^2)}$,

13. Prove that if $y = (1+x^2)^{\frac{1}{2}}$, then $y(dy/dx) = x$.

14. $\theta(x)$ is a function of x such that

$$d(\theta(px))/dx = p\theta(px).$$

Show that $y = \theta(p_1x)+\theta(p_2x)$ satisfies the equation

$$L\, d^2y/dx^2+R\, dy/dx+y/C = 0$$

if p_1 and p_2 are the roots of the quadratic equation $Lp^2+Rp+1/C = 0$.

***15.** The flux ϕ linking a coil of n turns situated in a field of flux density B and whose cross-sectional area is A is $BAn \cos\theta$, where θ is the angle between the axis of the coil and the field. If the coil rotates at a constant angular velocity w, show that the e.m.f e induced in the coil is $BAnw \sin wt$, given that $e = -d\phi/dt$.

Explain why the induced e.m.f e is alternating.

16. A body has a displacement which varies with time according to the equation $x = 10 \sin 20\pi t$. What is its velocity and acceleration when $t = 1/200$, and what is its maximum velocity and displacement?

17. The flux ϕ linking a circuit varies according to the equation $\phi = 200 \cos(100\pi t)$. What is the rate of change of flux when

(i) $t = 1/400$, (ii) $t = 0$, (iii) $t = 1/200$?

What is the maximum flux linkage?

18. A body moves so that its displacement from a fixed point at time t is given by the expression $0{\cdot}1 \cos \pi t$. Find expressions for its velocity and acceleration, and describe the motion.

†**19.** Explain why it is not legitimate to write $\delta y/\delta x = (\delta y/\delta z)(\delta z/\delta x)$ if $z = 5$ and $y = z^2$.

6.1.1. The procedure for differentiating the function of a function can be made even quicker. Rewriting the derivative of $(5-3x^2)^7$ in the form

$$7(5-3x^2)^6(-6x),$$

we can easily see that the first two factors (7 and $(5-3x^2)^6$) are constructed from the $(5-2x^2)$ term differentiated "as a whole"—that is, its original power (7) is brought down as a multiplying factor, and a new power one less (6) is given to it. The third factor $(-6x)$ is the derivative of the "inner function" or bracket contents. To take another example, we can obtain the derivative of, say, $(1+3x)^{10}$ by

(i) differentiating the outer function or bracket as a whole, giving $10(1+3x)^9$, and

(ii) multiplying by the derivative of the function inside the bracket—which in this case is 3.

The derivative of $(1+3x)^{10}$ is thus $10(1+3x)^9.3$, or $30(1+3x)^9$. Relating this to the substitution method, if we put z equal to $(1+3x)$, y will equal z^{10}. Then $dy/dz = 10z^9$, or $10(1+3x)^9$, and $dz/dx = 3$, so that

$$\frac{dy}{dx} = 10(1+3x)^9.3 = 30(1+3x)^9$$

as before.

This speedier technique can be put into general terms. Let us suppose that we have a function of x, say $\phi(x)$, raised to a power n. Abbreviating the function symbol to ϕ, we can see that its curve can be represented by the equation $y = \phi^n$. The gradient function of the curve is given by the value of dy/dx, and making the substitution $z = \phi$, this can be written as $(dy/dz)(dz/dx)$ or $nz^{n-1}(dz/dx)$. But $z = \phi$, and dy/dx can be written as ϕ', so that

$$\frac{dy}{dx} = n\phi^{n-1}\phi'.$$

Again it can be seen that the derivative is composed of two factors:

(i) the ϕ^n differentiated as a whole,
and (ii) the derivative ϕ' of the inner function.

The relationship

$$\boxed{\frac{d}{dx}(\phi^n) = n\phi^{n-1}\phi'} \qquad (6.4)$$

is an important one, and the technique of differentiating a bracket as a whole and then multiplying by the inside of the bracket differentiated mentally is one which will be found useful. It is far quicker to write $5(1+x^2)^4.2x$ for the derivative of $(1+x^2)^5$ than to substitute, but it should not be forgotten that $5(1+x^2)^4$ is dy/dz, and $2x$ is dz/dx.

The derivatives of two commonly occurring trigonometric functions—sec x and cosec x—can be obtained by employing equation (6.4). Thus

$$\frac{d}{dx}(\sec x) = \frac{d}{dx}\{(\cos x)^{-1}\}$$

$$= -1(\cos x)^{-2}.-\sin x$$

$$= \frac{\sin x}{\cos^2 x} = \frac{1}{\cos x}.\frac{\sin x}{\cos x} = \sec x \tan x,$$

identifying $\cos x$ with ϕ, and putting $n = -1$.

Similarly, $d(\text{cosec } x)/dx$ can be found to equal $-\text{cosec } x \cot x$, and these two important results are set out in a table for emphasis:

$d(\sec x)/dx = \sec x \tan x$
$d(\text{cosec } x)/dx = -\text{cosec } x \cot x$

Example 6.3

Write down the derivative of $\sqrt{(8-x^2)}$.

Let $\phi = 8-x^2$: the problem then reduces to finding the derivative of $\phi^{\frac{1}{2}}$. The derivative of $\phi^{\frac{1}{2}}$ is $\frac{1}{2}\phi^{-\frac{1}{2}}.\phi'$, and since $\phi' = -2x$, the derivative of $\sqrt{(8-x^2)}$ can be written as $\frac{1}{2}(8-x^2)^{-\frac{1}{2}}.-2x$, or

$$\frac{-x}{\sqrt{(8-x^2)}}.$$

Example 6.4

Find the derivative of $\sin^3 x$.

Let $\phi = \sin x$.

Then

$$\frac{d}{dx}(\sin^3 x) = \frac{d}{dx}(\phi^3) = 3\phi^2.\phi'.$$

But if $\phi = \sin x$,

$$\phi' \text{ or } \frac{d\phi}{dx} = \cos x;$$

thus

$$\frac{d}{dx}(\sin^3 x) = 3\sin^2 x \cos x.$$

The derivative of functions such as $\sin(x^3)$ can be obtained by making the substitution $z = x^3$. This gives $y = \sin z$, and then $dy/dz = \cos z$, and, from the substitution equation, $dz/dx = 3x^2$. Thus

$$\frac{dy}{dx} = 3x^2 \cos(x^3).$$

More generally, the derivative of $\sin \phi$, where ϕ is a function of x, can be found by letting $z = \phi$. Equation (6.3) then leads to a result which is usefully borne in mind, viz.

$$\frac{d}{dx}(\sin \phi) = \phi' \cos \phi. \qquad (6.5)$$

Once more it should be noticed that the inner function (the ϕ of $\cos \phi$) is unaltered in form, and ϕ' appears as a multiplying factor.

The most common use of equation (6.5) is exemplified by finding the derivative of a function such as $\sin 3x$. If we think of ϕ as $3x$, we obtain $d(\sin 3x)/dx = 3 \cos 3x$, and generalising this type of relation to an angle of ax, we can write:

$$\frac{d}{dx}(\sin ax) = a \cos ax$$

and

$$\frac{d}{dx}(\cos ax) = -a \sin ax.$$

Another situation in which equation (6.3) finds application arises when a function such as $\sin^3 (5x)$ is to be differentiated. If we replace $5x$ by θ, we can rewrite the derivative $d\{\sin^3(5x)\}/dx$ in the form $\{d(\sin^3\theta)/d\theta\}(d\theta/dz)$. Now $d\{\sin^3 \theta\}/d\theta = 3 \sin^2 \theta \cos \theta$, and since $d\theta/dx = 5$, we obtain the solution as $3 \sin^2 \theta . 5$, or $15 \sin^2(5x) \cos 5x$.

Again, we can see that

$$\frac{d}{dx}\{\sin^3 5x\} = \frac{d}{dx}(\phi^3) \quad \text{if} \quad \phi \equiv \sin 5x$$

$$= 3\phi^2 . \frac{d\phi}{dx}$$

$$= 3\phi^2 . 5 \cos 5x,$$

since

$$\frac{d\phi}{dx} = \frac{d}{dx}(\sin 5x) = 5 \cos 5x.$$

Example 6.5

Differentiate the following functions with respect to x:

(i) $(1+4x)^5$,
(ii) $(3-ax)^{10}$,
(iii) $\dfrac{1}{(2-x)^2}$,
(iv) $\sin 3x$,
(v) $\sin (x^4)$,
(vi) $\sin \sqrt{(\frac{1}{2}x)}$,
(vii) $\sec^2 x$.

(i) Bringing the power 5 down in front of the bracket, and reducing the power to 4 gives $5(1+4x)^4$; this has to be multiplied by the derivative of the contents of the bracket. As the derivative of $(1+4x)$ is simply 4, we get the derivative of the whole function as $5(1+4x)^4.4$, or $20(1+4x)^4$.

By substitution, the reasoning would proceed as follows: Let $z = 1+4x$; then setting y equal to the given function gives the equation $y = z^5$. Then

$$\frac{dy}{dz} = 5z^4, \quad \text{and} \quad \frac{dz}{dx} = 4.$$

So

$$\frac{dy}{dx} = \frac{dy}{dz} \cdot \frac{dz}{dx}$$

$$= 5(1+4x)^4.4$$

$$= 20(1+4x)^4.$$

(ii) In considering the derivative of $(3-ax)^{10}$, care must be exercised over the sign of the derivative of the function inside the bracket. This is $-a$, and not just a, so we get

$$\frac{dy}{dx} = 10(3-ax)^9.(-a)$$

$$= -10a(3-ax)^4.$$

(iii) $$\frac{d}{dx}\left(\frac{1}{2-x}\right)^2 = \frac{d}{dx}(2-x)^{-2}$$

$$= -2(2-x)^{-3}.-1 = \frac{2}{(2-x)^3}.$$

(iv) $\dfrac{d}{dx}(\sin 3x) = 3 \cos 3x$, the multiplying factor 3 being the derivative of the angle $3x$.

(v) $$\frac{d}{dx}(\sin x^4) = \cos (x^4) \frac{d}{dx}(x^4)$$

$$= 4x^3 \cos (x^4).$$

Notice that the angle is unchanged.

(Or let $z = x^4$; then $y = \sin z$,

$$\frac{dz}{dx} = 4x^3, \text{ and } \frac{dy}{dz} = \cos z;$$

thus $\dfrac{dy}{dx} = \cos z.4x^3 = 4x^3 \cos x^4.)$

(vi) $\dfrac{d}{dx}[\sqrt{\sin(\frac{1}{2}x)}] = \dfrac{d}{dx}\{(\sin\frac{1}{2}x)^{\frac{1}{2}}\}$

$= \frac{1}{2}\frac{1}{2}x)^{-\frac{1}{2}}.\ \frac{1}{2}\cos\frac{1}{2}x$

$= \dfrac{\cos\frac{1}{2}x}{4\sqrt{\sin\frac{1}{2}x}}.$

(vii) $\dfrac{d}{dx}[\sec^2 x] = \dfrac{d}{dx}\{(\sec x)^2\}$

$= 2(\sec x)^1.\sec x \tan x = 2\sec^2 x \tan x.$

EXERCISE 6b

Differentiate the following expressions with respect to x:

1. $(3-x)^4$.
2. $(1+3x^2)^{-6}$.
3. $(1+\sin x)^{10}$.
4. $3\sin 8x$.
5. $\frac{1}{3}\cos 15x$.
6. $\sin\frac{1}{4}x$.
7. $\cos(3x-1)$.
8. $3\left(x-\dfrac{1}{x}\right)^8$.
9. $\dfrac{5}{(1+x^2)}$.
10. $\dfrac{1}{9-5x}$.
11. $4\sec\frac{1}{2}x$.
12. $\cos(3-7x)$.
13. $\sin(x-\frac{1}{2}\pi)$.
14. $(\sin x)^2$.
15. $\dfrac{7}{(1+\cos x)}$.
16. $2\sqrt{\sin x}$.
17. $\cos\left(\dfrac{x-1}{8}\right)$.
18. $\frac{1}{4}\sin(4-2\pi x)$.
19. $(\sin 2x)^{-1}$.

Find the differential coefficients of:

20. $(2+3x)^5$.
21. $\left(1+\dfrac{1}{x}\right)^7$.
22. $(\sin x+\cos x)^{20}$.
23. $(1+\sqrt{x})^4$.
24. $\dfrac{1}{(1-x^2)}$.
25. $3\sqrt{(1+x^2)}$.
26. $3(\cos x^4)$.
27. $\dfrac{1}{\sqrt{(\cos 2x)}}$.
28. $1/(9+x)$.
29. $1/(14-x^4)^{\frac{1}{4}}$.
30. $3+8\sec^2 x$.
31. $-7\sin^3 x$.
32. $\operatorname{cosec}\frac{1}{2}x$.
33. $-5(1-x^4)^4$.
34. $\sqrt{\{1+\sqrt{(5x)}\}^3}$.
35. $\cos(7x^2+5)$.
36. $\cos\sqrt{x}$.
37. $\sqrt{(\sin 3x)}$.

38. Differentiate the following with respect to the variable p:

(i) $p^{-\frac{1}{7}}$.
(ii) $(a-bp)^{20}$.
(iii) $(2-7p)^8$.
(iv) $\left(p^2-\dfrac{1}{p^2}\right)^3$.
(v) $1+5\sec 3p$.
(vi) $\sin(3p^2)$.
(vii) $1/(1-2p)$.
(viii) $\cos(p^2)$.
(ix) $3-\sin^2 p$.
(x) $\cos\sqrt{p}$.
(xi) $\sqrt{(\cos p)}$.
(xii) $(3+p^3)^{-4}$.
(xiii) $-\operatorname{cosec}^2 p$.
(xiv) $(1+4p)^3$.
(xv) $6/(1-\sin p)$.

39. Write down the derivatives of the following:

(i) $x^{18\frac{1}{2}}$.
(ii) $(3x^2)^4$.
(iii) $(\sqrt{x})^{-2}$.
(iv) $\sin^3 x$.
(v) $1/(1-\sin^2 x)$.
(vi) $4x(1-x^2)^2$.
(vii) $\sec 3x/\cos 3x$.
(viii) $5\left(1-\dfrac{1}{x^3}\right)^7$.
(ix) $3x\sqrt{(1-x)}$.
(x) $7\operatorname{cosec} 5x$.
(xi) $\sin^2(x^2)$.
(xii) $(\cos x+\sin x)^{-2}$.
(xiii) $\sqrt{(7-6x-x^2)}$.
(xiv) $\sin^4 5x$.
(xv) $\sqrt{(1-x^2)}$.
(xvi) $\sqrt{(1-2x-x^2)}$.
(xvii) $(8+5x+3x^2)^3$.

40. The height h m of the tide t hours after noon on a certain day at a certain place is given by the formula $h = 7+5\sin(4\pi t/25)$. Find:

(i) the rate in m/hr at which the tide is rising at 5 minutes past 2 on this day, giving your answer correct to two places of decimals;
(ii) the time of the high tide on the next day and the value of h at this time;
(iii) the difference between the first two times after noon at which the tide is 10 m, giving your answer to the nearest minute.

(NUJMB)

41. The motion of a particle is described by the equation

$$s = 10\cos(5t-\tfrac{1}{4}\pi).$$

What are the maximum values of the displacement s, the velocity and the acceleration of the particle? At what times t do these maxima occur? What are the minimum values of these quantities?

42. If $y = \sin(xn+t)$, show that $d^2y/dx^2 = -n^2y$. Show also that if $z = \sin(nx+t)+a/n^2$, then $d^2z/dx^2 = -n^2z+a$.

43. Find the maximum value of $\{x-\sqrt{(4x^2+3)}\}$.

†**44.** Differentiate $x^6(1+1/x^2)^3$ with respect to x.

†**45.** Show that dy/dx is equal to $1/(dx/dy)$. Assume that

$$\lim_{\delta x=0}\left\{\frac{1}{f(\delta x)}\right\} = \frac{1}{\lim_{\delta x=0}\{f(\delta x)\}}.$$

†**46.** Given that $d[\xi(z)]/dz = 1/(1+z^2)$, where $\xi(z)$ is a particular form of function, find the derivatives of $\xi(3x)$, $\xi(\sin x)$ and $\xi(x^2)$.

†**47.** Show that if $p = \mathrm{d}y/\mathrm{d}x$, the derivative $\mathrm{d}^2y/\mathrm{d}x^2$ can be written as $\mathrm{d}p/\mathrm{d}x$ or $p(\mathrm{d}p/\mathrm{d}y)$.

†**48.** Show that the equation of the tangent to the ellipse

$$\frac{x^2}{a^2}+\frac{y^2}{b^2} = 1 \quad \text{at} \quad (x_1, y_1) \quad \text{is} \quad \frac{xx_1}{a^2}+\frac{yy_1}{b^2} = 1.$$

***49.** The optical path of a ray of light travelling in a medium of absolute refractive index μ is μx, where x is the actual length of the path in space.

A point X is situated in air (whose refractive index is 1) at a distance a from a boundary between the air and water (whose refractive index is μ). A second point Y is situated in the water at a distance b from the interface between the water and the air, at a distance c to the right of X (see Fig. 6.1). A ray of light

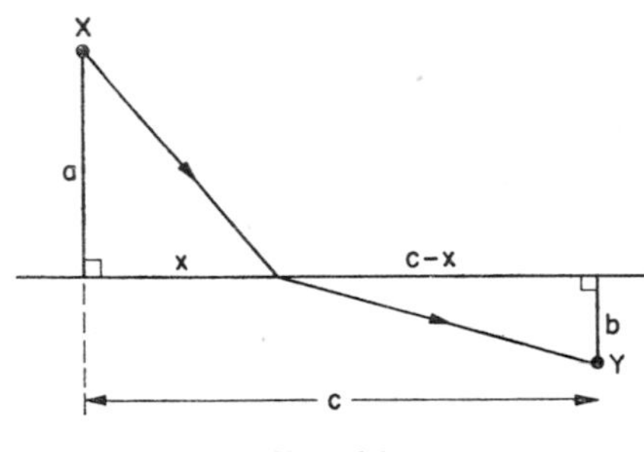

FIG. 6.1.

travels from X to Y striking the interface at a distance x to the right of X as shown in the figure. Using the principle that the optical path of such a ray is a minimum, find the relation between a, b, c and x. Explain the meaning of the equation $\sin i/\sin r = \mu$, and deduce this from your condition.

†**50.** Prove that the base radius r of a right circular cone of fixed volume V is given by the equation $\pi r^3 = 3V\sqrt{2}$, if the length of the slant edge is a minimum.

6.1.2. There is, in fact, no need to introduce any other variable such as y or z when finding the derivative of an expression like $(9+4x^2)^5$. We have seen that the derivative of a compound function of a general form $f\{\phi(x)\}$ can be obtained by making the substitution $z = \phi(x)$, and then evaluating

$$\frac{\mathrm{d}}{\mathrm{d}z}\{f(z)\}\,\frac{\mathrm{d}z}{\mathrm{d}x}.$$

Now this expression summarises the method of finding the derivative *with respect to x* of the function $f\{\phi(x)\}$ or $f(z)$, i.e.

$$\frac{\mathrm{d}}{\mathrm{d}x}\{(z)\} - \frac{\mathrm{d}}{\mathrm{d}z}\{f(z)\}\frac{\mathrm{d}z}{\mathrm{d}x}. \qquad (6.6)$$

Put this way, we are merely expressing equation (6.3) in a slightly different way, and we can paraphrase both equations by saying that the derivative of a function of one variable z with respect to another variable x is equivalent to the product of its own derivative $f'(z)$ and $\mathrm{d}z/\mathrm{d}x$. That is, to differentiate a function $f(z)$ with respect to x, differentiate it with respect to z and multiply the derivative so obtained by $\mathrm{d}z/\mathrm{d}x$. Thus:

$$\frac{\mathrm{d}}{\mathrm{d}x}(9z^2) = \frac{\mathrm{d}}{\mathrm{d}z}(9z^2)\,\frac{\mathrm{d}z}{\mathrm{d}x} = 18z\,\frac{\mathrm{d}z}{\mathrm{d}x},$$

$$\frac{\mathrm{d}}{\mathrm{d}x}(\sin z) = \frac{\mathrm{d}}{\mathrm{d}z}(\sin z)\,\frac{\mathrm{d}z}{\mathrm{d}x} = \cos z\,\frac{\mathrm{d}z}{\mathrm{d}x},$$

$$\frac{\mathrm{d}}{\mathrm{d}t}(8x^4-7) = \frac{\mathrm{d}}{\mathrm{d}x}(8x^4-7)\,\frac{\mathrm{d}x}{\mathrm{d}t} = 32x^3\,\frac{\mathrm{d}x}{\mathrm{d}t}.$$

After a little practice, the middle term can be omitted, and we can write the derivative directly in the product form, e.g.

$$\frac{\mathrm{d}}{\mathrm{d}y}(4x^3) = 12x^2\,\frac{\mathrm{d}y}{\mathrm{d}x},$$

or

$$\frac{\mathrm{d}}{\mathrm{d}x}(\cos t) = -\sin t\,\frac{\mathrm{d}t}{\mathrm{d}x}.$$

An application of the relationship expressed in equation (6.6) is commonly found in problems involving rates of change. The situation which often arises occurs in a three-variable problem, where it is useful to differentiate an equation relating two variables throughout with respect to the third variable. For example, if we know that a sphere is being inflated at a rate of 10 cm³/s, the rate of increase of its radius r when $r = 20$ cm can be found by differentiating the equation relating the volume V to the radius r throughout with respect to the time t. The equation relating V to r

is, as usual, $V = \frac{3}{4}\pi r^3$, and we can therefore write that

$$\frac{d}{dt}(V) = \frac{d}{dt}\left(\frac{4}{3}\pi r^3\right).$$

Now $d(V)/dt$ is simply the time rate of increase of the volume dV/dt, and we can evaluate the right-hand side of the equation by using the expression given in equation (6.6):

$$\frac{d}{dt}\left(\frac{4}{3}\pi r^3\right) = \frac{d}{dr}\left(\frac{4}{3}\pi r^3\right)\frac{dr}{dt},$$

where the differentiation of the right-hand side has now been rephrased in terms of differentiation with respect to r, the variable of which it is a function.

Then

$$\frac{dV}{dt} = 4\pi r^2 \frac{dr}{dt},$$

and knowing that $dV/dt = 10$ and $r = 20$, dr/dt can be immediately evaluated as

$$\frac{10}{4\pi(20)^2} \quad \text{or} \quad \frac{1}{160\pi}\,\text{cm/s}.$$

Example 6.6

By differentiating the equation

$$s^2+5s = 8t^3-7$$

throughout with respect to t, find an expression for ds/dt.

Writing the equation in the operative form

$$\frac{d}{dt}(s^2)+\frac{d}{dt}(5s) = \frac{d}{dt}(8t^3-5),$$

we can proceed to the equation

$$2s\frac{ds}{dt}+5\frac{ds}{dt} = 24t^2.$$

Thus
$$(2s+5)\frac{ds}{dt} = 24t^2,$$

and
$$\frac{ds}{dt} = \frac{24t^2}{2s+5}.$$

Example 6.7

Differentiate the equation $x^2+y^2 = 25$ throughout with respect to x, and hence find the equation of the tangent of the curve $x^2+y^2 = 25$ at the point (3, 4).

On differentiating the equation $x^2+y^2 = 25$ with respect to x, we obtain

$$2x+2y\frac{dy}{dx} = 0,$$

and the gradient function of the curve is thus $dy/dx = -x/y$.

At (3, 4) the gradient can then be evaluated as $-\frac{3}{4}$, so the tangent to the curve at this point has the equation

$$(y-4) = -\tfrac{3}{4}(x-3),$$

i.e.

$$4y-16 = -3x+9,$$

or

$$3x+4y-25 = 0.$$

EXERCISE 6c

1. Differentiate the following equations with respect to x:

(i) $x^2+y^2 = 9$,
(ii) $5x^2-9y^2 = 1$,
(iii) $x+11y^2 = 7$,
(iv) $x+y^2-8y = 5$,
(v) $x(x-3)+y(y+2)-18 = 0$.

2. Find the gradient functions of the curves:

(i) $3x^2-2y^2 = 6$,
(ii) $x(x+1)-y(y-1) = 5$,
(iii) $x^2+6x+y^2-3y-9 = 0$.

3. Explain geometrically why the gradient of the radius from the centre of the circle $x^2+y^2 = a^2$ to a point (x, y) on its circumference is y/x, and deduce that the gradient of the tangent at that point is $-x/y$. Also derive this latter fact by differentiating the equation of the circle.

4. Differentiate the equation $3x-5y+7 = 0$ throughout with respect to x, and hence show that the gradient of $3x-5y+7 = 0$ at any point is independent of the coordinates of the point. Interpret your results geometrically.

5. Find the equations of the tangents to the following curves at the points indicated:

(i) $x^2+y^2 = 25$, $(-3, -4)$,
(ii) $2x^2-y^2 = 2$, $(3, -4)$,
(iii) $5x-y^2+3y = 0$, $(2, 5)$,
(iv) $3x^2-2x+4y^2-2y-21 = 0$, $(1, -2)$.

6. Find expressions for $\mathrm{d}t/\mathrm{d}x$ from the equations

(i) $\sin x+\cos t = 0$, (ii) $x^4-t^4+t^2-3 = 0$.

7. The area of a circular ink blot is increasing at the rate of 1 cm²/s. Find the rate of increase of the radius r when

(i) $r = 0{\cdot}1$ cm, (ii) $r = 0{\cdot}5$ cm, (iii) $r = 1$ cm.

What is the radius when $\dot{r} = 0{\cdot}1$ cm/s?

8. At what rate is the volume of a spherical balloon decreasing when the radius of the balloon is decreasing at the rate of 0·6 cm/hr and when the radius equals 10 cm?

9. The end P of a rod PQ of length 5 cm slides along a horizontal bar as shown in Fig. 6.2, the other end Q sliding along a vertical bar joined to the horizontal one at O. Find the relation between a and b,

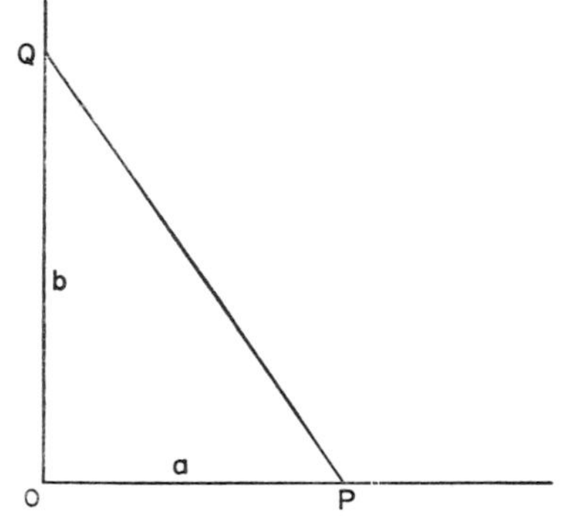

FIG. 6.2.

where a and b are the distances of P and Q, respectively, from O. Hence find the relationship between the rate of change of a and that of b.

If $\mathrm{d}a/\mathrm{d}t = 1$ cm/s, what is the value of $\mathrm{d}b/\mathrm{d}t$ when $a = 3$ cm?

10. The cross-section of a trough is an isosceles triangle with its vertex downwards. The height of the triangle is 2 m, its base is 2·5 m, and the length of the trough is 6 m. Water runs into the trough at the rate of 9 m³/min. Prove that, when the depth of water is x m, the volume of water is $\frac{15}{4}x^2$ m³, and find in m/min the rate at which the level is rising after $1\frac{1}{2}$ minutes. (OC)

***11.** The energy E stored in a capacitor whose capacity is C is $Q^2/2C$, where Q is the charge on the plates. If Q is constant, write down an expression for $\mathrm{d}E/\mathrm{d}x$, where x is the distance between the plates, and by using the relation $C = \varepsilon_0 A/x$, where A is the area of the capacitor plate, find an expression for the force F between the plates, given that $F = -\mathrm{d}E/\mathrm{d}x$. ($\varepsilon_0$ can be treated as a constant.)

Is this a force of attraction or repulsion? Relate your answer to the sign of your expression for F.

***12.** If an object is placed at a distance u from a lens of focal length f, the image of the object formed by the lens would be at a distance v from the lens, where $1/u+1/v = 1/f$. Find the rate of change of v with respect to u. Hence show that if the object is moved *away* from the lens through a small distance δu, the image moves *towards* the lens through a distance $v^2/u^2.\delta u$.

Explain why the magnification of a short object lying along the axis is v^2/u^2, and discuss whether or not the image is distorted.

13. A man is walking away from a street lamp at 1 m/s. If the man is 2 m high and the lamp is 3 m above the ground, find the rate at which the length of his shadow is increasing. (OC)

14. The volume (V m³) of water in a given vessel is given by $V = 2x^2+2x$, where x is the depth of the water in metres. If water is entering the vessel at a constant rate of 0·03 m³/s, find the rate at which the depth of water is increasing when $x = \frac{3}{4}$. Express your answer in metres per minute. (OC)

15. At an instant when the radius is 6 cm, the area of a circle is increasing at the rate of 20 cm² per second. Find the rate of increase of the radius at this instant, giving your answer correct to two significant figures. (OC)

16. A ladder, of length 10 m, is placed with its lower end, A, on horizontal ground and its upper end, B, in contact with a vertical wall. The end A slides directly away from the wall with a constant speed of 2 m/s. Find the rate of descent of B at the instant when A is 6 m from the wall. (OC)

***17.** The energy E stored in a system composed of a cell of e.m.f V connected permanently to a capacitor of capacity C is $BV-\frac{1}{2}CV^2$, where B is a constant. If $C = \varepsilon_0 A/x$, where A and x are the area of and distance between the capacitor plates respectively, find an expression for $\mathrm{d}E/\mathrm{d}x$, and by using the relation $F = -\mathrm{d}E/\mathrm{d}x$, find an expression for the force F between the plates. (ε_0 can be treated as a constant.)

Deduce from the sign of your answer whether this is a force of attraction or repulsion.

18. The volume V of a solid cube increases uniformly at the rate of 8 cm³/s. Find the rate of increase of its surface area S when the area of a face is 9 cm². (Start by finding the relationship between S and V, and differentiate it throughout with respect to time t.)

19. The surface area of a liquid in a hemispherical bowl is S when its depth is x. If a small volume δV of water is added, the depth of the liquid increases by δx, and the surface area by δS. Explain why

$$S\delta x \leqslant \delta V \leqslant (S+\delta S)\,\delta x,$$

and deduce that $\mathrm{d}V/\mathrm{d}x = S$.

If the bowl is being filled at a rate of 10 cm^3/s, find the rate of increase of depth when the surface area of water is 20 cm^2.

*20. A ray of light enters a spherical water drop making an angle i with the normal at the point incidence. It is reflected off the inside of the water surface once, and emerges into the air again. Show that the ray is deviated through an angle D, where $D = 180+2i-4r$, where $\sin i = \mu \sin r$, and where μ is the refractive index of water.

Show that a stationary value of D occurs when $dr/di = \frac{1}{2}$, and by differentiating the equation $\sin i = \mu \sin r$, show that $\mu \cos r = 2 \cos i$. Hence show that $\cos i = \sqrt{(\mu^2-1)/3}$.

*21. Find the angle of deviation of a ray which enters a spherical water drop at an angle i with the normal, but which suffers two internal reflections instead of one before re-emerging (see question 20). Show that the angle of deviation has a stationary value when $\cos i = \sqrt{(\mu^2-1)/8}$.

22. The surface area of a liquid in a hemispherical bowl of radius a is S when its depth is x. Show that $S = \pi x(2a-x)$. Hence find the rate at which the surface area is increasing when x is 2 cm and increasing at the rate of 0·2 cm/s, given that the radius of the bowl is 5 cm.

23. The rate at which a block of ice melts is proportional to the surface area S. Write down an equation expressing this fact, and find the rate of change of the length of an edge of a cubical block of ice, given that the constant of proportionality between the rate of decrease of volume in cm^3/s and the surface area in cm^2 is 10^{-3} (cm s^{-1}).

*24. The flux density on the axis of a magnet of moment M at a distance x from one end is approximately $2M/x^3$.

Write down an expression for the total flux linking a ring of radius r placed symmetrically with its plane horizontal and perpendicular to the vertical axis of the magnet, and at a distance x from it. If the magnet is allowed to fall towards the ring, the flux linking the circuit will change. Derive an expression for the rate at which the flux changes in terms of the velocity of the magnet.

25. The abscissa of the point P increases at the rate of b cm/s, and P moves along the curve $y = 4ax^2$. If P is initially at $(a, 4x^3)$, find an expression for the rate at which its ordinate is increasing after t seconds.

26. A hemispherical basin of radius 60 cm contains water which runs out at a rate of 20 cm^3/s. Prove that the rate at which the depth of water is falling when the basin is full is three-quarters of that when the depth is 30 cm. (When a cap of height x is cut from a sphere of radius r, the volume is $\frac{1}{3}\pi(3rx^2-x^3)$.) (OC)

27. The rate of escape of air from a spherical balloon is proportional to the area of the hole through which it escapes. If this area is a constant fraction of the total surface area of the balloon as the latter reduces its size, show that the radius decreases at a constant rate as the balloon shrinks.

28. The volume of water in a vessel is given by the formula $V = 3\pi x^3/16$, where x is the depth of the water and V is its volume. When the depth is 30 cm, the water is running into the vessel at the rate of 4·8 cm^3/min. Find, by the method of "small increases", the approximate increase in the depth of the water during the next 10 seconds. Give your answer as a decimal of a cm, correct to two significant figures. (Take π to be 3·142.) (OC)

29. A cylindrical jar, of radius a cm, stands on a table. If water is poured into the jar at a rate of k cm^3/s, at what rate is the depth of water in the jar increasing? (L)

30. A vessel is constructed so that the volume of water contained in it is

$$\frac{\pi}{192}(x^3+24x^2+192x) \text{ cm}^3,$$

when the depth is x cm.

What is the rate of increase of volume per unit increase of x when (i) $x = 2$, (ii) $x = 4$?

How many times faster does the surface rise when $x = 2$ than when $x = 4$, if water is poured in at a constant rate? (OC)

31. The surface area of a liquid contained in a jar is S. Show that the depth of the liquid increases by δx if a small volume δV is added, where $\delta V \doteqdot S\delta x$. Hence show that

(i) the rate of increase of the volume with respect to x is S,
(ii) the time rate of increase of the volume is S times as great as that of the depth of the liquid.

32. A block of ice in the form of a cube, whose edge is 2 metres, begins melting and its volume decreases at a constant rate, the block remaining cubical. If the rate of melting is such that the edge measures 1 metre after 16 hours, find

(i) the length of edge after 16 hours,
(ii) the rate at which the length of the edge is decreasing at this time. (NUJMB)

33. Given that $y = (x^2-1)^n$, where n is a positive integer, prove that

$$(x^2-1)\frac{dy}{dx}-2nxy = 0. \quad \text{(OC)}$$

34. A vessel is such that when the depth of water in it is x cm the volume of water is $x^2(1+\frac{1}{3}x)$ cm^3. What is then the area of the surface of the water? If water is poured in at the rate of 10 cm^3/min, at what rate is the level rising when the volume is 18 cm^3? At what rate is the area of the surface of the water then increasing?

6.2. THE PRODUCT RULE

The differentiation of a function such as $x^2 \sin x$ presents a new problem in that no substitution will be sufficient to reduce the function to a form which is differentiable with the rules and techniques already derived. It is in fact necessary to develop from first principles a special technique which will deal with this "product" type function.

To generalise the reasoning, replace the first factor in the example quoted (x^2) by the symbol u, where u is an abbreviation for $u(x)$ and stands for a function of x. Replace the second factor ($\sin x$) by v, which stands

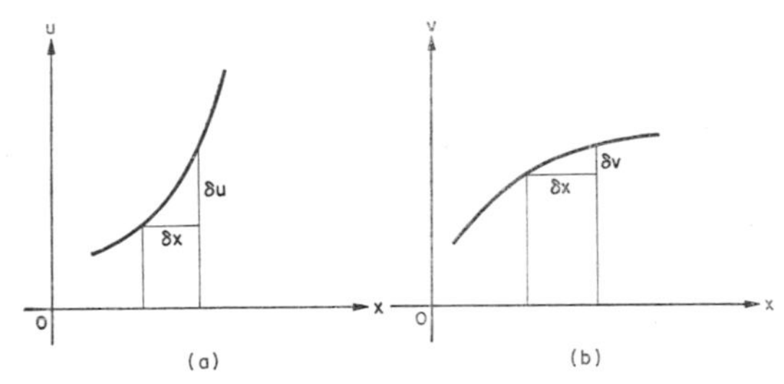

FIG. 6.3. (a) If u is a function of x, a change of δx in x will lead to a change of δu in u; (b) if v is a function of x, a change of δx in x will lead to a change of δv in v.

for another function $v(x)$ of x. The problem then generalises to finding an expression for the derivative of a product of functions uv.

If x increases by δx, the resulting changes in the values of u and v can be denoted by δu and δv* (Fig. 6.3). Denoting the product uv by a third function symbol y, we can write

$$y = uv,$$

and

$$y+\delta y = (u+\delta u)(v+\delta v),$$

where δy is the change in y associated with the change δx in x (Fig. 6.4). Subtracting these equations,

$$\delta y = (u+\delta u)(v+\delta v)-uv,$$

* These can, of course, be zero.

and multiplying out the brackets,

$$\delta y = u\,\delta v+v\,\delta u+\delta u\,\delta v. \qquad (6.7)$$

The original problem was to obtain the derivative of y (i.e. uv) with respect to x, and dividing equation (6.7) throughout by δx, we obtain

$$\frac{\delta y}{\delta x} = u\frac{\delta v}{\delta x}+v\frac{\delta u}{\delta x}+\frac{\delta u}{\delta x}\delta v.$$

The derivative of y with respect to x equals the limiting value of this ratio $\delta y/\delta x$ at $\delta x = 0$, so we can write

$$\frac{dy}{dx} = \lim_{\delta x=0}\left\{u\left(\frac{\delta v}{\delta x}\right)+v\left(\frac{\delta u}{\delta x}\right)+\left(\frac{\delta u}{\delta x}\right)\delta v\right\}.$$

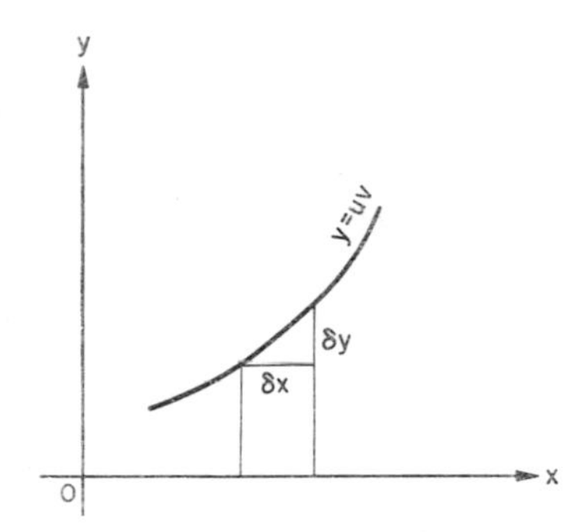

FIG. 6.4. Similarly, if $y = uv$, y will be a function of x, and a change of δx in x can be supposed to lead to a change of δy in y.

Assuming that the limit of a sum equals the sum of the separate limits,

$$\frac{dy}{dx} = \lim_{\delta x=0}\left(u\frac{\delta v}{\delta x}\right)+\lim_{\delta x=0}\left(v\frac{\delta u}{\delta x}\right)+\lim_{\delta x=0}\left(\frac{\delta u}{\delta x}\delta v\right).$$

Now the limit theorem quoted on p. 143 states that the limit of a product is equal to the product of the limits, and thus

$$\frac{dy}{dx} = \lim_{\delta x=0}(u).\lim_{\delta x=0}\left(\frac{\delta v}{\delta x}\right)+\lim_{\delta x=0}(v).\lim_{\delta x=0}\left(\frac{\delta u}{\delta x}\right)+\lim_{\delta x=0}\left(\frac{\delta u}{\delta x}\right).\lim_{\delta x=0}(\delta v).$$

Replacing the limiting values of the delta ratios by the derivatives, and noting that the value of u and v for any particular value of x is not changed by setting δx equal to zero, the equation

$$\frac{dy}{dx} = u\frac{dv}{dx}+v\frac{du}{dx}+\frac{du}{dx}\cdot \lim_{\delta x=0} (\delta v)$$

can be written. And, finally, if the relationship between x and the function $v(x)$ is continuous at the point in question, $\delta v = 0$ if $\delta x = 0$. The last term therefore disappears, and the relationship

$$\boxed{\frac{dy}{dx} = u\frac{dv}{dx}+v\frac{du}{dx}} \qquad (6.8)$$

remains. This relationship expresses the law for obtaining the derivative of a product of two functions: such a derivative is the sum of two terms, the first being the product of one of the functions (say u) and the derivative of the other (dv/dx), the second being the product of the derivative of the first function (du/dx) and the other (undifferentiated) function (v).

This can be written in an alternative form by replacing du/dx and dv/dx with the symbols u' and v', and, doing this, the expression

$$y' = uv'+u'v$$

is obtained.

Example 6.8

Differentiate with respect to x: (i) $x^2 \sin x$; (ii) $\cos x \sin x$; (iii) $x\sqrt{(1-x^2)}$.

(i) Denoting x^2 by u and $\sin x$ by v, we can write the derivatives du/dx and dv/dx as $2x$ and $\cos x$. But

$$\frac{d}{dx}(uv) = u\frac{dv}{dx}+v\frac{du}{dx},$$

and thus in this case

$$\frac{d}{dx}(x^2 \sin x) = x^2 \cos x+2x \sin x$$

$$= x(x \cos x+2 \sin x).$$

(ii) Let $u = \cos x$ and $v = \sin x$. It follows that $u' = -\sin x$ and $v' = \cos x$, and thus

$$\frac{d}{dx}(\cos x . \sin x) = \cos x . \cos x + \sin x . -\sin x$$

$$= \cos^2 x - \sin^2 x.$$

(iii) We can write that

$$\frac{d}{dx}\{x\sqrt{(1-x^2)}\}$$

$$= x\frac{d}{dx}\{\sqrt{(1-x^2)}\} + \sqrt{(1-x^2)}\frac{d}{dx}(x).$$

Since

$$\frac{d}{dx}\{\sqrt{(1-x^2)}\} = \frac{d}{dx}(1-x^2)^{\frac{1}{2}}$$

$$= \tfrac{1}{2}(1-x^2)^{-\frac{1}{2}} . -2x = \frac{-x}{1-x^2},$$

we have $\frac{d}{dx}\{x\sqrt{(1-x^2)}\}$

$$= x\frac{-x}{\sqrt{(1-x^2)}} + \sqrt{(1-x^2)}.1$$

$$= \frac{-x^2}{\sqrt{(1-x^2)}} + \sqrt{(1-x^2)}$$

$$= \frac{-x^2+(1-x^2)}{\sqrt{(1-x^2)}} = \frac{1-2x^2}{\sqrt{(1-x^2)}}.$$

EXERCISE 6d

Differentiate the following functions with respect to x:

1. $x \sin x$.
2. $x^3 \cos x$.
3. $3x \sin 2x$.
4. $x^2 \sin \frac{1}{2}x$.
5. $\cos^2 x \sin x$.
6. $\sin^2 x \cos x$.
7. $\sin^2 4x \cos 4x$.
8. $x\sqrt{(1-x)}$.
9. $x^4(1-x^3)^7$.
10. $x(1+x)^3$.
11. $(5x+4)^6(3x-8)^4$.
12. $x(\sin x - \cos^2 x)$.
13. $\sin^3 x \cos^2 3x$.
14. $\sin \frac{1}{2}x \sec \frac{1}{2}x$.
15. $\cos x \sqrt{(\operatorname{cosec} x)}$.
16. $(1-x^3)(1+x^2)$.
17. $\sin ax \cos ax$.
18. $\frac{3x}{(1-5x^2)}$ (write as $3x(1-5x^2)^{-\frac{1}{2}}$).
19. $(x^2-7) \sin^2 x$.
20. $x^3(1-7x)^{10}$.

21. What is the differential coefficient with respect to R of $E^2R(R+r)^{-2}$? For what value of R is the differential coefficient zero?

22. Sketch the graph of $y = x\sqrt{(1-x)}$.

23. Find the velocity and acceleration of a particle whose displacement s varies with time t according to the equation $s = 5 \sin 2t \sin 3t$.

24. Find the stationary values of $x^8(1-x^2)^5$.

25. By writing $\tan x$ as $\sin x \sec x$, find an expression for $\mathrm{d}(\tan x)/\mathrm{d}x$. What is $\mathrm{d}(\cot x)/\mathrm{d}x$?

26. Write expressions for

$$\text{(i)}\ \frac{\mathrm{d}}{\mathrm{d}t}(\sin \omega t \cos \omega t),\ \text{(ii)}\ \frac{\mathrm{d}}{\mathrm{d}t}(\tfrac{1}{2}\sin 2\omega t).$$

Why are the results the same?

27. Find an expression for $\mathrm{d}(\sin 3\theta \cos 2\theta)/\mathrm{d}\theta$ by

(i) using the product rule,
(ii) expressing $(\sin 3\theta \cos 2\theta)$ in terms of 5θ and θ, and then differentiating.

Show that your answers are the same.

28. Find the rate of change of z with respect to w if $z = w^3(3-w^2)^4$.

29. Form an equation whose solution gives the values of θ for which the function $\theta \cos \theta$ takes its maxima and minima (θ being measured in radians). Prove that this equation has a root between $\frac{1}{4}\pi$ and 1, and use an accurate graph to locate this root as closely as you can. Hence find the maximum value of $\theta \cos \theta$ for acute angles. (SU)

30. The height of a cone remain constant at 4 cm. The radius of the base increases by 0·1 cm/s. When the radius is 10 cm, calculate the rate at which

(a) the volume of the cone is increasing;
(b) the curved surface area A is increasing.

(Assume that $A = \pi r\sqrt{(r^2+4^2)}$ where r is the radius of the base.)

Leave your answers in terms of π. Surds need not be evaluated. (C)

31. Water is being poured at the rate of 5 cm³/s into a hollow vessel in the form of a right circular cone with vertex downwards. When the level of water in the vessel is 2 cm the rate of increase of depth is 1 cm/s. Find the rate of increase of depth of water when it is 3 cm deep. (OC)

32. $ABCD$ is a square field and the length of the diagonal is $2a$ km. A man starts to walk from A straight across to the opposite corner C at a speed of 4 km/h. At the same instant a second man starts to walk from B straight across to D at a speed of 3 km/h. Show that after t hours ($t \leqslant \frac{1}{2}a$) their distance apart is

$$\left\{(a-3t)^2+(a-4t)^2\right\}^{\frac{1}{2}}.$$

Prove that their least distance apart is $a/5$. Find how far each man is from the centre of the square at this moment. (OC)

33. What is the area of the greatest rectangle which can be fitted in a quadrant of the circle of radius a?

34. A rod of length l is to be cut into three bits to form an isosceles triangle. Prove that the area enclosed is greatest when the triangle is equilateral.

35. The displacement x of a body from a fixed point is given by the expression $t \sin(\frac{1}{2}\pi t)$. Derive an expression for its velocity and acceleration, given that t is the time, and describe the motion exhibited by the body.

36. Find the derivative of $\sqrt{(x \sin x)}$, and find the limiting value of this derivative as

$$\text{(i)}\ x \to 0+, \quad \text{(ii)}\ x \to 0-.$$

What can be said about the value of the derivative at $x = 0$?

37. Show that the product of three positive numbers a, x and y, where $a+x+y = 3$, is greatest when $x = y$ if the value of a is fixed. Deduce that the maximum value of axy is 1.

38. The ends A, B, of a rod AB of length 4 dm move on two perpendicular lines Ox, Oy. At time t seconds $(0 \leqslant t \leqslant 2)$, the position of A is given by $OA = 4t-t^2$. Prove that A starts from O with velocity 4 dm/s, and that B reaches O when $t = 2$.

Prove also that $OB = (2-t)\sqrt{(4+4t-t^2)}$, and show that when B reaches O its velocity is $-2\sqrt{2}$ dm/s. (OC)

39. Prove that, if u and v are functions of x, then

$$\frac{\mathrm{d}}{\mathrm{d}x}(uv) = u\frac{\mathrm{d}v}{\mathrm{d}x}+v\frac{\mathrm{d}u}{\mathrm{d}x}.$$

Differentiate with respect to x:

(i) $(x-1)(x^3+1)$,
(ii) $\cos x + x \sin x$, (OC)
(iii) $x^2\sqrt{(2-3x)}$.

40. A solid consists of a circular cylinder of radius r and height h, joined to a right circular cone whose base-radius and height are both r, the flat base of the cone being in contact with one of the circular ends of the cylinder. The total area of the solid is S. Prove that its volume is

$$r\{3S-\pi r^2(1+3\sqrt{2})\}/6.$$

(The curved surface of the circular cone is $\pi r^2\sqrt{2}$; the volume is $\frac{1}{3}\pi r^3$.)

Hence show that, for a given total surface, the volume of the solid is greatest when $h = r\sqrt{2}$. (OC)

41. A man in a boat on one side of a canal 30 m broad wishes to get as quickly as possible to a point 50 m along the canal bank from the point opposite to him. He can row at 1·5 m/s and run at 3 m/s. In what direction should he steer, and how long will he take?

Obtain the answers to the same questions if the point to be reached is (i) 80 m, (ii) 12·5 m along the bank. (SU)

*42. The minimum angle of deviation D of a ray of light passing through a prism of angle A is given by the equation

$$\mu = \frac{\sin\{(A+D)/2\}}{\sin \frac{1}{2}A},$$

where μ is the refractive index of the material of which the prism is made. Derive an expression for the rate of change of μ with D for a prism of angle 60°.

If $\mu = 1{\cdot}65$, and $\mathrm{d}\mu/\mathrm{d}\lambda = -925\ \mathrm{cm}^{-1}$, where λ is the wavelength of the light passing through the prism, find a value for the dispersive power $dD/\mathrm{d}\lambda$ for the prism when light of this particular wavelength is used.

43. Prove that, if $y = x^p \sin qx$, then

$$x^2 \frac{\mathrm{d}^2y}{\mathrm{d}x^2} - 2px\frac{\mathrm{d}y}{\mathrm{d}x} + \{p(p+1)+q^2x^2\}y = 0.$$

(SU, part question)

†**44.** Find the gradient function of $y = x \operatorname{cosec} x$. Show that for $0 \leqslant x \leqslant \pi/2$, the gradient function is positive. At what value of x does y begin to decrease? Sketch the graph of $x \operatorname{cosec} x$.

†**45.** A cone of semi-vertical angle θ is inscribed in a sphere of radius a. Prove that the volume of the cone is $\frac{8}{3}\pi a^3 \sin^2\theta \cos^4\theta$, and find the area of the curved surface.

Prove that if a is fixed and θ allowed to vary, the maximum volume of the cone is 8/27 of the volume of the sphere.

Prove further that if the volume of the cone is a maximum, the area of its curved surface is also a maximum. (OC)

†**46.** Find the maximum and minimum values of the function $y = (4x+1)(x-1)^4$.

A beam of rectangular cross-section is to be cut from a cylindrical log of diameter d. The stiffness of such a beam is proportional to xy^3, where x is the breadth and y the depth of the section. Find the cross-sectional area of the beam

(i) of greatest volume,
(ii) of greatest stiffness,

that can be cut from the log. (OC)

†**47.** Writing x^{p+1} as the product $x.x^p$, find an expression for the derivative of x^{p+1}, given that the derivative of x^p is px^{p-1}, and the product rule. Hence deduce that the expression px^{p-1} gives the derivative of x^p for $p = n+1$ if it gives it for $p = n$. Hence show that the expression px^{p-1} is valid for all integral values of p.

†**48.** Sketch the graph of $(1-x)^3/(1-2x)$.

†**49.** Derive an expression for $\mathrm{d}(uvw)/\mathrm{d}x$, where u, v and w are functions of x. The expression should be written in terms of u, v, w, $\mathrm{d}u/\mathrm{d}v$, $\mathrm{d}v/\mathrm{d}x$ and $\mathrm{d}w/\mathrm{d}x$ only. Use your result to find $\mathrm{d}\{x(1+2x)^4(1-x^2)^5\}/\mathrm{d}x$.

†**50.** Find from first principles an expression for the derivative of a ratio of functions f/g, where f and g are both functions of x, the variable with respect to which the derivative is to be found.

†**51.** The probability of gaining r successes in n trials is

$$\frac{n!}{(n-r)!\,r!}p^r(1-p)^{n-r}.$$

What must p equal if the chance of getting r successes in n trials is to be as high as possible?

†**52.** Show, from first principles, that

$$\frac{\mathrm{d}}{\mathrm{d}x}\{f(x)+g(x)\} = f'(x)+g'(x).$$

On what limit theorem does your proof rest?

6.3. QUOTIENTS

If it is necessary to find the derivative of a function expressed in the form of a ratio (e.g. $x/(1-x)$), it is often easier to use a technique slightly different from that described in the last section. Expressing the quotient of a function as u/v, where, as before, u and v denote functions of x, we can derive an expression for the derivative of a quotient analogous to that of the derivative of a product. Carrying out a similar procedure, we can write that:

$$\text{if} \qquad y = \frac{u}{v},$$

$$y+\delta y = \frac{u+\delta u}{v+\delta v},$$

and by subtracting these two expressions,

$$\delta y = \frac{u+\delta u}{v+\delta v} - \frac{u}{v}$$

$$= \frac{uv+v\,\delta u - uv - u\,\delta v}{v(v+\delta v)}$$

$$= \frac{v\,\delta u - u\,\delta v}{v(v+\delta v)}.$$

Dividing both sides of the equation by δx,

$$\frac{\delta y}{\delta x} = \frac{v\,(\delta u/\delta x) - u(\delta v/\delta x)}{v(v+\delta v)},$$

and, in the limit, at $\delta x = 0$ (assuming the relationships between x, u, v and y to be continuous),

$$\frac{\mathrm{d}y}{\mathrm{d}x} = \frac{v(\mathrm{d}u/\mathrm{d}x) - u(\mathrm{d}v/\mathrm{d}x)}{v^2},$$

the δv term in the denominator disappearing when δx is set to zero. Replacing y by u/v, the derivative of a quotient is thus given by the equation

$$\frac{\mathrm{d}}{\mathrm{d}x}\left(\frac{u}{v}\right) = \frac{v(\mathrm{d}u/\mathrm{d}x) - u(\mathrm{d}v/\mathrm{d}x)}{v^2}. \quad (6.9)$$

The advantage of this expression over that derived for the product (eqn. (6.8)) lies in the easier manipulation of the denominator. Writing a function such as $(x+1)/(x+2)$ in the form $(x+1)(x+2)^{-1}$, and using the product rule, leads inevitably to two terms which will involve factors of $(x+2)^{-1}$ and $(x+2)^{-2}$ respectively. The collection of these terms over a common denominator presents some hindrance, the more so if the denominator function is more complicated. The use of equation (6.9) avoids the necessity for any such steps. An important feature which should be noted is that the first term in the numerator of (6.9) is composed of the derivative of the numerator multiplied by the unchanged denominator, and since the two numerator terms are separated by a difference sign, the order is important.

An immediate consequence of equation (6.9) is that we can readily derive expressions for the derivatives of the remaining commonly occurring trig functions. For example, the derivative of $\tan x$ can be found by writing $\tan x$ as $\sin x/\cos x$. Identifying $\sin x$ with u and $\cos x$ with v, and using equation (6.9), we obtain

$$\begin{aligned}\frac{\mathrm{d}}{\mathrm{d}x}\left(\frac{\sin x}{\cos x}\right) &= \frac{\cos x \cos x - \sin x(-\sin x)}{\cos^2 x} \\ &= \frac{\cos^2 x + \sin^2 x}{\cos^2 x} \\ &= \frac{1}{\cos^2 x} \\ &= \sec^2 x.\end{aligned}$$

Similarly, $\mathrm{d}(\cot x)/\mathrm{d}x = \mathrm{d}(\cos x/\sin x)/\mathrm{d}x$
$= -\operatorname{cosec}^2 x.$

The derivatives of the important trigonometric functions are now collected for reference, and are as follows:

Function	Derivative
$\sin x$	$\cos x$
$\cos x$	$-\sin x$
$\tan x$	$\sec^2 x$
$\sec x$	$\sec x.\tan x$
$\operatorname{cosec} x$	$-\operatorname{cosec} x.\cot x$
$\cot x$	$-\operatorname{cosec}^2 x$

The following points are usefully borne in mind:

(i) the derivatives of the trig functions beginning with the letters *CO* are preceded by a minus sign;
(ii) if the angle x is replaced by ax, or $(ax-b)$, the derivative is multiplied by a, and the angle remains unchanged;
(iii) if the angle x is replaced by a function $\phi(x)$, the derivative is multiplied by $\phi'(x)$, and the angle remains unchanged as suggested in the general rule for differentiation of functions of a function (see eqn. (6.5)).

Example 6.9

Differentiate with respect to x:

(i) $\dfrac{x^2}{1+x}$, (ii) $\dfrac{x}{(1-x^2)}$, (iii) $\dfrac{\sin x}{\sin x+\cos x}$.

(i) Defining u and v by the equations $u = x^2$ and $v = 1+x$, we require the derivative of u/v. Since $\mathrm{d}u/\mathrm{d}x = 2x$, and $\mathrm{d}v/\mathrm{d}x = 1$, we can write

$$\begin{aligned}\frac{\mathrm{d}y}{\mathrm{d}x} &= \frac{(1+x)\,2x - x^2.1}{(1+x)^2} \\ &= \frac{1+x^2}{(1+x)^2}.\end{aligned}$$

(ii) Let $u = x$, and $v = \sqrt{(1-x^2)}$. Differentiating with respect to x, $u' = 1$, and $v' = \frac{1}{2}(1-x^2)^{-\frac{1}{2}}.-2x = -x/\sqrt{(1-x^2)}$.

Thus $$\frac{dy}{dx} = \frac{\sqrt{(1-x^2)}.1 - x\{-x/\sqrt{(1-x^2)}\}}{1-x^2}$$

$$= \frac{1-x^2+x^2}{(1-x^2)^{\frac{3}{2}}}$$

$$= \frac{1}{(1-x^2)^{\frac{3}{2}}}.$$

(iii) Let $u = \sin x$, and $v = \sin x + \cos x$. Differentiating, $u' = \cos x$, $v' = \cos x - \sin x$. Thus

$$\frac{d}{dx}\left(\frac{\sin x}{\sin x + \cos x}\right)$$

$$= \frac{(\sin x+\cos x)\cos x - \{\sin x(\cos x - \sin x)\}}{(\sin x+\cos x)^2}$$

$$= \frac{\cos^2 x + \sin^2 x}{(\sin x + \cos x)^2}$$

$$= \frac{1}{(\sin x+\cos x)^2}, \text{ since } \cos^2 x + \sin^2 x = 1.$$

EXERCISE 6e

Differentiate the following functions with respect to x:

1. $\dfrac{x^2}{(1+x)^3}$.
2. $\dfrac{4-x^2}{x(1-x)}$.
3. $\dfrac{1}{(3-x^2)}$.
4. $\dfrac{x^2}{x-1}$.
5. $\dfrac{\sin x}{(2+\cos x)^2}$.
6. $\dfrac{1-x^3}{1+x^3}$.
7. $\dfrac{\sin x}{\cos^3 x}$.
8. $\dfrac{1}{\cos 3x}$.
9. $\dfrac{x^3}{(2x^2-1)^2}$.
10. $\dfrac{x-2}{x^4}$.

11. Find the maximum and minimum values of $x^2/(x+1)$.

12. The velocity v of a particle varies with the time, t, as given by the equation $v = t/(3+t^2)$.

Sketch the curve representing this variation, marking in any stationary values or turning points.

13. Find the differential coefficients with respect to x of:

(i) $\dfrac{x-1}{(1+x^2)}$, (ii) $\dfrac{1+x}{x(1-x)}$.

14. Writing $\cot x$ as $\cos x/\sin x$, and regarding the expression as a ratio of functions, show that $d(\cot x)/dx = -\text{cosec}^2 x$, assuming only that $d(\sin x)/dx = \cos x$ and $d(\cos x)/dx = -\sin x$. Sketch the graphs of the functions involved, and show how the graph of $-\text{cosec}^2 x$ describes the variation in the gradient of the graph of $\cot x$.

15. Find the derivatives of $\sec x$ and $\text{cosec}\, x$ using a method similar to that suggested in question 14.

16. Differentiate with respect to t the function $\{a \tan(t/a)/t\}$.

17. Differentiate x^3 from first principles.

Differentiate with respect to x:

$$\frac{1}{1-x^3},\ x(1+x)^{\frac{1}{2}},\ \tan^2 x. \quad \text{(OC)}$$

***18.** The ratio R of two resistances in the gaps of a metre bridge circuit is given by the value of $x/(100-x)$, where x is a distance of the bridge balance point from one end. Find an expression for the rate of change of R with respect to x, and show that a small change δR in the value of the ratio causes a change δx in the balance point, where $\delta R = (100/(100-x)^2)\,\delta x$.

Show also that the percentage change in R is $100/(100-x)$ times the percentage change in x, and deduce that for a given δx, the value of $\delta R/R$ is a minimum when $x = 50$.

***19.** The power P delivered to a resistance R by a battery of e.m.f. E and internal resistance r is given by the equation $P = E^2R/(R+r)^2$. Find the relationship that must exist between R and r if the power delivered to the external resistance R is a maximum. (r is to be treated as a constant.)

20. The tractive force T exerted by an engine on the aeroplane it propels equals the power P it develops divided by the velocity v with which the aeroplane is travelling relative to the wind. If the fuel consumed in unit time equals kP, where k is a constant, and the wind resistance equals Cv^2, where C is another constant, show that the speed (relative to the wind) at which a journey of distance d should be covered so that the least amount of fuel is used is $\frac{3}{2}u$, where u is the velocity of the wind relative to the ground.

21. Prove that, if $f'(a) = 0$ and $f''(a)$ is positive, then the graph of the function $f(x)$ has a minimum at the point whose abscissa is a.

A variable isosceles triangle is circumscribed about a circle of given radius. Prove that the area of the triangle is a minimum (and not a maximum) when the triangle is equilateral. (OC)

22. Prove that if $f'(a) = 0$ and $f''(a)$ is negative, then the graph of the function of $f(x)$ has a maximum at the point whose abscissa is a.

Given that $f'(x) = (x-a)F(x)$ and $F(a) < 0$, prove that $f(x)$ has a maximum value when $x = a$.

Find the coordinates of the turning points of the function $(15+10x)/(4+x^2)$, distinguishing between maximum and minimum values.

Sketch the graph of the function for all values of x. (OC)

***23.** A load of resistance Z connected across the secondary windings of a transformer presents an effective resistance of nZ in the primary circuit, where n is a constant depending on the construction of the transformer. If the primary circuit is driven by a source of resistance R and e.m.f. E, the power delivered to the primary is given by

$$\frac{E^2nZ}{(R+nZ)^2},$$

and for a perfectly efficient transformer, this is the energy delivered to the secondary circuit. Show that this energy is a maximum if $n = R/Z$.

Given that n is the square of the ratio of the turns in the primary and secondary windings, calculate the value of this ratio so that the maximum power possible is delivered to a loudspeaker of resistance 15 ohms from a valve of effective impedance 1500 ohms.

24. Prove that, if u and v are functions of x, then

$$\frac{d}{dx}\left(\frac{u}{v}\right) = \frac{1}{v^2}\left\{v\frac{du}{dx} - u\frac{dv}{dx}\right\}.$$

Differentiate with respect to x:

$$\frac{x}{x^2+1}, \quad \frac{\sqrt{(1-x)}}{x}, \quad \sec x + \tan x. \qquad \text{(OC)}$$

25. Prove from first principles that

$$\frac{d}{dx}\left(\frac{1}{x}\right) = \frac{-1}{x^2}.$$

Differentiate the following functions with respect to x:

$$\text{(i)}\ \frac{x^2}{1+x}, \quad \text{(ii)}\ \frac{1}{(2x+1)^3}, \quad \text{(iii)}\ x^3 \tan x.$$

26. If $y = 2 \sin x + \tan x$, prove that $d^2y/dx^2 = 2 \sin x(\sec^3 x - 1)$. Show that for $0 \leqslant x \leqslant \frac{1}{2}\pi$ the gradient of the function is greater than 3, and that for $\frac{1}{2}\pi < x < \pi$ the function has a turning point and is zero for a value of x other than π. Sketch these two branches of the graph of the function. (SU)

***27.** A magnetic field of strength B acting perpendicularly to the earth's field B_0 at the centre of a tangent galvanometer causes a deflection θ, where $B = B_0 \tan \theta$. Show that a small change δB in the value of B produces a small change $\delta\theta$ in θ, where

$$\frac{\delta B}{B} = \frac{2\delta\theta}{\sin 2\theta}.$$

Hence deduce that the change in θ for a given percentage change in B is a maximum when $\theta = 45°$.

28. A flagpole of height 2 m stands on top of a building of height 10 m. A man whose eyes are 2 m above the ground sees the flagpole from a horizontal distance of x m. Show that the angle θ subtended by the flagpole at his eye is given by the equation

$$\tan \theta = \frac{2x}{x^2+80}.$$

Find the value of x for which θ is a maximum. (SU)

29. Prove that the greatest possible volume for a right pyramid on a square base whose sloping faces have a total area of A cm^2 is

$$\frac{A^{\frac{3}{2}}}{2^{\frac{1}{2}}.3^{\frac{7}{4}}}\ \text{cm}^3.$$

†**30.** $ABCD$ is a trapezium with AB horizontal and with BC and AD vertical. $AB = 4$ m, $BC = 7/3$ m, $AD = 16/3$ m. E is a point in AB such that $AE = 5/3$ m. The figure represents a vertical section of a room with a sloping ceiling, and with a light fixed to E. A man of height 2 m stands against the wall BC and then walks at a steady speed of $\frac{2}{3}$ m/s towards the light. Prove that at time t seconds from the start the speed of the highest point of his shadow is

$$\frac{2}{(6-t)^2}\ \text{m/s}$$

if t lies between 0 and 3, and find the corresponding result if t lies between 3 and 6. (SU)

†**31.** If $y = \sin \sqrt{(1+x)}$ and y_1 and y_2 denote the first and second differential coefficients of y with respect to x, prove that $4y_2(1+x)+2y_1+y = 0$. (SU, part question)

***32.** The magnetic field B on the axis of a coil of n turns and radius a is given by the expression $B = \mu_0 na^2 i/2(a^2+x^2)^{\frac{3}{2}}$, where i is the current flowing in the coil, x is the distance between the point and the plane of the coil, and μ_0 is a constant. Show that $d^2B/dx^2 = 0$ when $x = \frac{1}{2}a$.

***33.** The output V of a radio tuner varies with the frequency f of the input signal according to the equation $V = Af/B(B^2+C^2f^4)$. Sketch a graph showing how V varies with the value of f, and calculate for what ranges of f the value of V is more than 90% of the maximum value of V.

†**34.** The magnetic field on the axis of a coil of n turns of mean radius a at a point whose distance is x from the plane of the coil is

$$\frac{\mu_0 na^2 i}{2(a^2+x^2)^{\frac{3}{2}}},$$

where i is the current flowing in the coil, and where μ_0 is a constant. If two equal coils are placed coaxially at a distance apart equal to their radius, show that if B is the total field at any point on that common axis, then dB/dx and d^2B/dx^2 are both zero at the point mid-way between them.

Explain the significance of these results in terms of the variation of the magnetic field with position along the axis.

†**35.** Prove that if $f'(a) = 0$ and $f''(a)$ is negative, then the graph of the function $f(x)$ has a maximum at the point whose abscissa is a.

Prove that the function

$$y = \frac{\sin x \cdot \cos x}{1+2 \sin x+2 \cos x}$$

has turning points in the range $0 \leqslant x \leqslant 2\pi$ when $x=\frac{1}{4}\pi$ and $x = \frac{5}{4}\pi$, distinguishing between maximum and minimum values. Prove that the tangents at the origin and at the point $(\frac{1}{2}\pi, 0)$ meet at a point whose abscissa is $\frac{1}{4}\pi$. (OC)

†**36.** Prove that the function

$$\frac{1-\cos 2x}{\sqrt{(4+3 \cos 2x)}}$$

has turning-points when $x = 0$ and $x = \frac{1}{2}\pi$. Find the maximum and minimum values of the function.

Find the equation of the tangent at the point whose abscissa is $\frac{1}{4}\pi$.

Sketch the curve for values of x from 0 to $\frac{1}{2}\pi$. (OC)

†**37.** (i) Prove that if $y = \dfrac{\sqrt{x}+\sqrt{(x-1)}}{\sqrt{x}-\sqrt{(x-1)}}$, then

$$\frac{1}{y}\cdot\frac{dy}{dx} = \frac{1}{\sqrt{(x^2-x)}}.$$

(ii) Prove that, if a and b are positive, the minimum value of

$$\frac{a^2}{\sin^2 \theta}+\frac{b^2}{\cos^2 \theta} \quad \text{is} \quad (a+b)^2. \qquad \text{(SU)}$$

†**38.** Find the maximum and the minimum values of the function

$$\frac{x}{x^2+3x+1}.$$

Sketch the graph of the function. (SU)

†**39.** (i) Prove that the least value of $4 \sec \theta - 3 \tan \theta$ for $0 < \theta < \frac{1}{2}\pi$ is $\sqrt{7}$.

(ii) A body is made up of a hemisphere radius r with its plane face joined to one of the plane faces of a circular cylinder radius r and length x. Prove that, if the volume of the body is 45π cm³, the least value of the total surface area is 45π cm². (OC)

†**40.** Sketch the graph of $(x+2)/(x^2-3)$.

†**41.** An isosceles triangle has area A and perimeter P, and its vertical angle is 2θ. Express the height in terms of A and θ, and hence prove that $P^2 \tan \theta = 4A(\tan \theta+\sec \theta)^2$.

Prove that, for a given area, the perimeter is least when $\theta = 30°$.

(For full marks it is necessary to show that the value of the perimeter is a minimum and not a maximum.) (SU)

6.4. DIFFERENTIATION OF RELATIONSHIPS EXPRESSED IN AN IMPLICIT FORM

The differentiation of relationships expressed in an implicit form involves a technique similar to that discussed in Section 6.1 and summarised in equation (6.6). We have seen in earlier chapters that the gradient of a curve $y = f(x)$ is obtained simply by differentiating and finding the derivative of $f(x)$. But if we have a curve or relation such as $x^3+y^3 = 3y^2$, we cannot obtain an expression for dy/dx in this fashion without manipulation to rearrange the equation in an explicit form for y.

We can, however, differentiate the implicit expression throughout with respect to x: giving, for example, the equation

$$\frac{d}{dx}(x^3)+\frac{d}{dx}(y^3) = \frac{d}{dx}(3y).$$

Hence,

$$3x^2+3y^2\frac{dy}{dx} = 3\frac{dy}{dx},$$

and

$$3(1-y^2)\frac{dy}{dx} = 3x^2,$$

whence

$$\frac{dy}{dx} = \frac{x^2}{1-y^2}.$$

This, provided we do not mind having the variable y appearing in the expression, is the gradient function of $x^3+y^3 = 3y$.

We can deal with even more complicated relationships using the same method. If we have, for example, the curve $x^2+xy+y^2 = 16$ (Fig. 6.5), we can obtain its gradient function by evaluating the equation

$$\frac{d}{dx}(x^2)+\frac{d}{dx}(xy)+\frac{d}{dx}(y^2) = \frac{d}{dx}(16). \qquad (6.10)$$

The only new feature introduced by this equation is the term $d(xy)/dx$, and expanding this by the product rule (eqn. (6.8)), we can rewrite it in the form $yd(x)/dx+xd(y)/dx$. This can then be developed as $y.1+x\,\mathrm{d}y/\mathrm{d}x$, and we can return to equation (6.10), writing it in its entirety in the form

$$2x+\left(y+x\frac{\mathrm{d}y}{\mathrm{d}x}\right)+2y\frac{\mathrm{d}y}{\mathrm{d}x}=0.$$

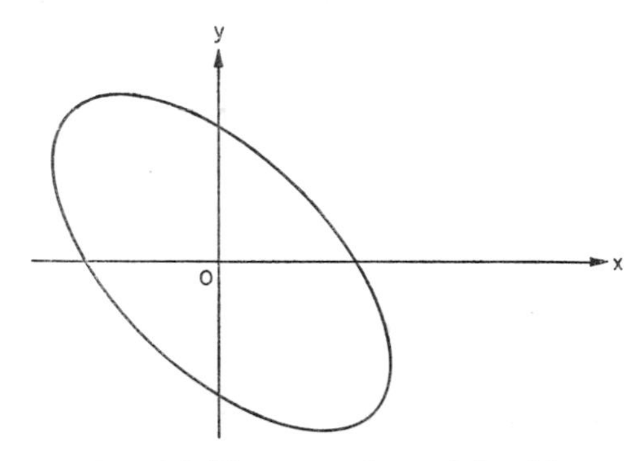

FIG. 6.5. The curve $x^2+xy+y^2=16$.

Thus, by rearrangement

$$(x+2y)\frac{\mathrm{d}y}{\mathrm{d}x}=-2x-y,$$

and the gradient function of $x^2+xy+y^2=16$ is

$$\frac{-(2x+y)}{x+2y}.$$

Example 6.10

Obtain an expression for the gradient function of the curve $y^2+xy+x^2=9$, and determine the gradient at the points (0, 3), (0, −3), (3, 0) and (−3, 0). At what points is the curve parallel to the x-axis? Sketch the graph.

Differentiating the equation throughout with respect to x, we obtain

$$2y\frac{\mathrm{d}y}{\mathrm{d}x}+x\left(1\frac{\mathrm{d}y}{\mathrm{d}x}\right)+1.y+2x=0.$$

Collecting the terms,

$$(2y+x)\frac{\mathrm{d}y}{\mathrm{d}x}=-(y+2x),$$

and thus

$$\frac{\mathrm{d}y}{\mathrm{d}x}=-\frac{(y+2x)}{(2y+x)}.$$

Substituting the values of the x- and the y-coordinates of the points in question, we obtain the gradients at these points as

$$\begin{array}{ll} (0, 3): & y'=-3/6=-\tfrac{1}{2}, \\ (0, -3): & y'=-(-3)/(-6)=-\tfrac{1}{2}, \\ (3, 0): & y'=-6/3=-2, \\ (-3, 0): & y'=-(-6)/(-3)=2. \end{array}$$

The curve is parallel to the x-axis when the gradient is 0, and this condition obtains when the numerator of the gradient function equals zero. Thus $y+2x=0$ or $y=-2x$, and substituting into the original equation for the curve, we obtain the condition that

$$(-2x)^2+x(-2x)+x^2=9$$

or

$$4x^2-2x^2+x^2=9,$$

i.e.

$$3x^2=9,$$

so that $x=+\sqrt{3}$ or $-\sqrt{3}$. The corresponding values of y at these points are $-2\sqrt{3}$ and $+2\sqrt{3}$.

Using the information derived, the graph can be sketched and appears as in Fig. 6.6.

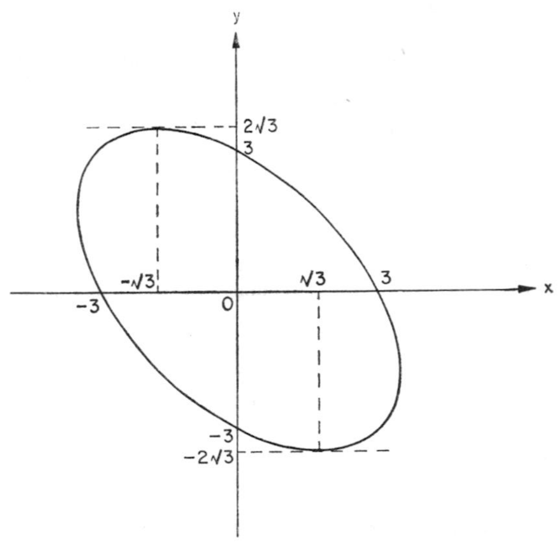

FIG. 6.6. The curve $x^2+xy+y^2=9$.

Exercise 6f

Differentiate the following equations with respect to x:

1. $x^2-3xy+y^2=4$. **3.** $x^3y+5x^2y^2=11$.
2. $x^4+x^2y^2-8y^4=0$. **4.** $y\sin^2 x=3$.
5. $\sin y\cos x+x\tan y=4$.

6. The relationship between the displacement of a particle from a fixed point and the time is expressed by the equation $s^2-3st+2t^2=0$, where s is the displacement and t the time. Determine the velocity and the acceleration of the particle when

(i) $s=2$, (ii) $t=3$.

7. Two variables are related by the equation $p^2+q^2=5p^2q^2$. What is the rate of change of p with respect to q?

8. Derive an expression for the value of $\delta(xy)$ in terms of x, y, δx and δy.

9. Write down an expression for

$$\text{(i) } \frac{d}{dp}(3p^2q^3),\quad \text{(ii) } \frac{d}{dt}\{x\sin\omega t\},$$

given that p, q, x and t are all variables.

10. Differentiate the following functions with respect to x:

$$\text{(i) } \frac{x^2}{x-1},\qquad \text{(ii) } \frac{(1+x^2)^{\frac{1}{2}}-x}{(1+x^2)^{\frac{1}{2}}+x}.$$

Prove that if $y=x\sin x$, then

$$x^2\frac{d^2y}{dx^2}-2x\frac{dy}{dx}+(2+x^2)y=0. \quad \text{(OC)}$$

11. (i) If $\sin y=\tan x$, find dy/dx in terms of x.
(ii) If $x^3+y^3=3axy$, find dy/dx in terms of x and y; and prove that dy/dx cannot be equal to -1 for finite values of x and y unless $x=y$. (SU, part question)

12. By differentiating the equation $x(dy/dx)=\sqrt{(1-x^2)}$, show that

$$x\frac{d^2y}{dx^2}+\frac{dy}{dx}+\frac{x}{\sqrt{(1-x^2)}}=0.$$

13. If the volume of a cone remains constant while the radius of its base is increasing at the rate of 1% per second, find the percentage rate per second at which its height is diminishing. (OC)

***14.** The bulk modulus of a gas is defined as the quantity $-v(dp/dv)$, where v is the volume of the gas and p its pressure. Find an expression for the bulk modulus if

(i) pv is a constant,
(ii) pv^γ is a constant, where $\gamma=1{\cdot}4$.

†15. A circle of radius b and centre at $(a, 0)$ has the equation $(x-a)^2+y^2=b^2$. The distance d of any point P on its circumference from the origin O is given by the equation $d^2=x^2+y^2$, and this can be written as a function of x only by substituting from the equation of the circle: $d^2=b^2+2ax-a^2$. Differentiating this expression for d^2 with respect to x gives $2d(dd/dx)=2a$, and thus $(dd/dx)=0$ only if $a=0$. Discuss the geometrical implications of this line of reasoning, and explain why it appears that no point on the circle is nearest the origin.

16. If $x\cos y=\sin x$, prove that

$$\frac{dy}{dx}=\frac{\cos y(\cos y-\cos x)}{\sin x\sin y}. \quad \text{(C)}$$

6.5. INVERSE TRIGONOMETRIC FUNCTIONS

The equation $\sin y=x$ can be written in the alternative form $y=\sin^{-1}x$, where the $\sin^{-1}x$ is not understood to mean $1/\sin x$ but the angle whose sine is x*. As such, the equation is explicit in form, but the derivative dy/dx is best obtained by differentiating the equation $\sin y=x$ with respect to x throughout. This gives

$$\cos y\frac{dy}{dx}=1,$$

and thus the value of dy/dx can be expressed as $1/\cos y$ or $\sec y$. We can therefore write that

$$\frac{d}{dx}(\sin^{-1}x)=\sec y,\quad \text{where}\quad y=\sin x.$$

It is often convenient, however, to have the expression for the derivative of $\sin^{-1}x$ written in terms of x rather than in terms of another variable y. This can be achieved by noting that if $\sin y=x$, $\cos y=\sqrt{(1-x^2)}$ (since for all values of y, $\sin^2 y+\cos^2 y=1$, and thus $\cos^2 y=1-\sin^2 y$). Using this

* There are, of course, an infinite number of angles with the same sine values. In general the terms such as $\sin^{-1}x$ will be used to represent the angle between $-\pi/2$ and $+\pi/2$ whose sine takes the value in question: this angle is called the principal value. When it is desired to include (for generality) other values, the terms will be written with a capital (e.g. $\text{Sin}^{-1}x$).

substitution,

$$\frac{d}{dx}(\sin^{-1} x) = \frac{1}{\sqrt{(1-x^2)}}$$

follows immediately

The derivative of $\sin^{-1} x$ is one which is needed fairly frequently in practice, and some saving of time is achieved if this result is committed to memory. Also important are the corresponding relationships of the other main trigonometrical functions, which results can be derived as follows:

(i) if $y = \cos^{-1} x$, $\cos y = x$. Differentiating both sides of the equation with respect to x, we obtain

$$-\sin y \frac{dy}{dx} = 1.$$

Thus $dy/dx = -1/\sin y$ or $-\operatorname{cosec} y$, and writing $\sin y$ as $\sqrt{(1-\cos^2 y)}$ or $\sqrt{(1-x^2)}$, we can see that dy/dx is equal to $-1/\sqrt{(1-x^2)}$. Thus

$$\frac{d}{dx}(\cos^{-1} x) = -\frac{1}{\sqrt{(1-x^2)}}.$$

That the derivative of $\cos^{-1} x$ is negative fits in with the general rule of thumb that the derivative of any simple trigonometric function beginning with *CO* is negative in sign.

(ii) if $y = \tan^{-1} x$, $\tan y = x$. Differentiating the equation throughout with respect to x, we obtain $\sec^2 y \, dy/dx = 1$, and thus $dy/dx = 1/\sec^2 y$. Now $\sec^2 y = 1+\tan^2 y = 1+x^2$, and thus $dy/dx = 1/(1+x^2)$. Replacing the y by $\tan^{-1} x$, we get

$$\frac{d}{dx}(\tan^{-1} x) = \frac{1}{1+x^2}.$$

The derivatives of functions such as $\sin^{-1}(x/2)$ or $\tan^{-1}(x/3)$ can be obtained in similar fashion specifically.

(iii) if $y = \sin^{-1}(x/2)$, $\sin y = x/2$. Differentiating, $\cos y(dy/dx) = \frac{1}{2}$, and thus

$$\frac{dy}{dx} = \frac{1}{2\cos y} = \frac{1}{2\sqrt{(1-\frac{1}{4}x^2)}} = \frac{1}{\sqrt{(4-x^2)}}.$$

$$\text{Thus } \frac{d}{dx}\left(\sin^{-1}\frac{x}{2}\right) = \frac{1}{(4-x^2)}.$$

(iv) if $y = \tan^{-1}(x/3)$, $\tan y = \frac{1}{3}x$. Thus $\sec^2 y \, (dy/dx) = \frac{1}{3}$, and

$$\frac{dy}{dx} = \frac{1}{3\sec^2 y} = \frac{1}{3(1+x^2/9)} = \frac{3}{9+x^2}.$$

$$\text{Thus } \frac{d}{dx}\left(\tan^{-1}\frac{x}{3}\right) = \frac{3}{9+x^2}.$$

Replacing the 2 or 3 of the last two examples by a, where a stands for any constant, the general rules

$$\frac{d}{dx}\left(\sin^{-1}\frac{x}{a}\right) = \frac{1}{\sqrt{(a^2-x^2)}}$$

$$\frac{d}{dx}\left(\cos^{-1}\frac{x}{a}\right) = \frac{-1}{\sqrt{(a^2-x^2)}}$$

and

$$\frac{d}{dx}\left(\tan^{-1}\frac{x}{a}\right) = \frac{a}{a^2+x^2}$$

can be obtained.

EXERCISE 6g

Differentiate the following with respect to x:

1. $\sin^{-1}\frac{x}{3}$.
2. $\sin^{-1}\frac{x}{4}$.
3. $2\sin^{-1}\frac{x}{5}$.
4. $\cos^{-1}\frac{x}{4}$.
5. $\cos^{-1}\frac{x}{7}$.
6. $\tan^{-1}\frac{x}{4}$.
7. $3\cos^{-1}\frac{x}{9}$.
8. $5\tan^{-1}\frac{1}{2}x$.
9. $x^2\sin^{-1}x$.
10. $\frac{1}{x}\tan^{-1}x$.
11. $(3x+7)\cos^{-1}x$.
12. $\frac{\tan^{-1}x}{1+x^2}$.
13. $\frac{\sin^{-1}\frac{1}{2}x}{\sqrt{(4-x^2)}}$.
14. $\sqrt{(9-x^2)}.\cos^{-1}\left(\frac{x}{3}\right)$.

15. By writing $\sin^{-1}(2x)$ as $\sin^{-1}(x/\frac{1}{2})$, find the derivative of $\sin^{-1}2x$ using the standard results on p. 163. Verify the correctness of your answer by finding dy/dx for the equation $\sin y = 2x$.

16. Differentiate with respect to t:

(i) $\tan^{-1}5t$, (ii) $\cos^{-1}3t$, (iii) $\sin^{-1}2t$, (iv) $\sin^{-1}(3t+\pi)$, (v) $5\tan^{-1}(5t-\frac{1}{2}\pi)$.

17. Write down the differential coefficients with respect to x of:

(i) $(\tan^{-1}x)^2$, (ii) $\sin^{-1}2x/\cos^{-1}3x$, (iii) $\text{cosec}^{-1}4x$, (iv) $\cot^{-1}x$.

18. Find expressions for dz/dt and d^2z/dt^2 given that $z = t\sin^{-1}t$.

19. Find the value of d^2y/dx^2 when $x = \frac{1}{2}\sqrt{3}$ and $y = 2\sin^{-1}x+3\cos^{-1}x$. (OC, part question)

†20. What are the gradient functions of the graphs of $\sin^{-1}x$ and $\tan^{-1}x$? Sketch the graphs, and relate them to those of $\sin^{-1}\frac{1}{2}x$ and $\tan^{-1}\frac{1}{2}x$.

†21. Write down an expression for the value of $d(\tan^{-1}x^2)/dx$ and $d(\tan^{-1}(\tan x))/dx$.

†22. Write down general expressions for the derivatives of $\sin^{-1}\phi$, $\cos^{-1}\phi$ and $\tan^{-1}\phi$, where ϕ represents a function of x.

†23. Given that $y = \sin^{-1}x$, prove that

$$(1-x^2)\frac{d^2y}{dx^2}-x\frac{dy}{dx} = 0.$$

Hence prove that

$$(1-x^2)y_{n+2}-(2n+1)xy_{n+1}-n^2y_n = 0,$$

where y_n denotes d^ny/dx^n.

†24. Differentiate

$$\{\cot^{-1}x+\cot^{-1}(1/x)\}$$

with respect to x.

Is $\{\cot^{-1}x+\cot^{-1}(1/x)\}$ a constant value function?

†25. Find from first principles the differential coefficient of $\cot x$. How could the sign of the result be predicted? If $y = \cot^{-1}x$, find dy/dx and prove that

$$\frac{d^2y}{dx^2} = 2\sin^3 y.\cos y. \quad \text{(SU)}$$

†26. Differentiate with respect to x:

(i) $\tan^{-1}(\sin^2 x)$, (ii) $\tan^{-1}\frac{(1-x^2)}{x^2}$.

(OC, part question)

†27. If $y = \frac{\sin^{-1}x}{\sqrt{(1-x^2)}}$, show that

$$(1-x^2)\frac{dy}{dx} = xy+1.$$

†28. (i) If $y^2 = ax^2+b$, prove that

$$x\left\{\left(\frac{dy}{dx}\right)^2+y\frac{d^2y}{dx^2}\right\} = y\frac{dy}{dx}.$$

(ii) If $y = \sin^{-1}x/a$, prove that

$$\frac{d^2y}{dx^2} = x\left(\frac{dy}{dx}\right)^3. \quad \text{(SU)}$$

6.6. DIFFERENTIATION WITH RESPECT TO A PARAMETER

The relationship between two variables is often expressed in terms of a third variable. This requires two equations which express each of the related variables in terms of the third; these equations are known as *parametric equations*, and the third variable is called the *parameter*. Thus the parametric equations $y = 3t$ and $x = 6t$ express y and x in terms of the parameter t, and in this simple case, it is easily seen that the relationship between x and y can be expressed in the single equation $y = x/2$.

The consequence of some importance following from the use of two parametric equations is that the position of a point on a plane defined by two variable axes can be expressed with the use of only one number. If, for example, $y = 3t$ and $x = 6t$, the position of the point $t = 2$ is immediately defined as (12, 6), using the conventional notation. It is thus possible to express the position of a point using either one equation and two coordinates, or two equations and one parametric value.

Parametric relationships are useful in situations such as that arising when the motion of a particle is compounded from two perpendicular motions. Thus if the velocity of a particle in the x direction is 6, the displacement x after a time t is given by the equation $x = 6t$, and if, superimposed on this, there is a velocity of 3 in the y direction, the displacement y after a time t is given by the equation $y = 3t$. The net result is that the path $2y = x$ is traced out by the particle in question.

Other parametric relationships may be more involved; the equation pair $y^2 = t^2-3t$ and $x = t^3-7t+4$ does not lend itself readily to elimination of t. As a result, it is worth developing a technique for differentiating parametric equations independently if the value of the rate of change of y with respect to x is required.

For two related variables x and y, the value of derivative dy/dx is defined as the value of

$$\lim_{\delta x=0}\left(\frac{\delta y}{\delta x}\right).$$

If now a third (parametric) variable—say t—is introduced, we can write the ratio $\delta y/\delta x$ as $(\delta y/\delta t)(\delta t/\delta x)$, or, more conveniently, $(\delta y/\delta t)/(\delta x/\delta t)$. If the relationship between y, x and t is continuous, as $\delta x \to 0$, $\delta t \to 0$. Thus

$$\frac{dy}{dx} = \lim_{\delta x=0}\left(\frac{\delta y}{\delta x}\right) = \lim_{\delta t=0}\left(\frac{\delta y/\delta t}{\delta x/\delta t}\right) = \frac{\lim_{\delta t=0}\left(\dfrac{\delta y}{\delta t}\right)}{\lim_{\delta t=0}\left(\dfrac{\delta x}{\delta t}\right)}$$

$$= \frac{dy}{dt}\Big/\frac{dx}{dt},$$

assuming that $\lim (a/b) = \lim (a)/\lim (b)$. This final equation is the desired result, and the rate of change of y with respect to x can be found in terms of t by evaluating the ratio of the derivative of y with respect to t to that of x with respect to t.

It was explained in Section 4.9 that the symbols $\dot{x}$, $\dot{y}$ and $\ddot{x}$, etc., represent the time derivatives dx/dt, dy/dt and d^2x/dt^2. Identifying the parameter t of the previous paragraphs with time, we can now see that an alternative form of $dy/dx = (dy/dx)/(dx/dt)$ is $dy/dx = \dot{y}/\dot{x}$.

Example 6.11

Find the rate of change of y with respect to x in terms of (i) t, (ii) x, given the parametric equations $x = t^3$ and $y = 3t^2-7$.

Differentiating the parametric equations with respect to t,

$$\frac{dx}{dt} = 3t^2 \quad\text{and}\quad \frac{dy}{dt} = 6t.$$

Thus

$$\frac{dy}{dx} = \frac{dy}{dt}\Big/\frac{dx}{dt} = \frac{6t}{3t^2} = \frac{2}{t}.$$

Substituting $t = x^{\frac{1}{3}}$ from the equation for x, the rate of change of y with respect to x in terms of x is obtained as

$$\frac{dy}{dx} = \frac{2}{\sqrt[3]{x}}.$$

Answer: (i) $2/t$, (ii) $2/\sqrt[3]{x}$.

Exercise 6h

Find dy/dx in terms of t for the following parametric equations:

1. $x = at^2$, $y = 2at$.
2. $x = a+t$, $y = a \sin t$.
3. $x = a \sin t$, $y = b \cos t$.
4. $x = \sqrt{t}$, $y = t-3$.
5. $x = \sec 2t$, $y = \cos 2t$.
6. $x = \sin^{-1} t$, $y = \cos^{-1} t$.
7. $x = t \tan t$, $y = (1/t) \cot t$.

8. Find the equations relating x to y for the parametric equations in questions 1–7.

9. Find dy/dx in terms of x for questions 1–7.

10. Given that $x = 4t^3+2t$, $y = 3t^2-t$, find dy/dx in terms of t.

Differentiate the following functions with respect to x:

$$\text{(i) } x^2(1-3x)^{\frac{1}{2}}, \quad \text{(ii) } \frac{x^3}{1+2x}, \quad \text{(iii) } \sin^3 2x.$$

(OC)

11. Show that $\dfrac{d}{dt}\left(\dfrac{\dot{y}}{\dot{x}}\right) = \dfrac{\dot{x}\ddot{y}-\ddot{x}\dot{y}}{\dot{x}^2}$,

and that $\dfrac{d}{dx}\left(\dfrac{y}{x}\right) = \dfrac{d^2y}{dx^2} = \dfrac{\dot{x}\ddot{y}-\ddot{x}\dot{y}}{\dot{x}^3}$.

12. Using the results of question 11, find expressions for d^2y/dx^2 if

(i) $x = at^2$ $\quad y = 2at$,
(ii) $x = a \sin 3t$, $\quad y = b \cos 3t$,
(iii) $x = \sin^{-1}(t/3)$, $\quad y = \cos^{-1}(t/3)$.

13. Find an expression for d^2y/dx^2 if $x = t^3, y = t^4$. Check your solution by obtaining an expression relating y and x directly, and differentiating this with respect to x.

14. Express the gradient function of the curve defined by the equations $x = at^2$, $y = a(t^3-t^2)$ in terms of

(i) the parameter t,
(ii) x,

given that a is a constant.

Find the stationary values of y.

15. The average change of a persons weight with his age when he is 13 years old is 11 kg per year, and the average rate of change of height with age is 13 mm per year. What is the average rate of change of weight with height?

16. Differentiate the following functions with respect to x:

$$\text{(i) } \frac{\sqrt{(1-x^2)}}{x^3}, \quad \text{(ii) } \frac{\sin x}{1+\cos x}.$$

If $x = 3t+t^3$ and $y = 3-t^{\frac{5}{2}}$ express dy/dx in terms of t and prove that, when $d^2y/dx^2 = 0$, then x has one of the values $0, \pm 6\sqrt{3}$. (OC)

17. Prove that, if u and v are functions of x, then

$$\frac{d}{dx}(uv) = v\frac{du}{dx}+u\frac{dv}{dx}.$$

Differentiate the following with respect to x:

$$\text{(i) } (1+x)^3 \tan 3x, \quad \text{(ii) } \frac{5+3\sin x}{3+5\sin x}.$$

If $x = \dfrac{2+t}{1+2t}$, $y = \dfrac{3+2t}{t}$,

prove that $\dfrac{dy}{dx} = \dfrac{(1+2t)^2}{t^2}$,

and find the value of d^2y/dx^2, when $x = 0$. (OC)

18. Differentiate with respect to x:

$$\text{(i) } x+\sqrt{(1+x^2)}, \quad \text{(ii) } \frac{2+7\cos x}{3+5\cos x}.$$

A curve is given by the equations $x = a\cos^3\theta$, $y = a\sin^3\theta$. Show that $dy/dx = -\tan\theta$ and find the value of d^2y/dx^2 at the point $\theta = \frac{1}{4}\pi$. (OC)

19. A curve is given by $x = a(2\cos\theta+\cos 2\theta)$, $y = a(2\sin\theta-\sin 2\theta)$. Prove that the radius of curvature ϱ at any point has magnitude $8a\sin(3\theta/2)$, given that the formula for the radius of curvature is

$$\varrho = \frac{(\dot{x}^2+\dot{y}^2)^{\frac{3}{2}}}{\dot{x}\ddot{y}-\dot{y}\ddot{x}},$$

where dots denote differentiation with respect to the parameter. (OC modified)

20. A point P of a curve is given by the parametric equations $x = a\cos^3 t$, $y = a\sin^3 t$. Prove that the equation of the tangent at P is

$$x\sin t+y\cos t = a\sin t\cos t.$$

This tangent meets the axis of x at T, ON is the perpendicular from the origin O onto PT, and Q is the mid-point of PN.

Prove that $OQ = QT = \frac{1}{2}a$. (SU)

21. (i) Differentiate $\sqrt{(3-4\sin^2\theta)}$ with respect to θ.

(ii) Find dy/dx in terms of t if

$$x = \frac{3(1+t^2)}{1-t^2} \quad \text{and} \quad y = \frac{8t}{1-t^2},$$

simplifying your answer. (C, part question)

22. Write each of the expressions $2+2t-t^2$, $2+2t+t^2$ in the form $a\pm(t+b)^2$ (where a and b are numbers to be determined in each case). Hence find, for each of the expressions, whether it has a maximum or a minimum value, and if so what these values are.

Find also the maximum and minimum values of

$$\frac{2+2t-t^2}{2+2t+t^2}.$$

What information do these results give about the graph represented by the parametric equations $x = 2+2t-t^2$, $y = 2+2t+t^2$? Draw a sketch of this graph. (SU)

23. Differentiate $\cos x$ from first principles.

If $x = a(\theta-\sin\theta)$, $y = a(1-\cos\theta)$, find dy/dx in terms of θ and show that $1+(dy/dx)^2 = 2a/y$. (SU)

24. If $x = \sin\theta$, and $y = \cos\theta$,

(a) prove that $d(y^3)/dx = -\frac{3}{2}\sin 2\theta$;
(b) obtain d^2y/dx^2 in terms of θ. (SU, part question)

†**25.** Given that $v = u^3$ and $u = \cos t$, write down dv/dt in term of t.

The parametric equations of a curve are $x = a\cos^3 t$, $y = a\sin^3 t$.

Write down dx/dt and dy/dt and hence show that $dy/dx = -\tan t$. Deduce the equation of a tangent of the curve at the point t, and simplify the equation.

Show that the length of the perpendicular from the point $(a, 0)$ to this tangent is $|q|$, where $q = a(\sin t - \frac{1}{2}\sin 2t)$. Calculate dq/dt, and hence find the maximum length of the perpendicular. (NUJMB)

†**26.** The position at any instant of a particle moving in a circle with a constant angular velocity ω can be described by the parametric equations $x = r\cos\theta$, $y = r\sin\theta$, where r is the radius of the circle, θ is the angle that the radius vector from the centre of the circle to the particle makes with the direction of the x-axis, and $d\theta/dt = \omega$ (Fig. 6.7) By differentiating these equations with respect to the time t,

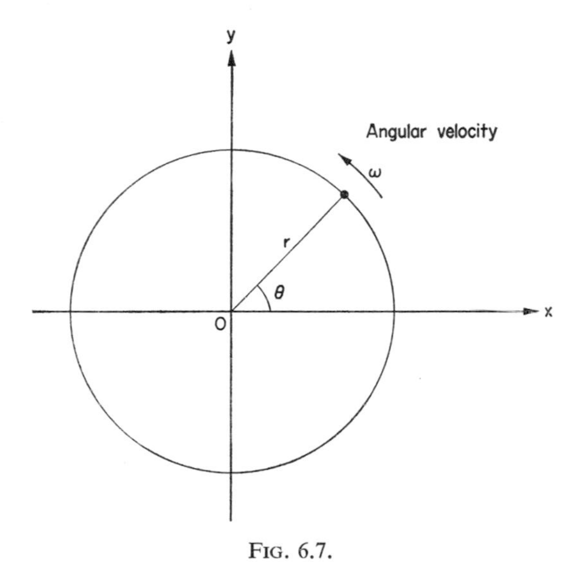

FIG. 6.7.

show that the components of the particle's velocity parallel to the x- and y-axes are $-r\omega\sin\theta$ and $r\omega\cos\theta$, respectively, and hence show that the velocity of the particle is $r\omega$ at $(\pi/2+\theta)$ to the x-axis (i.e. along the tangent). By differentiating again, find the x and y components of the particle's acceleration, and show that this is equal to $r\omega^2$ at an angle of $(\pi+\theta)$ with the x-axis (i.e. along the radius and directed towards the centre).

†**27.** If the angular velocity of the motion described in question 26 is accelerated at a rate denoted by $\ddot{\theta}$, show that the acceleration of the particle can be resolved into $r\dot{\theta}^2$ along the radius, and $r\ddot{\theta}$ along the tangent.

†**28.** Find the magnitude of the velocity of a particle whose position at any instant is described by the equations $x = r\cos\theta$, $y = r\sin\theta$, if r, θ and $d\theta/dt$ are all changing with time.

†**29.** Find the velocity of a particle moving according to the parametric equations $x = a\cos^3\theta$, $y = a\sin^3\theta$, where θ is changing at a constant rate ω.

SUMMARY

If y is a function of z, and z is a function of x, y will be a function of x, and

$$\frac{dy}{dx} = \frac{dy}{dz}\cdot\frac{dz}{dx}.$$

Writing y as $f(g(x))$ and substituting z for $g(x)$ gives

$$\frac{d}{dx}\{f(g(x)\} = \frac{d}{dz}\{f(z)\}\cdot\frac{d}{dx}\{g(x)\}.$$

This relationship can be expressed in alternative forms: for example, the derivative of a function ϕ of x raised to a power n (i.e. ϕ^n) is $n\phi^{n-1}.\ \phi'$, and the derivative of $\sin\phi$ is $\phi'\cos\phi$. Two important consequences of this latter relation are:

$$\frac{d}{dx}(\sin ax) = a\cos ax,$$

$$\frac{d}{dx}(\cos ax) = -a\sin ax.$$

Again, the derivative of a function of a variable z with respect to another variable x can be written down using the relation

$$\frac{d}{dx}\{f(z)\} = \frac{d}{dz}\{f(z)\}\cdot\frac{dz}{dx},$$

and an epuation can be differentiated throughout with respect to any variable.

The derivative of a product of two functions u and v can be evaluated by forming the sum $u(dv/dx)+v(du/dx)$; that of a quotient u/v can be found by evaluating $\{v(du/dx)-u(dv/dx)\}v^2$. The value of a derivative such as $d(x^2y)/dx$ is thus seen to be $\{2xy+x^2(dy/dx)\}$ by a combination of the product rule and the principles outlined above.

Important derivatives of inverse trigonometric functions are:

$$\frac{d}{dx}\{\sin^{-1}(x/a)\} = \frac{1}{\sqrt{(a^2-x^2)}},$$

$$\frac{d}{dx}\{\cos^{-1}(x/a)\} = \frac{-1}{\sqrt{(a^2-x^2)}},$$

$$\frac{d}{dx}\{\tan^{-1}(x/a)\} = \frac{a}{a^2+x^2}.$$

If y and x are expressed as functions of a parameter t, the derivative $\mathrm{d}y/\mathrm{d}x$ can be evaluated by determining the value of the quotient $(\mathrm{d}y/\mathrm{d}t)/(\mathrm{d}x/\mathrm{d}t)$; or, in abbreviated notation, $\dot{y}/\dot{x}$.

MISCELLANEOUS EXERCISE 6

*1. Show that $\dfrac{\mathrm{d}x}{\mathrm{d}y}\Big/\dfrac{\mathrm{d}p}{\mathrm{d}q} = \dfrac{\mathrm{d}x}{\mathrm{d}p}\cdot\dfrac{\mathrm{d}q}{\mathrm{d}y}$.

A ray of light enters a prism of angle A making an angle i with the normal at the point of incidence. The ray continues into the prism at an angle r with the normal, where $\sin i = \mu \sin r$, and where μ is the refractive index of the material of which the prism is made. The ray then strikes the opposite face of the prism at an angle r' with the normal at the point of incidence, and emerges at an angle i' with that normal, where $\sin i' = \mu \sin r'$. Derive expressions for $\mathrm{d}i/\mathrm{d}r$ and $\mathrm{d}i'/\mathrm{d}r'$, and by dividing, show that

$$\frac{\mathrm{d}i'}{\mathrm{d}i} = \frac{\cos i.\cos r'}{\cos r.\cos i'}\cdot\frac{\mathrm{d}r'}{\mathrm{d}r}.$$

Show further that $A = r+r'$, and hence that $\mathrm{d}r'/\mathrm{d}r = -1$.

Deduce that $\dfrac{\mathrm{d}i'}{\mathrm{d}i} = -\dfrac{\cos i.\cos r'}{\cos r.\cos i'}$.

*2. Show that the deviation D of the ray passing through the prism described above is $i+i'-A$. Differentiate this expression with respect to i, and show that D is a minimum when $i = i'$ and $r = r'$. You may use the results of question 1 if you wish.

3. If $y = (\sin^{-1}x)^2$, prove that

$$(1-x^2)\frac{\mathrm{d}^2y}{\mathrm{d}x^2} = x\frac{\mathrm{d}y}{\mathrm{d}x}+2.$$

Differentiate $\tan^{-1}\{x\sqrt{2}/(1-x^2)\}$ with respect to x. (SU, part question)

4. A point moves on the x-axis and at time t its position is given by $x = t(t^2-6t+12)$.

Show that its velocity at the origin is 12, and that its velocity decreases to zero at the point where $x = 8$, and thereafter continues to increase.

Prove that, if v is the velocity and a the acceleration at any point, then $a^2 = 12v$. (OC)

5. A man wishes to cross a square ploughed field from one corner to the opposite corner in the shortest possible time. He can walk at 10 km/h when he keeps to the edge of the field and at 6 km/h when he walks on the ploughed land. Prove that he will achieve his object if he first walks along the edge for a distance equal to one-quarter of the length of the side and then crosses over direct. (OC)

6. Given that $y^2 = (uv)$, where u and v are functions of x, prove that

$$4y^3\frac{\mathrm{d}^2y}{\mathrm{d}x^2} = 2y^2\left(u\frac{\mathrm{d}^2v}{\mathrm{d}x^2}+v\frac{\mathrm{d}^2u}{\mathrm{d}x^2}+2\frac{\mathrm{d}u}{\mathrm{d}x}\cdot\frac{\mathrm{d}v}{\mathrm{d}x}\right) - \left(u\frac{\mathrm{d}v}{\mathrm{d}x}+v\frac{\mathrm{d}u}{\mathrm{d}x}\right)^2. \quad \text{(OC)}$$

7. Differentiate $y = (a/x)\tan(x/a)$ and $z = \tan^{-1}\{(a/x)\tan(x/a)\}$ with respect to x. (OC)

8. Given that $\cos x(\mathrm{d}^2y/\mathrm{d}x^2) + \sin x(\mathrm{d}y/\mathrm{d}x) - y\cos^3 x = \sin x\cos^3 x$ and $z = \sin x$, prove that $(\mathrm{d}^2y/\mathrm{d}z^2) - y = z$. (OC)

9. Differentiate the following functions with respect to x:

(i) $\dfrac{1}{\cos\sqrt{x}}$, (ii) $\dfrac{1+x}{x(1-x)}$, (iii) $\tan^{-1}\sqrt{(ax+b)}$. (OC)

10. Differentiate the following functions with respect to z:

(i) $\dfrac{(1-z)^2}{1+z}$, (iii) $\dfrac{z^2}{(1+z)^3}$,

(ii) $z^2\tan^{-1}z^2$, (iv) $\tan^{-1}\left(\dfrac{\cos z}{1+\sin z}\right)$. (OC)

11. Prove from first principles that $\mathrm{d}(\tan x)/\mathrm{d}x = \sec^2 x$. (You can assume that $\sin\theta/\theta \to 1$ as $\theta \to 0$.)

Differentiate the following functions of x with respect to x:

(i) $\sqrt{\{x(1+x^2)\}}$, (ii) $\dfrac{\sin x}{(2+\cos x)^2}$.

Prove that, if $y^2+ay+b = x$, then

$$\frac{\mathrm{d}^2y}{\mathrm{d}x^2}+2\left(\frac{\mathrm{d}y}{\mathrm{d}x}\right)^3 = 0. \quad \text{(OC)}$$

12. The area of a triangle is to be calculated from the formula $\Delta = \frac{1}{2}bc\sin A$. In a certain triangle, b, c and A are measured, A being 45°; while c is measured accurately, errors of 1% and 2% respectively, are made in measuring b and A. Prove that the maximum percentage error in the calculated value of Δ is $2\frac{4}{7}$% (take $\pi = 3\frac{1}{7}$).

(Hint: Assume that b, c, A and Δ become $b+\delta b$, $c+\delta c$, $A+\delta A$ and $\Delta+\delta\Delta$, and find an expression for $\delta\Delta$.) (OC)

13. AB is a diameter of a given circle of radius a; CD is a variable chord parallel to AB. Express the perimeter, P, of the trapezium $ACDB$, in terms of the angle AOC ($\equiv \theta$), O being the centre of the circle. Prove that P is a maximum (and not a minimum) when $\theta = 60°$. (OC)

14. In the triangle ABC, the sides AB, AC are equal and contain an angle 2θ. The circumscribed circle of the triangle has radius R. Show that the sum of the lengths of the perpendiculars from A, B, C to the opposite sides of the triangle is $2R(1+4\sin\theta-\sin^2\theta-4\sin^3\theta)$.

If R is constant and θ varies, show that this sum has only one stationary value, and that this value is a maximum. (NUJMB)

15. An engine A is approaching an open level crossing $\frac{1}{2}$ km directly ahead at 30 km/h. At the same instant a car B travelling on a straight road at right angles to the railway track is $\frac{3}{4}$ km from the crossing and approaching it at 50 km/h. If the speed of neither is altered, show that the distance between the car and the engine is least after approximately 56 seconds and find this least distance.

Find the initial angular velocity, in radians/minute, of the straight line joining A and B. (L)

***16.** A slab of dielectric is suspended between two parallel conducting plates a distance t apart, so that x of its length is between the plates. The total length of the plates is a, their breadth is b, and the capacity C of this system is

$$\frac{\varepsilon_0\{ab+xb(\varepsilon-1)\}}{t},$$

where ε and ε_0 are constants, assuming that the dielectric just fits between the plates. If the two plates are connected to a source of potential difference of magnitude V, the energy E of the system is $(A-\frac{1}{2}CV^2)$, where A is a constant. Show that the dielectric experiences a force tending to pull it into the space between the plates whatever the value of x.

17. If in question 16 the plates are charged and the system isolated, the energy of the system becomes B/C, where B is a constant. Find the force acting on the dielectric under these conditions if $\varepsilon > 1$. What happens if $\varepsilon \leqslant 1$?

18. A tree trunk is in the form of a frustum of a right circular cone, the radii of the end faces being a and b respectively ($a > b$) and the distance between these faces being l. A log in the form of a right circular cylinder is to be cut out of the trunk, the axis of the cylinder being perpendicular to the end faces of the frustum. Show that, if $b < 2a/3$, the volume of the log is a maximum when its length is

$$\frac{al}{3(a-b)}.$$

If $b > 2a/3$, what is the length of the log when its volume is as great as possible? (L)

19. A searchlight is placed on level ground 100 m from the nearest point of a straight vertical wall. Its beam is directed horizontally on to the nearest point of the wall and is then rotated in the horizontal plane at 10° per second. Show that the point of illumination moves with increasing velocity as the beam turns through 90°. Find its velocity and acceleration after 3 seconds. (SU)

20. A particle is moving on a straight line, and its distance x from the fixed point O on the line at time t is given by

$$x = a(1+\cos^2 t).$$

Show that the acceleration of the particle is $6a-4x$. Find the values of x at the points where the velocity of the particle is (i) zero, (ii) a maximum. (OC)

21. Prove that $\mathrm{d}(3\sin x+\sin 3x)/\mathrm{d}x = 6\cos x\cos 2x$.

For what ranges of values of x between 0 and π is $(3\sin x+\sin 3x)$ an increasing function of x?

Sketch a *rough* graph of $y = 3\sin x+\sin 3x$ for values of x from 0 to π, indicating particularly the positions of the turning values. (OC)

22. Prove that, if u and v are functions of x, then

$$\frac{\mathrm{d}}{\mathrm{d}x}(uv) = u\frac{\mathrm{d}v}{\mathrm{d}x}+v\frac{\mathrm{d}u}{\mathrm{d}x}.$$

Differentiate with respect to x:

(i) $(x+1)(x^2+1)$, (iii) $x^2\sqrt{(1-2x)}$.

(ii) $\sin x - x\cos x$, (OC)

23. Find from first principles the differential coefficients of x^2 and $1/x$.

Differentiate with respect to x:

(i) $\sqrt{(x^2+1)}$, (ii) $\dfrac{x^2}{1-x}$, (iii) $\sin x\cos x$.

(OC)

24. A particle moves along the axis of x and at the end of a time t its position is given by $x = 5+4\sin 2t+3\cos 2t$. Prove that its acceleration is $20-4x$. Prove that its velocity is zero when $x = 0$ and $x = 10$, and that it moves backwards and forwards between these two points. (OC)

25. Prove from first principles that $\mathrm{d}(x^n)/\mathrm{d}x = nx^{n-1}$, where n is a positive integer.

Differentiate with respect to x:

(i) $\dfrac{(x+1)^2}{x}$, (ii) $(1+3x)^{\frac{1}{2}}$, (iii) $\cos^2 x$. (OC)

26. Find and distinguish between the turning points of $(x-1)^3(12x^2-9x-43)$.

27. A hollow right circular cone is of height h, and semi-vertical angle θ. Prove that the radius of the largest sphere which can be placed entirely within the cone is $(h\sin\theta)/(1+\sin\theta)$. Prove further that the ratio of the volume of the sphere to that of the cone is greatest when $\sin\theta = \frac{1}{3}$, and is then equal to $\frac{1}{2}$. (SU)

***28.** Show that the minimum distance between an object and its image formed by a lens of focal length f is $4f$. (Let x be the distance between the object and its image, and use the relationship $1/u+1/(x-u) = 1/f$.)

***29.** A battery of e.m.f. 3 volts and internal resistance 5 ohms is connected to a load of resistance R ohms. Write down an expression for the power developed in R, and find the value of R for which the power is a maximum.

30. If $y = \sin\sqrt{(1+x)}$, and y_1 and y_2 denote the first and second differential coefficients of y with respect to x, prove that

$$4y_2(1+x)+2y_1+y = 0. \quad \text{(SU, part question)}$$

31. A rectangular parcel of square cross-section is such that the sum of its length and girth is 6 dm. If its length is x dm, show that its volume, V dm^3, is given by $16V = x(6-x)^2$. Find the maximum volume of a parcel of this shape, and prove that your result is, in fact, a maximum. (The girth of the parcel is the perimeter of its square cross-section.) (OC)

32. Find the equations of the tangent to the curve $27y^2 = 4x^3$ at the point $(3t^2, 2t^3)$. Find also the equation of the normal at $(3u^2, 2u^3)$. (OC)

33. Prove from first principles that if $f(u)$ is a function of u and $u(x)$ is a function of x,

$$\frac{\mathrm{d}}{\mathrm{d}x}f\{u(x)\} = \frac{\mathrm{d}f}{\mathrm{d}u}\cdot\frac{\mathrm{d}u}{\mathrm{d}x}. \qquad \text{(OC)}$$

34. A particle moves along a cycloid so that its coordinates at time t are given by $x = a(t-\sin t)$, $y = a(1-\cos t)$. Obtain the speed v of the particle at time t,

(i) expressed in terms of t,
(ii) expressed in terms of y.

Sketch the graphs

(i) of x against y,
(ii) of v against t,
(iii) of v against y,

all as t varies in the range from 0 to 4π. (OC)

35 In the same diagram, sketch the graphs of the functions

$$\frac{(x+1)(x-3)}{x-4} \quad \text{and} \quad \frac{-2}{x},$$

for values of x from -3 to $+4$. Prove that the abscissa of the point of intersection of the two graphs is the root of the equation $x^3-2x^2-x-8 = 0$. Prove (do not merely derive the result from your graph) that the maximum ordinate of the first graph within the range -3 and $+3$ for x is $6-2\sqrt{5}$. (OC)

36. Prove from first principles that $\mathrm{d}(1/x^3)/\mathrm{d}x = -3/x^4$.

Differentiate $(x-1)/(1+x^2)$ with respect to x.

Prove that, if $y = \dfrac{\cos x - \sin x}{\cos x + \sin x}$, then

$$\frac{\mathrm{d}^2y}{\mathrm{d}x^2}+2y\frac{\mathrm{d}y}{\mathrm{d}x} = 0. \qquad \text{(OC)}$$

37. Find the condition for a maximum of $f(x)$ when $x = a$ if $F(x)$ is a function such that $f'(x) = (x-a)F(x)$ and $F(a) \neq 0$.

A variable isosceles triangle has a constant perimeter $2c$. If x denotes one of the equal sides, express the area of the triangle in terms of x and c. Prove that the area is a maximum (and not a minimum) when the triangle is equilateral. (OC)

38. If $y = f(u)$ and $u = g(x)$, prove from first principles that

$$\frac{\mathrm{d}y}{\mathrm{d}x} = \frac{\mathrm{d}f}{\mathrm{d}u}\cdot\frac{\mathrm{d}g}{\mathrm{d}x}.$$

Gas flow sin a cylindrical pipe of variable cross-section. The speed of the gas is measured by the "Mach" number M, which is defined as the ratio of the actual speed to the speed of sound. If $S(x)$ is the area of the cross-section at a distance x along the pipe, the Mach number may be calculated from the formula

$$MS(x) = C\{1+AM^2\}^{(n+1)/2A},$$

where C and A are positive constants. Prove that at the smallest cross-section either the speed of the gas has a stationary value or the speed of sound is reached. (OC)

39. Find the gradient of the tangent at any point on the curve whose parametric equations are $x = a\sec t$, $y = b\tan^3 t$ $(a, b > 0)$, and prove that it cannot exceed b/a.

Sketch the curve, paying attention to the shape at the points $(a, 0)$ and $(-a, 0)$. Indicate the correspondence between parts of the curve and the four possible quadrants for the parameter t. (OC)

40. Prove that, if $y = x-(1-x^2)^{\frac{1}{2}}\sin^{-1}x$, then $(1-x^2)\mathrm{d}y/\mathrm{d}x = x(x-y)$. Use this result to evaluate

$$\frac{\mathrm{d}y}{\mathrm{d}x},\quad \frac{\mathrm{d}^2y}{\mathrm{d}x^2},\quad \text{and} \quad \frac{\mathrm{d}^3y}{\mathrm{d}x^3}$$

at $x = 0$. (OC)

41. Prove that the equations of the tangent and normal at the point $P(ct, c/t)$ on the rectangular hyperbola $xy = c^2$ are, respectively, $x+t^2y = 2ct$, $t^3x-ty = c(t^4-1)$.

The normal at P meets the hyperbola again at $R(cT, c/T)$. Prove that $t^3T = -1$.

The tangent at P meets the y-axis at Q. Find the area of the triangle PQR in terms of t and prove that the area is a minimum when $t = \pm 1$. (OC)

42. Find the values of A and B for which $y = A\sin x+B\cos x$ satisfies the differential equation

$$\frac{\mathrm{d}^2y}{\mathrm{d}x^2}+4\frac{\mathrm{d}y}{\mathrm{d}x}+3y = \sin x. \qquad \text{(OC)}$$

43. Given that $y = (\sec x+\tan x)^p$ where p is constant, prove that $\cos x(\mathrm{d}y/\mathrm{d}x) = py$. (OC)

44. Find an expression for the derivative of $\sec^2 x$, and show that this is zero when $x = n\pi$, where n is an integer. Show that the values of $\sec^2 x$ at these points are relative minima, and explain why there are no maxima. State values of x for which the function $\sec^2 x$ is discontinuous.

45. A cup is in the form of a hollow hemisphere of radius 4 cm with its top horizontal. Water is poured into it at the rate of 5 cm^3/s. At what rate is the surface rising when the depth of water in the cup is 3 cm? (The volume of a segment of height h of a sphere of radius r is $\frac{1}{3}\pi h^2(3r-h)$.) (L)

46. If y is a function of x, show that the increment δy in y caused by a small increment δx in x is approximately equal to $(dy/dx)\delta x$.

Find the "difference" in the "cosec" tables, corresponding to the increase in angle from 45° to 45° 0·5′. (SU)

47. A tower of height h m stands on level ground a distance a m due north of a point O on a road running due east–west. A balloon is sent vertically from the top of the tower, with constant speed u m/s, commencing its ascent at the same instant as a man starts to walk along the road from O, due east, at v m/s. Obtain an expression for the tangent of the angle of elevation of the balloon from the man at time t seconds. Prove that the angle of elevation is a maximum (and not a minimum) after time a^2u/hv^2 seconds. (C)

48. A solid is formed by placing a hemisphere of radius x cm on one end of a cylinder radius x cm and height 12 cm. Express the volume V as a function of x. If R_1 and R_2 are the rates of increase of V with x when $x = 9$ cm and when $x = 10\frac{1}{2}$ cm, show that $5R_1 = 4R_2$. (OC)

49. The base angles of an isosceles triangle are each 2θ. Prove that the ratio of the area of the inscribed circle to the area of the triangle is $\frac{1}{2}\pi \tan\theta(1-\tan^2\theta)$, and deduce that the ratio is greatest when the triangle is equilateral.

Obtain also the ratio of the inscribed sphere to the volume of a right circular cone whose generators make an angle 2θ with the base, and find for what value of θ the ratio is greatest. (SU)

50. Sketch the graphs of

$$\text{(i) } y = \sqrt{x^3-x}, \quad \text{(ii) } y = \sqrt{(1-x^{\frac{2}{3}})^3}.$$

For what range of values of x are the functional relationships defined? Consider carefully the variation in the gradient of the curves.

51. Show that the function $y = (1-x)^4(2x^2-2x-3)$ has stationary values at the points where $x = -\frac{1}{2}$, 1, and $\frac{5}{3}$. Find the sign of dy/dx

$$\text{(i) when } -\tfrac{1}{2} < x < 1, \quad \text{(ii) when } 1 < x < \tfrac{5}{3},$$

and determine which of the stationary values of y is a maximum and which a minimum. (OC)

52. When $y = \cos 5\theta$ and $x = \cos\theta$, express dy/dx in terms of θ and show that

$$(1-x^2)\frac{d^2y}{dx^2} - x\frac{dy}{dx} + 25y = 0. \quad \text{(OC)}$$

53. Show that for the curve $ax^2+2hxy+by^2 = 1$, where a, h and b are constants,

$$\frac{d^2y}{dx^2} = \frac{h^2-ab}{(hx+by)^3} \quad \text{(OC)}$$

54. Find the coordinates of the turning points of the function $(3\cos x)/(2-\sin x)$ between the values 0 and 2π of x, distinguishing between maximum and minimum values. Sketch the graph of the function between $x = 0$ and $x = 2\pi$.

Prove that the tangents at the points where the function is zero intersect at the point $(\frac{3}{4}\pi, -\frac{3}{4}\pi)$. (OC)

55. An open tank is to be constructed with a square horizontal base and vertical sides. The capacity of the tank is to be 2000 m³. The cost of the material for the sides is £2 per m², and for the base is £1 per m². Prove that the minimum cost of the material involved is £1200, and give the corresponding dimensions of the tank. (OC)

56. If $p = dy/dx$, prove that $d^2y/dx^2 = p(dp/dy)$.

57. A man wishes to cross a river 100 m wide in order to get to a point A 200 m further along the opposite bank. If the boat in which the man crosses the stream travels at 4 km/h, the stream flows at 3 km/h and the man can walk at 5 km/h on land, where on the opposite bank should he land so that he gets to A in the least possible time if A is (i) upstream, (ii) downstream from his starting position?

58. Assuming that the relation

$$\frac{d}{dx}[\{u(x)\}^n] = n\{u(x)\}^{n-1}.u'(x)$$

is correct for a particular value N of n, derive an expression for $d[\{u(x)\}^{N+1}]/dx$ using the product rule.

Show that this is the same as the original expression if $n = N+1$.

***59.** A ray of light from a source X strikes a mirror at P and is reflected to a point Y (Fig. 6.8). If Q and R are the feet of the perpendiculars from X and Y to the mirror, $QP = x$, $PR = (b-x)$ and $XQ = YR = a$. Find an expression for the distance travelled by the ray of light along the path XPY. Find the value of x for which this path length has a stationary value, and hence deduce the second law of reflection.

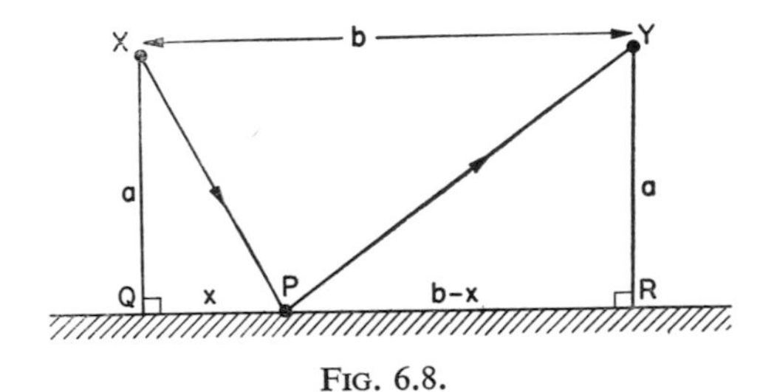

Fig. 6.8.

60. Sketch the graphs of

$$y = \frac{x(x-1)}{(x+2)},$$

and

$$y = \frac{x-2}{x(x+3)},$$

and state the values of x for which the functional relationships are not defined.

PART 3

THE INTEGRAL CALCULUS

CHAPTER 7

Introduction to Integration

7.1. THE CONCEPT OF AREA

It was shown in Section 2.2 that the area of a rectangle under a horizontal line velocity–time graph is numerically equal to the distance covered in the interval from the origin to the time t, provided that unity scaling is used (cf. Fig. 7.1). This fact is a consequence of the meaning

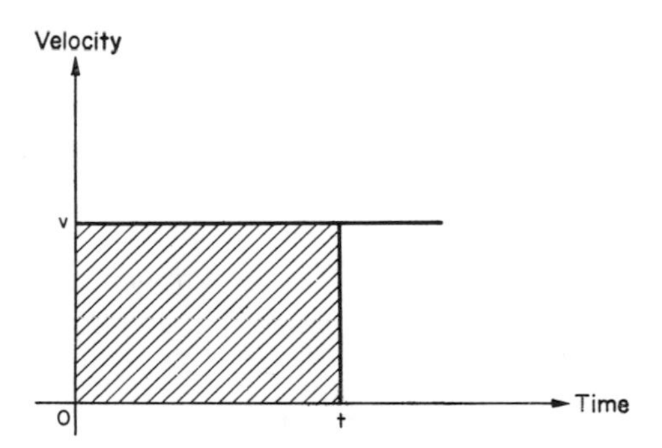

FIG. 7.1. The area between the unity scale horizontal velocity–time graph and the time axis gives the distance covered by the body whose motion is described by the graph.

of the "area" of a rectangle, which can be conveniently defined as the product of its length and breadth (Fig. 7.2). This definition, of course, is used almost as an axiom in everyday speech. The concepts of area and the integral of a function are closely related, and to lay a solid foundation for the ideas of the integral calculus to be shown later, we will first

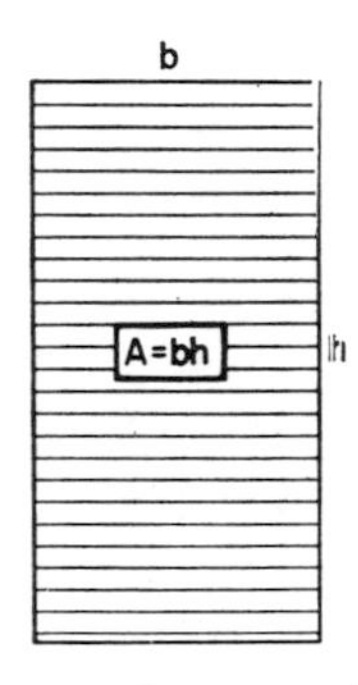

FIG. 7.2. The area of a rectangle is defined to be the breadth multiplied by the length.

consider the concept of the area of irregular surfaces.

Consider the area of a right-angled triangle. That this is equal to half the product of its base and height is well known: the derivation

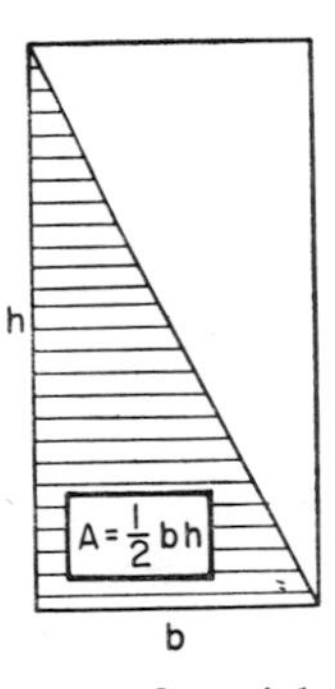

FIG. 7.3. The area of a right-angled triangle must be given by the product of half its base and its height if area possesses the property of being additive.

comes directly from the fact that the diagonal of a rectangle divides the figure into two equal parts (Fig. 7.3). If we endow area with an additive property, each triangle must have an area half that of the rectangle, and the result follows as a corollary of the definition of the area of the rectangle. Triangles without a right angle have areas given by the same expression (half the base multiplied by the height), and this corollary can be derived using Euclidean geometry or shearing theorems.

What then of the area of a circle? Here some new ideas have to be introduced, and it is

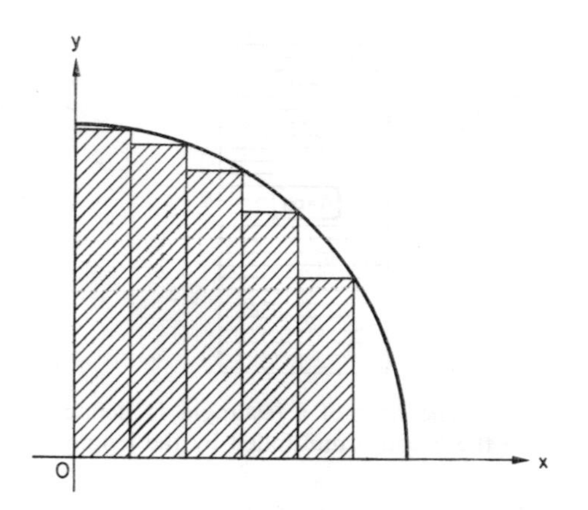

FIG. 7.4. The area of the five shaded rectangles is less than the area of the quadrant.

instructive to pursue them in some detail. If we take just one quadrant of the circle, we can construct a number of rectangles (say five) between a bounding radius and the circumference (Fig. 7.4), and either by measurement or by calculation, we can write down the values of the areas of these rectangles. Now since we have defined area to have a cumulative property, we can evaluate the total area of the rectangles by direct addition. We can also see that it will be possible to construct more (smaller) rectangles in the spaces we have between the rectangles and the circumference (i.e. in the unshaded areas of Fig. 7.4), and we can deduce that the total area of the five rectangles

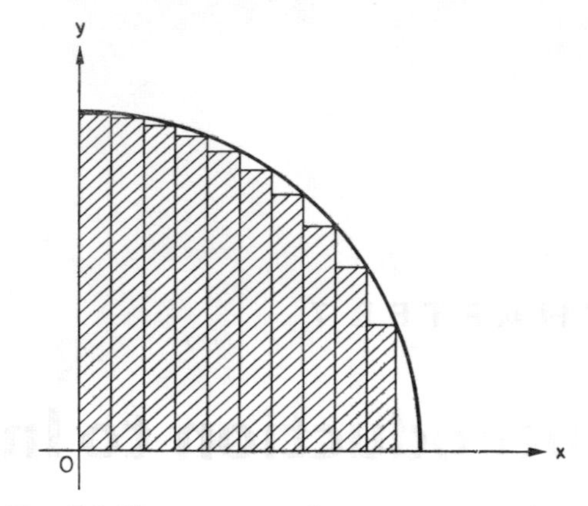

FIG. 7.5. If more rectangles are constructed, the (unshaded) area between the rectangles and the arc decreases.

is less than the area of the quadrant of the circle. By constructing a larger number of rectangles, the difference between the area of the quadrant and the total area of the rectangles is reduced (Fig. 7.5), and it would be quite sensible to suggest that we define the area of the quadrant as the limit of the total area of the number of rectangles constructed inside the quadrant as their number increases without bound (Fig. 7.6). (The method of "counting squares" described in Section 2.2 can be seen to be a procedure for finding an area under a curve which is a development of this definition. It will be recalled that an allowance

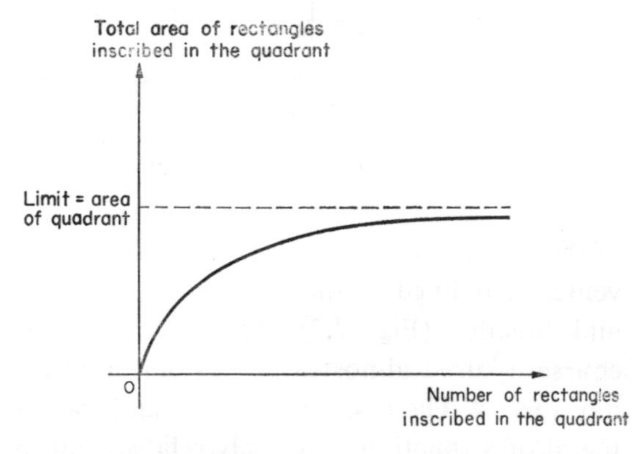

FIG. 7.6. The area of the rectangles constructed inside the quadrant approaches a limit as their number increases without bound.

can be made for the area between the graph and the complete squares of graph paper by counting any square with over half its area beneath the curve as wholly included, and any square with less than half its area as wholly excluded.)

It is, however, usual to adopt a definition of the area under a curve which is slightly different to that suggested above. So far, we have obtained an approximate value of the area of the quadrant of a circle by constructing rectangles on a number of ordinates drawn from a radius to its circumference, and evaluating the total area of the rectangles. The ordinates on which these "inner" rectangles are constructed are those subsequent to the first or radial ordinate, and as a result we obtain an estimate of the area which is always too small. Now we could equally well have started by constructing "outer" rectangles on the first and subsequent ordinates (Fig. 7.7),

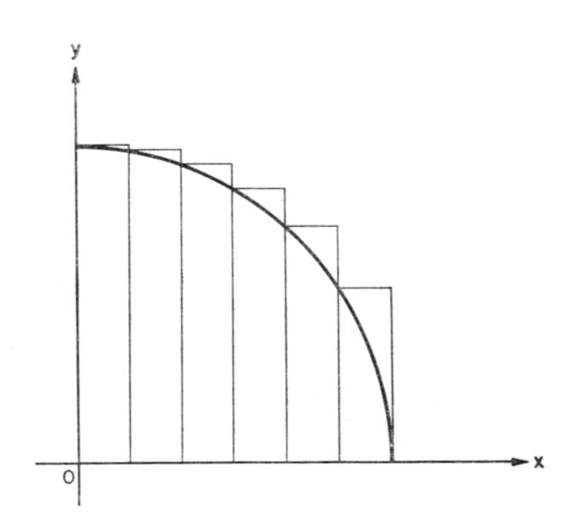

FIG. 7.7. Construction of rectangles on the first and subsequent ordinates leads to an estimate of the area of the quadrant which errs on the high side.

and the estimate of the area of the quadrant would then have erred on the high side. By combining both methods, we can obtain two numbers between which the true area of the quadrant must lie and, by constructing a sufficiently large number of rectangles, we can make these upper and lower estimates or (preferably) *bounds* of the area as close as we like. To be able to specify two bounds between which the true area must lie is quite an advance on specifying only one lower or one upper bound, and, of course, it is likely that the average of the bounds will be a better approximation to the value of the true area than any extrapolation of a curve such as that in Fig. 7.6.

We can now proceed to construct a definition for the "area" of the quadrant of the circle. If we imagine that n rectangles are constructed on ordinates inside and outside the quadrant as in Figs. 7.4 and 7.7, the total areas of which give a lower and an upper bound to the area of the quadrant, then if a number A can be found such that

$$L \leqslant A \leqslant U,$$

where L and U are the lower and upper bounds to the area, and where this inequality holds for all values of n, no matter how large, A is defined to be the area of the quadrant. It will be clear from what has been said that if an accurate estimate of A is to be obtained from an evaluation of the bounds, n must be large—that is, many rectangles will have to be constructed. To estimate the value of A by constructing a large number of rectangles is equivalent in a way to drawing the circle or curve containing the area on graph paper with small squares. It is interesting to ponder that no matter how small the squares, the area contained by a circle or a part of a circle will never enclose whole squares only, and there will always be bits "left over". To express the area of these bits as fractions of the smaller squares is equivalent to dividing the graph paper squares into even smaller ones, and this will give only a more accurate but still inexact solution. In the end, it is possible to define the area under a curve only as a limit—the common limit of the total area of the inner and outer rectangles as their number increases without bound (Fig. 7.8).

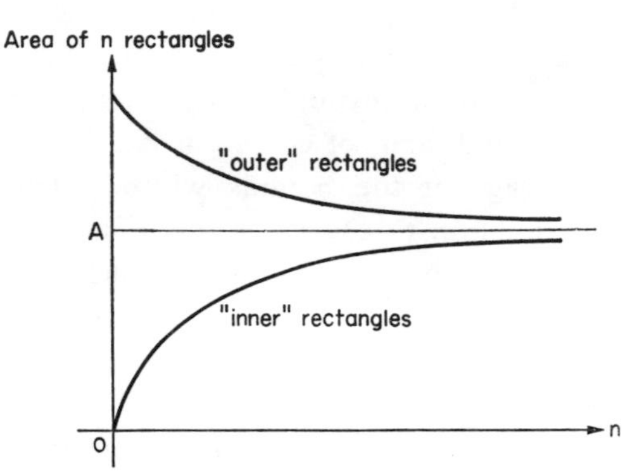

FIG. 7.8. The area of the quadrant (A) is defined as that number which lies between the area of n "outer" rectangles (i.e. those constructed on the first and subsequent ordinates) and n "inner" rectangles (i.e. those constructed on the second and subsequent ordinates) for all values of n.

Example 7.1

Draw a unity scale graph of $y = \frac{1}{2}(x^2+1)$ for $0 \leqslant x \leqslant 2$. Construct five equally spaced ordinates from the x-axis to the curve, and by direct measurement, calculate the areas of the outer and inner rectangles which can be constructed on the ordinates and the y-axis. Hence state an upper and lower bound to the value of the area enclosed between the curve and the x-axis between $x = 0$ and $x = 2$.

A table of values for plotting the sample points can be constructed from the equation $y = \frac{1}{2}(x^2+1)$, giving

x	0	$\frac{1}{2}$	1	$1\frac{1}{2}$	2
y	$\frac{1}{2}$	$\frac{5}{8}$	1	$1\frac{5}{8}$	$2\frac{1}{2}$

The points are then plotted and joined, giving the curve shown in Fig. 7.9a.

The five equally spaced ordinates are constructed at $x = 0{\cdot}4$, $0{\cdot}8$, $1{\cdot}2$, $1{\cdot}6$ and $2{\cdot}0$ (Fig. 7.9b), and by measurement, their lengths can be found as 0·58 cm, 0·82 cm, 1·22 cm, 1·78 cm and 2·5 cm. The breadth of each rectangle is

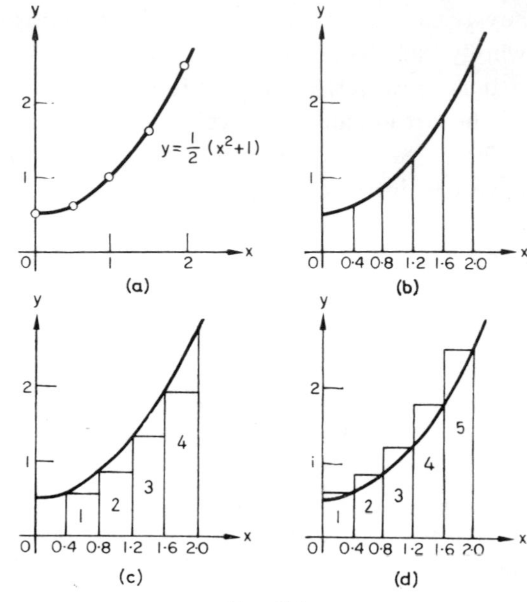

FIG. 7.9.

0·4 cm, so the area of the four inner rectangles (Fig. 7.9c) can be calculated:

$$A_1 = 0{\cdot}4 \times 0{\cdot}58 = 0{\cdot}232 \text{ cm}^2$$
$$A_2 = 0{\cdot}4 \times 0{\cdot}82 = 0{\cdot}328 \text{ cm}^2$$
$$A_3 = 0{\cdot}4 \times 1{\cdot}22 = 0{\cdot}488 \text{ cm}^2$$
$$A_4 = 0{\cdot}4 \times 1{\cdot}78 = 0{\cdot}712 \text{ cm}^2$$

The total area of the four rectangles is the sum of these; i.e. 1·76 cm².

Similarly the area of the five outer rectangles are (Fig. 7.9d):

$$B_1 = 0{\cdot}4 \times 0{\cdot}58 = 0{\cdot}232 \text{ cm}^2$$
$$B_2 = 0{\cdot}4 \times 0{\cdot}82 = 0{\cdot}328 \text{ cm}^2$$
$$B_3 = 0{\cdot}4 \times 1{\cdot}22 = 0{\cdot}488 \text{ cm}^2$$
$$B_4 = 0{\cdot}4 \times 1{\cdot}78 = 0{\cdot}712 \text{ cm}^2$$
$$B_5 = 0{\cdot}4 \times 2{\cdot}5 \;\; = 1{\cdot}000 \text{ cm}^2$$

and thus their total area is the value of this sum or 2·76 cm².

Thus if A is the area under the curve between $x = 0$ and $x = 2$, we can state that

$$1{\cdot}76 \leqslant A \leqslant 2{\cdot}76 \quad (\text{cm}^2).$$

EXERCISE 7a

1. Draw the upper right (or first) quadrant of a circle of radius 4 cm on a piece of graph paper with ten small squares to the centimetre. Construct inside the quadrant an ordinate PQ from the horizontal radius to the circumference 2 cm from the centre. Complete a rectangle on this ordinate inside the circle, another on the vertical radius and PQ produced and a third on PQ outside the circle as shown in Fig. 7.10a. By direct measurement evaluate the area of the two outer rectangles and the one inner rectangle.

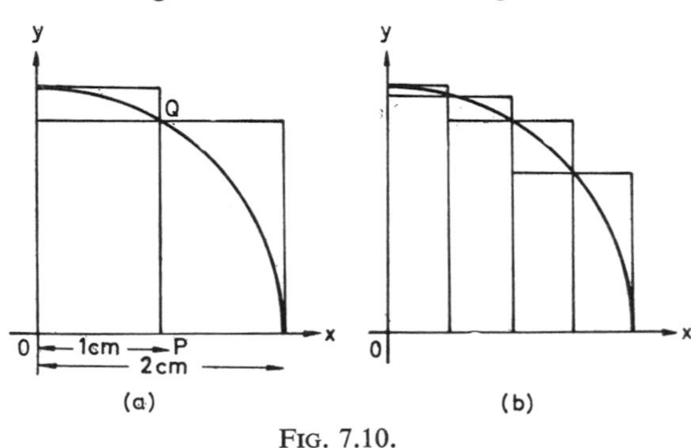

FIG. 7.10.

By further construction, divide the area inside the quadrant into three rectangles, constructing four rectangles outside the quadrant as in Fig. 7.10b, and by measurement calculate their total areas. Finally divide each rectangle into two again, and evaluate a lower and an upper bound of the area of the quadrant of the circle.

Plot the results on a graph, showing how the upper and lower bounds of the area vary with the number of constructed rectangles, and using as the independent variable the number of outer rectangles. By extrapolation estimate the common limit of these two bounds as the number of outer rectangles increases without bound.

Count the number of small squares enclosed by the circle, making allowances for parts of squares enclosed, and using the fact that each small square has an area of 0·01 cm², estimate the area of the quadrant of the circle by this method.

Compare your answers with the correct value π cm².

2. Draw a large-scale graph of the curve given by the equation $y = x(1-x)$ in the range $0 \leqslant x \leqslant 1$. By constructing rectangles inside and outside the curve, estimate the area enclosed by the curve.

3. Draw a large scale graph of the function $y = x^2$ in the range $0 \leqslant x \leqslant 2$. By constructing rectangles, estimate the area between the curve and the x-axis between $x = 0$ and $x = 2$.

4. Draw a sketch of the graph of $y = \sqrt{x}$ for $0 \leqslant x \leqslant 4$. If ten outer rectangles with equal bases are constructed on this curve between these limits, what will be the base of each if 1 cm is used to represent unity on both the y- and the x-axes?

Write down an expression for the height of each ordinate, and with the use of a slide rule or tables calculate these heights. Hence calculate an upper and a lower bound to the area between the curve and the x-axis.

5. Given that the area under the curve $y = x^3$ between $x = 0$ and $x = 2$ is 4 cm² if 1 cm on the x- and y-axes represents unity, write down the areas under the curves

(a) $y = 2x^3$ (d) $y = ax^3$
(b) $y = 5x^3$ (e) $y = x^3+2$
(c) $y = \frac{1}{2}x^3$ (f) $y = ax^3+b$

between the same two limits.

6. An "annulus" is the name given to the ring-shaped strip formed between two concentric circles of different radii (Fig. 7.11). Imagine such an annulus cut

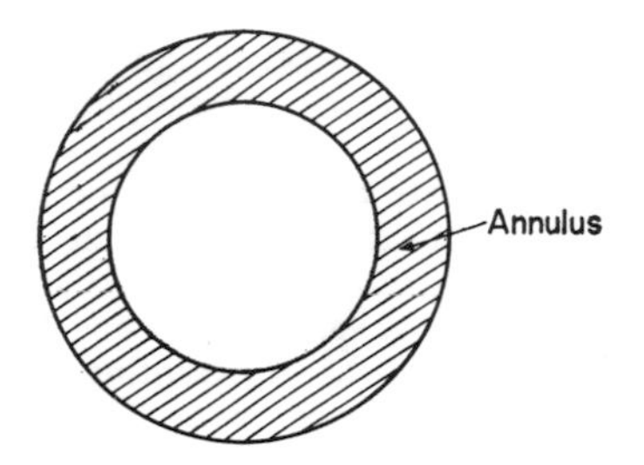

FIG. 7.11.

along a radius and "straightened out" into a trapezium. Write down the lengths of the parallel sides of the trapezium if the radii of the circles from which the annulus was formed were a and $(a+h)$. Hence deduce that the area of the trapezium is $\pi h(2a+h)$.

7. Two variables x and y are related so that corresponding pairs of values are as follows:

y	0	0·1	0·9	3	4·6	5	4·1
x	0	1	2	3	4	5	6

Draw a sketch graph of the relationship between them, treating x as the independent variable, and by constructing rectangles on the ordinates given above, find an upper and a lower bound to the area between the ordinates at $x = 0$ and $x = 6$.

8. Explain why the concept of the area under a graph of $1/x$ between $x = 0$ and $x = a$, where a is any positive number, is meaningless.

†**9.** Discuss whether or not the graph of a function with a discontinuity can contain an area between itself and the independent variable axis.

†**10.** Draw the graph of $y = x^2-2x-8$ for values of x between $x = -4$ and $x = +5$.

Use your graph

(i) to find the range of values of x for which y is positive;
(ii) to find the range values of x for which x^2-2x is less than $+4$;
(iii) to solve the quadratic equation $x^2-2x-13 = 0$;
(iv) to find the area contained between the graph and the x-axis. (SU)

*11. A piece of thin elastic, 1 m long, has one end attached to a hook and the other to a lump of metal of mass 6 kg. When it hangs in equilibrium from the hook, the metal stretches the elastic by an additional 3 m. Draw a graph showing the tension in the elastic at various extensions, and deduce from it the energy of the elastic when it is extended by 8 m.

The hook is 10 m above the floor, and the metal is held in contact with the hook. When the metal is released, it falls vertically, so that the elastic stretches. When the total length of the elastic is 5 m, it snaps. With what velocity will the metal hit the floor? Assume that the elastic obeys Hooke's Law until it snaps. (SU)

†12. Four graphs of a certain function are drawn: the first (A) is drawn with cm unity scaling, the second (B) with mm unity scaling, the third (C) with n cm along the independent variable axis and 1 cm along the dependent variable axis representing unit change, and the fourth (D) with n cm along both axes representing unit change. If the area contained between the graph, the independent variable axis and two particular ordinates on graph A is 39 cm², what is the value of the area enclosed by the same lines and curves on

(i) graph A in mm²,
(ii) graph B in mm²,
(iii) graph C in cm²,
(iv) graph D in cm²?

7.2. THE TRAPEZIUM RULE

7.2.1. It is usual to construct the rectangles used to estimate the area under a curve on equal bases: this is clearly the most suitable arrangement for convenience. Accepting this convention, we can rewrite the ideas of Section 7.1 in a more concise way by noting first that if n equal based rectangles are constructed under a portion of a curve ranging from $x = a$ to $x = b$ (Fig. 7.12), then the breadth of each will be $(b-a)/n$. Denoting this common width of the rectangles by the letter h, and

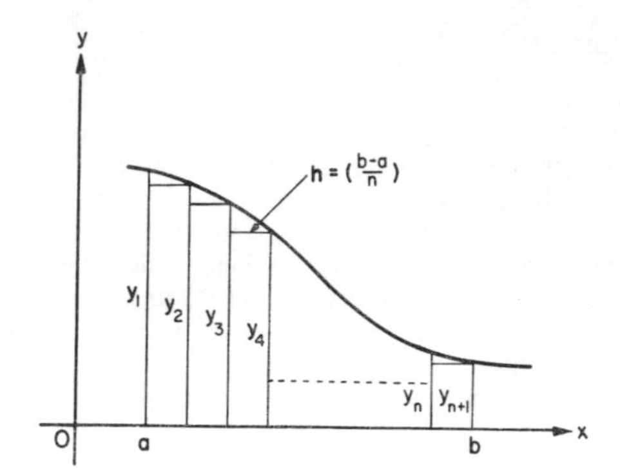

FIG. 7.12. The breadth (h) of each rectangle is $(b-a)/n$.

the length of the ordinates bounding the rectangles by y_1, y_2, y_3 and so on up to y_{n+1}, we can then write down the expression for the lower bound of the area as

$$hy_2+hy_3+hy_4+\ldots+hy_{n+1}. \quad (7.1)$$

This can be factorised and written more concisely in the form

$$h(y_2+y_3+y_4+\ldots+y_{n+1}). \quad (7.2)$$

Referring now to Fig. 7.13, we can write an expression corresponding to this for the combined area of the outer rectangles, namely

$$h(y_1+y_2+y_3+\ldots+y_n). \quad (7.3)$$

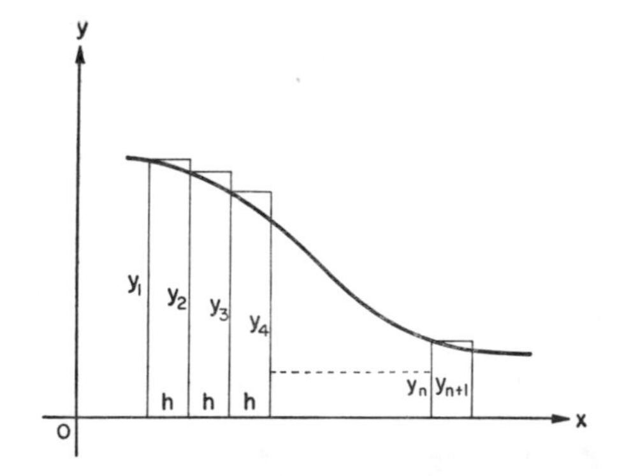

FIG. 7.13. The area of the outer rectangles is $h(y_1+y_2+\ldots+y_n)$.

Now once the number of rectangles to be constructed has been decided, it will be possible to calculate the width of each. And knowing the value of x at the lower boundary of the area (a), the ordinates at the edges of the rectangles are easily seen to have abscissae of $(a+h)$, $(a+2h)$, $(a+3h)$, and so on up to (b). From these values, which are also the abscissae of the points on the curve at the upper ends of the ordinates, the lengths of the ordinates from the x-axis to the curve can be calculated from the equation of the curve. Thus without drawing accurately any graph, it is possible to determine the values of y_1, y_2, y_3 and so on up to y_{n+1}, and thus upper and lower bounds to the area can be calculated without direct measurement.

Example 7.2

Calculate the values of the upper and lower bounds to the area under the curve $y = 16-x^4$ in the range $0 \leqslant x \leqslant 2$, using eight rectangles.

The curve is an inverted quartic curve (Fig. 7.14), crossing the y-axis at 16, and the x-axis at 2. If eight rectangles are constructed, the base of each will be $\frac{1}{4}$ (units of x). The

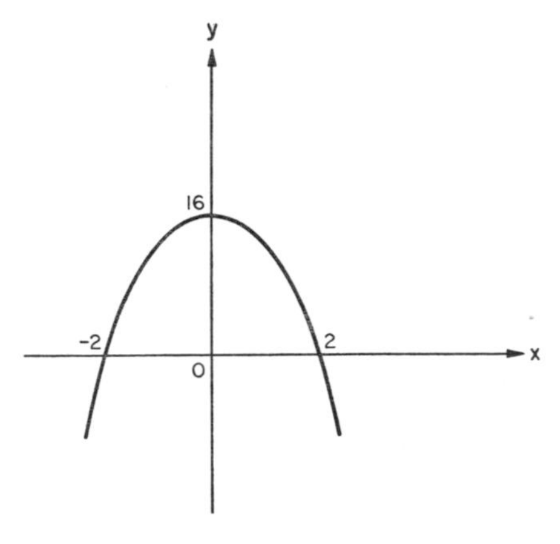

FIG. 7.14.

boundaries of the rectangles thus occur at $x = 0, \frac{1}{4}, \frac{1}{2}, \frac{3}{4}, 1, 1\frac{1}{4}, 1\frac{1}{2}, 1\frac{3}{4}$ and 2. The heights of the ordinates at these points are thus

$$\begin{aligned}
y_1 &= 16-0^4 &&= 16\\
y_2 &= 16-(0{\cdot}25)^4 &&= 16{\cdot}00\\
y_3 &= 16-(0{\cdot}5)^4 &&= 15{\cdot}94\\
y_4 &= 16-(0{\cdot}75)^4 &&= 15{\cdot}68\\
y_5 &= 16-(1{\cdot}0)^4 &&= 15\\
y_6 &= 16-(1{\cdot}25)^4 &&= 13.56\\
y_7 &= 16-(1{\cdot}5)^4 &&= 10{\cdot}92\\
y_8 &= 16-(1{\cdot}75)^4 &&= 6{\cdot}64\\
y_9 &= 16-(2{\cdot}0)^4 &&= 0
\end{aligned}$$

The lower bound is equal to $h(y_2+y_3+y_4+\ldots+y_9)$, and since the width of each rectangle is $\frac{1}{4}$, this equals

$$\tfrac{1}{4}(16{\cdot}0+15{\cdot}94+15{\cdot}68+15{\cdot}0+13{\cdot}56+10{\cdot}94+6{\cdot}64+0) = \tfrac{1}{4}(92{\cdot}76) = \underline{23{\cdot}19.}$$

Similarly, the upper bound

$$\begin{aligned}
&= h(y_1+y_2+y_3+\ldots+y_8)\\
&= \tfrac{1}{4}(16+16{\cdot}0+15{\cdot}94+15{\cdot}68+\ldots+6{\cdot}64)\\
&= \tfrac{1}{4}(109{\cdot}74)\\
&= \underline{27{\cdot}44.}
\end{aligned}$$

Thus, if A is the area under the curve,

$$23{\cdot}2 < A < 27{\cdot}4.$$

(It is supposed that the graphs are drawn with unity scaling: if centimetre unity scaling is used, A will be in cm².)

EXERCISE 7b

Calculate upper and lower bounds to the area under the following curves between the ordinates indicated, using the number of rectangles given:

1. $y\ \sqrt{x}$; $x = 0$ and $x = 9$; nine rectangles.

2. $y = 3x^2$; $x = 0$ and $x = 2$; five rectangles.

3. $y = \sin x$; $x = 0$ and $x = \frac{1}{2}\pi$; six rectangles

4. $y = \dfrac{1}{x}$; $x = 1$ and $x = 5$; eight rectangles.

5. $y = \dfrac{1}{x^2}$; $x = 1$ and $x = 9$; eight rectangles.

6. A function of x varies with x as given in the table below. Sketch its graph, and using the six ordinates given, calculate lower and upper bounds to the area under the curve between the ordinates $x = 0$ and $x = 5$ if 2 cm is used to represent 20 units of

$f(x)$ and 2 units of x along the axes. Repeat for the second function $g(x)$.

x	0	1	2	3	4	5
$f(x)$	5	15	30	50	75	105
$g(x)$	0	10	25	45	70	100

7. A function $g(t)$ of a variable t varies with the value of t as shown in the table below. Sketch its graph, and using a suitable number of ordinates, calculate lower and upper bounds of the area under the curve between ordinates

(i) $t = 0$ and $t = 20$,
(ii) $t = 20$ and $t = 40$,
(iii) $t = 0$ and $t = 40$,

if 1 cm is used to represent 50 units of $g(t)$ and 10 units of t.

t	0	5	10	15	20	25	30	35	40
$g(t)$	0	20	35	45	50	50	40	20	0

If $g(t)$ is a velocity measured in cm/s, and t is in seconds, between what limits does the distance travelled by the body whose motion is described by the curve in the (a) first 20 seconds, (b) second 20 seconds of motion, (c) altogether, lie?

8. The velocity of a body varies with the time as follows:

Time (s)	0	1	2	3	4	5
Velocity (km/h)	0	5·2	10·8	16·1	21·2	28·8

Time (s)	6	7	8	9	10
Velocity (km/h)	34·1	42·1	49·8	55·7	60·6

Make an estimate of the total distance travelled by the body in the 10 seconds.

7.2.2. It will probably have occurred to the reader by now that a closer estimate of the area under a curve can be obtained by averaging a corresponding pair of upper and lower bounds. This is in fact the case, and although the size of the error in the estimate is not known, it is often useful to have a more precise guess of the area in question.

An analytical expression for the average of a pair of upper and lower bounds can be obtained by adding equations (7.2) and (7.3), and dividing by 2; this gives the expression

$$\tfrac{1}{2}h(y_1+2y_2+2y_3+2y_4 \ldots +2y_n+y_{n+1}) \qquad (7.4)$$

as an approximation for the area A under the curve between the ordinates in question. Writing the corresponding equation for A in the form

$$A \simeq h[\tfrac{1}{2}(y_1+y_2)+\tfrac{1}{2}(y_2+y_3)+\tfrac{1}{2}(y_3+y_4)+ \ldots + \tfrac{1}{2}(y_n+y_{n+1})]$$

it becomes clear that the evaluation of equation (7.4) is equivalent to summing the areas of the trapezia drawn between the ordinates $y_1, y_2, \ldots, y_{n+1}$ (Fig. 7.15), and this is again apparent when it is recognised that the mean of the areas of a pair of inner and outer rectangles is equal to the area of the trapezium drawn on the same ordinates (Fig. 7.16). For

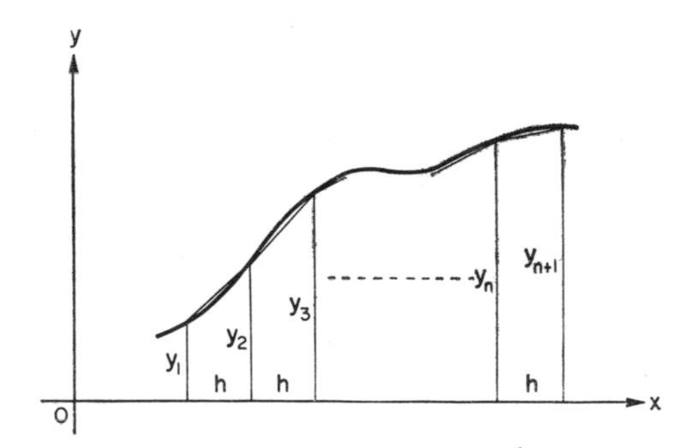

FIG. 7.15. The evaluation of $h\{\frac{1}{2}(y_1+y_2)+\frac{1}{2}(y_2+y_3)+ \ldots +\frac{1}{2}(y_n+y_{n+1})\}$ is equivalent to finding the area of the trapezia constructed on the ordinates y_1, y_2, etc., to y_{n+1}.

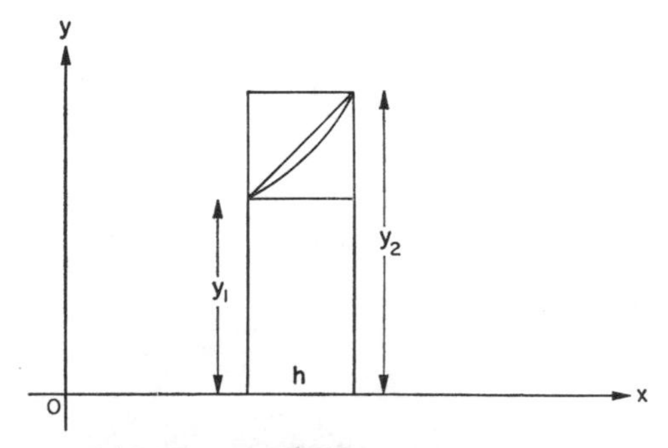

FIG. 7.16. The area $\frac{1}{2}h(y_1+y_2)$ of the trapezium is the average of the areas hy_1 of the inner rectangle and hy_2 of the outer rectangle.

this reason, the expression of equation (7.4) is known as the "*Trapezium Rule*".

In practice the trapezium rule is more usefully expressed in the form

$$A \simeq \tfrac{1}{2}h[(y_1+y_{n+1})+2(y_2+y_3\ldots+y_n)], \quad (7.5)$$

so that in finding the lengths of the ordinates between the boundaries, there is no need to double each one individually. The sum of the boundary ordinates y_1 and y_{n+1} is found, as is the sum of y_2 to y_n, and this latter sum can then be doubled as a whole.

A little reflection will show that the use of the trapezium rule is equivalent to approximating to the curve bounding the area by linear portions (Fig. 7.17); the degree of ap-

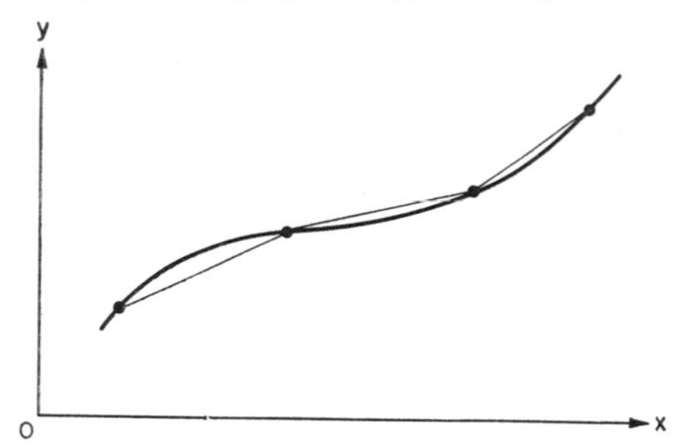

FIG. 7.17. Using the trapezium rule is equivalent to replacing the curve by a set of straight line approximations.

proximation inherent in the method clearly depends on the departure of the curve from linearity between the ordinates, and this becomes less as the spacing between the ordinates is reduced. In general, the method is less tedious and more accurate than the method of counting squares described on p. 42, and given an indefinite amount of labour, equation (7.5) can be used to obtain an answer correct to any specified degree of accuracy.

Example 7.3

Find the area enclosed between the curve $y = 2+x^2$, the ordinates $x = 0$ and $x = 3$, and the x-axis on a cm unity scale graph. Use the trapezium rule and 6 rectangles.

Using six rectangles as instructed, ordinates must be constructed at intervals of $(3-0)/6$ or $\frac{1}{2}$ cm. The ordinates (see Fig. 7.18) thus occur

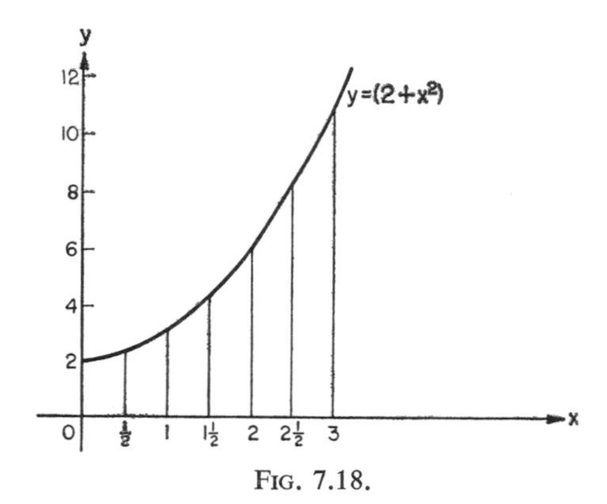

FIG. 7.18.

at $x = 0, \frac{1}{2}, 1, 1\frac{1}{2}, 2, 2\frac{1}{2}, 3$. Evaluating and writing down the heights of these ordinates in a convenient tabular form, we obtain:

$y_1 = 2+(0)^2 = 2$	$y_2 = 2+(\frac{1}{2})^2 = 2{\cdot}25$
	$y_3 = 2+(1)^2 = 3{\cdot}0$
	$y_4 = 2+(1\frac{1}{2})^2 = 4{\cdot}25$
	$y_5 = 2+(2)^2 = 6{\cdot}0$
$y_7 = 2+(3)^2 = 11$	$y_6 = 2+(2\frac{1}{2})^2 = 8{\cdot}25$
$y_1+y_7 = 13$	$y_2+y_3+y_4+y_5+y_6 = 23{\cdot}75.$

$$\begin{aligned} A &\simeq \tfrac{1}{2}h[(y_1+y_7)+2(y_2+y_3+\ldots y_6)] \\ &= \tfrac{1}{2}\cdot\tfrac{1}{2}[13+2(23{\cdot}75)] \\ &= \tfrac{1}{4}[13+47{\cdot}5] \\ &= \tfrac{1}{4}[60{\cdot}5] \\ &= 15{\cdot}1 \text{ cm}^2. \end{aligned}$$

(The accurate answer will be shown later to equal 15 cm².)

EXERCISE 7c

Estimate the value of the areas enclosed between the x-axes, the curves and the ordinates indicated, using the trapezium rule and the number of trapezia suggested. Assume that the graphs are drawn with cm unity scaling.

1. $y = 6x^2$; $x = 0$ and $x = 1$; 10.
2. $y = \tan x$; $x = 0$ and $x = \frac{1}{4}\pi$; 5.
3. $y = (25-x^2)$; $x = 0$ and $x = 5$; 5.
4. $y^2 = (25-x^2)$; $x = 0$ and $x = 5$; 5. Compare the value you obtain in this example with the exact value.

5. Obtain three estimates of the area enclosed by the curve $y = x^2$, the x-axis and ordinates at $x = 0$ and $x = 10$ using in turn five, ten and twenty trapezia. Represent your results graphically, and estimate the value of the limit of the result obtained by using the trapezium rule as the number of trapezia increases without bound. Assume that the curve is plotted on a cm unity scale graph.

†6. Estimate the area enclosed between the x-axis, the line $x = 1\frac{1}{2}\pi$ and the curve $y = \sin x$, using the trapezium rule and six trapezia. State carefully the scaling you assume, and calculate the area on a unity scale graph using

(i) degrees,
(ii) radians,

as the unit on the x-axis.

7. The speed of two different cars starting from rest varies with time as follows:

Time (s)	0	2	4	6	8	10	12
v_1 (m/s)	0	3	6	10	13	15	18
v_2 (m/s)	0	2	5	9	12	14	17

Compare the distances travelled by the two cars in the first 6 seconds of their motion using the trapezium rule, and estimate how much ahead of the slower car the faster one will be at the end of 12 seconds.

8. The rate of change of a variable y with respect to x varies with the value of x as follows:

dy/dx	0	0·1	0·3	0·8	1·3	1·0	0·4	0·2	0·2
x	0	1	2	3	4	5	6	7	8

Draw a sketch graph of this variation, and find the value of y when x is equal to

(i) 2, (iii) 6,
(ii) 4, (iv) 8,

using the trapezium rule and given that $y = 0$ and $x = 0$.

9. Draw a graph of the straight line $y = 1+x$ between $x = 0$ and $x = 10$. Divide the trapezium formed by this line, $x = 0$, $x = 10$ and the x-axis into ten equal strips by constructing ordinates at $x = 1, 2, 3, \ldots, 9$. By considering the total area of the rectangles formed under the line by adjacent pairs of ordinates, show that the area of the trapezium cannot be less than 55.

Show also that the area of the trapezium cannot be greater than 65, and by imagining it to be divided into 100 equal strips, show that its area A satisfies the inequality $59{\cdot}5 \leqslant A \leqslant 60{\cdot}5$.

Show that if the area is divided into n equal strips,

$$60-\frac{50}{n} \leqslant A \leqslant 60+\frac{50}{n},$$

and deduce the area of the trapezium to be 60.

10. By a method similar to that described in question 9, show that the area of a trapezium formed by two parallel lines of length a and b a distance h apart is $\frac{1}{2}(a+b)h$.

11. The velocity of a car starting from rest varies with time as follows:

Time (s)	0	2	4	6	8	10	12	14
Velocity (m/s)	0	5	8	12	15	19	24	27

Calculate the approximate distance travelled in these first 14 seconds.

12. The commission on the sale of an article costing 60p is 10% for the first 10,000 articles sold, 15% for the second 10,000, 20% for the next 5000 and 25% for any sales over 25,000. Sketch the graph of the rate of commission in new pence per article against the number of articles sold, and assuming 1 cm is used to represent 1000 articles on the sales axis, and 1p along the rate of commission axis, calculate the *exact* area under the graph between ordinates at zero sales and (i) 8000, (ii) 14,000, (iii) 22,000, (iv) 34,000 articles sold.

Why is this area a measure of the total commission earned? Calculate the conversion factor from square centimetres of graph paper into new pence commission, and hence calculate the commission earned by selling (a) 8000, (b) 14,000, (c) 22,000, (d) 34,000 articles.

13. Sketch the graph of $y = 2x^2$ for $0 \leqslant x \leqslant 4$. If 1 cm is used to represent 2 units of y and $\frac{1}{2}$ unit of x along the axes, what are the upper and lower bounds of the area under the curve between the ordinates at $x = 0$ and $x = 4$ if rectangles of base (a) 1 cm, (b) $\frac{1}{2}$ cm, (c) $\frac{1}{4}$ cm are constructed? Make an estimate of the actual value of the area.

14. Repeat question 13 for the graph of $y = \sqrt{x}$ in the range $0 \leqslant x \leqslant 16$, using 1 cm to represent unit change in y and 2 units change in x.

†15. Draw a graph of the curve $y = 1+x^2$ for $0 \leqslant x \leqslant 10$, and divide the area formed by the curve, $x = 0$, $x = 10$ and $y = 0$ into ten equal vertical strips.

If A is the area between the curve and the x-axis, show that $295 \leqslant A \leqslant 395$.

Show also that the bounds on A can be brought nearer together by dividing the area into n equal strips, and that if this is done,

$$\frac{10}{n}\sum_{r=0}^{n-1}\left(1+\frac{100r^2}{n^2}\right) \leqslant A \leqslant \frac{10}{n}\sum_{r=1}^{n}\left(1+\frac{100r^2}{n^2}\right)$$

or

$$343\frac{1}{3}-\frac{500}{n}+\frac{500}{3n^2} \leqslant A \leqslant 343\frac{1}{3}+\frac{500}{n}+\frac{500}{3n^2}.$$

Hence find the value of A.

$$\left(\sum_{r=1}^{n} r^2 = n(n+1)(2n+1)/6).\right)$$

7.3. THE DEFINITE INTEGRAL

It will have been appreciated that the numerical value of the area under a curve on a piece of graph paper depends largely on the scales employed for the axes of the graph, and more important than the area itself is the mathematical or physical quantity it represents. The number given by the area under a curve on a unity scale graph can be identified with the "*integral*" of the function represented by the curve over the range of values of the independent variable in question. (To avoid unnecessary repetition, it will be assumed in this section that all unity scale graphs are drawn using 1 cm rather than 1 mm for the units of the variables.)

The first and the simplest definition of an integral applies to a constant value function. If we have a function of time which has a constant value v over an interval of time t, the integral of the function over that interval is defined to have a value $v \times t$. Considering the graphical representation of this part of the function on a graph with unity scaling (Fig. 7.19), we can see that the numerical

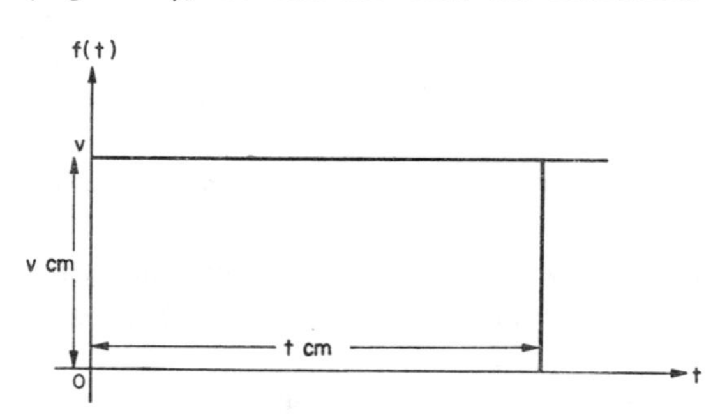

FIG. 7.19. The integral of a constant value function $f(t)$ between 0 and t is defined to be the product vt, where $v = f(t)$: and thus the integral has the same numerical magnitude as the area enclosed between the ordinates at 0 and t on a unity scale graph of the function $f(t)$.

value of the area under the portion of the graph being considered is equal to the value of the integral. Or, to take a numerical example, if we have a function whose value is constant and equal to 4 over a range of independent variable x from 2 to 5 (Fig. 7.20), the integral of the function over that range is (4×3) or 12, and, of course, 12 cm² is the area under the graph if unity scaling is em-

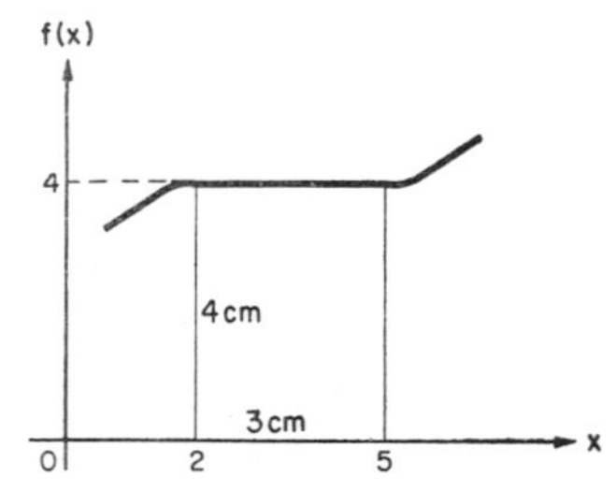

FIG. 7.20. Similarly the integral of $f(x) = 4$ between $x = 2$ and $x = 5$ is (4×3) or 12.

ployed. The interpretations which can be given to the value of such an integral will be seen later: for the moment the fact that the value of the integral of velocity over a range of time is equal to the change in displacement gives an illustration of the kind of use to which it can be put.

If we require the integral of a function which varies in value as the independent variable changes, our definition (which applies only to constant value functions) will no longer suffice. We extend our ideas as follows.

Suppose the function we are considering varies in the manner shown in Fig. 7.21. If we

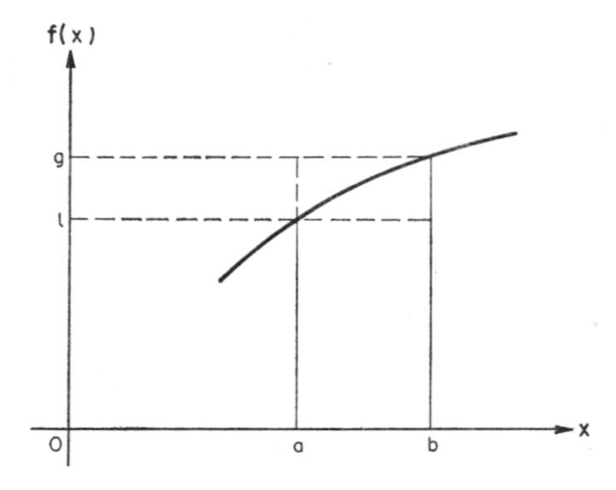

FIG. 7.21. If g and l are the greatest and least values of $f(x)$ between a and b, the integral I of $f(x)$ between a and b is defined to lie between that of $g(x)$ and $l(x)$, or $l(b-a) \leqslant I \leqslant g(b-a)$.

wish to evaluate the integral of this function between $x = a$ and $x = b$, we proceed by first specifying two bounds between which it must lie. To do this, consider two constant value functions $g(x)$ and $l(x)$ whose values g and l throughout the interval $x = a$ to $x = b$ equal the greatest and least values respectively that $f(x)$ attains in the interval (Fig. 7.21). Now we have defined the integral of $g(x)$ as the product $g(b-a)$ over the interval from $x = a$ to $x = b$, and it follows that the magnitude of the integral is the same as the area under a unity scale graph of $g(x)$ between two ordinates at $x = a$ and $x = b$, and it is logical to extend this association and specify that, since the area under the graph of $f(x)$ in the interval from $x = a$ to $x = b$ is less than that under the graph of $g(x)$, the integral of $f(x)$ throughout the interval is less than the integral of $g(x)$ or $g(b-a)$. And similarly, since the area under $f(x)$ is greater than that under $l(x)$, we specify that the integral of $f(x)$ between $x = a$ and $x = b$ is greater than that of $l(x)$ between the same two limits. Thus without yet defining a value for the integral of $f(x)$, we have specified that its value shall lie between $l(b-a)$ and $g(b-a)$, where l and g are the least and greatest values of the function in the interval.

Now let us suppose that we divide the interval between $x = a$ and $x = b$ into n small intervals of size δx (where, if these small intervals are equal, $n\,\delta x = b-a$), and let us make a similar specification for the values of the integrals of $f(x)$ in each interval—namely, that the integral of $f(x)$ over any one of the intervals δx must lie between $l\,\delta x$ and $g\,\delta x$, where l and g are the least and greatest values the function takes in the interval. If we then endow the integral with a cumulative property—the property with which we endowed area—we can add together the integrals $l\,\delta x$ and $g\,\delta x$ for each interval, and say that the sum of the products $l\,\delta x$ and $g\,\delta x$ for all the intervals into which a range is divided specify a lower and upper bound to the value of the integral of $f(x)$ throughout that range.

Graphically we are maintaining the association of the integral with the area on a unity scale graph: just as the sum of the areas $(l_1\,\delta x+l_2\,\delta x+\ldots+l_n\,\delta x)$ is less than the area under the graph of $f(x)$, which in turn is less than the sum of the areas $(g_1\,\delta x+g_2\,\delta x+\ldots+g_n\,\delta x)$, so we define the sum of the integrals $(l_1\,\delta x+l_2\,\delta x+\ldots+l_n\,\delta x)$ to be less than the integral of $f(x)$, and the integral of $f(x)$ to be less than the sum of the integrals $(g_1\,\delta x+g_2\,\delta x+\ldots+g_n\,\delta x)$.

We finally take the same step as we did in defining the area under a curve, and define the integral I of any function $f(x)$ between the values a and b as that number which always lies between the two sums of the integrals of the least value and greatest value functions, no matter how large n. As n increases without bound, the two integral sums become more and more nearly equal, and the definition just suggested enables us to assign a precise value to the integral, just as a precise value is assigned to the area under a curve by an identical definition. In fact, the consequence of the definition we have just developed for the integral of any varying function is that *the numerical magnitude of the area under a unity scale graph of the function between ordinates at* $x = a$ *and* $x = b$ *is equal to the value of the integral of that function over the same range (Fig. 7.22).*

We can therefore determine the value of the integral of a function over any range by evaluating the corresponding area under a unity scale graph of that function.

It it important to be able to derive the units or dimensions of the physical quantity represented by the integral. The area under a curve is, of course, measured in units such as cm^2, and it is only the *value* of the integral which equals the *value* of the area (e.g. with cm unity scaling, an area of 13 cm² corresponds to an integral whose value is 13). The *unit* of the integral is derivable from the fact that the integral is defined as a product of the two variables involved: and the unit of the integral

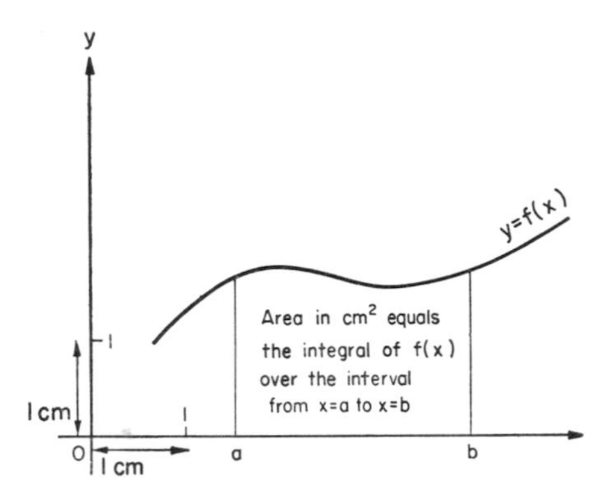

FIG. 7.22. The numerical magnitude of the area under a unity scale graph of $f(x)$ between $x = a$ and $x = b$ is equal to the value of the integral of $f(x)$ over the same range.

is then given by the product of the units of the variables. Thus if a graph is one of a dependent variable measured in cm/s against an independent variable measured in seconds, the unit of the integral is (cm/s)×(s)—i.e. cm. Or if the dependent variable is measured in new pence/cm and the independent variable in min/new pence, the unit of the integral will be p min/cm.

As in the case of the derivative, a special sign is employed to denote the integral of a function over a range. In view of the fact that the idea of the value of the integral can be approached by considering the limiting value of a sum (as we shall see in Section 8.3), an old form (∫) of the letter S is employed. The symbol ∫ is called the ***integral sign***, and the integral of a function $f(x)$ of **a** variable x over a range between $x = a$ and $x = b$ is denoted by the symbol

$$\int_a^b f(x)\,dx.$$

The symbol is read "the integral of $f(x)$ with respect to x between ($x =$, understood) a and b", and the numbers a and b are called the *limits of integration*, integration being the name given to the process which obtains the value of the integral. The final dx may seem superfluous, but in fact it is an important computational aid, as will become apparent later. Thus, the value of

$$\int_5^8 3\,dx$$

is numerically equal to the area under the unity scale graph of the constant value function 3 between $x = 5$ and $x = 8$ (Fig. 7.23), and this equals 9. Similarly, the value of $\int_0^4 x\,dx$ is numerically equal to the area under

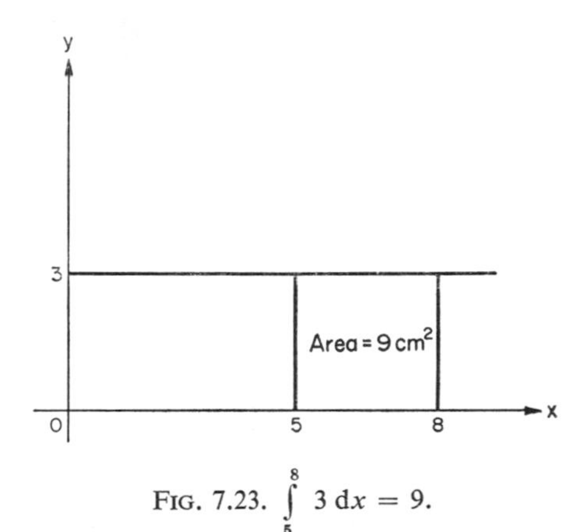

FIG. 7.23. $\int_5^8 3\,dx = 9.$

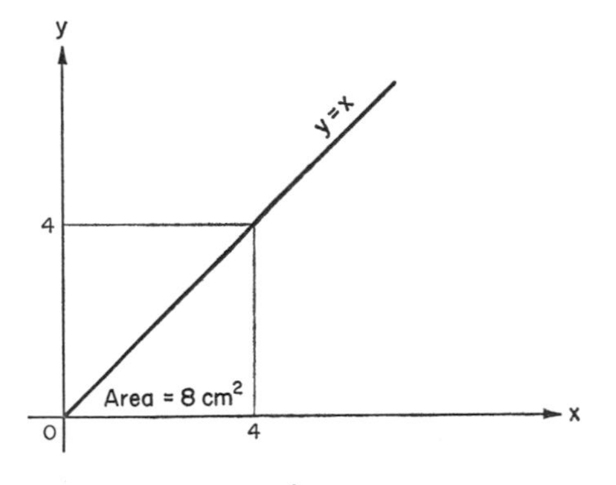

FIG. 7.24. $\int_0^4 x\,dx = 8.$

the graph of $y = x$ between the values $x = 0$ and $x = 4$ (Fig. 7.24), which equals $\frac{1}{2}.4.4$, or 8.

Example 7.4

Find the values of

(a) $\int_1^2 3\,dx$, (c) $\int_2^5 a\,dx$,

(b) $\int_0^4 3\,dx$, (d) $\int_9^2 x\,dx$.

(a) The value of the integral of 3 with respect to x can be identified with the area under the graph of $y = 3$ between $x = 1$ and $x = 2$, since the area under a y–x curve between these limits can be identified with $\int y\,dx$, and the y

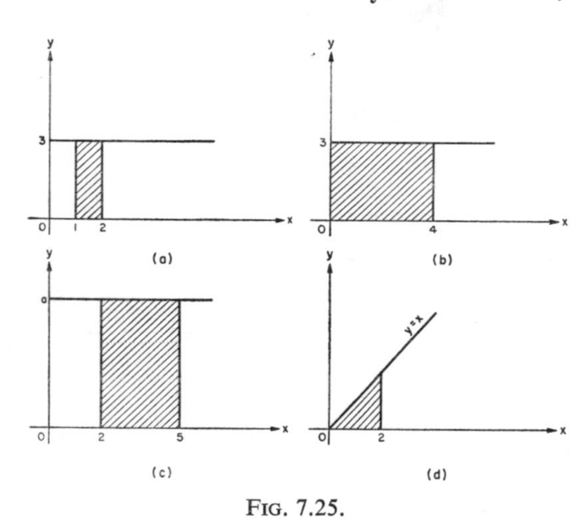

FIG. 7.25.

in this case is replaced by 3 (Fig. 7.25a). If the graph uses centimetre unity scaling, the area is 3 cm^2; and thus

$$\int_1^2 3\,dx = 3$$

(b) Similarly, the value of $\int_0^4 3\,dx$ can be identified with the shaded area in Fig. 7.25b and thus is equal to 12.

(c) The graph of $y = a$, where a is a constant, is also horizontal, and thus

$$\int_2^5 a\,dx = 3a \quad \text{(Fig 7.25c).}$$

(d) The graph of $y = x$ is a straight line of gradient 1, and the area which can be identified with $\int_0^2 x\,dx$ is thus triangular (Fig. 7.25d). Numerically the area of the triangle between the origin and the ordinate $x = 2$ equals $\frac{1}{2}$(base$\times$height), i.e. $\frac{1}{2}(2\times 2)$, and thus $\int_0^2 x\,dx$ equals 2.

EXERCISE 7d

Write down the values of:

1. $\int_2^3 3\,dx$.
2. $\int_0^1 2\,dx$.
3. $\int_1^2 1\,dx$.
4. $\int_5^8 7\,dx$.
5. $\int_{27}^{45} a\,dx$.
6. $\int_{100}^{500} 18\,dx$.
7. $\int_{-1}^{0} 2\,dx$.
8. $\int_{-2}^{+2} 5\,dx$.
9. $\int_0^4 x\,dx$.
10. $\int_1^4 3x\,dx$.
11. $\int_2^5 (x+7)\,dx$.
12. $\int_8^{10} (x-1)\,dx$.
13. $\int_5^6 (x-5)\,dx$.
14. $\int_2^3 4\,dx + \int_3^5 4\,dx$.
15. $\left\{\int_0^2 + \int_2^4\right\}(x\,dx)$.
16. $\int_0^6 5x\,dx + \int_8^{10} 5x\,dx$.
17. $\dfrac{\int_1^2 5x\,dx}{\int_2^3 5x\,dx}$.
18. $\int_0^5 3x\,dx - \int_3^5 3x\,dx$.
19. $\left\{\int_0^1 x\,dx\right\}\left\{\int_1^2 x\,dx\right\}$.
20. $\dfrac{\int_1^2 x\,dx}{\int_1^2 5\,dx}$
21. $\int_0^3 (2x+1)\,dx$.
22. $\int_1^5 (7x-1)\,dx$.

23. Show that $\int_0^2 3\,dx + \int_2^4 3\,dx = \int_0^4 3\,dx$.

†**24.** Show that $\int_a^b f(y)\,dy + \int_b^c f(y)\,dy = \int_a^c f(y)\,dy$,

and that $\int_a^c g(q)\,dq - \int_a^b g(q)\,dq = \int_b^c g(q)\,dq$.

†**25.** Explain why the integral of the expression for the variation in the velocity of a body with respect to the time gives an expression for the displacement, and why the integral of the acceleration of the body with respect to time gives an expression for the velocity.

†**26.** The velocity of a body is equal to $2t$, where t is the time. How far does it travel during the interval between $t = 3$ and $t = 5$?

†**27.** A step function of the variable x is denoted by $s(x)$, and is defined by the following equations: $s(x) = 2$ in the range $0 \leqslant x < 1$; $s(x) = 1$ in the range

$1 \leqslant x < 3$, and $s(x) = 3$ in the range $3 \leqslant x \leqslant 6$.

Write down the values of:

(i) $\int_0^1 s(x)dx$; (ii) $\int_0^2 s(x)dx$; (iii) $\int_0^6 s(x)dx$.

†**28.** A step function of a variable x is denoted by $t(x)$, and defined by the following equations:

$$\begin{aligned} t(x) &= 2 \text{ for } -1 \leqslant x \leqslant 1; \\ t(x) &= 4 \text{ for } 1 < x \leqslant 2; \\ t(x) &= 2 \text{ for } 2 < x < 4; \\ t(4) &= 6; \\ \text{and } t(x) &= 1 \text{ for } 4 < x \leqslant 5. \end{aligned}$$

Write down the values of:

(i) $\int_0^1 t(x)dx$; (iv) $\int_0^4 t(x)dx$;

(ii) $\int_{-1}^{+1} t(x)dx$; (v) $\int_0^5 t(x)dx$.

(iii) $\int_{-1}^{3 \cdot 2} t(x)dx$;

Would the value of any of these integrals have been changed if $t(4) = -2$?

†**29.** Write down the values of:

(i) $\int_0^3 [x]\,dx$; (ii) $\int_0^{4 \cdot 6} \{x-[x]\}\,dx$,

where $[x]$ denotes a function which takes the greatest integer with a value less than or equal to the value of x.

†**30.** What is the value of $\int_0^n [x]\,dx$ if n is positive and

(i) integral, (ii) non-integral?

†**31.** Discuss the meaning of the following statements:

(i) The integral of a function over any range is defined in terms of a series of least and greatest value functions which approximate more and more to the original function as their number increases.

(ii) The integral of a function over any range is defined as the limiting value of the sum of the integrals of a set of least value functions as their number increases without bound.

†**32.** Explain why $\int_a^5 g(x)\,dx > \int_a^5 f(x)\,dx$ if $g(x)$ is greater than $f(x)$ for all values of x between a and b.

†**33.** Explain why the expressions $\int_{-1}^{1} \frac{1}{x}\,dx$ and $\int_1^2 \frac{1}{x-1}\,dx$ are meaningless.

34. As a car travels along a short stretch of road from A to B the total resistance R N to its motion varies with its distance s m from A, the values being given by the following table:

s (m)	0	20	40	60	80	100	120
R (N)	250	350	415	445	440	400	325

The mass of the car is 800 kg. When it reaches A it is going at 15 m/s, and the engine is then cut off, so that the car free-wheels from A to B. Find

(i) the work done against the resistance,

(ii) the speed with which the car arrives at B.

State the units in which the answers are measured. (SU)

35. A body of mass 20 kg, initially at rest, is acted on by a variable force P N whose direction is constant and whose magnitude is shown, in the following table, at various distances s m from the starting-point.

s	0	2	4	6	8	10	12	14	16	m
P	4·0	4·2	4·7	5·3	5·0	4·3	3·0	1·8	0	N

Assuming that no other forces act on the body, find

(i) the total work done,

(ii) the final velocity. (SU)

7.4. TWO PROPERTIES OF THE INTEGRAL

$$\text{(i)} \int_a^b f(x)\,dx = \int_a^b f(u)\,du.$$

Translated into words, this equation simply means that the area between the curve $y = f(x)$ and the x-axis of a y-x graph is the same as that between the curve $y = f(u)$ and the u-axis on the corresponding y-u graph (both graphs employing unity scaling) (Fig. 7.26). The form of the functional relationship

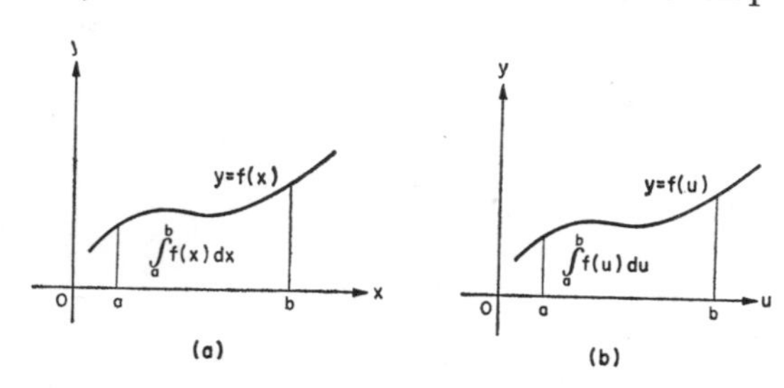

FIG. 7.26. (a) The area between $x = a$ and $x = b$ under a graph of $y = f(x)$ is the same as that (b) between $u = a$ and $u = b$ under a graph of $y = f(u)$, where f denotes the same function or relation.

between the dependent and independent variables in the two cases is the same, since the same symbol f is used, and the graphs therefore have exactly the same shapes. Thus the areas under them between the values a and b of the independent variable are equal, and the integrals are therefore also equal.*

$$\text{(ii)} \int_a^l f(x)\,dx + \int_b^c f(x)\,dx = \int_a^c f(x)\,dx.$$

This equality expresses the fact that the area under a curve between the boundaries

$$x = a \quad \text{and} \quad x = b\,(\int_a^b f(x)\,dx)$$

* This theorem thus makes the point that the only quantities which affect the value of an integral are the range of integration (i.e. a to b) and the form of the functions involved: the letter which actually appears plays no essential part in the definition or value of the integral.

when added to the area under that curve between

$$x = b \quad \text{and} \quad x = c\,(\int_b^c f(x)\,dx)$$

equals the area between the boundaries drawn at

$$x = a \quad \text{and} \quad x = c\,(\int_a^c f(x)\,dx) \quad \text{(Fig. 7.27).}$$

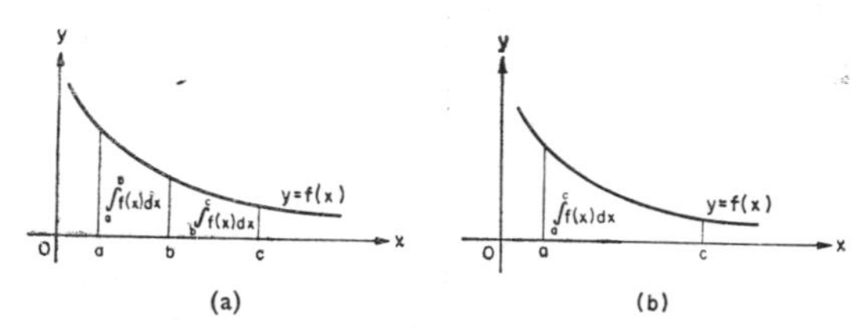

FIG. 7.27. (a) The sum of the areas between $x = a$ and $x = b$, and $x = b$ and $x = c$ is the same as that (b) between $x = a$ and $x = c$.

7.5. THE SIGN OF THE INTEGRAL

Defining the integral of a constant value function y over a range of independent variable x as the product xy means that the integral of a function with a negative value is to be accorded a negative sign. (Thus the integral of the constant value -2 over a range of independent variable 4 is -8.) In terms of the graphical representation of the value of the integral, this means that if we are to maintain the numerical equality between an integral and the corresponding area between a unity scale graph of the integral and the independent variable axis, any area lying below the independent variable axis must be given a negative sign (see Fig. 7.28).

In general this will cause little trouble, but care must be exercised when evaluating the area under a curve which crosses the independent variable axis (as that, for example, in Fig. 7.29), for since the area P is to be accorded a positive value, and that labelled Q is to be accorded a negative value, if the areas are

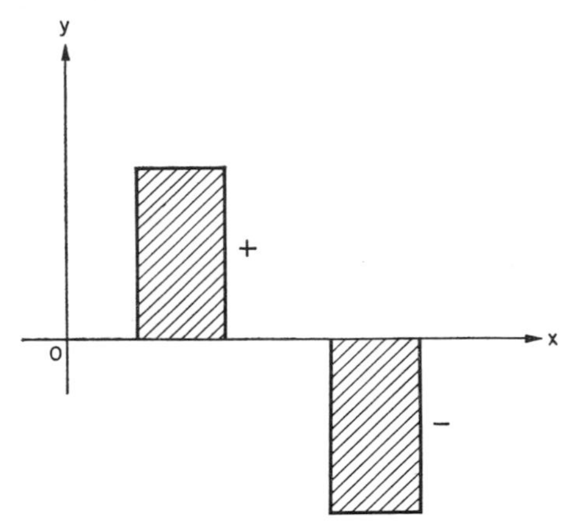

FIG. 7.28. An area below the independent variable axis is to be given a negative value.

added directly they will cancel each other wholly or to some extent. Thus the value of

$$\int_a^b f(x)\,dx$$

for this particular function may be positive, zero, or negative, and will give the difference between the absolute values of P and Q rather than the actual sum of P and Q. If the absolute value of $(P+Q)$ is required,

$$\int_a^c f(x)\,dx \quad \text{and} \quad \int_c^b f(x)\,dx,$$

where c is the value of the independent variable where the curve of the function crosses

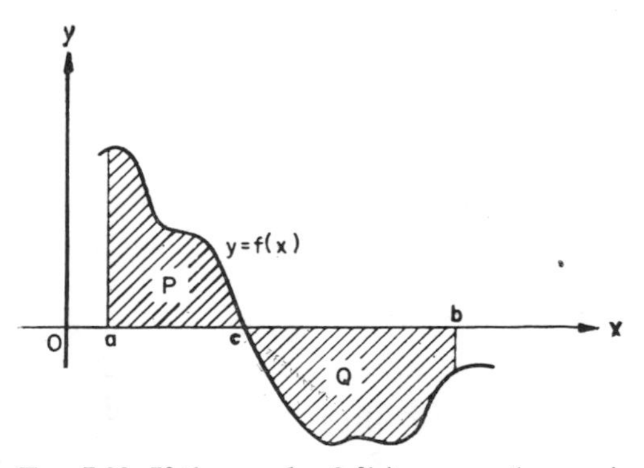

FIG. 7.29. If the graph of $f(x)$ crosses the x-axis between a and b, $\int_a^b f(x)\,dx = P-Q$, and not $P+Q$.

the axis, must be obtained separately, and then added, discounting any negative sign associated with a. Often, however, the difference between the values of the areas is the quantity which is required.

Example 7.5

Determine the values of

$$\int_0^2 x\,dx, \quad \int_{-1}^2 x\,dx \quad \text{and} \quad \int_{-1}^2 x\,dx.$$

A sketch of the line $y = x$ is shown in Fig. 7.30, and it can be readily determined that the area between the origin, the ordinate $x = 2$

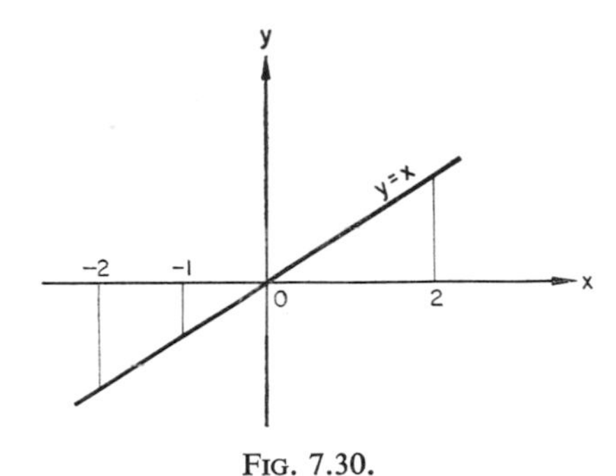

FIG. 7.30.

and the line $y = x$ is 2 (the area of a triangle being given by the product of half its base and its height). Thus

$$\int_0^2 x\,dx = 2.$$

To find the value of

$$\int_{-1}^2 x\,dx$$

we must add the triangle below the axis between the origin and $x = -1$. The area of this triangle is $\frac{1}{2}$, and since it lies below the x-axis, its contribution to the total area between $x = -1$ and $x = 2$ must detract from the area between $x = 0$ and $x = 2$. Thus

$$\int_{-1}^2 x\,dx = 1\tfrac{1}{2}.$$

We can similarly think of the value of

$$\int_{-2}^{2} x\,dx$$

as the sum of

$$\int_{-2}^{0} x\,dx \quad \text{and} \quad \int_{0}^{2} x\,dx,$$

and identifying the value of the two integrals with the area of the triangles between the origin and $x = -2$ and that of the triangle between the origin and $x = +2$, we can write that

$$\int_{-2}^{2} x\,dx = 2-2$$

$$= 0.$$

7.6. THE INDEFINITE INTEGRAL*

If we consider the area enclosed between two ordinates $u = b$, $u = x$, the u-axis and a unity scale graph of a function $f(u)$ (Fig. 7.31),

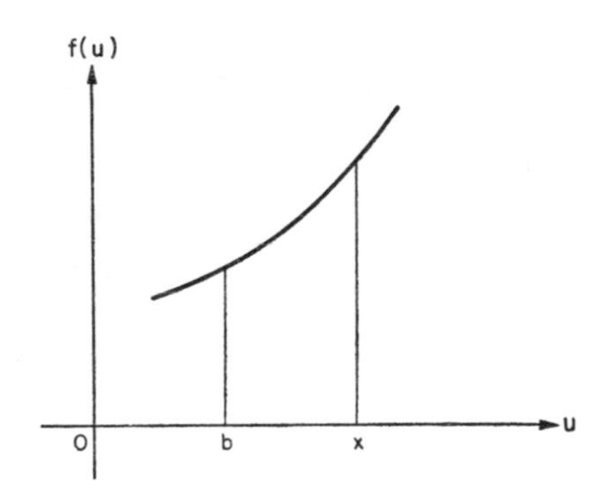

FIG. 7.31. The area enclosed between the curve and the u-axis depends on the value of x.

we can say that the value of the enclosed area depends only on the form of $f(u)$ and the values chosen for b and x. If we then regard the position of the lower boundary ordinate $u = b$ as fixed, the value of the enclosed area will depend only on $f(u)$ and x, and, for any particular form of $f(u)$, the enclosed area will depend solely on the value assigned to x.

Let us take a simple example and consider the area under the unity scale graph of the constant value function $f(u) = 3$ (Fig. 7.32a).

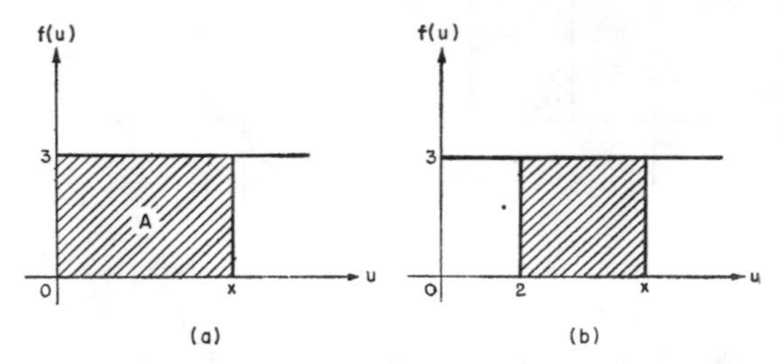

FIG. 7.32. (a) The indefinite integral is a function which describes how the area A depends on the value of the upper boundary x. (b) If the lower boundary is moved to $u = 2$, the area enclosed is only $(3x-6)$.

If we fix a lower boundary at $u = 0$, the area A between this ordinate, the graph and $u = x$ will be by $A = 3x$, or, in other words, the function describing how the size of the enclosed area varies with x is $3x$.

Now the area we are considering has a value equal to that of the integral

$$\int_{0}^{x} 3\,du,$$

and since the upper limit associated with the integral has deliberately been left unspecified, the function $3x$ which expresses the dependence of the integral on the value of the upper limit can be referred as the indefinite integral of the function $f(u)$.

There is one other point to be made. If we had chosen to fix the lower boundary, to the area under consideration at, say, $u = 2$ instead of $u = 0$ (Fig. 7.32b), the function expressing the relationship between the enclosed area and x would have been $(3x \ -6)$. This new function differs from the old only in that it is 6 less, and it is important to note that the manner in which x is involved remains unchanged. We can generalise this conclusion as follows.

*Some readers may prefer to read Section 8.1 first.

Let us suppose that we have a function $f(u)$ whose graph has the form shown in Fig. 7.33. If we fix a lower boundary to the enclosed area at $u = b$ for the moment, we can suppose that

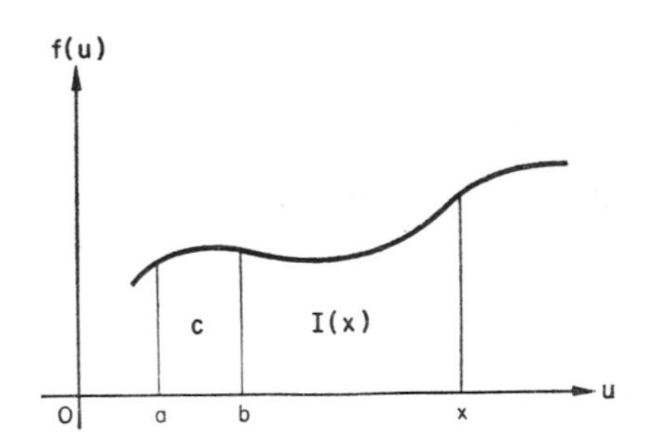

FIG. 7.33. If the lower boundary is changed from b to a, the indefinite integral increases by an amount c. It is unchanged as a function of the upper boundary x.

there is some function $f(x)$ which describes how the value of this area varies with the value of x. In terms of an indefinite integral,

$$\int_b^x f(u)\, du = I(x).$$

If we now change the lower boundary to $u = a$ (Fig. 7.33), we can consider the enclosed area as the sum of the areas enclosed by the u-axis, the graph and the pairs of ordinates $u = a$ and $u = b$, and $u = b$ and $u = x$. In the integral notation

$$\int_a^x f(u)\, du = \int_a^b f(u)\, du + \int_b^x f(u)\, du \quad (7.6)$$

(cf. Section 7.4 (ii)).

The area between $u = a$ and $u = b$ has, for any value of x, a certain value, and this value is constant even if x is allowed to vary. If we denote its value by c, equation (7.6) can be rewritten in the form

$$\int_a^x f(u)\, du = \int_b^x f(u)\, du + c,$$

or, replacing $\int_b^x f(u)\, du$ by $I(x)$,

$$\int_a^x f(u)\, du = I(x) + c. \quad (7.7)$$

The effect of changing the lower boundary thus alters the indefinite integral by a numerical quantity independent of x.

The implications of equation (7.7) are that for any function $f(u)$ there exists an indefinite integral function $I(x)$ which describes how the value of the integral

$$\int_a^x f(u)\, du$$

varies with the value of the upper limit x. Until the form of $f(u)$ is known, and until the value of the lower limit a is chosen, the precise value of the integral cannot be determined. Altering the value of a does not change the form of the dependence of the integral on x; it merely increases or decreases its value by a magnitude which is the same for all values of x.

In the next chapter we shall discuss some of the techniques which enable the form of $I(x)$ to be found once the form of $f(u)$ is specified. In the meanwhile it will be sufficient to content ourselves with the observations that:

(i) if the form of $I(x)$ is to be found by deriving an expression for the area contained under a unity scale graph of $f(u)$, it is convenient to choose the position of the lower boundary ordinate $u = a$ so that the form of $I(x)$ is as simple as possible (and if this is done, the arbitrary nature of the choice is fully represented by the inclusion of the undefined constant c in the final expression);

(ii) since the undefined constant c added to the expression for $I(x)$ entirely allows for any variation in the value assigned to the lower boundary abscissa a, we need not write two indefinite constants in our expression

$$\int_a^x f(u)\, du = I(x) + c,$$

and can therefore omit the a altogether;

(iii) if we choose to consider functions of such variables as t, z or even x, the preceding comments apply equally well, and we can therefore write

$$\int^{x} f(t)\,dt = \int^{x} f(z)\,dz = \int^{x} f(x)\,dx$$
$$= I(x)+c;$$

and (iv) since the letter chosen to represent the (variable) upper limit x is indicated by the letter chosen for the variable in the expression for $I(x)$ (for if we were considering an expression for $\int^{y} f(u)\,du$, we would write $I(y)$ to represent the integral function), we need not write it twice, and the equation $\int f(u)du = I(x)+c$ is sufficiently unambiguous to commend itself for general use.

To summarise the suggestions of this section, we can finally define the indefinite integral of a function $f(u)$ as that function $I(x)$ which expresses the manner in which the value of the indefinite integral

$$\int_{a}^{x} f(u)\,du$$

varies with x, a being regarded as fixed in value. To allow for the change in the value of the integral

$$\int_{a}^{x} f(u)\,du$$

which occurs if a is changed, we must add an undefined constant c to any expression we write for

$$\int_{a}^{x} f(u)\,du,$$

and since there is no need to write either the indefinite lower limit a or the indefinite upper limit x on the integral sign, we can—without any loss of generality—write the equation

$$\int f(u)\,du = I(x)+c$$

as a definition of the indefinite integral function $I(x)$. It must be borne in mind that the equation implies that the lower limit to the integral is not specified until the value of c is defined, and that the letter representing the (variable) upper limit is x.

Example 7.6

Find an expression for the indefinite integral $\int u\,du$.

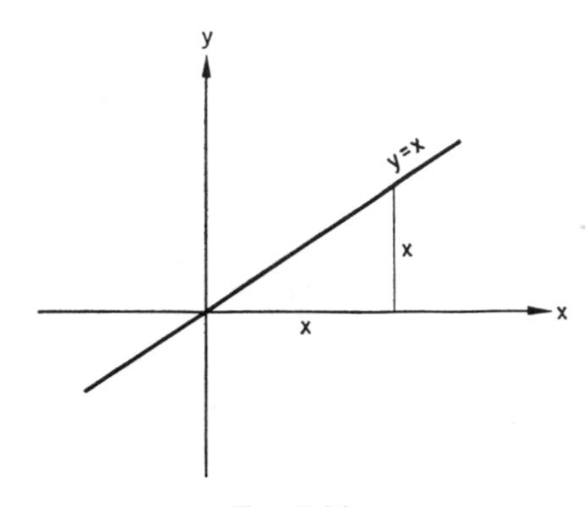

FIG. 7.34.

Referring to Fig. 7.34, we can see that the value of

$$\int_{0}^{x} u\,du$$

is given by the area of the triangle between the origin and the ordinate at x. Since this is $\frac{1}{2}x^2$, we can write an expression for the indefinite integral by adding a constant to allow for the indefinite position of the lower boundary, and thus obtain the expression $\frac{1}{2}x^2+c$.

7.7. FOUR MORE PROPERTIES OF THE INTEGRAL

(iii) $$\int_{a}^{b} f(u)\,du = -\int_{b}^{a} f(u)\,du.$$

Interchanging the limits on a definite integral reverses the sign of its value. Suppose, in indefinite form, that $\int f(u) = I(x)+c$. Then

$$\int_{a}^{b} f(u)\,du$$

is equal to the difference between the values of $I(x)+c$ at $x = b$ and $x = a$, i.e.

$$\int_a^b f(u)\,du = I(b)-I(a),$$

Similarly,

$$\int_b^a f(u)\,du = I(a)-I(b),$$

and by comparison the result follows.

(iv) If we are considering a function whose graph is symmetrical about the y-axis, then

$$\int_0^a f(x)\,dx = \int_{-a}^0 f(x)\,dx.$$

The result follows immediately by considering the magnitudes and signs of the areas under the symmetrical graph for the two ranges of integration (Fig. 7.35).

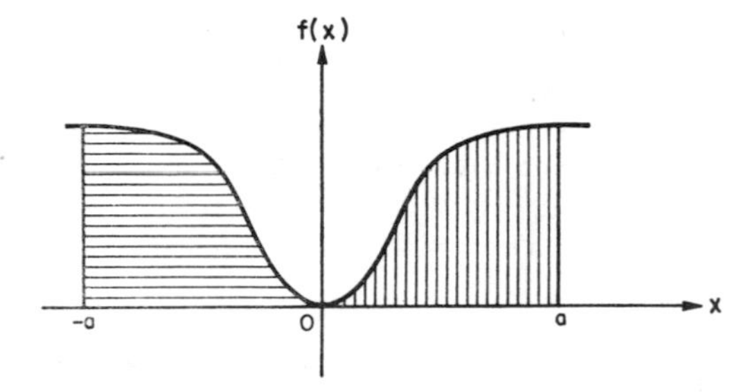

FIG. 7.35. The two shaded areas are equal, since the graph is symmetrical about the y-axis. They are also both positive, since they lie above the x-axis.

A function which is symmetrical about the y-axis {i.e. one for which $f(x) = f(-x)$} is called an *even* function, and it follows that

$$\int_{-a}^a f(x)\,dx = 2\int_0^a f(x)\,dx.$$

(v) If we are considering a function whose graph is symmetrical about the origin, i.e. an *odd* function, or one for which $f(x) = -f(-x)$, then (see Fig. 7.36)

$$\int_0^a f(x)\,dx = -\int_{-a}^0 f(x)\,dx,$$

and

$$\int_{-a}^a f(x)\,dx = 0.$$

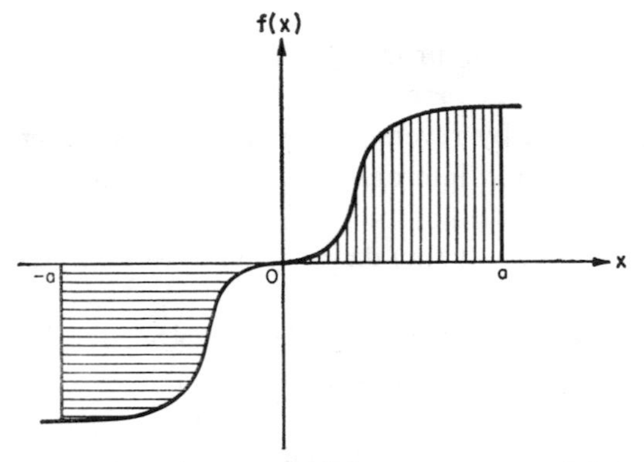

FIG. 7.36. The two shaded areas are equal but opposite in sign. Thus $\int_{-a}^{a} f(x)\,dx = 0$.

(vi) For any function defined in the range $0 \leqslant x \leqslant a$,

$$\int_0^a f(x)\,dx = \int_0^a f(a-x)\,dx.$$

This property can be produced by recognising that the graph of $f(a-x)$ is the mirror image of the graph of $f(x)$ in the line $x = \frac{1}{2}a$ (Fig. 7.37). The area to the left of $x = \frac{1}{2}a$

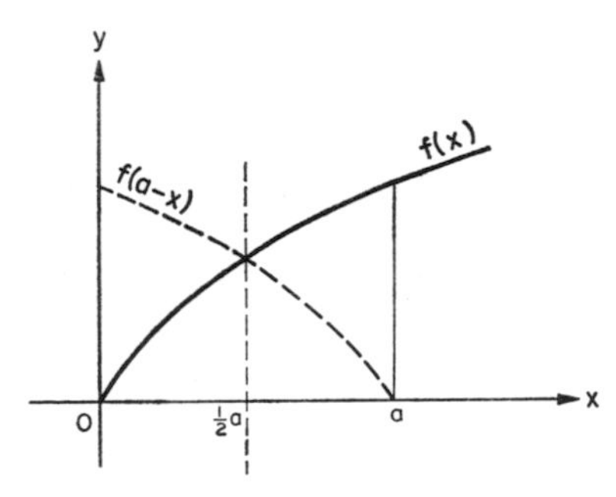

FIG. 7.37. The graph of $f(a-x)$ is the mirror image of that of $f(x)$ in the line $x = \frac{1}{2}a$.

under $f(x)$ is thus identical to that to the right of $x = \frac{1}{2}a$ under $f(a-x)$, and that to the right of $x = \frac{1}{2}a$ under $f(x)$ is identical to that to the left of $x = \frac{1}{2}a$ under $f(a-x)$.

EXAMPLE 7.7

Show that $\int_0^2 f(x)\,dx - \int_3^2 f(x)\,dx = \int_0^3 f(x)\,dx$.

Using the property described in Section 7.7 (iii), we can rewrite the left-hand side of the equation by reversing the sign of the second integral and interchanging the limits, i.e.

$$\int_0^2 f(x)\,dx - \int_3^2 f(x)\,dx = \int_0^2 f(x)\,dx + \int_2^3 f(x)\,dx.$$

Using now the second relation in Section 7.4, we rewrite the right-hand side as

$$\int_0^3 f(x)\,dx,$$

and the result is therefore proved.

EXERCISE 7e

Evaluate the following integrals:

1. $\int_0^2 (-2)\,dx$. 8. $\int_{-3}^5 (2x-1)\,dx$. 15. $\int_0^5 \sqrt{(25-x^2)}\,dx$.

2. $\int_0^a (-a)\,dx$. 9. $\int_{-1}^1 (x-3)\,dx$. 16. $\int_0^3 \sqrt{(9-q^2)}\,dq$.

3. $\int_{-2}^0 (-2)\,dx$. 10. $\int_0^x x\,dx$. 17. $\int_0 3\,dx$.

4. $\int_{-a}^0 a\,dx$. 11. $\int_1^x x\,dx$. 18. $\int^0 5\,dx$.

5. $\int_{-b}^0 x\,dx$. 12. $\int_{-x}^x x\,dx$. 19. $\int x\,dx$.

6. $\int_{-2}^0 2x\,dx$. 13. $\int_{-x}^x (-5)\,dx$.

7. $\int_{-2}^{+2} x\,dx$. 14. $\int_{-x}^{+x} x^3\,dx$.

20. Write down an expression for the indefinite integral $\int^x 5\,du$. For what value of x is this function zero? Hence deduce the value of the constant of integration if the lower limit of the integral is chosen to be at

(i) $u = -1$, (ii) $u = -2$, (iii) $u = 0$, (iv) $u = 1$, (v) $u = -a$.

21. Write down expressions for

(i) $\int_0^x u\,du$; (ii) $\int_a^x u\,du$; (iii) $\int_{-a}^x u\,du$.

Find an expression for the indefinite integral of x with respect to x, writing the constant of integration in terms of a, the lower limit of integration being

(iv) 0, (v) a, (vi) $-a$.

22. The lower boundary of the indefinite integral $\int x\,dx$ is at $x = a$. Write down an expression for the constant of integration c, and also expressions for the values of

$$\int^3 x\,dx \quad \text{and} \quad \int^4 x\,dx.$$

Explain why the difference between these last two expressions give the value of

$$\int_3^4 x\,dx.$$

†23. Show that

$$\int_{-a}^a x^n\,dx = 0 \text{ if } n \text{ is odd,}$$

$$\text{and } \int_{-a}^a x^n\,dx = 2\int_0^a x^n\,dx \text{ if } n \text{ is even.}$$

†24. An "odd" function is a function such that $g(-x) = -g(x)$: an "even" function is a function such that $f(-x) = f(x)$. Show that

$$\int_{-a}^a p(x)\,dx = 0 \text{ for an odd function,}$$

and

$$\int_{-a}^a p(x)\,dx = 2\int_0^a p(x)\,dx \text{ for an even function.}$$

†25. Show that

$$\int_{-a}^b p(x)\,dx = \int_0^b p(x)\,dx + \int_0^a p(x)\,dx$$

if $p(x)$ is even, and that

$$\int_{-a}^b p(x)\,dx = \int_a^b p(x)\,dx \quad \text{if} \quad p(x) \text{ is odd.}$$

†26. Show that

(i) $\int_{-\pi/2}^{\pi/2} \sin x\,dx = 0$, (ii) $\int_0^\pi \cos x\,dx = 0$,

(iii) $\int_0^\pi \sin x\,dx = 2\int_0^{\pi/2} \sin x\,dx$.

†27. Find the integral values of n and m for which

(i) $\int_0^{n\pi} \sin x\,dx$, (ii) $\int_0^{n\pi/2} \cos x\,dx$,

(iii) $\int_{n\pi}^{m\pi} \sin x \, dx$

are equal to zero.

†**28.** Find the values of

(i) $\int_3^5 p(x)\, dx$, (iii) $\int_{-2}^3 p(x)\, dx$,

(ii) $\int_{-2}^1 p(x)\, dx$, (iv) $\int_{-2}^5 p(x)\, dx$,

where $p(x)$ is the step function defined by the equations:

$$p(x) = 1 \text{ for } -2 \leqslant x < 1;$$
$$p(x) = -2 \text{ for } 1 \leqslant x < 3;$$
$$p(x) = 4 \text{ for } 3 \leqslant x \leqslant 5.$$

†**29.** Sketch the graph of the function $x-2[x]$, and find the values of

(i) $\int_0^3 \{x-2[x]\}\, dx$, (ii) $\int_0^{4.4} \{x-2[x]\}\, dx$

and (iii) $\int_0^a \{x-2[x]\}\, dx$.

(See question 29, p. 189.)

†**30.** Explain why the integral of the speed of a body over a certain range of time is equal to the distance the body travels.

†**31.** Sketch the graph of $y = x(x-1)(x-2)$, and explain why the area contained by it, the x-axis and the ordinates $x = 0$ and $x = 2$ is not zero, but the value of

$$\int_0^2 x(x-1)(x-2)\, dx$$

is.

†**32.** If $(1/u)\, du = \theta(x)$, explain why

$$\int_a^b (1/u)\, du = \theta(b) - \theta(a).$$

†**33.** The interest payable per annum on a sum of money invested increases with the sum according to the equation $y = 3+x/5$, where y equals the interest per annum in per cent, and x equals the sum invested in hundreds of pounds. Evaluate

$$\int_0^{10} \left(3+\frac{x}{5}\right) dx,$$

and hence find the interest payable per annum on £1000.

***34.** The coefficient of linear expansion y (per degree Celsius) of a metal is given by the equation $y = 0{\cdot}001+0{\cdot}00004x$, where x is the temperature in degrees Celsius. Calculate the total increase in length of a bar of the metal 1 cm long if its temperature increases from 0°C to 100°C.

†**35.** Show that

$$\int_a^b f(x)\, dx = 2\int_{\frac{1}{2}a}^{\frac{1}{2}b} f(2y)\, dy = 2\int_{\frac{1}{2}x}^{\frac{1}{2}b} f(2x)\, dx.$$

If we have a function $f(x)$, we can write that

$$\int_1^2 f(x)\, dx = \int_0^2 f(x)\, dx - \int_0^1 f(x)\, dx.$$

If we then rewrite

$$\int_0^2 f(x)\, dx \quad \text{as} \quad 2\int_0^1 f(2x)\, dx$$

on the right-hand side, and suppose that, for a particular $f(x)$, $f(2x) = \frac{1}{2}f(x)$, we can write

$$\int_1^2 f(x)\, dx = 2\int_0^1 f(2x)\, dx - \int_0^1 f(x)\, dx$$
$$= 2\int_0^1 \tfrac{1}{2}f(2x)\, dx - \int_0^1 f(x)\, dx.$$
$$= 0.$$

Now $f(2x) = \frac{1}{2}f(x)$ is true if $f(x) \equiv 1/x$, so

$$\int_1^2 \frac{1}{x}\, dx = 0;$$

or, in graphical terms, the area under the graph of $1/x$ between the ordinate $x = 1$ and $x = 2$ is zero. Discuss.

SUMMARY

The area of a rectangle is defined as the product of its breadth and length. The area of a number of rectangles is equal to the sum of their areas, and the area *under a curve* between two ordinates is that number A which satisfies the inequality

total area of "inner" rectangles $\leqslant A \leqslant$ total area of "outer" rectangles

however many rectangles are constructed between the two ordinates (Fig. 7.38).

The area A under a curve can be estimated by the *trapezium rule*:

$$A \simeq \tfrac{1}{2}h[(y_1+y_{n+1})+2(y_2+y_3+y_4 \ldots +y_n)],$$

where n trapezia are constructed on ordinates to the curve evenly spaced between the bound-

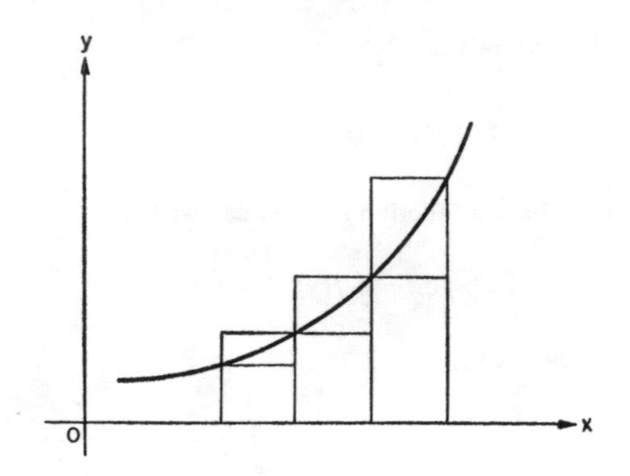

FIG. 7.38. The magnitude of the area under a curve is that number which always lies between the area of the rectangles constructed on the second and subsequent ordinates and that of the rectangles constructed on the first and subsequent ordinates.

aries of the area (Fig. 7.39). The letter h stands for the breadth of the trapezia, and y_1, y_2, etc., are the ordinate lengths which can be calculated from the abscissae values a, $(a+h)$, $(a+2h)$, etc.

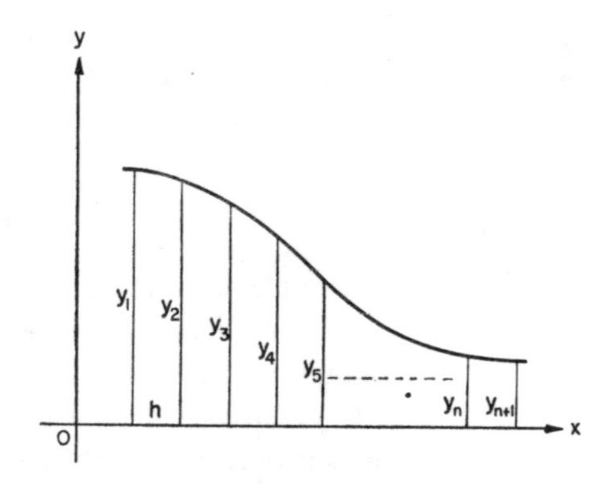

FIG. 7.39. The area under the curve or the integral of the function between a and b is approximately equal to

$$\tfrac{1}{2}h\{(y_1+y_{n+1})+2(y_2+y_3+\ldots+y_n)\}.$$

The *integral* of a function over a range can be associated with the area under the curve of the function on a unity scale graph between the same limits.

$$\int_a^b f(x)\,dx$$

stands for the value of the integral of $f(x)$ over the range a to b. Areas lying above the horizontal axis (and therefore integrals of functions which take positive values throughout the range of integration) are accorded a positive sign: those lying below the axis are accorded a negative sign.

The integral possesses the properties that:

(i) $\int_a^b f(x)\,dx = \int_a^b f(u)\,du$,

(ii) $\int_a^b f(x)\,dx + \int_b^c f(x)\,dx = \int_a^c f(x)\,dx$,

(iii) $\int_a^b f(x)\,dx = -\int_b^a f(x)\,dx$.

The function which expresses the variation of the value of the integral of a function $f(u)$ with the upper limit is called the *indefinite integral* of that function.

MISCELLANEOUS EXERCISE 7

1. Draw a graph of $y = x^3$ for the range $0 \leqslant x \leqslant 2$. Construct four rectangles under the curve, and determine their area. Compare this with the area of the five "outer" rectangles on the same bases, and hence give upper and lower estimates of the area enclosed by the curve, the x-axis and the ordinate at $x = 2$.

2. Repeat question 1 for the curve $y = x^4$ in the range $0 \leqslant x \leqslant 2$.

3. Explain why the area enclosed by the curve $y = ax^n$, the x-axis and the ordinate $x=b$ is a times as great as the area enclosed by the curve $y = x^n$, the x-axis and the ordinate $x = b$. Assume $n > 1$, $a > 0$.

4. Show that the area δA between two circles of radii r and $(r+\delta r)$ is given by the equation $\delta A=\pi\,\delta r(2r+\delta r)$. Hence show that the area A of a circle is related to its radius by the equation $dA/dr = 2\pi r$.

5. Draw a graph of $y=\sin^2 x$ for values of x between 0 and $\frac{1}{2}\pi$, using 4 cm to represent unity on the y-axis and 1 cm to represent 15° on the x-axis. Estimate the area under the curve by constructing rectangles and measuring their sides, and calculate what your results would have been had your scaling been

(i) 1 cm to represent unity on the y-axis,
(ii) 1 cm to represent unity on the y-axis and 1° on the x-axis,
(iii) 1 cm to represent unity on the y-axis and 1 radian on the x-axis.

6. A spherical shell of internal radius a and external radius $(a+\delta a)$ is flattened out onto a trapezoidal parallelepiped (Fig. 7.40). What will be the areas of the upper and lower faces, and the approximate volume of the solid?

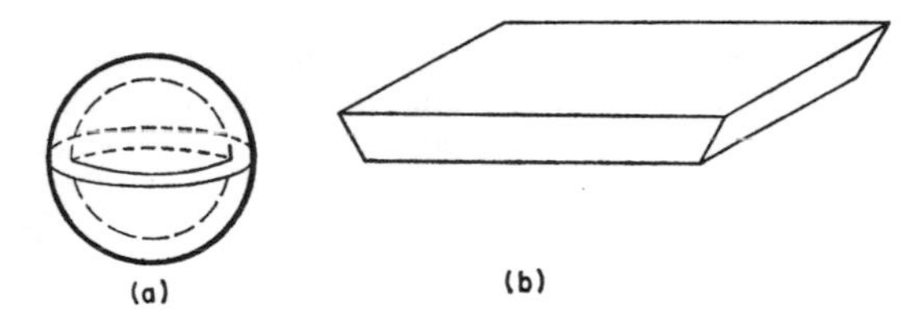

FIG. 7.40.

7. The rate of interest given on a sum of money invested in a deposit account varies during 1 year as follows:

Rate of interest (%) p.a.	$4\frac{1}{2}$	4	$3\frac{3}{4}$	$3\frac{1}{2}$
Number of days during which rate applied	105	51	37	136

What actual interest was received from the sum invested in the deposit throughout the year? Relate this to the area under a unity scale graph of the variation.

8. Estimate the area under the curve $y = 1/x$ drawn on a unity scale graph between the ordinates

(i) $x = 1$ and $x = 2$,
(ii) $x = 1$ and $x = 6$,
(iii) $x = 1$ and $x = 9$,

using the trapezium rule and five trapezia.

9. The velocity v cm/sec of a body varies according to the equation $v = 2\cos t$, where t is the time in seconds. Draw a sketch graph of the variation, and calculate the distance travelled between $t = 0$ and $t = \frac{1}{2}\pi$ seconds, using the trapezium rule. Radian measure should be employed along the time axis.

10. The speed of a car varies with time as shown in the following table:

Time (s)	0	2	4	6	8	10
Speed (m/s)	0	3	8	13	17	19

How far does the car travel in the first 10 seconds starting from rest?

11. Two variables are related so that the corresponding pairs of their values are as follows:

Independent variable	0	0·1	0·2	0·3	0·4	0·5
Dependent variable	0	1·0	2·5	4·0	5·0	5·3

Using the trapezium rule, estimate the area under the curve of the dependent variable against the independent variable for values of the independent variable between 0 and 0·5.

12. Write down the values of the following integrals:

(i) $\int_{3}^{4} 8\,dx$, (ii) $\int_{11}^{20} 9\,dx$, (iii) $\int_{2}^{3} 17x\,dx$,

(iv) $\int_{0}^{1} (x+2)\,dx$, (v) $\int_{0}^{1} x\,dx$.

13. What are the values of:

(i) $\int_{-3}^{0} 5\,dx$, (ii) $\int_{-3}^{0} -5\,dx$, (iii) $\int_{-3}^{0} x\,dx$,

(iv) $\int_{-2}^{0} (x-1)\,dx$, (v) $\int_{-2}^{0} (1-x)\,dx$,

(vi) $\int_{-2}^{0} (x+1)\,dx$?

14. Evaluate the following integrals:

(i) $\int_{-2}^{1} x\,dx$, (ii) $\int_{-2}^{1} (-2)\,dx$ (iii) $\int_{-4}^{3} (x+1)\,dx$,

(iv) $\int_{-5}^{5} (1-x)\,dx$.

15. Write down the values of the following integrals:

(i) $\int_{0}^{2} -8\,dx$, (ii) $\int_{-2}^{0} -6\,dx$, (iii) $\int_{-4}^{-2} -5\,dx$,

(iv) $\int_{-4}^{2} -5\,dx$, (v) $\int_{-3}^{3} -3\,dx$, (vi) $\int_{-2}^{-1} +7\,dx$.

16. Write down the values of the following:

(i) $\int_{-3}^{0} 3x\,dx$, (ii) $\int_{0}^{2} -2x\,dx$, (iii) $\int_{-2}^{0} -5x\,dx$,

(iv) $\int_{-2}^{2} -b\,dx$, (v) $\int_{-5}^{-3} -x\,dx$, (vi) $\int_{6}^{8} (7-x)\,dx$,

(vii) $\int_{2}^{4} (4-3x)\,dx$, (viii) $\int_{-4}^{-2} (4-3x)\,dx$,

(ix) $\int_{-1}^{3} (4-3x)\,dx$, (x) $\int_{0}^{x} -x\,dx$.

17. Estimate the value of

$$\int_{0}^{3} y\,dx$$

given the following corresponding pairs of values of x and y:

y	0	2	5	9
x	0	1	2	3

18. Explain why

$$\int_0^x f(t)\,dt = \int_0^x f(x)\,dx,$$

and point out the meaning of the various x's.

19. Estimate the value of

$$\int_0^8 y\,dx$$

given the following corresponding pairs of values of y and x:

y	3	5	2	−2	−6
x	0	2	4	6	8

20. The rate of change of a function $f(x)$ with respect to x varies with the value of x as follows:

$f'(x)$	5	10	20	18	−4	−10	−12
x	0	5	10	15	20	25	30

Estimate the values of $f(10)$, $f(20)$, and $f(30)$, given that $f(0) = 0$.

21. Representing the independent variable of question 11 by the letter u, and the dependent variable by the function symbol $f(u)$, estimate the value of

$$\int_0^{0.4} f(u)\,du.$$

†**22.** Write down an expression for the value of

$$\int_0^a [x]\,dx, \text{ given that } a > 0.$$

†**23.** Calculate the area enclosed by 100 m of fencing if the shape of the area enclosed is a regular n-sided polygon where

(i) $n = 3$,
(ii) $n = 4$,
(iii) $n = 5$,
(iv) $n = 6$,
(v) $n = 10$,
(vi) $n = 20$,
(vii) $n = 50$.

Plot a graph of your results, and estimate the value of the limit of this area as $n \to \infty$. Comment on your result.

CHAPTER 8

Techniques of Integration

8.1. INTEGRATION THE REVERSE OF DIFFERENTIATION

The process of finding an expression for the indefinite integral of a function is called *integration*: and, as already explained, any expression for the indefinite integral of a function must include an arbitrary constant. The value of this constant can be associated with the area enclosed between the lower boundary chosen for the integral and the ordinate at which the integral function $I(x)$ equals 0 (Fig. 8.1). The function to be integrated $\{f(x)\}$ is

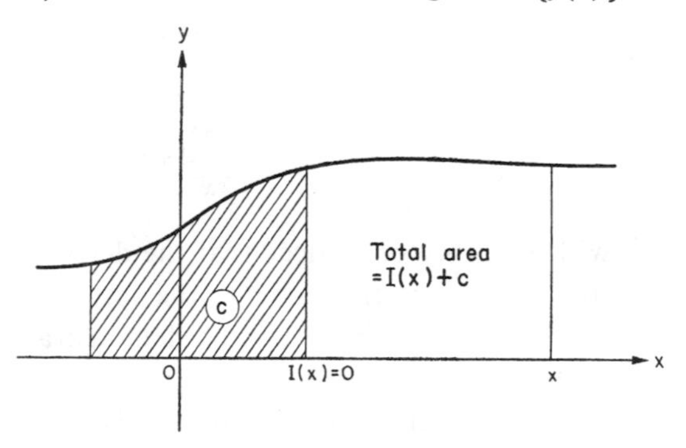

FIG. 8.1. The value of the constant of integration c represents the area between the ordinate chosen as the lower boundary of the indefinite integral and the ordinate for which $I(x) = 0$.

known as the *integrand*, and the relationship between the integrand $f(x)$ and the integral $I(x)$ can be expressed by the equation

$$\int f(x)\, dx = I(x) + c,$$

the value of c being indeterminate until further information about the value of the integral function for some particular value of x is forthcoming.

The form of the integral of a constant value function follows from definition. Equating the value of the integral with the area under the curve of the integrand on a unity scale graph enables us to derive an expression for the integral. Thus by considering the unity scale graph of the constant value function 3 in Fig. 8.2, we can see that the area enclosed between ordinates at 0 and that of the indefinite value x is rectangular in shape and has a value $3x$. Allowing for a change in the position of the ordinate $x = 0$ at the lower boundary, the

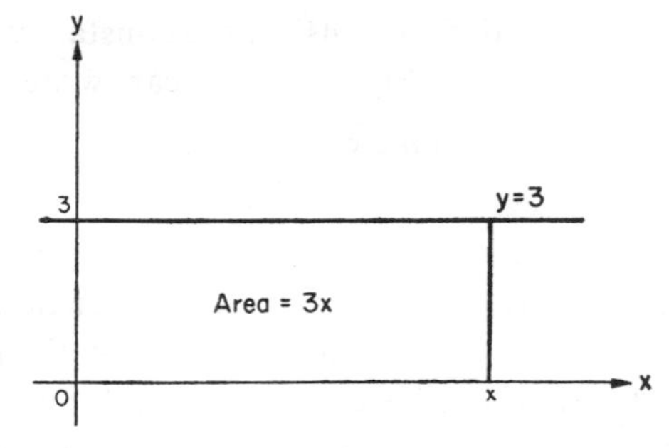

FIG. 8.2. The area under the constant value function 3 between ordinates at 0 and x is $3x$.

expression for the area with an indefinite lower boundary must be written as $(3x+c)$, and this can be identified with the value of the integral. Thus

$$\int 3\,dx = 3x+c.$$

The integral of the constant value function 7 can similarly be deduced as $(7x+c)$ (see

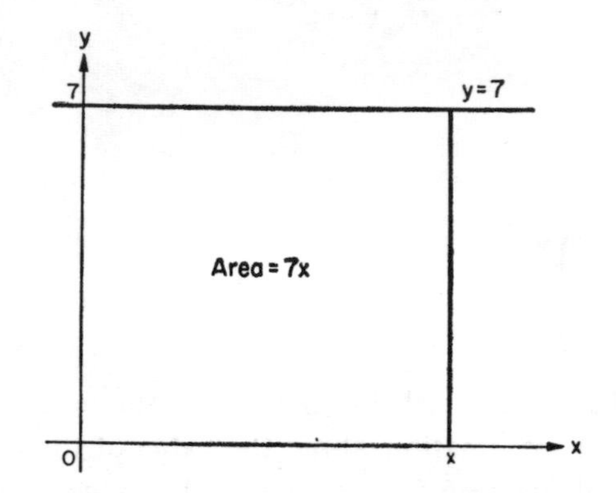

FIG. 8.3. Similarly the area under the constant value function 7 is $7x$...

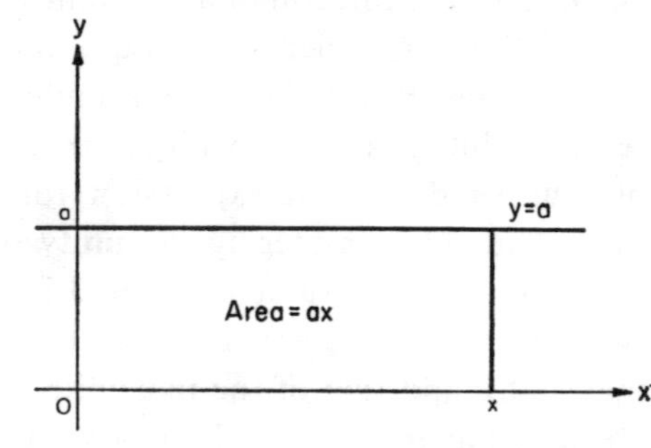

FIG. 8.4. ...and that under the constant value function a is ax.

Fig. 8.3), and if we consider the constant value function a (see Fig. 8.4), we can write that

$$\int a\,dx = ax+c.$$

The integral of the function x is not difficult to derive either. Starting from the origin (Fig. 8.5), the area of the right-angled triangle ending on the ordinate at x is given by the product of half the base and the height, i.e. $\frac{1}{2}x.x$ or $\frac{1}{2}x^2$. Thus

$$\int x\,dx = \frac{1}{2}x^2+c.$$

Similarly the integral of $2x$ (Fig. 8.6) can be seen to be given by the equation

$$\int 2x\,dx = x^2+c.$$

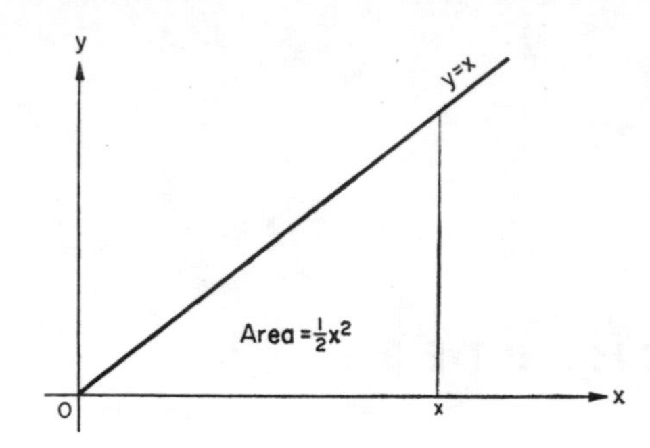

FIG. 8.5. The area of the triangle formed by $y = x$, the x-axis and an ordinate at x is $\frac{1}{2}x^2$...

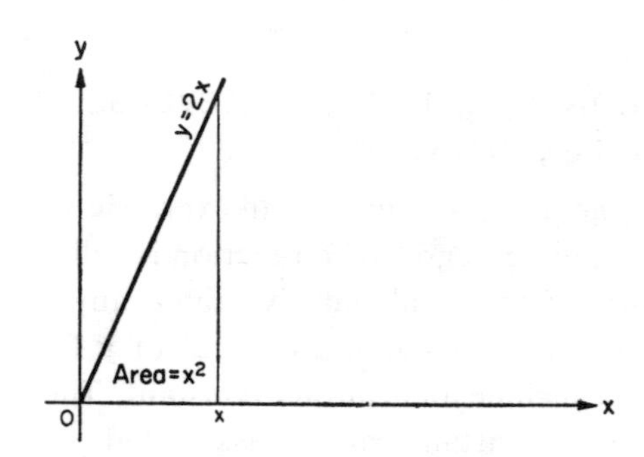

FIG. 8.6. ...and that under $y = 2x$ is x^2.

Collecting these results together, we obtain the table:

Integrand	*Integral*
3	$3x+c$
7	$7x+c$
a	$ax+c$
x	$\frac{1}{2}x^2+c$
$2x$	x^2+c

It will probably be noticed that the derivative of the integral is the same as the integrand: that this is generally so can be demonstrated as follows.

Consider a unity scale graph such as that shown in Fig. 8.7. Denoting the area between the y-axis and an ordinate at x by A (where A is a function of x, and is the indefinite integral of y), we can say that if a third ordinate is constructed at $(x+\delta x)$, the extra piece of area to be added to A can be denoted by δA, where $\delta A = A(x+\delta x)-A(x)$. And further, if we consider the rectangles constructed on ordinates

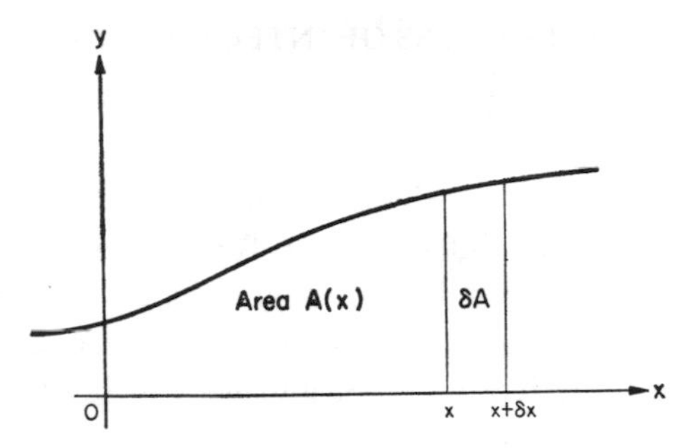

FIG. 8.7. The area between ordinates at x and $(x+\delta x)$ can be represented by δA...

of height l and g at x and $(x+\delta x)$, where l and g are the least and greatest values y or $f(x)$ takes in the interval from x to $x+\delta x$, we can write the inequality

$$l\,\delta x \leqslant \delta A \leqslant g\,\delta x \quad \text{(Fig. 8.8).}$$

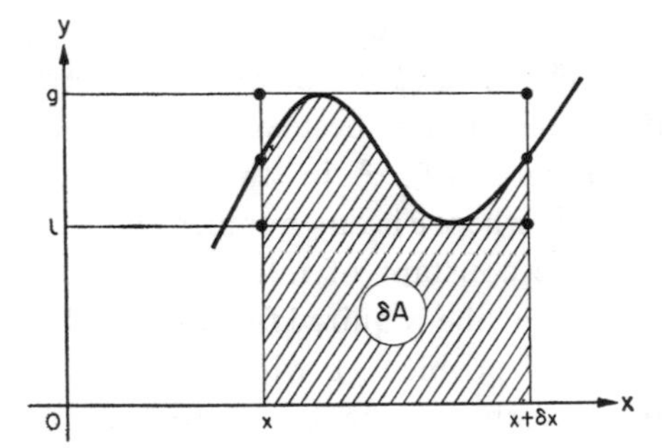

FIG. 8.8. ...and δA satisfies the inequality $l\,\delta x \leqslant A \leqslant g\,\delta x$.

Dividing this throughout by δx, we obtain

$$l \leqslant \frac{\delta A}{\delta x} \leqslant g,$$

and considering the behaviour of this relationship as $\delta x \to 0$ (when l and g both approach the value y or $f(x)$), we see that

$$y = f(x) = \frac{dA}{dx}.$$

Thus the derivative of A is equal to y, the function representing the shape of the curve.*

* This assumes that $f(x)$ is continuous at the point x in question. Another way of looking at this is to say that $A(x)$ must be differentiable at x (see also Exercise 8*a*, question 7).

We have already seen that A can be identified with the indefinite integral of the curve function. Generally, if

$$\int f(x)\,dx = I(x)+c,$$

then

$$\frac{dI}{dx} = f(x),$$

and the process of integration reduces to one of finding a function $I(x)$ which will give the integrand $f(x)$ when differentiated. In many cases this step of differentiating the supposed integral to check that it does in fact equal the integrand is an invaluable detector of error.

Example 8.1

Find the integral with respect to x of: (i) $6x$, (ii) $5x^4$, (iii) $10x^4$.

(1) The integral of $6x$ is a function whose derivative is $6x$. The derivative of x^2 is $2x$, and from this it is easy to proceed to the function $3x^2$ whose derivative is $6x$. Thus the integral of $6x$ is $3x^2+c$, or symbolically

$$\int 6x\,dx = 3x^2+c.$$

(2) A function whose derivative is equal to $5x^4$ must involve a power in x of 5 (since the power is reduced by 1 on differentiation), and the derivative of x^5 itself is $5x^4$. Thus

$$\int 5x^4\,dx = x^5+c.$$

(3) The derivative of x^5 is $5x^4$, and the derivative of $2x^5$ will thus equal $10x^4$, the integrand. Adding the indeterminate constant of integration,

$$\int 10x^4\,dx = 2x^5+c.$$

The integrals are thus $(3x^2+c)$, (x^5+c), and $(2x^5+c)$, where, of course, the constants c are not necessarily the same in the three cases.

Exercise 8a

1. A triangle is formed by the line $y = 2x$, the x-axis, and a line parallel to the y-axis whose abscissa is x. Write down the area A of the triangle in terms of x, and verify that $dA/dx = 2x$.

2. Show that the area bounded by the curve $y = x^3$, the x-axis, and a line parallel to the y-axis whose abscissa is x is given by the function whose first derivative with respect to x is x^3.

3. Find, by trial and error, functions which will give 1, x, x^2 and x^3 when differentiated with respect to x. What function will give x^n?

4. Which function when differentiated with respect to θ will give (i) 2θ, (ii) $\cos 2\theta$?

†**5.** Find the area bounded by the x-axis, the curve $y = x^2$ and the lines $x = 0$ and $x = 4$, on a unity scale graph.

†**6.** Find the area bounded by the x-axis, the curve $y = \cos x$, and the lines $x = \pi/6$ and $x = \pi/3$ on a unity scale graph.

†**7.** Draw a sketch of the graph of the function $f(x)$ for which $f(x) = 0$, when $x \leqslant 1$, and $f(x) = 1$, when $x > 1$.

Write down expressions or values for $\int_0^x f(x)\, dx$ if

(i) $x \leqslant 1$, (ii) $x > 1$,

and sketch the graph of the value of this integral for values of x between about -2 and $+3$.

Denoting the value of the integral by $I(x)$, discuss whether $I(x)$ is (a) continuous, (b) differentiable at $x = 1$.

†**8.** If $d\{\theta(\lambda x)\}/dx = \lambda\{\theta(\lambda x)\}$, explain why

$$\int \theta(\lambda x)\, dx = \{\theta(\lambda x)\}/\lambda + c.$$

†**9.** Discuss the following equation with particular reference to the graphs of $1/x^2$ and $-1/x$ for values of x between -1 and $+1$:

$$\int_{-1}^{1} \frac{dx}{x^2} = \left[-\frac{1}{x}\right]_{-1}^{1} = -1-(1)$$
$$= -2.$$

Denoting the integrand by $f(x)$, and $-1/x$ by $I(x)$, discuss whether the argument is fallacious because $f(x)$ is not continuous at $x = 0$, or because it is not differentiable at $x = 0$.

†**10.** Show that, if y is a function of x which decreases in value when x increases from x to $x+\delta x$, then the area δA contained by the graph of y, the x-axis, and the ordinates $y+\delta y$ and y at x and $x+\delta x$, respectively, satisfies the inequality

$$(y+\delta y)\, \delta x \leqslant \delta A \leqslant y\, \delta x,$$

and explain why $y = dA/dx$.

8.2. THE PROCESS OF INTEGRATION

It has just been established that the purpose of integration is to find a function whose derivative has a particular form. This particular form is the integrand or function which is to be integrated. To integrate x^2, therefore, is to carry out a process giving a function whose derivative is x^2, and we can see fairly easily that the power of x in the required function or integral must be one more than the two of the x^2, for the power will be reduced by one by the differentiation. If we assume that the integral of x^2 is x^3, and check the assumption by differentiation, we do not get the original integrand x^3, but $3x^2$. The function x^3 is therefore three times too large, and it is $\frac{1}{3}x^3$ which gives x^2 when differentiated. One integral of x^2 is thus $\frac{1}{3}x^3$, but this is not the only one, for $(\frac{1}{3}x^3+4)$ or $(\frac{1}{3}x^3-2)$ will also give x^2 when differentiated. The best we can do is to write the integral of x^2 in the indefinite form $(\frac{1}{3}x^3+c)$, where c is any number and is referred to as the constant of integration. This gives the form of the required function, and for any particular problem, c will have a particular value which can be easily found, or is immaterial.

Repeating this procedure for a few simple integrands, we can draw up the following table:

Integrand (function whose integral is required)	*Integral* (function which when differentiated gives the integrand)
x^1	$\frac{1}{2}x^2+c$
x^2	$\frac{1}{3}x^3+c$
x^3	$\frac{1}{4}x^4+c$
x^4	$\frac{1}{5}x^5+c$

It can be seen that in these cases the integral can be obtained from the integrand by making three changes:

(i) increase the power of x by one,
(ii) divide by the new power,
(iii) add the constant of integration.

The integral of x^n can therefore be deduced to be $x^{n+1}/(n+1)+c$, and the process represented diagrammatically by the three steps:

$$\left.\begin{array}{l}\text{(i) } x^n \rightarrow x^{n+1} \\ \text{(ii) divide, obtaining } \dfrac{x^{n+1}}{n+1} \\ \text{(iii) add } c\end{array}\right\} \begin{array}{l}\dfrac{x^{n+1}}{n+1}+c \\ = \int x^n\,dx.\end{array}$$

We will now consider some particular applications of these general rules in more detail:

(a) the integral of x is $\frac{1}{2}x^2+c$; x is understood to be x^1;

(b) the integral of ax^n, where a is any number, is a times the integral of x; thus $\int 6x\,\mathrm{d}x = 6\int x\,\mathrm{d}x = 6.\frac{1}{2}x^2+c = 3x^2+c$, and this can be checked by differentiation;

(c) the integral of 1 is $(x+c)$: this can again be checked quite easily by differentiation, and since $x^0 = 1$, this can be regarded as the special case when the power of x in the integrand is 0;

(d) from (ii) and (iii), the integral of any number b is $(bx+c)$;

(e) when n is a negative integer, the power of x in the integral is numerically smaller than it is in the integrand. A positive number added to a negative one makes the negative number less negative, e.g. $(-2)+(1) = (-1)$, so

$$\int \frac{1}{x^2}\,\mathrm{d}x = \int x^{-2}\,\mathrm{d}x = -x^{-1}+c = -\frac{1}{x}+c.$$

(f) if $n = -1$, $(x^0/0)+c$ is obtained from the general rule. The term $x^0/0$ is meaningless, and if we are to deal with the integral of a function such as x^{-1} we shall need to approach the problem differently. This will be done later, and for the moment the special case of $n = -1$ is excluded from the general rule.

(g) the result holds for fractional powers as well: if, for example, $n = -\frac{1}{2}$, we find that the integral of $1/x^{\frac{1}{2}}$ is $(2x^{\frac{1}{2}}+c)$, and this correctly gives the integrand when differentiated.

We can finally add two comments of general use, namely:

(h) the integral of a sum of terms is equal to the sum of the integrals of the separate terms. Thus the integral of (x^2+x) is $\frac{1}{3}x^3+\frac{1}{2}x^2+c$, there being no need to write two constants.

(i) the integral of a simple product can be obtained by multiplying it out, and then integrating the terms separately. For example, the integral of $x(x-2)$ is the same as the integral of (x^2-2x), which is $(\frac{1}{3}x^3-x^2+c)$.

To avoid any ambiguity which might arise if there is more than one variable, it is necessary to indicate with respect to which variable the integration is to be carried out. The integral of 7 with respect to x is $(7x+c)$, whereas its integral with respect to y is $(7y+c)$. The variable with respect to which the integration is being performed is the variable whose power or form in the integrand is to be altered, and this variable is indicated by the dx or dy following the integral sign and the integrand. Thus in a sense the dx can be taken as representing the phrase "with respect to x", and the whole expression $\int x\,\mathrm{d}x$ can be read "the integral of x with respect to x". In an analogous way to the d and dx of differentiation, the integral sign and the dx stand for a non-algebraic operation, and are not subject to the usual laws of algebra.

Example 8.2

Write expressions for:

(i) $\int (8x^3-3)\,\mathrm{d}x$,

(ii) $\displaystyle\int \frac{\mathrm{d}y}{y^3}$,

(iii) $\displaystyle\int \frac{z-1}{z^4}\,\mathrm{d}z$.

Increasing the power of the variable with respect to which the integration is to be carried out by one in each case, and dividing by the increased power as explained above, we obtain after rearranging:

(i) $\int (8x^3-3)dx = \int 8x^3\,dx - \int 3dx$
$= 8\int x^3\,dx - 3\int 1\,dx$
$= 2x^4-3x+c.$

(ii) $\displaystyle\int\frac{dy}{y^3} = \int y^{-3}\,dy = \frac{y^{-2}}{-2}+c = -\frac{2}{y^2}+c.$

(iii) $\displaystyle\int\frac{z-1}{z^4}\,dz = \int\frac{dz}{z^3}-\int\frac{dz}{z^4}$

(writing the integrand as two separate terms)

$$= \int z^{-3}\,dz - \int z^{-4}\,dz = \frac{z^{-2}}{-2}-\frac{z^{-3}}{-3}+c$$

$$= -\frac{2}{z^2}+\frac{3}{z^3}+c$$

$$= \frac{3-2z}{z^3}+c.$$

EXERCISE 8b

Integrate with respect to x:

1. x^5.
2. x^3.
3. x.
4. $8x$.
5. 3.
6. $2x$.
7. x^{-2}.
8. $\frac{2}{x^3}$.
9. $\frac{7}{x^5}$.
10. $x^{\frac{1}{4}}$.
11. $\frac{1}{\sqrt{x}}$.
12. $\frac{1}{x^{\frac{2}{3}}}$.
13. $\frac{8}{x^{\frac{1}{4}}}$.
14. $3x^{-\frac{3}{2}}$.
15. $\frac{2}{3}x^{-\frac{2}{3}}$.
16. $x(x-7)$.
17. $5(x+3)$.
18. $(x+1)^2$.
19. $(x-7)(2x+1)$.
20. $(x+2)\sqrt{x}$.
21. a.
22. ax^7.
23. $x^2+\frac{7}{x^2}$.
24. $\frac{3x-2}{\sqrt{x}}$.
25. $\frac{x^2+1}{x^{\frac{1}{2}}}$.
26. $\frac{x^3+x^2}{x^{10}}$.

Write expressions for:

27. $\int(3p+y)\,dp$.
28. $\int\frac{q^2+1}{q^4}\,dq$.
29. $\int z\sqrt{z}\,dz$.
30. $\int(at^2+b)^2\,dt$.

†31. What sets of functions give

(i) $\sin x$, (ii) $1/\sqrt{(9-x^2)}$, (iii) $6\sin^4 2x\cos^2 2x$,

when differentiated with respect to x?

†32. Write an expression for

$$\int\frac{6x}{(1+x^2)^4}\,dx.$$

†33. Find the area under the curve $y = x^8$ between $x = 0$ and $x = 2$.

†34. Evaluate: $\displaystyle\int_1^2\frac{1}{x^2}\,dx$ and $\displaystyle\int_0^2 x(1-x)^2\,dx$.

†35. The velocity of a particle is given by the equation $v = 5t+7$. Deduce an expression for the relationship between the distance s it travels between the time 0 and t.

†36. Integrate with respect to x

(a) $(2x-3)^2$, (b) $\sqrt{x}-\frac{1}{\sqrt{x}}$. (OC)

8.3. SOME STANDARD FORMS

The preceding two sections contain the explanation and operation of integration, and a little careful thought will show that the procedure for integration is the exact opposite of that for differentiation. The derivative of x^3 is $3x^2$; the integral of $3x^2$ is (x^3+c). The operation of differentiating results in the power of x^n being reduced by one; that of integrating increases it. Differentiation results in the *original* power *multiplying* the derivative; integration results in the *final* power *dividing* the integral. The derivative of the area equation is the curve equation; the integral of the curve equation is the area equation. A two-way relationship can be written:

$$x^n+c \underset{\text{Integrate}}{\overset{\text{Differentiate}}{\rightleftharpoons}} nx^{n-1},$$

where the upper line is read from left to right, and the lower line from right to left.

Function	*Derivative*	*Integrand*	*Integral*
x^{n+1}	$(n+1)x^n$	x^n	$\frac{x^{n+1}}{n+1}+c$
$\sin ax$	$a\cos ax$	$\cos ax$	$\frac{1}{a}\sin ax+c$
$\cos ax$	$-a\sin ax$	$\sin ax$	$-\frac{1}{a}\cos ax+c$
$\tan ax$	$a\sec^2 ax$	$\sec^2 ax$ or $1/\cos^2 ax$	$\frac{1}{a}\tan ax+c$
$\cot ax$	$-a\,\mathrm{cosec}^2\, ax$	$\mathrm{cosec}^2\, ax$ or $1/\sin^2 ax$	$-\frac{1}{a}\cot ax+c$
$\sin^{-1}\left(\frac{x}{a}\right)$	$\frac{1}{\sqrt{(a^2-x^2)}}$	$\frac{1}{\sqrt{(a^2-x^2)}}$	$\sin^{-1}\left(\frac{x}{a}\right)+c$
$\tan^{-1}\left(\frac{x}{a}\right)$	$\frac{a}{a^2+x^2}$	$\frac{1}{a^2+x^2}$	$\frac{1}{a}\tan^{-1}\left(\frac{x}{a}\right)+c$
ϕ^{n+1}	$(n+1)\phi'\phi^n$	$\phi'\phi^n$	$\frac{\phi^{n+1}}{n+1}+c$

If therefore we can differentiate any function, we can integrate the derivative, the integral being the original function with which we started, plus a constant. So knowing, for example, that $\mathrm{d}(\sin x)/\mathrm{d}x = \cos x$, we can say that $\int \cos x\,\mathrm{d}x = \sin x+c$. Drawing up a list of similar results we obtain columns 1 and 2 of the table above. The two columns are related by the operation of differentiation; if a function in column 2 is integrated, the function in column 1, together with a constant, will be obtained. Knowing this, the integrals of column 3 can be deduced, with the results shown in column 4. (Note that if the integrand is divided or multiplied by any number, the integral will also be multiplied or divided by that number, but this applies *only* to pure numbers, and not to functions of x.) For the last example, the symbol ϕ is to be taken to represent any simple function of x; an eye has to be developed for integrands of the type $\phi'\phi^n$, but this is quite easy with practice.

To take an example or two of this latter sort, consider the integrand $\sin x\cos^2 x$. If ϕ is set equal to $\cos x$, this integrand could be represented by the expression $-\phi'\phi^2$, where ϕ' is the derivative of ϕ. The integral is then $-\frac{1}{3}\phi^3+c$, or

$$-\tfrac{1}{3}\cos^3 x+c.$$

Similarly, if $x^2(1+x^3)^4$ is to be integrated, it can be made into an integrand of the same sort:

$$I = \int x^2(1+x^3)^4\,\mathrm{d}x$$
$$= \frac{1}{3}\int 3x^2(1+x^3)^4\,\mathrm{d}x.$$

Then if we put

$$\phi = (1+x^3),$$
$$I = \frac{1}{3}\int \phi'\phi^4\,\mathrm{d}x$$
$$= \tfrac{1}{3}\cdot\tfrac{1}{5}(1+x^3)^5+c$$
$$= \tfrac{1}{15}(1+x^3)^5+c\,.$$

Or again,

$$I = \int \frac{2x}{(1+x^2)^4}\,\mathrm{d}x$$
$$= \int 2x(1+x^2)^{-4}\,\mathrm{d}x$$
$$= -\tfrac{1}{3}(1+x^2)^{-3}+c,$$

putting $\phi = (1+x^2)$ and $n = -4$.

Note that the result does not hold for $n = -1$, and if the ϕ' *is not present* (as in $\int(1+x^2)^3\,dx$, for example), the integral cannot be written down in this way; to divide afterwards by the absent ϕ'

$$\left(\text{giving } \frac{(1+x^2)}{4}\cdot\frac{1}{2x} \text{ in this case}\right)$$

is not legitimate. The integrand must be *exactly* of the form $\phi'\phi^n$ as far as the x's are concerned; only numbers and signs can be adjusted after integration.

We can also express the general rule in words: if a simple function of x is raised to any power other than minus one and multiplied by its derivative, the integral of the product is the function raised to a power of one more than the original, plus the usual constant of integration.

Example 8.3

Write expressions for the integrals of x^8, $\sin 4x$, $1/\sqrt{(4-x^2)}$, $2\cos 7x$.

Using the table of standard results on p. 207, we obtain:

$\frac{1}{9}x^9$ as the integral of x^8,

$-\frac{1}{4}\cos 4x$ as the integral of $\sin 4x$,

$\sin^{-1}(x/2)$ as the integral of $1/\sqrt{(4-x^2)}$,

and $\frac{2}{7}\sin 7x$ as the integral of $2\cos 7x$, checking each result by differentiation.

Examples 8c

Integrate the following with respect to x:

1. $\sin 3x$.

2. $\cos \frac{1}{2}x$.

3. $2x(1+x^2)^4$.

4. $2x\sqrt{(1+x^2)}$.

5. $\sin x\cos^3 x$.

6. $\dfrac{1}{\sin^2 3x}$.

7. $\dfrac{x^2}{\sqrt{(1-x^3)}}$.

8. $\sec^2 x\tan^5 x$.

9. $\sin x/\cos^3 x$.

10. $\sec 5x\tan 5x$.

11. $(5+4x)^8$.

12. $\operatorname{cosec}^2 2x\cot^5 2x$.

13. $\operatorname{cosec} x\cot x$.

14. $\dfrac{1}{\sqrt{(9-x^2)}}$.

15. $\dfrac{x}{(16+x^2)^2}$.

16. $\dfrac{1}{16+x^2}$.

17. $\dfrac{x}{(16+x^2)^2}$.

18. $\dfrac{3}{\sqrt{(16-4x^2)}}$.

19. $\dfrac{1}{25+16x^2}$.

20. $x^4(1-2x^5)^{12}$.

21. $\dfrac{x}{\sqrt{(3-x^2)}}$.

22. $\sin x\cos^8 x$.

23. $\dfrac{1}{7+5x^2}$.

24. $\sec^2 x\tan x$.

†**25.** $\sin^3 x$.

†**26.** $\cos^2 x$.

†**27.** $\dfrac{1+\sin^2 x}{\cos^2 x}$.

†**28.** $\cos^3 3x$.

†**29.** $\sqrt{\left(\dfrac{1+x}{1-x}\right)}$.

†**30.** $\sin x\cos 2x$.

†**31.** $\tan^2 x$.

†**32.** $\sin^4 x$.

8.4. THE INTEGRAL AS THE LIMITING VALUE OF A SUM

In considering the concept of the area under a curve in Section 7.1, it was shown that the actual value of the area involved could be approached by evaluating a lower and upper bound. Referring to Fig. 8.9, it will be recalled

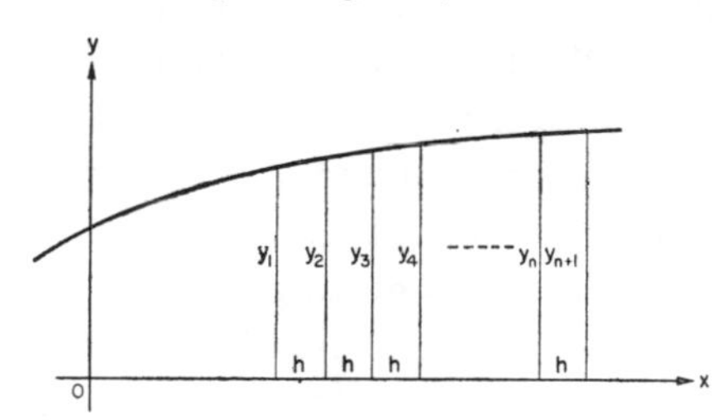

FIG. 8.9. The area under the curve bounded by y_1 and y_{n+1} lies between $h(y_1+y_2+\ldots+y_n)$ and $h(y_2+y_3+\ldots+y_{n+1})$.

that the value of these bounds can be written as

$$h(y_1+y_2+\ldots+y_n)$$

and

$$h(y_2+y_3+\ldots+y_{n+1}),$$

where h is the size of the (equal) spacing between the ordinates y_1 to y_{n+1}, and where y_1 and y_{n+1} define the boundaries of the area being considered. If we now construct a very large number of ordinates n distance δx apart, we can write the inequality

$$(y_1+y_2+\ldots+y_n)\,\delta x \leqslant A \leqslant (y_2+y_3+\ldots+y_{n+1})\,\delta x, \quad (8.1)$$

where A = area under curve between the ordinates y_1 and y_{n+1}.

If further we consider what happens as δx is made to approach zero (when more and more ordinates must be constructed), we can see that the two sums approach (and in fact define) the value A, i.e.

$$\lim_{\delta x=0} \{(y_1+y_2+\ldots+y_n)\,\delta x\} = A = \lim_{\delta x=0} \{(y_2+y_3+\ldots+y_{n+1})\,\delta x\}^*. \quad (8.2)$$

But we know that the value of A is equal to the definite integral of the curve function y between the ordinates y_1 and y_{n+1}, so we can write that

$$\lim_{\delta x=0}\{(y_1+y_2+\ldots+y_n)\,\delta x\} = \int_{x_1}^{x_{n+1}} y\,dx = \lim_{\delta x=0} \{(y_2+y_3+\ldots+y_{n+1})\delta x\},$$

where x_1 and x_{n+1} are the abscissae corresponding to the ordinates y_1 and y_{n+1} (Fig. 8.9).

This can be expressed more concisely. The Greek letter $\sum$ (sigma) is used in mathematics to stand for a sum, and the equality

$$\sum_{r=1}^{n} r = 1+2+3+\ldots+n$$

can be used as an expression of its definition. The term or set of terms after the $\sum$ sign is evaluated with $r = 1$ (written underneath the $\sum$ sign), and then added to the value of the expression when $r = 2$ (the variable is always increased by unity). This sum is then in turn added to the value of the expression when $r = 3$ (adding another 1), and so on until $r = n$, when the last addition is made. The notation can be used for other letters or expressions: a few examples will make the idea clear:

(i) $\sum_{x=1}^{n} 2x = 2.1+2.2+2.3+2.4+\ldots+2.n,$

(ii) $\sum_{r=1}^{n} x^r = x^1+x^2+x^3+\ldots+x^n,$

(iii) $\sum_{p=3}^{n+1} (p+4) = (3+4)+(4+4)+(5+4)+\ldots+(n+4)+\{(n+1)+4\}$

(note that the expression starts with $p = 3$ and ends with $p = n+1$).

(iv) $\sum_{r=1}^{n} y_r\,\delta x = y_1\,\delta x+y_2\,\delta x+y_3\,\delta x+\ldots+y_n\,\delta x.$

We can use this last equation to abbreviate our identification of the integral with the limiting value of a sum, for equation (8.1) can now be written as

$$\sum_{r=1}^{n} y_r\,\delta x \leqslant A \leqslant \sum_{r=2}^{n+1} y_r\,\delta x,$$

and equation (8.2) as

$$\lim_{\delta x=0}\sum_{r=1}^{n} y_r\,\delta x = \int_{x_1}^{x_{n+1}} y\,dx = \lim_{\delta x=0}\sum_{r=2}^{n+1} y_r\,\delta x.$$

Since we are considering the limiting values as $\delta x \to 0$, n is going to increase without bound, and the values of y_r will be little different from each other. Dropping the suffixes for the general case $y_r = y$ and $x_r = x$, we obtain the equation

$$\lim_{\delta x=0} \sum y\,\delta x = \int y\,dx, \quad (8.3)$$

where it is understood that the range of integration is the same as the range of summation,

* The use of the static limit implies that the limiting value of the sum as δx approaches zero through positive values equals that as δx approaches zero through negative values. If δx approaches zero through negative values, the second ordinate of the element is considered to lie to the left of the ordinate at x.

and where only one limit is quoted since

$$\lim_{\delta x=0} \sum_{r=1}^{n} y_r\,\delta x = \lim_{\delta x=0} \sum_{r=2}^{n+1} y_r\,\delta x$$

(cf. Fig. 8.10).

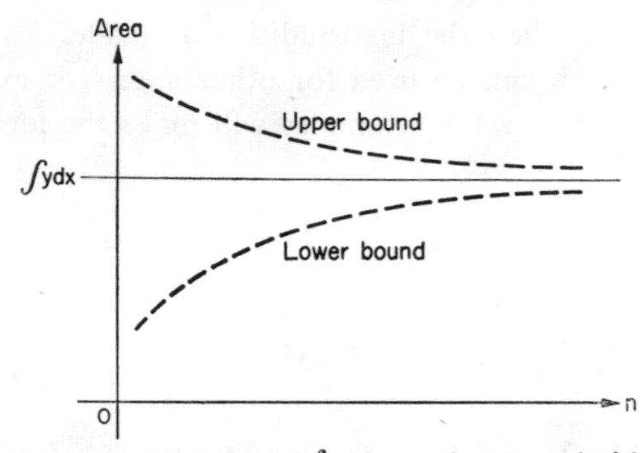

FIG. 8.10. The value of $\int y\,dx$ can be equated with that of the area under a unity scale graph, and is equal to the limiting values of both the upper and the lower bounds.

The integral of a function can therefore be associated with the limiting value of a sum: replacing y by $f(x)$, an alternative expression of equation (8.3) is

$$\lim_{\delta x=0} \sum f(x)\,\delta x = \int f(x)\,dx,$$

and in general it can be seen that the combination of symbols

$$\lim_{\delta x=0} \sum \ldots \delta x$$

is equivalent to $\int \ldots dx$. The importance of this idea will become apparent in Chapter 9.

Example 8.4

Write down the values of:

$$\text{(i)} \sum_{r=1}^{3} r^2; \quad \text{(ii)} \sum_{a=0}^{5} (2a+1).$$

(i) Expanding the abbreviated notation, $\sum_{r=1}^{3} r^2 = 1^2+2^2+3^2$, and this can be readily summed, giving the answer 14.

(ii) Similarly $\sum_{a=0}^{5} (2a+1) = 1+3+5+7+9+11 = 36.$

EXERCISE 8d

Evaluate the following sums:

1. $\sum_{x=1}^{4} x.$
2. $\sum_{x=1}^{3} 2x.$
3. $\sum_{x=1}^{3} x^2.$
4. $\sum_{x=2}^{5} 3x^2.$
5. $\sum_{x=1}^{6} (3+x).$
6. $\sum_{x=1}^{4} (7-x).$
7. $\sum_{r=1}^{5} 2^r.$
8. $\sum_{r=2}^{r=3} 3^r.$
9. $\sum_{r=1}^{3} p^3.$

†10. $\sum_{x=1}^{30} (5+3x).$

†11. $\sum_{x=1}^{50} (8+7x).$

†12. $\sum_{r=1}^{20} (2)^r.$

†13. $\sum_{r=0}^{11} 4.2^r.$

†14. $\sum_{r=1}^{n} \{a+(r-1)d\}.$

†15. $\sum_{m=0}^{n-1} ar^m.$

†16. $\lim_{n\to\infty} \sum_{r=0}^{n} (\tfrac{1}{2})^r$

†17. $\lim_{n\to\infty} \sum_{r=0}^{\infty} (\tfrac{1}{4})^r.$

†18. $\lim_{n\to\infty} \sum_{r=0}^{n} (1/10)^r.$

†19. $\lim_{n\to\infty} \dfrac{\sum_{r=0}^{n} (3+5r)}{\sum_{r=0}^{n} (4+7r)}.$

†20. $\lim_{n\to\infty} \dfrac{\sum_{r=0}^{n} (\tfrac{1}{3})^r}{\sum_{r=0}^{n} (\tfrac{1}{4})^r}.$

21. Write down expressions for

(i) $\lim_{\delta x=0} \sum \cos x\,\delta x,$

(ii) $\lim_{\delta x=0} \sum x^5\,\delta x,$

(iii) $\lim_{\delta y=0} \sum \dfrac{1}{y^2}\,\delta y.$

22. Give expressions for

(i) $\lim_{\delta x=0} \sum (1-x)\,\delta x,$

(ii) $\lim_{\delta x=0} \sum \sec^2 2x\,\delta x$

(iii) $\lim_{\delta p=0} \sum \sqrt{p}\,\delta p.$

†23. Which of the following statements are true?

(i) $\dfrac{\sum_{r=0}^{n} (a)^r}{\sum_{r=0}^{n} (b)^r} = \sum_{r=0}^{n} \left(\dfrac{a}{b}\right)^r,$

(ii) $\sum_{x=1}^{5} f(x) - \sum_{x=3}^{5} f(x) = \sum_{x=1}^{2} f(x),$

(iii) $\sum_{x=a}^{b} f(x) - \sum_{x=b}^{c} f(x) = \sum_{x=a}^{c} f(x),$ where $a<b<c,$

(iv) $\dfrac{\sum_{r=1}^{n} a^r}{\sum_{r=1}^{m} a^r} = \sum_{r=m}^{n} a^r,$

(v) $\lim_{n\to\infty} \dfrac{\sum_{r=1}^{n}(3+2r)}{\sum_{r=1}^{n}(3+4r)} = \dfrac{3}{3} = 1,$

(vi) $\lim_{n\to\infty} \dfrac{\sum_{r=0}^{n}\left(\frac{1}{a}\right)^r}{\sum_{r=0}^{n}\left(\frac{1}{b}\right)^r} = \dfrac{a(b-1)}{b(a-1)}.$

†**24.** By considering a graph of the function $y = 3$, show that

$$\sum_{x=0}^{x=5} 3\,\delta x = 15.$$

†**25.** Show that $\lim_{\delta x=0} \sum_{r=0}^{r=x} r\,\delta r = \frac{1}{2}x^2.$

†**26.** Write expressions for

(i) $\lim_{\delta x=0} \sum \sin x \cos^3 x\,\delta x,$

(ii) $\lim_{\delta x=0} \sum x\sqrt{(1+x^2)}\,\delta x,$

(iii) $\lim_{\delta x=0} \sum \dfrac{\sin^{-1} x}{\sqrt{(1-x^2)}}\,\delta x.$

***27.** Calculate the work done in pumping a liquid of density ϱ out of a cylindrical tank of height h and cross-section area A. (Consider first the work done in pumping out the liquid contained between two horizontal planes at depth of x and $x+\delta x$ from the top.)

***28.** The tension in a stretched elastic string is given by an expression of the form $\lambda x/a$, where x is the extension of the string from its natural length a, and where λ is a constant. Show that the work done in stretching the string through a further distance δx is $\lambda x \delta x/a$, and deduce that the total work done in stretching the string from its natural length to an extension X is $\lambda X^2/2a$.

29. Show that the area A under a unity scale graph of $y = x^2$ between the ordinates $x = 0$ and $x = 8$ is such that

$$\frac{8^3}{n^3}\sum_{r=0}^{n-1}(r)^2 \leqslant A \leqslant \frac{8^3}{n^3}\sum_{r=1}^{n}(r)^2.$$

By considering the behaviour of this inequality as $n \to \infty$, deduce that the area enclosed by the graph and the x-axis is $8^3/3$.

$$\left[\sum_{r=1}^{n} r^2 = n(n+1)(2n+1)/6.\right]$$

30. The surface area of a liquid of a hemispherical bowl of radius a is S when its depth is x. Show that $S = \pi x(2a-x)$. Show also that if the depth increases by δx, the extra volume δV of liquid added must be such that

$$S\,\delta x \leqslant \delta V \leqslant (S+\delta S)\,\delta x,$$

where δS is the increase in the surface area due to the added liquid. Deduce that the volume of liquid V in the bowl is given by $V = \int S\,dx$, and find an expression relating V to x.

***31.** The potential of a capacitor whose capacitance is C is Q/C, where Q is the charge on it. Explain why the work necessary to bring a further charge δQ from a very large distance is $Q.\delta Q/C$, and hence show that the total work done in giving the capacitor a charge Q is $Q^2/2C$.

32. Find the value of

$$\lim_{\delta x=0}\sum_{x=0}^{x=a} px^2\,\delta x$$

if x varies from 0 to a, and p is a constant.

***33.** The twist couple in a torsion wire is $k\theta$ when the angle of twist is from the natural position is θ. Show that the work done in twisting the wire through a further angle $\delta\theta$ is $k\theta\,\delta\theta$, and hence deduce that the energy stored in the wire at an angle of twist θ is $\frac{1}{2}k\theta^2$.

***34.** The tension in a wire stretched a distance x from its natural length is YAx/L, where Y is a constant, A the cross-sectional area, and L the natural unstretched length. Show that the energy stored in the wires when stretched through a total distance X from its natural length is $YAX^2/2L$.

***35.** The moment of inertia of an annular element of radius r and thickness δr in a disc rotating about an axis perpendicular to its plane and through its centre is $2\pi r^3\varrho\delta r$, where ϱ = the density per unit area. Show that the total moment of inertia of a solid disc about such an axis is $\frac{1}{2}Ma^2$, where a is its radius and M is its total mass.

***36.** The tension in a spring is $80x$ newtons, where x is its extension. Show that the work done in stretching it from an extension of 5 m to one of 10 m is 3000 J.

***37.** The magnetic flux density due to a short piece of wire of length δl carrying a current i amps at a point r m away is $\mu_0 i\,\delta l \sin\theta/4\pi r^2$, where θ is the angle between a line from the point to the mid-point of the length δl and the direction of δl, and μ_0 is a constant. Show that the flux density at the centre of a circular coil of n turns of radius a m is $\mu_0 ni/2a$.

38. (i) Show that if A is the area under the curve $y = x^2$ between $x = 0$ and $x = 5$, then

$$\frac{5^3}{n^3}\sum_{r=0}^{n-1} r^2 \leqslant A \leqslant \frac{5^3}{n^3}\sum_{r=1}^{n} r^2.$$

Consider the range between $x = 0$ and $x = 5$ to be divided into n equal parts.

(ii) Show that the application of the trapezium rule leads to the expression

$$\left.\frac{a^3}{2n}+\frac{a^3}{n^3}\sum_{r=1}^{n-1} r^2\right\}$$

for the area under $y = x^2$ between $x = 0$ and $x = a$.

†**39.** Show that

(i) $\displaystyle\int_a^b f(x)\,dx = \frac{1}{n}\int_{na}^{nb} f(y)\,dy$ if $y = nx$,

(ii) $\displaystyle\int_a^b g\{f(x)\}f'(x)\,dx = \int_{f(a)}^{f(b)} g(y)\,dy$ if $y = f(x)$.

†**40.** Use the results of question 39 to show that

$$\frac{1}{\sigma\sqrt{(2\pi)}}\int_{a\sigma+\bar{x}}^{b\sigma+\bar{x}} e^{-(x-\bar{x})^2/2\sigma^2}\,dx = \frac{1}{\sqrt{(2\pi)}}\int_a^b e^{-y^2/2}\,dy.$$

SUMMARY

The process of finding an expression for the indefinite integral of a function is called *integration*; the function to be integrated is the *integrand.*

Integration can be regarded as the converse of differentiation; the derivative of the integral function gives the integrand. A short list of standard integrals follows:

Integrand	*Integral*
x^n	$\dfrac{x^{n+1}}{n+1}+c \quad (n \neq -1)$
$\sin ax$	$-\dfrac{1}{a}\cos ax + c$
$\cos ax$	$\dfrac{1}{a}\sin ax + c$
$\sec^2 ax$	$\dfrac{1}{a}\tan ax + c$
$\operatorname{cosec}^2 ax$	$-\dfrac{1}{a}\cot ax + c$
$\dfrac{1}{\sqrt{(a^2-x^2)}}$	$\sin^{-1}\left(\dfrac{x}{a}\right)+c$
$\dfrac{1}{a^2+x^2}$	$\dfrac{1}{a}\tan^{-1}\dfrac{x}{a}$
$\phi'\phi^n$	$\dfrac{\phi^{n+1}}{n+1} \quad (n \neq -1)$

The value of an integral can also be identified with the limiting value of a sum: the symbol sets

$$\lim_{\delta x = 0}\sum \phi\,\delta x \quad \text{and} \quad \int \phi\,dx$$

are equivalent.

MISCELLANEOUS EXERCISE 8

Integrate the following functions with respect to x:

1. x^8.
2. x^4.
3. 4.
4. $-x$.
5. $-x^2$.
6. $-7x$.
7. $-8x^8$.
8. 7.
9. $7-9x^8$.
10. $2-x$.
11. $5+2x$.
12. $6+2x+9x^2$.
13. $12x^3-7$.
14. $\dfrac{1}{\sqrt{x^3}}$.
15. $x^{\frac{5}{2}}$.
16. $\dfrac{7}{2}x^{\frac{5}{2}}$.
17. $x^{\frac{1}{3}}$.
18. $3x^{\frac{5}{3}}$.
19. $6x^{\frac{1}{5}}$.
20. $4x^{\frac{4}{5}}$.
21. $9x^{1\frac{1}{4}}$.
22. x^{-3}.
23. $7x^{-8}$.
24. $\dfrac{1}{\sqrt{x}}$.
25. $\dfrac{3}{\sqrt[3]{x}}$.
26. $7x^{-2}+5x^{-3}$.
27. $-x^{-\frac{1}{2}}$.
28. $-x^{-\frac{3}{2}}-2x^{-2}$.
29. $\dfrac{p}{x^5}-\dfrac{q}{x^6}$.
30. $\dfrac{ax+b}{x^8}$.
31. $\dfrac{3}{x^3}-\dfrac{5}{x^5}$.
32. $\dfrac{(3x+2)}{\sqrt{x}}$.
33. $(2x-5)^2$.
34. $\frac{1}{7}$.
35. $\frac{1}{3}x^5-8$.
36. $\dfrac{1}{8x^7}$.
37. $\dfrac{9}{5x^4}$.
38. $\dfrac{12}{7x^6}$.
39. $\dfrac{11}{5\sqrt{x}}$.
40. $\dfrac{7x-3}{8\sqrt{x}}$.
41. $(2x^{-1})^2$.
42. $2(x^{-1})^2$.
43. $(\frac{1}{2}x^2)^{-2}$.
44. $x\left(\sqrt{x}-\dfrac{1}{\sqrt{x}}\right)^2$.
45. $(3x+1)(x-8)$.
46. $(7-3x)(1+x)$.
47. $\dfrac{(2-x)(1-x)}{x^4}$.

48. $\left(2x^{\frac{1}{4}}\right)^3$.

49. $\left(2x^{\frac{1}{4}}\right)^{-3}$.

50. $\sin 8x$.

51. $\cos 5x$.

52. $2 \cos 2x$.

53. $3 \sin 9x$.

54. $\sin \frac{1}{2}x$.

55. $7 \cos \frac{1}{2}x$.

56. $8 \cos \frac{1}{4}x$.

57. $11 \cos \frac{1}{5}x$.

58. $2x(1+x^2)^7$.

59. $x(1+x^2)^6$.

60. $x(1+x^2)^{\frac{1}{2}}$.

61. $x^2(1+x^3)^{-3}$.

62. $x(1-x^2)^{-8}$.

63. $\dfrac{7x}{(1-x^2)^4}$.

64. $\dfrac{2x}{\sqrt{(1-x^2)}}$.

65. $\dfrac{1}{\sqrt{(1-x^2)}}$.

66. $\dfrac{2}{\sqrt{(9-x^2)}}$.

67. $\dfrac{5}{\sqrt{(1-16x^2)}}$.

68. $\dfrac{3}{\sqrt{(8-x^2)}}$.

69. $\dfrac{2}{1+x^2}$.

70. $\dfrac{2x}{(1+x^2)^8}$.

71. $\dfrac{1}{1+4x^2}$.

72. $\dfrac{1}{9+x^2}$.

73. $\dfrac{7}{16+x^2}$.

74. $\dfrac{8}{1+16x^2}$.

75. $\dfrac{-5}{9+4x^2}$.

76. $\dfrac{12}{16+9x^2}$.

77. $\dfrac{6}{4+9x^2}$.

78. $\dfrac{2}{\sqrt{(25-16x^2)}}$.

79. $\dfrac{2x}{\sqrt{(25-16x^2)}}$.

80. $\dfrac{5}{\sqrt{(9-4x^2)}}$.

81. $\dfrac{5x}{\sqrt{(9-4x^2)}}$.

82. $7x\sqrt{(36-49x^2)}$.

83. $\dfrac{7x}{\sqrt{(36-49x^2)}}$.

84. $\dfrac{7}{36+49x^2}$.

85. $3 \sin 3x \cos^2 3x$.

86. $3 \sin 3x \cos^9 3x$.

87. $\dfrac{\sin 5x}{\cos^8 5x}$.

88. $\dfrac{\cos 4x}{\sin^{11} 4x}$.

89. $\cos 10x \sin^8 10x$.

90. $\sec^2 2x \tan^4 2x$.

91. $\operatorname{cosec}^2 \frac{1}{2}x \cot^7 \frac{1}{2}x$.

Write down expressions for the following:

92. $\int 7 \, dr$.

93. $\int (3r^2-8) \, dr$.

94. $\int q.q^{\frac{3}{2}} dq$.

95. $\int t^{\frac{1}{5}}.t^{\frac{1}{4}} \, dt$.

96. $\int (\sqrt{a})^3 da$.

97. $\displaystyle\int \frac{ds}{4s^5}$.

98. $\displaystyle 4\int \frac{db}{b^9}$.

99. $\int 2x(1+x^2)^4 \, dx$.

100. $\displaystyle\int \frac{27.dy}{(1+y^2)^3}$.

101. $-\int \sin\theta \cos^5\theta \, d\theta$.

102. $\int \cos\theta \sin^4 d\theta$.

103. $\displaystyle\int \frac{\sin\phi}{\cos^3\phi} d\phi$.

104. $\int \sec^2\alpha \, d\alpha$.

105. $\int \tan x \sec^3 x \, dx$.

106. $\int \sec^2 5x \tan 5x \, dx$.

107. $\int \operatorname{cosec}^2 3x \cot^8 3x \, dx$.

108. $\int \sin x \sqrt{(\cos x)} \, dx$.

109. $\displaystyle\int \frac{\sin x}{\sqrt{\cos x}} dx$.

110. $\int \cos 2\theta \sin 2\theta \, d\theta$.

Evaluate the following:

111. $\displaystyle\lim_{n\to\infty} \sum_{r=0}^{n} \left(\frac{1}{8}\right)^r$.

112. $\displaystyle\lim_{n\to\infty} \left\{ \frac{\sum_{r=0}^{n} \left(\frac{1}{5}\right)^r}{\sum_{r=0}^{n} \left(\frac{1}{10}\right)^r} \right\}$.

113. $\displaystyle\lim_{\delta x=0} \sum (x^2+1)^2 \, \delta x$.

†**114.** Integrate with respect to x:

(i) $\dfrac{\cos^{-1} x}{\sqrt{(1-x^2)}}$; (ii) $\dfrac{\sqrt{\{\sqrt{(\sin x)}\}} \cos x}{\sqrt{\sin x}}$.

†***115.** How much work is needed to wind up a cable of length L and constant linear density ϱ which starts by hanging freely from a winch?

†***116.** The force on a charge q_1 at a distance r from another charge q_2 is

$$\frac{q_1 q_2}{4\pi\varepsilon_0 r^2}.$$

Show that the work done in moving one of the charges through a distance δr towards the other is

$$\frac{q_1 q_2}{4\pi\varepsilon_0 r^2}.\delta r,$$

and hence show that

$$q_1 q_2 \frac{(b-a)}{4\pi\varepsilon_0 ab}$$

is the work necessary to bring q_1 from a point at a distance a from q_2 to a point at a distance b from it.

†***117.** Calculate the work done against gravity in raising a body from the earth's surface to a height h.

†**118.** (i) The integral of $\sin x \cos x$ can be written as $\frac{1}{2}\sin^2 x$ or $-\frac{1}{2}\cos^2 x$. How is it possible to have two different integrals?

(ii) The integral of $1/\sqrt{(1-x^2)}$ is written $\text{Sin}^{-1} x$. Is this any more correct than writing it $\sin^{-1}x$? (see p. 162).

CHAPTER 9

Applications of Integration

9.1. EVALUATION OF DEFINITE INTEGRALS

As will be seen in the subsequent sections of this chapter, many applications of the integral calculus involve the evaluation of the definite integral of a function. We will therefore turn first to this problem.

It has been explained that the value of the integral of a function over a certain range can be identified with the area under a unity scale graph of that function between ordinates constructed at the range boundaries (see Fig. 9.1).

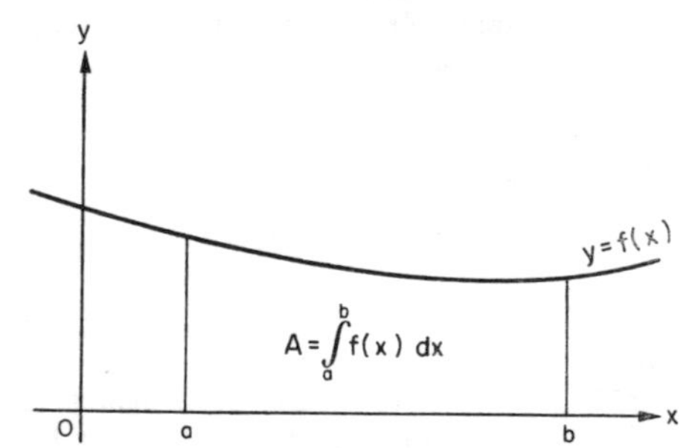

FIG. 9.1. The magnitude of the area between any two ordinates on a unity scale graph gives the value of the integral between those two limits.

We are now concerned with finding this area analytically.

The techniques shown in Chapter 8 enable an "indefinite" integral function to be derived from the integrand, and this integral function expresses the variation in the area under the integrand function curve as the upper boundary is moved. Thus to write that the integral of x^2 is $(\frac{1}{3}x^3+c)$ is to state that the area under the unity scale graph of $y = x^2$ increases as

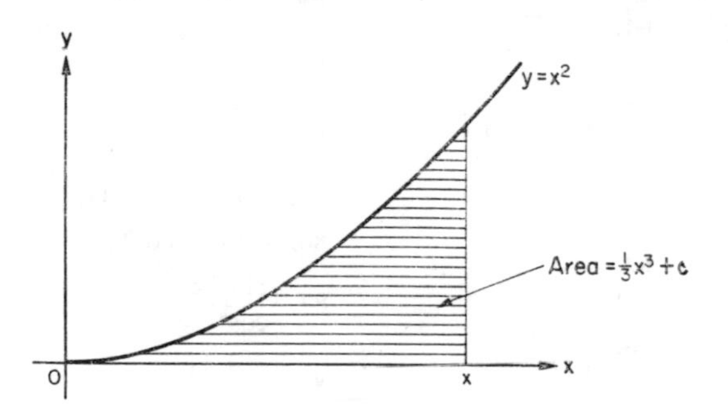

FIG. 9.2. The equation $\int x^2\, dx = \frac{1}{3}x^3+c$ indicates that the area under the curve $y = x^2$ increases with the cube of the upper boundary x.

the cube of the upper boundary (Fig. 9.2). We can then write that the area between an arbitrary lower boundary and an ordinate at $x = 3$ is equal to $(\frac{1}{3}.3^3+c)$ or $(9+c)$, where c depends on the position of the lower boundary (Fig. 9.3a). Again the area between the same lower boundary and an ordinate at $x = 6$ is $(\frac{1}{6}.6^3+c)$, or $(72+c)$ as shown in Fig. 9.3b, and by direct subtraction the area between two ordinates at $x = 3$ and $x = 6$ can be obtained as $[(72+c)-(9+c)]$, or 63 units (Fig. 9.3c). The value of this area can then be identified with the value of the integral of x^2 over the range $x = 3$ to $x = 6$, i.e.

$$\int_3^6 x^2\,dx = 63.$$

It should be noted that although the position of the lower boundary of the areas or integrals to $x = 3$ and $x = 6$ was not specified,

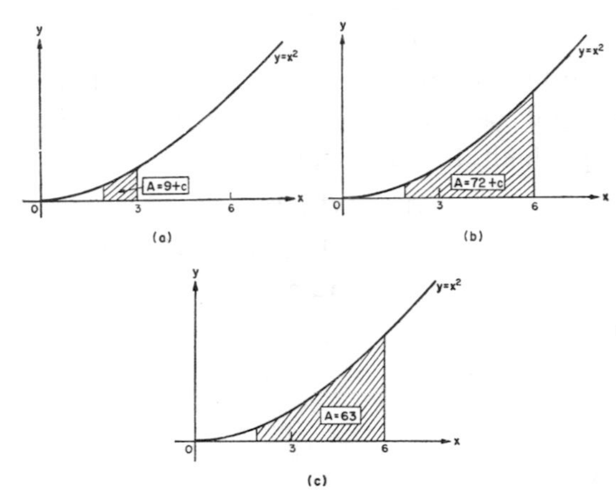

FIG. 9.3. (a) Knowing that the area between an arbitrary boundary and $x = 3$ is $(9+c)$, (b) and that that between the same arbitrary boundary and $x = 6$ is $(72+c)$, (c) we can deduce that the area between $x = 3$ and $x = 6$ is 63 by direct subtraction.

the constant of integration c which depends on its position will always disappear on subtraction, and this will always happen for *any* definite integral. It is therefore unnecessary to bring it into the evaluation of definite integrals, and for this reason it is often omitted altogether in such cases (see Example 9.1). In other cases, however, it often plays an important part, and must not be overlooked on any account. If it is necessary to assign a value to it, the value of the integral at some point must be known: this is often the case (especially at $x = 0$ or $t = 0$), and direct substitution will quickly lead to the value of c (see Example 9.2).

Example 9.1

Evaluate:

$$\text{(i)} \int_1^3 2x^3\,dx,$$

$$\text{(ii)} \int_0^{\pi/2} \cos x\,dx.$$

(i) The indefinite integral of $2x^3$ is $\frac{1}{2}x^4+c$, and so the value of the definite integral between $x = 1$ and $x = 3$ can be written in the conventional manner $[\frac{1}{2}x^4]_1^3$, or $[\frac{1}{2}x^4]_{x=1}^{x=3}$, where the square brackets indicate that the value of the "upper" limit is to be substituted into the expression inside the bracket, and then the value of the lower limit, and the two results subtracted. Unless two letters or variables are involved, it will often be unnecessary to state explicitly that $x = 3$ or $x = 1$, and the numbers 3 and 1 will suffice. Thus

$$\begin{aligned}\int_1^3 2x^3\,dx &= [\tfrac{1}{2}x^4]_1^3\\ &= [(\tfrac{1}{2}.3^4)-(\tfrac{1}{2}.1^4)]\\ &= \{\tfrac{1}{2}.81-\tfrac{1}{2}.1\}\\ &= 40,\end{aligned}$$

and the integral of $2x^3$ from $x = 1$ to $x = 3$ is 40. Note that the constant of integration can be omitted since we are dealing with a definite integral.

$$\begin{aligned}\text{(ii)} \int_0^{\pi/2} \cos x\,dx &= [\sin x]_0^{\pi/2}\\ &= \{\sin \tfrac{1}{2}\pi - \sin 0\}\\ &= 1-0\\ &= 1.\end{aligned}$$

Example 9.2

Find an expression for the area under the curve $y = 4x^3+2$ between an ordinate whose abscissa is 1 and one whose abscissa is indefinitely written as x.

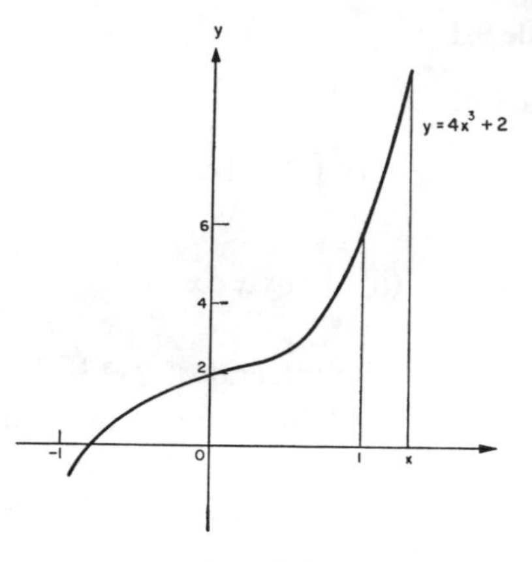

FIG. 9.4.

A sketch of the curve is shown in Fig. 9.4, and we can write that the area A under the curve is

$$\int (4x^3+2)\,dx$$

or

$$x^4+2x+c,$$

identifying the area with the integral. Now c depends on the position of the lower boundary to the area, and since in this case it is to be situated at $x = 1$, we can write that A must equal 0 when $x = 1$, for then no area is enclosed between the two ordinates (Fig. 9.5). If this is so, the expression for A gives

$$0 = 1^4+2.1+c,$$

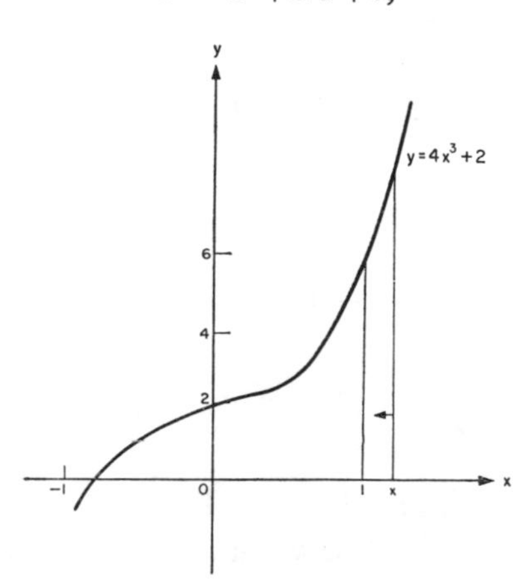

FIG. 9.5. The constant of integration is zero when x is put equal to the lower limit.

whence

$$c = -3.$$

Thus the required expression is

$$A = x^4+2x-3.$$

An alternative method for solving this problem is to write the expression for the area A as

$$\int_1^x (4x^3+2)\,dx,$$

regarding the area as the value of the definite integral between $x = 1$ and the indefinite upper limit. Integrating, this gives

$$\begin{aligned} A &= [x^4+2x]_1^x \\ &= (x^4+2x)-(1.1^4+2.1) \\ &= x^4+2x-3 \end{aligned}$$

as before.

EXERCISE 9a

Write down the values of:

1. $[x]_1^2$.
2. $[2x^2-3]_2^4$.
3. $[7x]_0^8$.
4. $[3x+1]_{-1}^1$.
5. $[\sin x]_0^{\pi/2}$.
6. $[4-2x]_{-2}^2$.
7. $[8-x]_{-2}^0$.
8. $\left[\frac{1}{x}\right]_{-4}^2$.

Evaluate the following definite integrals:

9. $\int_0^2 3x\,dx$.
10. $\int_0^8 (x-7)\,dx$.
11. $9\int_0^1 x^6\,dx$.
12. $\int_2^3 t^3\,dt$.
13. $\int_{-1}^1 (3p-7)\,dp$.
14. $\int_{-2}^0 q^2\,dq$.
15. $\int_1^5 \frac{1}{2}\,dx$.
16. $\int_2^4 \frac{2}{x^4}\,dx$.
17. $\int_{\pi/8}^{\pi/4} \cos 2x\,dx$.
18. $\int_{\frac{1}{4}}^{\frac{1}{2}} \frac{1}{p^3}\,dp$.
19. $\int_0^{\pi/12} \sin 6\theta\,d\theta$.
20. $\int_0^3 (x-3)^2\,dx$.

21. $\int_1^2 \frac{dt}{(t+1)^3}$.

22. $\int_0^4 \frac{dx}{(2x-7)^2}$.

23. $\int_{-1}^1 \frac{dp}{(3p+2)^2}$.

24. $\int_0^1 (3x+1)^4\, dx$.

25. $\int_0^1 \frac{dx}{\sqrt{(1-x^2)}}$.

26. $\int_0^3 \frac{dy}{\sqrt{(9-y^2)}}$.

27. $\int_0^2 \frac{dx}{4+x^2}$.

28. $\int_0^{\pi/12} \sec^2 3x\, dx$.

29. $\int_0^1 x(1+x^2)^4\, dx$.

30. $\int_0^1 \frac{x\, dx}{(1+x^2)^3}$.

31. $\int_0^{3/4} \frac{dt}{\sqrt{(9-4t^2)}}$

32. $\int_0^1 \frac{t\, dt}{\sqrt{(9-4t^2)t}}$.

33. $\int_{\frac{2}{3}}^1 \frac{x\, dx}{\sqrt{(16-9x^2)}}$.

34. $\int_0^{\pi/3} \sin x \cos^4 x\, dx$.

35. $\int_0^{\pi/4} \frac{\sin x}{\cos^3 x}\, dx$.

36. $\int_0^2 \frac{dx}{\sqrt{(16-x^2)}}$.

37. $\int_0^{\pi/4} \sec^2 x \tan^2 x\, dx$.

38. $\int_{\pi/6}^{\pi/2} \operatorname{cosec} x \cot x\, dx$.

†**39.** $\int_{\pi/3}^{\pi/6} \sin^2 \theta\, d\theta$.

†**40.** $\int_0^1 \frac{2x}{1+x^4}\, dx$.

†**41.** Derive an expression for the area under the curve $y = x^2+1/x^2$ between an ordinate where abscissa is (i) 1, (ii) a, and an ordinate whose abscissa is x. What are the restriction on the values of a and x?

†**42.** Write down an expression for $\int_{\pi/10}^{x} \cos 5x\, dx$.

†**43.** Find the area under the curve $y = 4x^2+8$ between $x = x_1$ and $x = x_2$.

†**44.** Find the equation of the hypotenuse AB of the right-angled triangle AOB in Fig. 9.6, and obtain the value of the area of the triangle by integration. Check your answer by using the normal rule for finding the area of a triangle.

†**45.** A parabola (whose equation is of the form $y = ax^2+bx+c$) passes through three points $(-h, y_1)$, $(0, y_2)$ and (h, y_3). Find its equation and find the area under the parabola between $x = -h$ and $x = +h$ in terms of h, y_1, y_2 and y_3 only. Give a sketch of the area.

†**46.** Estimate the value of $\int_0^5 f(x)\, dx$, where $f(0) = 0$,

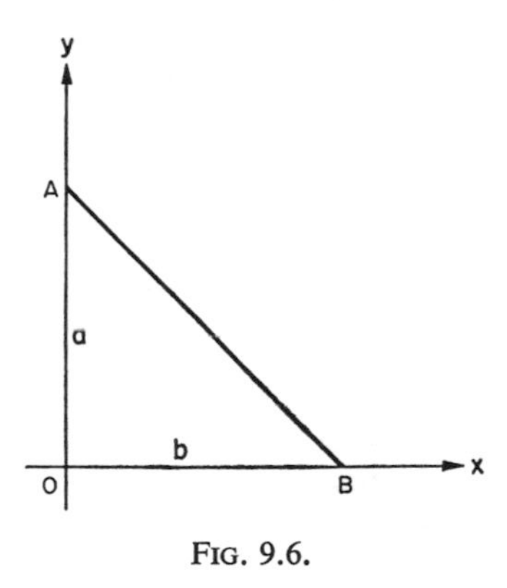

FIG. 9.6.

$f(1) = 0{\cdot}5$, $f(2) = 0{\cdot}9$, $f(3) = 1{\cdot}25$, $f(4) = 1{\cdot}5$ and $f(5) = 1{\cdot}7$.

†**47.** Estimate the value of $\int_1^{10} \log_{10} x\, dx$.

†**48.** Estimate the value of $\int_0^{\frac{1}{2}} \sin^{-1} x\, dx$.

49. Evaluate

(a) $\int_0^6 (x^3-x^2)\, dx$, (b) $\int_1^4 (x^{\frac{1}{2}}+x^{-\frac{1}{2}})\, dx$.

(OC)

50. Find

(a) $\int (5-3x)^2\, dx$, (b) $\int_1^{36} x^{-\frac{3}{2}}\, dx$.

(OC)

51. The tangent of the curve $y = 1-x^2$ at $x = \alpha$, where $0 < \alpha < 1$, meets the axes at P and Q. Prove that, as α varies, the minimum value of the area of the triangle OPQ is $2/\sqrt{3}$ times the area bounded by the axes and the part of the curve for which $0 \leqslant x \leqslant 1$. (SU)

***52.** The work done by a gas expanding from the volume v_1 to a volume v_2 is given by

$$\int_{v_1}^{v_2} p\, dv,$$

where p is the pressure and v the volume, at any instant. If $pv^{\gamma} = k$, calculate the work done by the gas expanding from a volume v to twice that volume, given that the gas occupies (2×10^2) m³ when $p = 10^5$ N m^{-2}, (2×10^{-2}) m³ being the initial volume and 10^5 N m^{-2} being the initial pressure. ($\gamma = 1{\cdot}4$.)

53. A function $f(x)$ is defined by the equations $f(0) = 0$, $f(x) = 1-(\frac{1}{2})^n$ for $1-(\frac{1}{2})^{n-1} < x \leqslant 1-(\frac{1}{2})^n$ for all positive integral values of n greater than or equal to 1, and $f(1) = 1$. Draw a sketch graph of the function, and by considering the area under a unity scale graph of the function, deduce the value of $\int_0^1 f(x)\, dx$.

54. Evaluate

$$\text{(a)} \int_0^{\pi/2} \sin x \cos 2x \, dx, \quad \text{(b)} \int_0^{\pi/4} \sin^2 x \, dx. \quad \text{(L)}$$

55. Evaluate

$$\int_1^2 x^2(3-x)\, dx. \quad \text{(C)}$$

56. Explain why the equation $\int (x-1)^{-2} dx = 1/(1-x)+c$ is invalid if the range of integration includes the value $x = 1$.

What is the value of c if the lower boundary of integration is $x = 0$, and the upper boundary x is less than 1?

57. Evaluate

$$\text{(a)} \int_1^2 \frac{x+1}{x^3} dx, \quad \text{(b)} \int_2^3 (x-2)(x-3)\, dx. \quad \text{(C)}$$

58. Find the value of $\int_0^1 dx/(x+1)$ by drawing a graph and counting squares.

59. What are the values of a and b for which the equation

$$\int_a^b \frac{1}{x^2} dx = \left(\frac{1}{a} - \frac{1}{b}\right) \text{ is valid?}$$

60. Explain why the equation

$$\int \frac{2}{x^3} dx = -\frac{2}{x^4} + c$$

must imply that the lower limit of integration is greater than zero or less than zero, or that the value of x must be greater than zero or less than zero, respectively.

61. Write an expression for $\int_a^b |x| \, dx$ if

(i) $b > a > 0$, (ii) $a < b < 0$, (iii) $a < 0 < b$.

62. Find

$$\text{(a)} \int x(x+1)(x-1)\, dx, \quad \text{(b)} \int_1^4 (x^2-4)\, dx. \quad \text{(OC)}$$

63. Evaluate

$$\text{(a)} \int_{-1}^{1} x(x^2+1) dx, \quad \text{(b)} \int_0^{\pi/4} \cos 2x \, dx. \quad \text{(L)}$$

64. Show that $\dfrac{d}{dx}\left\{\dfrac{x}{\sqrt{(1-x^2)}}\right\} = \dfrac{1}{(1-x^2)^{\frac{3}{2}}}$,

and hence evaluate

$$\int_0^{\frac{1}{2}} \frac{dx}{(1-x^2)^{\frac{3}{2}}}.$$

65. Given that $\int \phi' e^{\phi} \, dx = e^{\phi}$, find $\int \cos x \, e^{\sin x} \, dx$.

66. Write an expression for the value of $\int_0^a [x]\, dx$, given that a is a positive.

67. Evaluate $\displaystyle\int_{\pi/6}^{\pi/2} \frac{\cos y \, dy}{\sin^3 y}$.

†68. If we consider the value of the integral

$$I = \int_{2\cos\theta}^{1} \frac{a \, dx}{\sqrt{(1-a^2x^2)}}$$

we can evaluate it as

$$[\sin^{-1} ax]_{2\cos\theta}^{1} = \sin^{-1}(a) - \sin^{-1}(2a\cos\theta).$$

If then we make a take the value $\sin\theta$, we find that

$$I = \sin^{-1}(\sin\theta) - \sin^{-1}(\sin 2\theta) = -\theta.$$

If then we further set $\theta = \pi/3$, we have that $2\cos\theta = 1$, and

$$I = \int_1^1 \frac{a\, dx}{\sqrt{(1-a^2x^2)}} = 0,$$

since the limits on the integral we identical. Thus

$$0 = -\theta = -\pi/3.$$

and

$$\pi = 0.$$

Discuss.

†69. Find the values of

$$\int_0^{\pi/2} (\sin x + x \cos x)\, dx \quad \text{and} \quad \int_0^{\pi/2} \sin x \, dx,$$

and deduce the value of $\int_0^{\pi/2} x \cos x \, dx$.

†70. By writing an expression for $d(uv)/dx$, where u and v are functions of x, show that

$$\int_a^b uv' \, dx = [uv]_a^b + \int_a^b u'v \, dx.$$

Hence find the value of $\int_0^{\pi/2} x \cos x \, dx$.

9.2. CHANGES, RATES OF CHANGE AND SIMPLE DIFFERENTIAL EQUATIONS

Some examples of finding the values of a change in a dependent variable produced by a change in the independent variable have been given in Section 5.5; the method employed there is applicable only to small changes in either variable, and of course the whole equation $\delta y \triangleq (\mathrm{d}y/\mathrm{d}x)\,\delta x$ is an approximation anyway.

For larger changes it is necessary to derive a function relating the two variables from the function describing the variation in the rate of change of the one variable with respect to the other, i.e. obtain y as a function of x from the expression for $\mathrm{d}y/\mathrm{d}x$. This procedure is, of course, integration. (An alternative method is to use an approximate numerical method: such a technique is described as numerical integration, and use can be made of the trapezium rule or Simpson's rule (Section 9.5). This procedure is necessary whenever it is not possible to obtain or to express the integral function analytically.)

The problem, put in general terms, is thus to find an expression for the relationship between the variables y and x, given the form of the rate of change $\mathrm{d}y/\mathrm{d}x$. An obvious application of this procedure arises in dynamics—where, for example, we know that the acceleration of a body falling towards the centre of the earth is equal to 9·80 m/s²—that is, its velocity increases by 9·80 m/s every second. Thus the rate of change of velocity is equal to 9·80 m/s², or, expressing the same thing in the form of a *differential equation*, $\mathrm{d}v/\mathrm{d}t = 9{\cdot}80$. Now we can obtain an expression for v as a function of the time t directly by integration: for, referring to p. 202, we know that the derivative of the integral is the integrand. Thus in this case, since the time derivative of v is 9·80, the integral of 9·80 with respect to the time is v, or

$$v = \int 9{\cdot}80\,\mathrm{d}t + c$$
$$= 9{\cdot}80t + u,$$

and instead of the constant of integration c we write u to represent the velocity when $t = 0$. Thus from the differential equation $\mathrm{d}v/\mathrm{d}t = 9{\cdot}80$ we have obtained an equation which tells us how v varies with the time, and which enables the value of v at any instant to be calculated if the value of u is known.

It is important to realise the equivalence of the equations $\mathrm{d}y/\mathrm{d}x = f(x)$ and $y = \int f(x)\,\mathrm{d}x + c$. It will not waste time to demonstrate their equivalence in two slightly different ways:

(i) if we start with the relation $y = \int f(x)\,\mathrm{d}x$, and differentiate it throughout with respect to x, we are attempting to develop the equation

$$\frac{\mathrm{d}}{\mathrm{d}x}(y) = \frac{\mathrm{d}}{\mathrm{d}x}\left\{\int f(x)\,\mathrm{d}x\right\}.$$

Now
$$\frac{\mathrm{d}}{\mathrm{d}x}\left\{\int f(x)\,\mathrm{d}x\right\} = f(x)$$

(Section 8.1), and so we obtain the equivalent equation $\mathrm{d}y/\mathrm{d}x = f(x)$ directly.

(ii) if we start with the equation $\mathrm{d}y/\mathrm{d}x = f(x)$ and integrate both sides with respect to x, we obtain as an expression of our intention the equation

$$\int \frac{\mathrm{d}y}{\mathrm{d}x}\,\mathrm{d}x = \int f(x)\,\mathrm{d}x + c;$$

and rewriting the left-hand side successively in the forms

$$\lim_{\delta x=0}\sum\left(\frac{\mathrm{d}y}{\mathrm{d}x}\right)\delta x \quad\text{or}\quad \lim_{\delta x=0}\sum\left(\lim_{\delta x=0}\frac{\delta y}{\delta x}\right)\delta x$$
$$= \lim_{\delta x=0}\sum\left(\frac{\delta y}{\delta x}\,\delta x\right)$$
$$= \lim_{\delta y=0}\sum(\delta y)$$

(assuming that $\delta y \to 0$ as $\delta x \to 0$, which implies that y must be a continuous function of x), we can see that

$$\lim_{\delta y=0}\sum \delta y = \int f(x)\,\mathrm{d}x + c;$$

or, reverting entirely to integrals,

$$\int 1.\mathrm{d}y = \int f(x)\,\mathrm{d}x + c.$$

Now in general $\int 1.\mathrm{d}y = y + k$, but since we already have one constant of integration (c) in our equation, there will be no need to introduce a second. The final result of integrating the equation $\mathrm{d}y/\mathrm{d}x = f(x)$ with respect to x can therefore be expressed by the equation

$$y = \int f(x)\,\mathrm{d}x + c.$$

The manipulation which leads from $\mathrm{d}y/\mathrm{d}x = f(x)$ to $y = \int f(x)\,\mathrm{d}x + c$ can be summarised in just two steps:

(i) integrate both sides of the equation with respect to x, obtaining

$$\int \frac{\mathrm{d}y}{\mathrm{d}x}\,\mathrm{d}x = \int f(x)\,\mathrm{d}x;$$

(ii) replace the left-hand side by the equivalent term $\int \mathrm{d}y$ or $\int 1.\mathrm{d}y$, and integrate with respect to y immediately, obtaining

$$y = \int f(x)\,\mathrm{d}x + c,$$

where c represents the difference between the constants of integration of the two sides.

This example can be extended further. Velocity is rate of change of displacement with time, i.e. $v = \mathrm{d}s/\mathrm{d}t$. In the example used above we can therefore write $\mathrm{d}s/\mathrm{d}t = 9{\cdot}80t + u$, and again by integration, we can obtain an expression for the displacement s, namely $s = 4{\cdot}90t^2 + ut + c$, where c is the value of s when $t = 0$. These equations for v and s can then be used to determine the change in values of v or s during an interval of any length.

An example in electrical theory arises from the relationship between electrical current and the flow of charge in the circuit: the size of the current i flowing in the circuit is equal to the rate of flow of charge, and in the notation of calculus, this relationship can be written in the form $i = \mathrm{d}q/\mathrm{d}t$, where the variables q and t stand for charge and time respectively. If now the charge accumulates on a capacitor (see Fig. 9.7), the expression for the charge at any time t on that capacitor can be obtained by integrating the expression for the current, for if $\mathrm{d}q/\mathrm{d}t = i$, then $q = \int i\,\mathrm{d}t$.

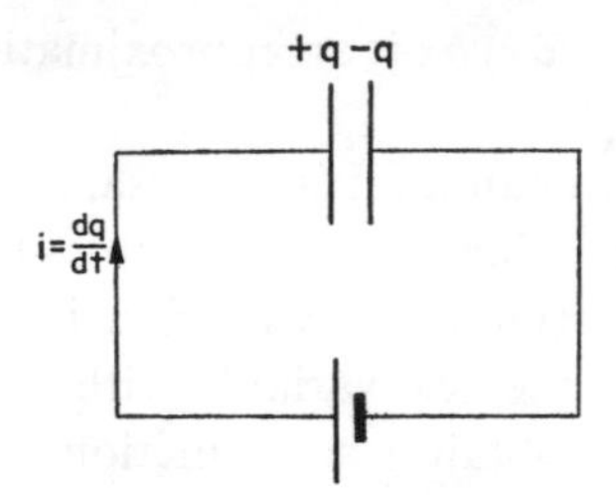

FIG. 9.7. The current i and the charge q in an electrical circuit are related according to the equation $i = \mathrm{d}q/\mathrm{d}t$.

Example 9.3

The velocity v of a body rolling down an inclined plane is given by the equation $v = 6t$, where v is the velocity t seconds after it is released from the top. Derive an expression for its displacement from the top t seconds after the start.

Since $v = \mathrm{d}s/\mathrm{d}t$, where s = the displacement from the top, we can write that $\mathrm{d}s/\mathrm{d}t = 6t$. Integrating this equation with respect to t, we can write that

$$s = \int 6t\,\mathrm{d}t$$
$$= 3t^2 + c,$$

where c is the constant of integration. Now when $t = 0$ (i.e. at the start), $s = 0$, and by substitution into the equation, c must equal 0.

Thus $s = 3t^2$ is the required expression.

Example 9.4

The alternating current i flowing in a circuit containing a capacitor (Fig. 9.8) is given by the equation $i = 2\sin(100\pi t)$. If the charge on the capacitor is zero when $t = 0$ (i.e. the

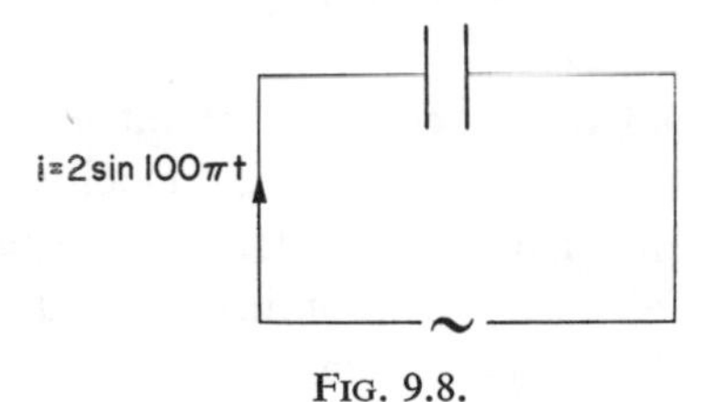

FIG. 9.8.

moment of switching on), derive an expression for the charge on the capacitor t seconds after connecting the supply.

Since $i = 2 \sin(100\pi t)$, we can write that $dq/dt = 2 \sin(100\pi t)$. Integrating the equation with respect to t, we find that

$$q = \int 2 \sin(100\pi t)\, dt$$
$$= -\frac{1}{50\pi}\cos(100\pi t)+c.$$

Now when $t = 0$, $q = 0$ (given), and by substitution into the integrated equation,

$$0 = -\frac{1}{50\pi}+c,$$

since $\cos 0° = 1$.

Thus $c = +1/50\pi$, and the required expression is

$$q = \frac{1}{50\pi}\{1-\cos(100\pi t)\}.$$

Example 9.5

The rate of change of a variable p with respect to a variable q is equal to $20q^2+8q$. By how much does p change when q changes from 1 to 3?

That $\frac{dp}{dq} = 20q^2+8q$ is given:

and integrating with respect to q,

$$p = \int (20q^2+8q)\, dq$$
$$= \frac{20}{3}q^3+4q^2+c.$$

Thus the change in p when q changes from 1 to 3 is equal to

$$\left\{\left(\frac{20}{3}.3^3+4.3^2+c\right)-\left(\frac{20}{3}.1^3+4.1^2+c\right)\right\}$$

$$= 180+36+c-\frac{20}{3}-4-c$$

$$= 205\tfrac{1}{3}.$$

Example 9.6

If $dy/dx = \log_{10} x$, by what does y change when x changes from 1 to 2?

The techniques described so far are insufficient to solve the equation $y = \int \log_{10}x\, dx$, and it will therefore be necessary to resort to a numerical method. Rewriting the equation in the definite form

$$y = \int_1^2 \log_{10}x\, dx,$$

we can use the trapezium rule

$$y = \tfrac{1}{2}h[g_1+2(g_2+g_3+\ldots+g_n)+g_{n+1}],$$

where $g = \log_{10} x$, $h = (2-1)/n$ and n is the number of elements to be taken between $x = 1$ and $x = 2$ (see Fig. 9.9 and p. 183).

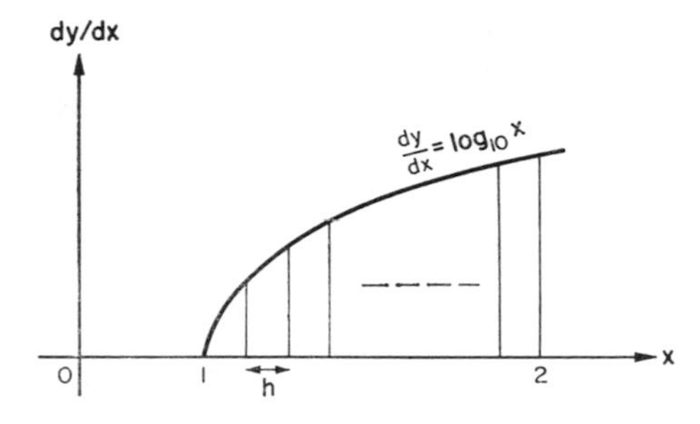

FIG. 9.9. The value of $\int_1^2 \log_{10} x\, dx$ has to be found by determining the area under $dy/dx = \log_{10} x$.

Using five elements, $h = 0{\cdot}2$, and

$$g_2 = \log_{10}1{\cdot}2 = 0{\cdot}079 \quad g_1 = \log_{10}1 = 0{\cdot}000$$
$$g_3 = \log_{10}1{\cdot}4 = 0{\cdot}146$$
$$g_4 = \log_{10}1{\cdot}6 = 0{\cdot}204$$
$$g_5 = \log_{10}1{\cdot}8 = 0{\cdot}256 \quad g_6 = \log_{10}2 = 0{\cdot}301$$

$$\sum_{r=2}^{5} g_r = 0{\cdot}685$$

$$2\sum_{r=2}^{5} g_r = 1{\cdot}37, \quad g_1+g_6 = 0{\cdot}301,$$

and so

$$y = \tfrac{1}{2}(0{\cdot}2)[0{\cdot}301+1{\cdot}37]$$
$$= (0{\cdot}1)(1{\cdot}671)$$
$$= 0{\cdot}167.$$

EXERCISE 9b

***1.** An electron accelerates in a vacuum between two electrodes at a rate of Ve/dM m/s², where $V =$ the potential difference between electrodes, $e =$ the charge on the electron, $d =$ the distance between the electrodes, and $M =$ the mass of the electron. Show that after a time t seconds the electron will be travelling with a velocity Vet/dM m/s, and that it will have travelled $Vet^2/2dM$ m. How long does it take to reach the other electrode? How fast is it then travelling? What is its kinetic energy when it strikes the positive electrode?

2. The rate of change of the area A of a circle with respect to its radius r is equal to the circumference. Write an expression for dA/dr, and derive an expression for A from it.

3. The rate of change of the volume V of a sphere with respect to its radius is $4\pi r^2$, where r is the radius of the sphere. Write an expression for dV/dr, and derive from it an expression for V.

4. Show that the distance travelled in t seconds by a body starting from rest and accelerating at a constant rate a is $\frac{1}{2}at^2$. What would the result be if the body started with a velocity u?

***5.** The electric field E existing in any region is equal to the rate of change of the potential with respect to distance (x) with the sign reversed: $E = -dV/dx$, where V equals the potential at a point. What is the potential difference between two points a cm apart in a region where the field

(i) is constant,
(ii) increases as the first power of the distance from the point of higher potential?

In both cases the field at the point of lower potential is 10.

Solve the following differential equations, given the particular values of y when $x = 0$:

6. (i) $\dfrac{dy}{dx} = 5x$; 0. (ii) $\dfrac{dy}{dx} = 3-8x$; 0.

7. (i) $\dfrac{dy}{dx} = \sin 2x$; 1. (ii) $\dfrac{dy}{dx} = (x-2)(1-x)$; 2.

8. (i) $\dfrac{dy}{dx} = \dfrac{1}{\sqrt{(1-x^2)}}$; 0.

(ii) $\dfrac{dy}{dx} = \sec x \tan x$; 0.

9. (i) $y' = \dfrac{1}{4+x^2}$; 2. (ii) $y' = \dfrac{1}{\sqrt{(9-4x^2)}}$; 5.

10. (i) $y' = \dfrac{x}{\sqrt{(9-x^2)}}$; 3. (ii) $y' = 2x(1+x^2)^7$; 0.

11. (i) $y' = \cos 2x \sin^9 2x$; $\frac{1}{2}$.

(ii) $y' = \sec^2 5x \tan^4 5x$; 2.

12. (i) $y' = x \cos x^2$; 0. (ii) $y' = \sec^4 x \tan x$; 0.

13. (i) $\dfrac{dy}{dx} = 7-5\sqrt{x}$; 0.

(ii) $\dfrac{dy}{dx} = 4 \cos\left(8x+\dfrac{\pi}{6}\right)$; $\frac{1}{2}$.

14. Solve the equations:

(i) $\dfrac{dx}{dt} = \sqrt{3t}$,

(ii) $\dfrac{dx}{dt} = \dfrac{1}{(1+2t)^4}$

given that $x = 0$ when $t = 0$ in both cases.

15. Solve the differential equation $d^2y/dx^2 = -5x$, given that $dy/dx = 1$ and $y = 0$ when $x = 0$.

16. Given that $y' = \frac{1}{2}$ and $y = 1$ when $x = 0$, solve the differential equation $y'' = 2 \sin(4x+\pi/3)$.

17. The velocity of a body t seconds after starting from rest is equal to $0{\cdot}2t$ m/s. Find how far it goes in 10 seconds from its starting-point.

18. Find how far a body starting from rest travels in 2 seconds given that its velocity t seconds after the start is $\frac{1}{2}t^2$.

19. A particle moves along a straight line OX and its velocity, v m/s, t seconds after leaving 0, is given by $v = 4+3t-t^2$.

Calculate

(i) the velocity when $t = 4$,
(ii) at what times the value of v is 21·6 km/h,
(iii) at what time the acceleration is zero,
(iv) the distance travelled on the 3rd second. (SU)

20. A particle moves in a straight line so that its acceleration time t seconds from the beginning of the motion is $6(4-t)$ m/s². Find the velocity with which the particle must begin to move if it is to describe 94 m in the first 2 seconds of its motion. Show that in this case the particle will come to instantaneous rest after a further 7 seconds and find how far the particle will then be from its starting-point. (NUJMB)

21. At all points on a certain curve it is known that $d^2y/dx^2 = 6x-2$. If $dy/dx = -1$ and $y = 1$ when $x = 0$, prove by integration that the equation of the curve is $y = x^3-x^2-x+1$.

Find the stationary points on the curve and determine, for each point, whether y has a maximum or a minimum value. (OC)

22. The slope of a curve at any point (x, y) is given by $dy/dx = 2x-1$. The curve passes through the point (4, 6); find the equation of the tangent at that point.

Find also the equation of the curve, and the coordinates of the point at which the tangent is parallel to the x-axis. (OC)

23. A particle moves along a straight line so that if a m/s² is the acceleration at time t seconds, then

$da/dt = 12$; at the end of 1 second the distance from the origin is 4 m, the velocity is 4 m/s and the acceleration is 4 m/s^2. Find the distance of the particle from the origin, its velocity and acceleration when $t = 2$. Show also that, if v m/s is the velocity at 5 seconds, then $24v = a^2+80$. (NUJMB)

24. Solve the equation $d(v^2)/dx = -2x$ for v^2, given that $v = 4$ when $x = 0$.

25. For a certain curve $d^2y/dx^2 = 6x-4$ and y has a minimum value 5 when $x = 1$. Find the equation of the curve and the maximum value of y. (L)

26. A rocket is launched with an initial speed of 100 m/s. For the first 16 seconds its acceleration, t seconds after launching, is $(75+10t-t^2)$ m per second per second. After 16 seconds, it travels with constant speed.

For this initial 16 seconds, calculate (a) the maximum acceleration, (b) the time taken to attain the greatest speed, (c) the greatest speed attained. (C)

27. A conical vessel full of liquid is held vertex downwards, and with the base horizontal, and at a given instant the tip is pierced so that liquid runs out. It is noted that the rate at which the volume of liquid in the vessel decreases is proportional to the depth x of liquid in the vessel. Prove that $x(dx/dt) = -k$, where k is a positive constant.

If the original depth of liquid in the vessel was h, and if time T was required to empty the vessel, prove that at time $\frac{1}{2}T$ the depth of liquid was $\frac{1}{2}h\sqrt{2}$.

Find at what time the depth of liquid was $\frac{1}{2}h$. (C)

28. At any point on a curve the product of the slope of the curve and the square of the abscissa of the point is 2. If the curve passes through the point $x = 1$, $y = -1$, find its equation. (L)

29. The gradient at any point (x, y) on a certain curve is $-9/x^2$. If the curve passes through the point (3, 6), find its equation, and find also the equation of the tangent and normal at the point (3, 6).

***30.** A particle of mass m experiences an accelerating force which is proportional to its displacement from a fixed point, and directed away from the fixed point. Write down an equation for the second-order derivative of the displacement of the particle from the fixed point, given that the constant of proportionality between the distance and the force acting on the particle is k.

***31.** A particle of mass m experiences a force of mn^2x, where n is a constant and x is the distance of the particle from a fixed point. The direction of the force is towards the fixed point. Write down an equation of motion for the particle, and show that it is satisfied by an expression of the form $x = a \sin nt + b \cos nt$, where a and b are constants.

***32.** A particle of mass m experiences a force of n^2x, where x is the distance of the particle from a fixed point, and n is a constant. If the force is directed towards the fixed point, and the particle also experiences a retarding force due to friction of magnitude k times its velocity, write down an equation representing the motion of the particle.

33. The equation of motion of a particle is $(d^2x/dt^2)+2x\,(dx/dt)-7x = 0$, where x is the displacement of the particle from a fixed point, and t is the time. Show that its maximum speed is 3·5.

***34.** The heat conducted through a cross-section of a uniform circular bar in one second is given by the expression $-kA(d\theta/dx)$, where k is a constant, A is the cross-sectional area of the bar, and $d\theta/dx$ the temperature gradient at the point whose distance is x from one end, θ being the temperature at the point. If the bar is lagged, the heat conducted through any cross-section along the bar is constant, and equal to that conducted out of the cold end. Denoting this by Q, show that the temperature varies along the length according to the equation $\theta = \theta_0-(Q/kA)x$, where θ_0 is the temperature at the hot end, and where x is measured from the hot end.

35. A body is moving in a straight line and t seconds after passing a point A in the line its velocity is $(3t^2+4)$ m/s. Find the distance moved by the body while its velocity increases from 31 to 112 m/s. Calculate the acceleration of the body when it has moved 5 m from A. (L)

†**36.** Solve the following differential equations, given that y takes the value indicated after the equation when $x = 0$:

(i) $$\left(\frac{dy}{dx}\right)^2 = \frac{1}{25-9x^2}\,;\ 1.$$

(ii) $$\frac{dy}{dx} = \sin x(1+\cos x)^8;\ 0.$$

(iii) $$y' = \frac{\sin x}{(1+\cos x)^2}\,;\ 0.$$

(iv) $$y' = \frac{\sin x \cos x}{(1+\cos^2 x)^4}\,;\ 0.$$

***37.** An electron travelling with a velocity of 4×10^7 m/s enters an electric field of strength 30 V/m. If the field is perpendicular to the initial direction of motion, find an expression for the velocity of the electron in the direction of the field t seconds after it enters the field, given that the charge on the electron $e = 1{\cdot}6\times10^{-19}$ C, and its mass $m = 9{\cdot}1\times10^{-31}$ kg.

Hence deduce an expression for the distance travelled by the particle in the direction of the field t seconds after entering it, and by combining this with an expression for the distance travelled by the electron in a direction perpendicular to the field, find an equation for the path of the particle in the field.

***38.** Show that

$$\text{(i)} \int P\,ds = \int P\frac{ds}{dv}\,dv, \quad \text{(ii)} \ \frac{dv}{dt} = v\frac{dv}{ds}.$$

Hence show that the work done by the constant force P acting on a body accelerating from rest is

$\frac{1}{2}mv^2$, where v is the final velocity of the body. (The work done by a force P moving along its line of action is $\int P.ds$, and by Newton's Second Law the acceleration a of a body of mass m is related to the force producing the acceleration by the equation $P = ma$.)

†**39.** If $y = vx$, where v is a function of x, find an expression for dy/dx. Using this expression, convert the differential equation $x(dy/dx) - y = x^3$ into a differential equation relating v with x. Solve the equation by integration, and find y in terms of x.

†**40.** Solve the equation $x(dy/dx) - y = x^3\sqrt{(1+x^2)}$ using a method similar to that suggested in question 39.

†**41.** Given that l, w, E, z and I are constants, and that y and dy/dx are zero when $x = 1$, solve the equation $EI(d^2y/dx^2) = (3wlx)/z - wx^2/z$ for y.

For what value of x is y a maximum?

†**42.** A gramophone turntable is rotating at $33\frac{1}{3}$ r.p.m. On switching off it slows down at a rate of 40 degrees/s^2. Write down a differential equation for the angular velocity of the turntable t seconds after switching off, solve it, and state the range of time values for which your solution is correct. Through what angle does the table turn while coming to rest?

***43.** The e.m.f. applied to an inductance of size L must be equal to $L\,dI/dt$, where I is the current flowing in the circuit and t is the time. Find an expression for I if an e.m.f. $E_0 \cos 2\pi ft$ is applied to the inductance, and $I = 0$ when $t = 0$. (f may be taken as a constant equal to the frequency of alternation of the applied e.m.f.) Discuss what is meant by the statements

(i) the current lags behind the voltage,
(ii) the impedance of an inductance is proportional to the frequency of the applied e.m.f.

9.3. MEAN VALUES

The speed of a car travelling between two towns A and B will not have a constant value due to the presence of other traffic and the nature of the road. The average speed which the car maintains between A and B is often a more important piece of knowledge than that of the precise form of the fluctuations in speed during the journey. Similarly, the average value of the current flowing in a circuit carrying alternating current is as important a statistic as those describing the way in which the current fluctuates; and the average power produced by a car engine is often just as important as the manner in which the engine produces the power.

Pursuing this example of engine power further, we can devise as follows a method of calculating the average or mean value of the fluctuations if the form of them is known. Let us suppose that the power delivered by the engine fluctuates in the simple manner shown in Fig. 9.10. Each vertical rectangle or boost of power is produced by one cylinder when the fuel in

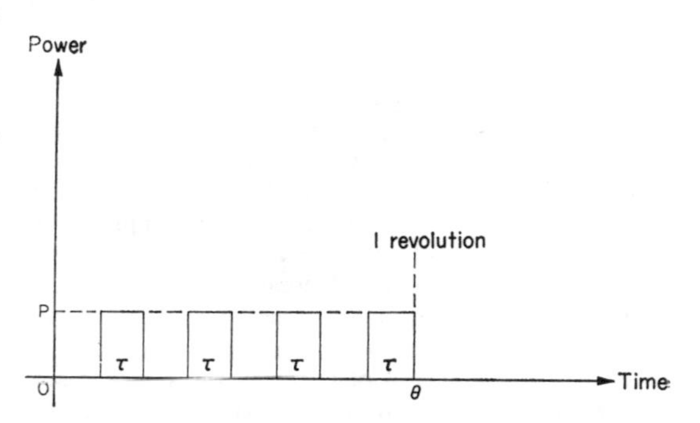

FIG. 9.10. The power developed by an engine comes in bursts.

it burns or explodes. Now power is defined as the rate of production of work, and thus the product of a steady power and the time for which it is delivered is equal to the work delivered in that interval. This quantity is equal to the integral of the power with respect to time over the interval in question, which in turn is numerically equal to the area under the power-time graph if a unity scale graph is used. The total area of the four rectangles which are produced in one revolution in Fig. 9.10 thus gives the total work produced per engine revolution. Since the mean power during one revolution is sensibly defined as the total work produced during that revolution divided by the time taken for the revolution, we can write that this mean power is given by the fraction $4P\tau/\theta$, where τ is the duration of each power burst, θ the time taken for one complete revolution of the engine, and P that power developed during the firing of each cylinder.

If instead of a fluctuating supply of power we have a steady supply whose magnitude equals $4P\tau/\theta$, the total work supplied in the time θ would be the same (Fig. 9.11). Expressed in graphical terms, this means that the

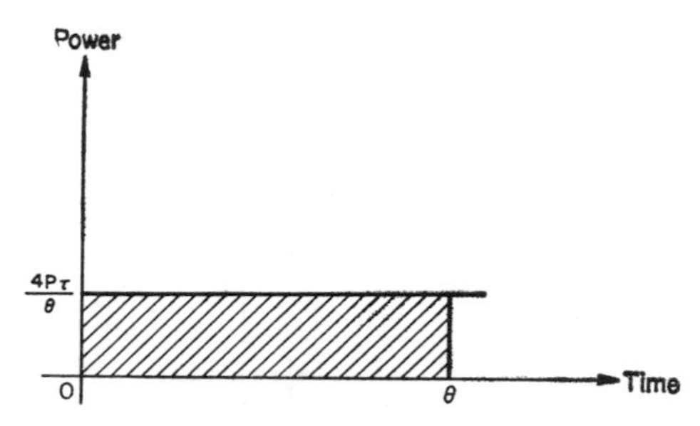

FIG. 9.11. The total work delivered in a time θ by a steady supply of power of magnitude $4P\tau/\theta$ is the area under the graph, and this is equal to $4P\tau$.

area of the rectangle formed when the mean power is delivered for a time θ is equal to the area under the graph of the fluctuating power supply for the same interval. We can thus define the mean power as that constant power which contains the same area under its graph as the fluctuating power over the same time interval (Fig. 9.12).

Translating these ideas into terms of integrals, we can say that for a fluctuating power supply, the total work produced during an interval of time from θ_1 to θ_2 is equal to

$$\int_{\theta_1}^{\theta_2} P\,\mathrm{d}t,$$

where P is the power developed and is a function of the time t, and the mean power that this represents is that steady power which will deliver the same amount of work in the same interval. Denoting the mean power by $\bar{P}$, this latter quantity is given by the expression

$$\int_{\theta_1}^{\theta_2} \bar{P}\,\mathrm{d}t,$$

and since $\bar{P}$ is by definition constant, it can be taken outside the integral sign, giving

$$\bar{P}\int_{\theta_1}^{\theta_2} 1.\mathrm{d}t \quad \text{or} \quad \bar{P}(\theta_2-\theta_1).$$

Thus

$$\bar{P} = \frac{\int_{\theta_1}^{\theta_2} P\,\mathrm{d}t}{\theta_2-\theta_1},$$

and $\bar{P}$ is equal to the height of the horizontal line above the time axis which contains the same area between two given ordinates as the graph of the fluctuating variable.

Generalising the results, we can say that the mean value $\bar{y}$ of a function $f(x)$ over interval x_1 to x_2 is given by the equation

$$\bar{y} = \frac{\int_{x_1}^{x_2} f(x)\,\mathrm{d}x}{x_2-x_1}, \tag{9.1}$$

and the result can be expressed graphically as in Fig. 9.13.

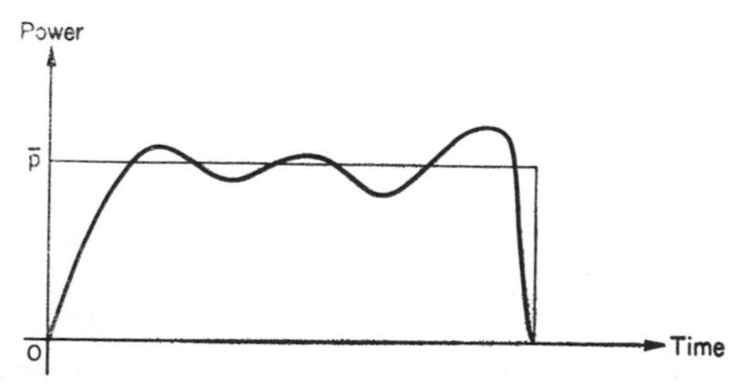

FIG. 9.12. The mean power delivered by a fluctuating supply is the unvarying power which contains the same area under its graph as that contained under the graph of the fluctuating supply for the same time interval.

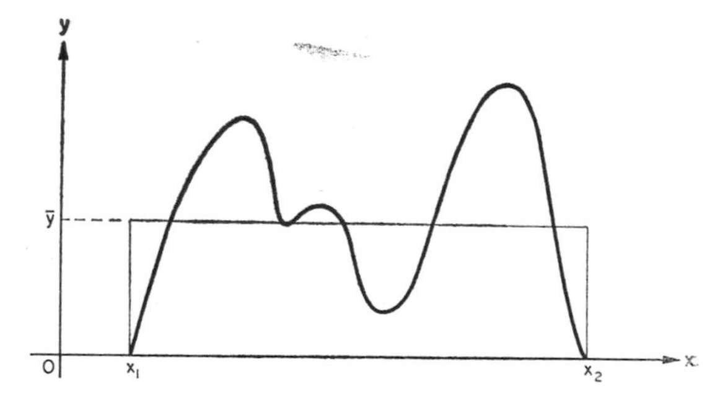

FIG. 9.13. The mean value of y ($\bar{y}$) between x_1 and x_2 is such that the area under the curve of y between x_1 and x_2 equals the area of the rectangle $\bar{y}(x_1-x_2)$.

Example 9.7

Find the mean value of the function x^2 between $x = 0$ and $x = 3$.

The graph of the function is shown in Fig. 9.14, and the area under the curve between $x = 0$ and $x = 3$ is equal to

$$\int_0^3 x^2\,dx \quad \text{or} \quad [\tfrac{1}{3}x^3]_0^3 \quad \text{or} \quad 9.$$

This must equal the area of the rectangle of height $\bar{y}$ constructed on the same base, where $\bar{y}$ is the mean value of the function over the interval, and thus $3\bar{y} = 9$. The mean value $\bar{y}$ of the function x^2 for the interval from $x = 0$ to $x = 3$ is thus 3.

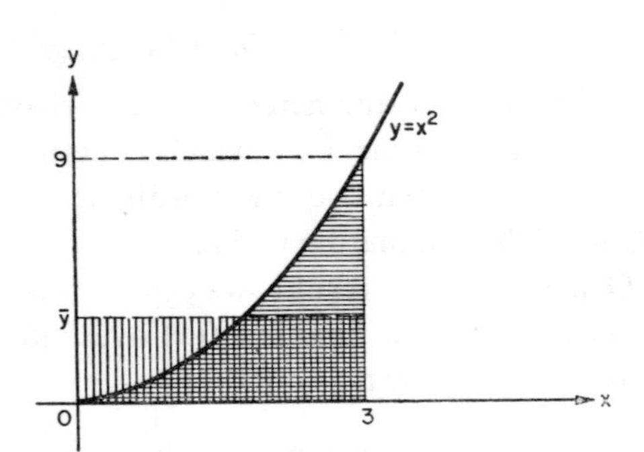

FIG. 9.14.

(This can also be calculated analytically, without reference to a graph of the function, from equation 9.1:

$$\bar{y} = \int_0^3 f(x)\,dx/(x_2 - x_1) = \frac{\int_0^3 x^2\,dx}{(3-0)}$$

$$= \frac{[\frac{1}{3}x^3]_0^3}{3}$$

$$= \tfrac{9}{3} = 3).$$

EXERCISE 9c

1. Find the mean values of the functions whose graphs are shown in Fig. 9.15 over the interval from $t = 0$ to $t = 5$.

2. Find the mean values of the functions

(i) $5x$, (ii) $5x+7$, (iii) x^3,

in the interval $0 \leqslant x \leqslant 4$.

3. Find the mean values of

(i) $\sin\theta$,

(ii) $\cos\theta$,

(iii) $\cos^2\theta$,

(iv) $\sin 2\theta$,

(v) $\sin\theta\sin 2\theta$,

in the interval $0 \leqslant \theta \leqslant \pi/2$.

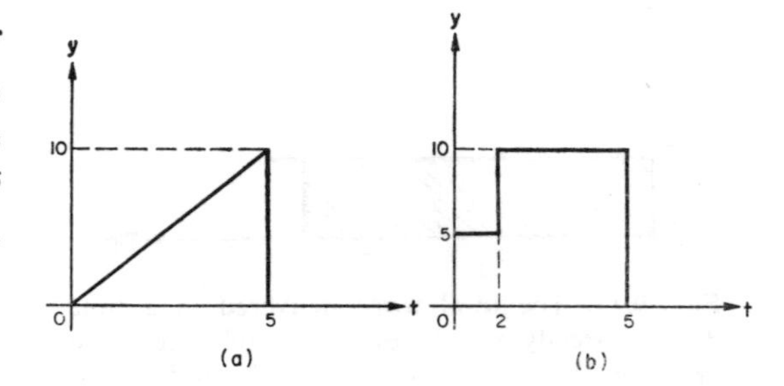

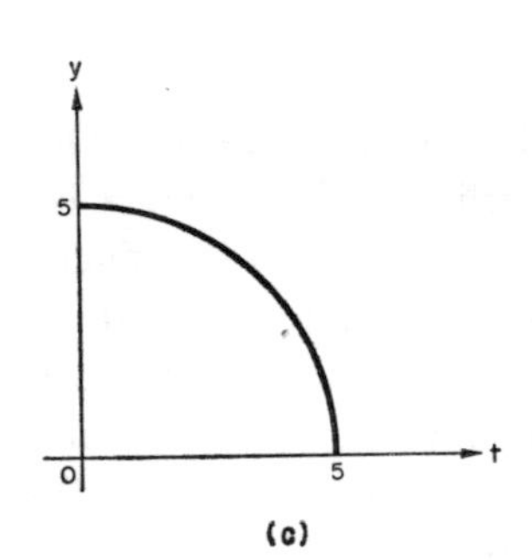

FIG. 9.15.

4. Find the mean values of the following functions in the ranges indicated:

(i) $(1+5x)^4$, $0 \leqslant x \leqslant 1/5$;

(ii) $x(1+x^2)^3$, $0 \leqslant x \leqslant 1$;

(iii) $\dfrac{1}{\sqrt{(4-9x^2)}}$, $0 \leqslant x \leqslant 1/3$;

(iv) $\dfrac{9}{4-9x^2}$, $0 \leqslant x \leqslant 1/3$;

(v) $\dfrac{2}{4+25x^2}$, $0 \leqslant x \leqslant 2/5$.

5. Find the mean values of the following functions in the ranges indicated:

(i) $(1-t)$, $-3 \leqslant t \leqslant -1$;

(ii) $7-3t^2$, $-2 \leqslant t \leqslant -1$;

(iii) $t\sqrt{(1-t^2)}$, $-1 \leqslant t \leqslant 0$.

6. The speed of a locomotive varies according to the equation $v = t(20-t)$ where $0 \leqslant t \leqslant 20$, and where the speed v is in km/h and the time t in minutes after the start of the motion. Sketch the graph of this variation, and find the mean velocity for the journey. How far does the locomotive travel between its stops?

7. The torque produced by an engine varies regularly every two seconds according to the following table:

Torque	10	20	25	20	50	40	20	15	10
Time	0	0·25	0·5	0·75	1·0	1·25	1·5	1·75	2

What is the mean torque delivered?

***8.** The power delivered in an electrical circuit varies according to $E_0^2/Z \sin wt \sin (wt-\varepsilon)$, where E_0, Z, w and ε are constants. What is the mean power delivered? (The total work delivered in one cycle should be calculated.)

9. If the Bank Rate changes according to the following table during a non-leap year, what will be the average rate for the year?

Date	1st Jan.	4th Apr.	1st June	1st Nov.
Bank Rate	4%	6%	7%	5%

10. The speed of a car varies as follows. Starting from rest, the car accelerates uniformly to 100 km/h in $\frac{1}{2}$ minute and then travels at this speed for $2\frac{1}{2}$ minutes. It then retards uniformly to 40 km/h for $\frac{1}{2}$ minute, and travels at 40 km/h for another $\frac{1}{2}$ minute. It accelerates again to 120 km/h, taking $\frac{1}{2}$ minute to reach this speed, and after travelling for $\frac{1}{2}$ minute at 120 km/h, it slows uniformly to a halt taking 1 minute to stop. What is its average speed for the journey, and how far does it travel?

11. Why is it not true to say that if a journey from A to B is covered at an average speed of 60 km/h and the return journey at an average speed of 40 km/h, the average speed for both journeys is 50 km/h?

12. Estimate the mean value of $9-x^2$ over the range $0 \leqslant x \leqslant 2$.

13. Estimate the mean value of the function f of t in the range $0 \leqslant t \leqslant 10$, given that $f(0) = 2, f(2) = 6, f(4) = 5, f(6) = 2, f(8) = 3, f(10) = 5$.

14. Two variables, x and y, are related so that corresponding pairs of values are as follows:

x	-4	-2	0	2	4
y	1	2	-2	-1	3

Estimate the mean value of y over the range $-4 \leqslant x \leqslant 4$.

***15.** A direct current meter gives a deflection θ proportional to the current i flowing through it. (Thus $\theta = ki$, where k is a constant.) If a fluctuating current of the form $i = i_0 \sin wt$ passes through it, what is the average value of the deflection over (i) half a cycle; (ii) the whole cycle?

An alternating current meter gives a deflection which is proportional to the square of the current (i.e. $\theta = ki^2$). What average deflection does this meter show?

***16.** The displacement x of a particle undergoing simple harmonic motion varies with time according to the equation $x = a \sin wt$, where a and w are constants. Show that its velocity at any instant is $wa \cos wt$, and find the average value of its

(i) velocity, (ii) speed, (iii) kinetic energy

(its mass is m) during one cycle.

The force acting on the particle at any moment is mf, where f is its acceleration. Show that this can be rewritten as $-mw^2x$, and show that the potential energy of the body relative to the centre of oscillation when at a distance x from the centre of oscillation is $-\frac{1}{2}mw^2x^2$.

Hence show that the average value of the potential energy over a cycle is the same as that of the kinetic energy.

†**17.** The graph of Fig. 9.16 describes the result of an experiment which measured the length of a large number of safety matches. The ordinate f of a

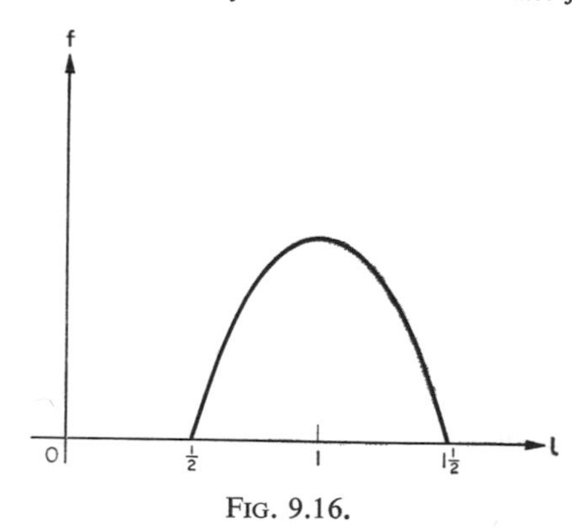

FIG. 9.16.

point on the curve describes the number of matches in the sample with a length equal to the abscissa l of the point. The mean length of the matches in the sample is given by the equation

$$\bar{l} = \frac{\int l f(l)\, dl}{\int f(l)\, dl},$$

where l is the match length, and $f(l)$ the expression describing the variation in the frequency as a function of length, and where the integral is to be taken over the complete range of match lengths.

If $f(l) \equiv 100(8l-3-4l^2)$, find the mean length of the matches.

†**18.** Show that the mean value of $\cos\theta$ is 0 for the range $0 \leqslant \theta \leqslant \pi$ by means of symmetry, and hence that the mean values of $\sin^2\theta$ and $\cos^2\theta$ over the range $0 \leqslant \theta \leqslant \pi/2$ are $\frac{1}{2}$. Deduce that the mean values of $\sin^2\theta$ and $\cos^2\theta$ over the range $0 \leqslant \theta \leqslant \pi$ are also $\frac{1}{2}$.

†**19.** A step function $s(x)$ is defined as follows:

$s(x) = 1$ for $-2 \leqslant x \leqslant 0$, $s(x) = 2$ for $0 < x < 1$.

$s(x) = -1$ for $1 \leqslant x \leqslant 5$.

What is the mean value of $s(x)$ over the range

(i) $-2 \leqslant x \leqslant 1$, (ii) $-2 \leqslant x \leqslant 5$,

(iii) $0 \leqslant x \leqslant 4$?

†**20.** Find the mean value of $\log_{10} x$ between $x = 1$ and $x = 6$.

***21.** Find the average value of a rectified alternating current whose variation is expressed by the equation $I = 50 \sin(100\pi t)$, where t is the time, and assuming that the rectification is (i) half wave, (ii) full wave.

What is its root mean square value?

***22.** A plane coil of 50 turns of mean area 0·025 m² is fitted with a commutator and rotates at 2000 rev/min about a diameter lying across a uniform magnetic field of flux density 5×10^{-3} T. The resistance of the coil is 4 ohms, and a load of resistance 6 ohms is connected across it. Find

(i) the mean e.m.f. generated,
(ii) the mean current flowing,
(iii) the mean power delivered to the load.

How could this latter quantity be increased?

9.4. AREAS

It is often important to know the value of the surface area of a body, and the area of a lamina or sheet with regular edges can be calculated quite easily by determining the area under a unity scale graph with the same boundaries as the lamina.

Occasionally the lamina whose area is to be found is defined by the y-axis and a curve whose equation is written explicitly for x (Fig. 9.17). We can proceed in these cases as follows. Construct an element of width δy parallel to the x-axis and contained in the required area (Fig. 9.18). The area of the element lies between $x\,\delta y$ and $(x+\delta x)\,\delta y$, where x and $x+\delta x$ are the abscissae of the points whose ordinates

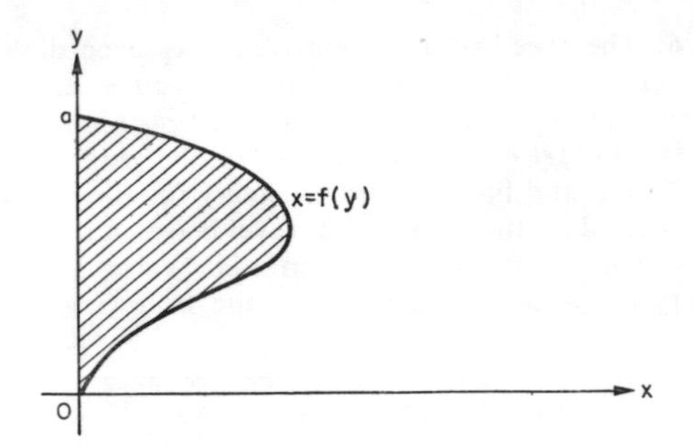

FIG. 9.17. The area of a lamina defined by the curve $y = f(x)$ and the y-axis can be found by integration.

are y and $y+\delta y$. Then the required area A can be written in terms of the sum of a number of these elements, or specifically

$$\sum x\,\delta y \leqslant A \leqslant \sum (x+\delta x)\,\delta y.$$

Taking the limits as $\delta y \to 0$ we obtain

$$\lim_{\delta y=0} \sum x\,\delta y = A,$$

or, more usefully,

$$\int x\,dy = A,$$

where the range over which the integration is to be carried out is that defined by the boundaries of the shaded area ($y = 0$ and $y = a$ in Fig. 9.17: and the limits of integration are the values of y at these boundaries, since the integration is to be carried out with respect to y).

The variable x cannot be integrated directly with respect to y: the problem of changing the variable of integration will be dealt with later, and for the moment we will consider only the

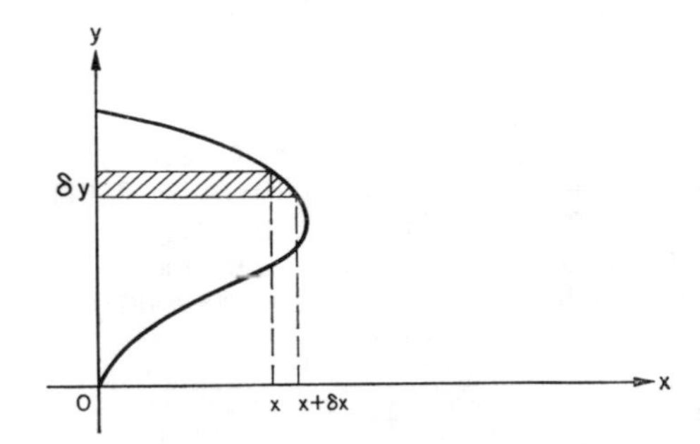

FIG. 9.18. The area of the shaded element lies between $x\,\delta y$ and $(x+\delta x)\,\delta y$.

cases where the integrand x can be replaced by a function of y: the integration can be carried out immediately x has been replaced by $f(y)$ as demonstrated in Example 9.9.

The evaluation of the area of a lamina bounded by two curves need not be any more complicated. If the upper boundary of the lamina in Fig. 9.19 is described by the equation $y = f(x)$, the area contained by this curve, the x-axis and the ordinate $x = a$ is

$$\int_0^a f(x)\,\mathrm{d}x.$$

Denoting the value of this by A, we can write the value of the area under $y = g(x)$ between the same two limits as

$$B = \int_0^a g(x)\,\mathrm{d}x,$$

and the area enclosed between the two curves can be written as

$$A - B = \int_0^a f(x)\,\mathrm{d}x - \int_0^a g(x)\,\mathrm{d}x,$$

or

$$\int_0^a \{f(x) - g(x)\}\,\mathrm{d}x$$

by direct subtraction (Fig. 9.20).

Many other examples can be solved by geometrical considerations (see Example 9.11).

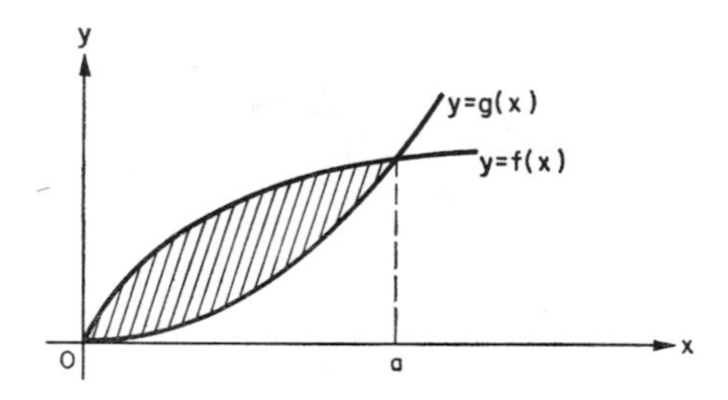

FIG. 9.19. The area of a lamina bounded by two curves can also be determined.

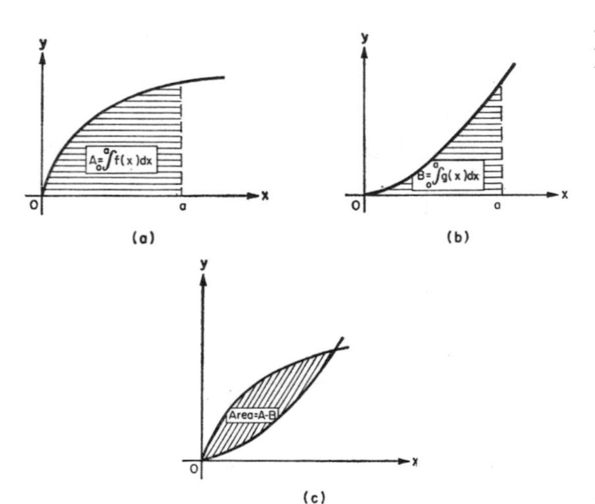

FIG. 9.20. The area between $y = f(x)$ and $y = g(x)$ is given by $\int_0^a \{f(x) - g(x)\}\,\mathrm{d}x$, assuming that the curves meet at the origin and at $x = a$, and that $f(x) > g(x)$ for all x between 0 and a.

Example 9.8

The x-axis and the curve $y = x(4-x)$ for $0 \leqslant x \leqslant 4$ describe the shape of a lamina. If y and x are measured in centimetres, calculate the area of the lamina.

A sketch of the curve is shown in Fig. 9.21, and the area is easily seen to equal $\int_0^4 x(4-x)\,\mathrm{d}x$. Integrating, the area A is found to be given by the equation

$$A = \left[2x^2 - \frac{x^3}{3}\right]_0^4 = \left(2\cdot 16 - \frac{64}{3}\right) - (0)$$

$$= 32 - \frac{64}{3}$$

$$= \frac{32}{3}\ \mathrm{cm}^2.$$

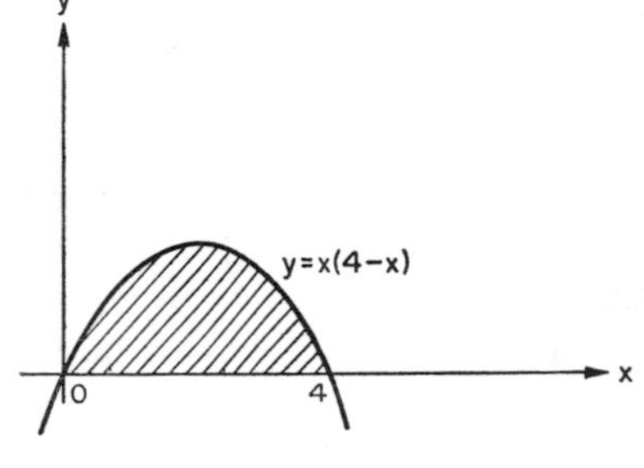

FIG. 9.21.

Example 9.9

Find the area enclosed by the y-axis, the curve $y = \sin^{-1} x$, and the lines $x = 0$ and $y = \pi/6$.

The required area is shaded in Fig. 9.22, and its value can be written as

$$\lim_{\delta y=0} \sum_{0}^{\frac{1}{6}\pi} x\,\delta y,$$

or

$$\int_0^{\pi/6} x\,\mathrm{d}y \quad \text{(the lower limit is 0 since } y = 0 \text{ when } x = 0).$$

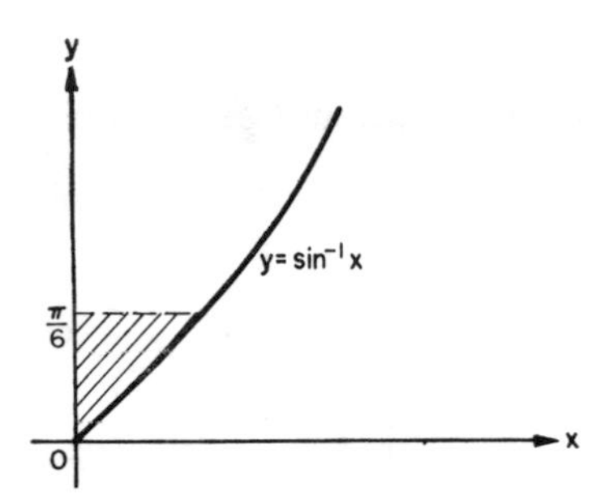

FIG. 9.22.

Now if $y = \sin^{-1} x$, $x = \sin y$, so

$$A = \int_0^{\pi/6} \sin y\,\mathrm{d}y.$$

Integrating,

$$\begin{aligned} A &= [-\cos y]_0^{\pi/6} \\ &= \tfrac{1}{2}\sqrt{3}-(-1) \\ &= 1 - \tfrac{1}{2}\sqrt{3} \\ &= 0{\cdot}134 \end{aligned}$$

Example 9.10

Find the area enclosed between the curves $y = x^2$ and $y = \sqrt{x}$.

The required area is shaded in Fig. 9.23: the two curves intersect at $x = 0$ and $x = 1$. The area under the curve $y = x^2$ between $x = 0$ and the intersection can be written as

$$\int_0^1 x^2\,\mathrm{d}x,$$

and that under $y = \sqrt{x}$ between the same two limits as

$$\int_0^1 \sqrt{x}\,\mathrm{d}x.$$

Thus the area enclosed between the curves is equal to*

$$\int_0^1 (\sqrt{x}-x^2)\,\mathrm{d}x,$$

and integrating, this equals $[\frac{2}{3}x^{3/2} - \frac{1}{3}x^3]_0^1$,

i.e. $(\frac{2}{3} - \frac{1}{3})$

or $\frac{1}{3}$.

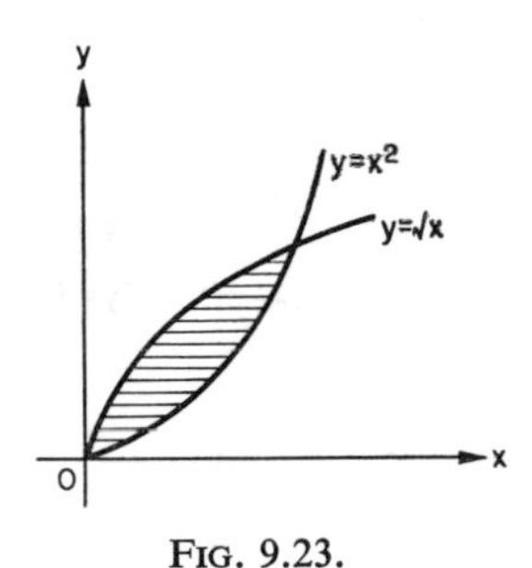

FIG. 9.23.

Example 9.11

Find the area enclosed by $y = 2$ and the portion of the curve, $y = 2+\sin^2 x$ which lies between $x = 0$ and $x = \pi$.

The shaded area of Fig. 9.24 can be seen to

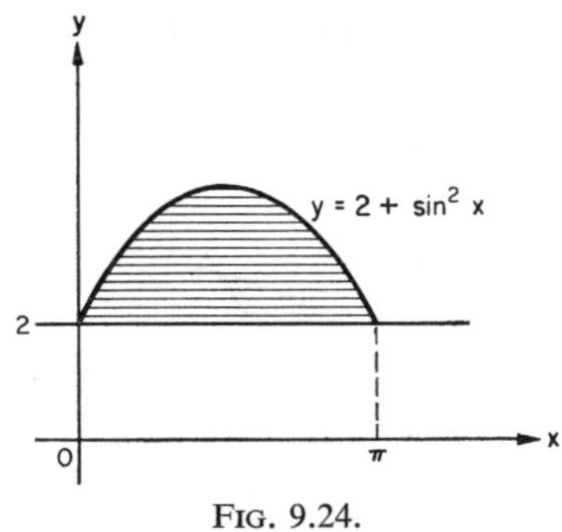

FIG. 9.24.

* The functions are subtracted in the order $(\sqrt{x}-x^2)$ since it can be seen that the area under $y = \sqrt{x}$ between $x = 0$ and $x = 1$ is greater than that under $y = x^2$ between the same two limits.

be that under the curve $y = 2+\sin^2 x$ less the rectangle bounded by $x = 0$, $y = 2$, $x = \pi$, and $y = 0$. Thus the required area can be written as

$$\int_0^{\pi} (2+\sin^2 x)\,dx - \pi.2$$

(since the area of the rectangle equals $\pi.2$), and this can be evaluated as $[2x+\frac{1}{2}x-\frac{1}{4}\sin 2x]_0^{\pi}-2\pi$ (using the substitution $\sin^2 x = \frac{1}{2}(1-\cos 2x)$ before integrating)

$$= (\tfrac{5}{2}\pi - \tfrac{1}{2}.0)-(0)-2\pi$$
$$= \tfrac{1}{2}\pi.$$

EXERCISE 9d

Find the area enclosed by the following curves and lines:

1. $y = x^2$, $y = 0$, $x = 2$.

2. $y = \sqrt{x}$, $y = 0$, $x = 4$.

3. $y = x(2-x)$, $y = 0$.

4. $y = x^2$, $x = 0$, $y = 4$.

5. $y = x^{3/2}$, $x = 0$, $y = 8$.

6. $y = \cos^{-1} x$, $x = 0$, $y = \pi/3$.

7. $2y(1+y^2)^3 = x$, $x = 0$, $y = 1$.

8. $y = \sqrt{(2-x)}$, $x = 0$.

9. $y\sqrt{(1-y^2)} = x$, $y = 0$, $y = 1$.

10. $y = x \sin x^2$, $y = 0$, $x = \frac{1}{2}\sqrt{\pi}$.

11. $y = \dfrac{25x}{(2+3x^2)^2}$, $y = 0$, $x = 1$.

12. $2y = \sin^{-1} 3x$, $x = 0$, $y = \frac{1}{4}\pi$.

13. $y = \sin x \cos^2 x$, $y = 0$, $x = 0$.

†14. $y=2-2\sin^2 x$, $y = 0$, $x = 0$.

Find the area enclosed by the following pairs of curves:

15. $y = x^2$, $y = x$.

16. $y = x^3$, $y = x^4$.

17. $y = x(4-x)$, $y = x^2$.

†18. $y = 8/x^2$, $y = x$ and $y = 8x$ $(x > 0)$.

†19. $y = 1$, $y = \sqrt{(2-1/x^2)}$, $x = 1$, $y = 0$.

†20. $y = 1$, $y = 2\sin^2 x$, $y = 2$.

21. Find the area enclosed by the curve $4x^2+y-12 = 0$ and the x-axis by finding the limiting value of the sum of the areas of small elements parallel to the (i) x-axis, (ii) y-axis.

22. Find the total area enclosed between

(i) $y = x(x-3)$ and $y = x(3-x)$,
(ii) $y = x^2$, $4y = x^3$.

23. Find the area enclosed between the curves $y = x^2+3x-4$ and $y = 8+x-x^2$.

24. Find the area enclosed between $y = x$ and $y = \frac{1}{2}x^2$, and find the areas of the two parts into which this area is divided by $x = 1$.

25. A square is formed by the two axes of coordinates and the perpendiculars drawn to them from the point (7, 7). Sketch, preferably on squared paper, that part of the curve

$$10y = x^2+5x-14$$

which lies within the square.

Calculate the areas of the two parts into which the square is divided by the curve. (OC)

26. Determine the areas of the two parts into which the square in question 25 is divided by the curve by counting the squares of graph paper lying within the two parts.

27. Plot the points on the curve $y = 3x^2-x^3$ for which $x = -1, 0, 1, 2, 3$ and sketch this portion of the curve. Calculate the gradients of the curve at the point $(-1, 4)$ and find, also, at which point on this portion of the curve the gradient has its greatest positive value.

Find the area bounded by the curve and the part of the x axis from $x = 0$ to $x = 3$. (OC)

28. The gradient of a curve which passes through the point $x = 3$, $y = 1$ is given by $dy/dx = x^2-4x+3$.

Find the equation of the curve and the area enclosed by the curve, the maximum and minimum ordinates and the x-axis. (L)

29. A curve whose equation is of the form $y = ax^2+bx+c$, where a, b and c are constants, passes through the point (3, 18) and crosses the y-axis at right angles at the point (0, 24). Find the equation of the curve and show that the total area in the first quadrant between the curve and the tangent to the curve at the point (3, 18) is $16\frac{1}{2}$ square units. (NUJMB)

30. Find the coordinates of the points of inflexion on the graph of $y = 1/(1+x^2)$.

Prove that the area bounded by the curve and the normals at its point of inflexion is

$$\frac{\pi}{3}-\frac{11\sqrt{3}}{54}. \qquad \text{(SU)}$$

31. Prove that there are points of inflexion on the curve $y = 1/(3+x^2)$ at the points where $x = \pm 1$.

Obtain the equations of the normals to the curve at these points, and prove that they meet at $(0, -7\frac{3}{4})$.

Prove also that the area bounded by these normals and the part of the curve between the points of inflexion is

$$\frac{\pi}{3\sqrt{3}}+\frac{15}{2}. \quad \text{(SU)}$$

32. By considering the graph of $1-\cos 2x$ for values of x between 0 and $\pi/2$, deduce that the value of

$$\int_0^{\pi/2} (1-\cos 2x)\, dx \text{ is } \pi/2.$$

Hence deduce the average value of $\sin^2 x$ for values of x between 0 and 2π.

33. Describe with reference to Fig. 9.25 the values of the following integrals:

$$\text{(i) } \int_0^a f(x)\, dx, \quad \text{(iv) } \int_0^c |f(x)|\, dx,$$

$$\text{(ii) } \int_0^b f(x)\, dx, \quad \text{(v) } \left|\int_a^b f(x)\, dx\right|.$$

$$\text{(iii) } \int_0^c f(x)\, dx.$$

Express your answers in terms of P, Q and R.

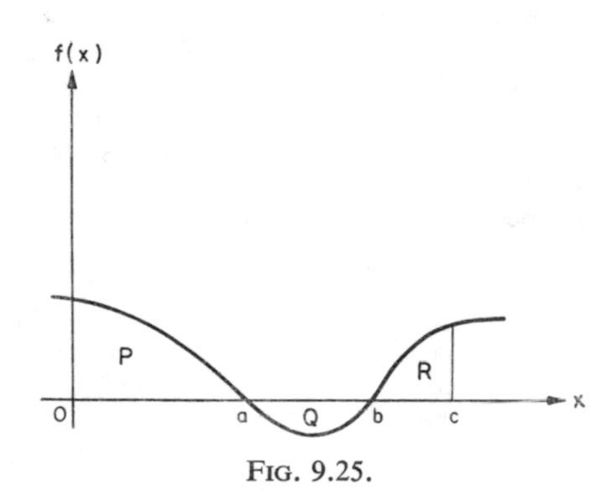

FIG. 9.25.

34. Find the area enclosed by the curve $y = \sin x$, the x-axis and the following pairs of ordinates:

(i) $x = -\frac{1}{2}\pi,\ x = 0$,

(ii) $x = -\frac{1}{2}\pi,\ x = \frac{1}{3}\pi$,

(iii) $x = -\pi,\ x = 0$.

35. Find the area enclosed by the curve $y = x^2$ and the following pairs of ordinates:

(i) $x = -2,\ x = 0$, (ii) $x = -2,\ x = 2$

36. Find the area enclosed by the curve $y = (x-5)(x-3)$, the x-axis and the ordinates:

(i) $x = 3,\ x = 5$, (iii) $x = 5,\ x = 6$,
(ii) $x = 1,\ x = 3$, (iv) $x = -1,\ x = 5$.

Give a labelled sketch of each of the areas.

37. What is the value of the area enclosed by the curve $x = (y-2)(y-3)$ and the y-axis? Account for its sign.

38. Find the area enclosed by the curve $9x^2+y-25 = 0$, the y-axis and the x-axis by taking an element of area parallel to

(i) the y-axis, and (ii) the x-axis.

39. Find the areas enclosed between the following curves and lines:

(i) $y = (x-1)^2,\ y = 1$,
(ii) $y^2 = x,\ x = 0,\ y = 4$,
(iii) $y = \sin x,\ y = \frac{1}{2},\ x = 0$,
(iv) $y = x^2,\ y = x$,
(v) $y = \sqrt[3]{x},\ y = \sqrt{x}$.

40. What is the area enclosed between the curves $y = \sqrt[m]{x}$ and $y = \sqrt[n]{x}$? Assume that $m > n$.

41. Find the area enclosed by the curve $y = \sqrt{(x+4)}$ and the lines $x = -4$ and $y = 2$.

42. What is the area enclosed by the curve $x^2+y^2=9$, and the lines $x = 0$ and $y = 0$?

9.5. NUMERICAL INTEGRATION

A method which enables us to evaluate a definite integral for a function whose integral is not easily obtained by analytic means was described in Section 7.2 (the trapezium rule). The use of an expression slightly different to that of equation (7.5), and known as Simpson's rule, in general yields a more accurate result.

One derivation of Simpson's rule requires that we start by giving particular attention to three neighbouring points on the graph of the function whose integral is to be found. Instead of using these three points to define two trapezia as we did in Section 7.2, we use them to define a parabola whose equation has the form $y = ax^2+bx+c$. By finding the values of a, b and c we can determine the exact area enclosed by the parabola, the x-axis and the ordinates to the two extreme points by integration, and this gives a comparatively accurate estimate of the area bounded by the actual curve through the three points, the same two bounding ordinates and the x-axis.

With reference to Fig. 9.26, let us suppose that the curve $APBQCRD$ is the graph of the

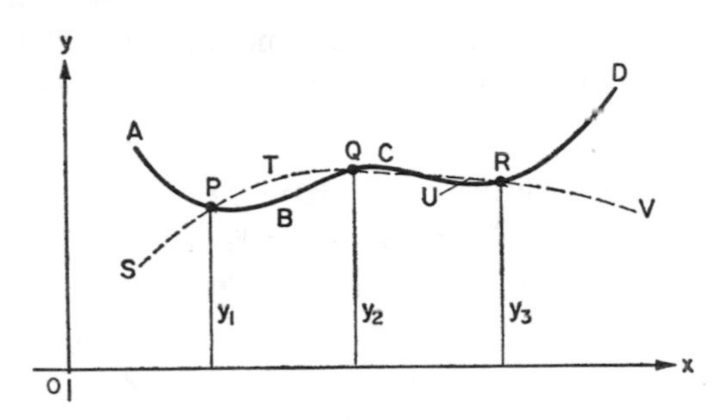

FIG. 9.26. We can obtain an approximate value for the area under *PBQCR* by fitting a parabola *SPTQURV* to pass through *PQR* and by finding the area *PTQUR* exactly.

integrand in the region involved in the integral whose value is to be estimated. *P* and *R* are the points whose abscissae are equal to the limits of the required integral, and *Q* is chosen so that it lies midway between *P* and *R*. We then "fit" a parabola *SPTQURV* to pass through *P*, *Q* and *R*, and determine the area under *PTQUR* by direct integration.

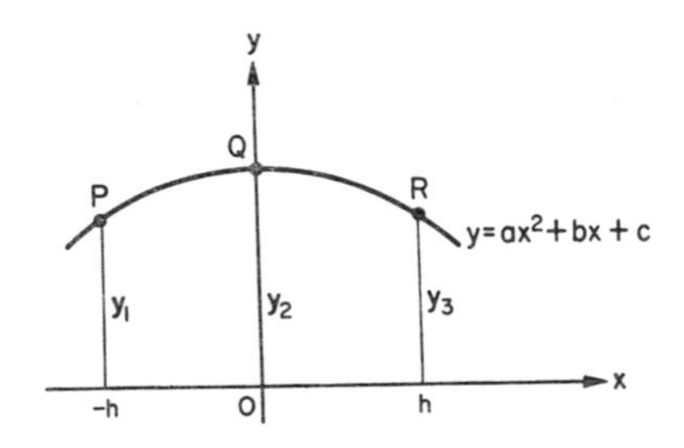

FIG. 9.27. To find the equation of the parabola through *PQR* it is convenient to choose a new *y*-axis to go through *Q*.

To derive the equation of the parabola *SPTQURV* it is convenient to set up a new *y*-axis which passes through *Q* (Fig. 9.27). The ordinates of *P*, *Q* and *R* will then be unchanged, and if we denote the difference between the abscissae of *P* and *R* by $2h$, we can write the abscissae of *P*, *Q* and *R* as $-h$, 0 and $+h$, respectively. The area under the parabola can then be written as the integral

$$\int_{-h}^{h} (ax^2+bx+c)\,dx,$$

and the values of a, b and c can be found once the coordinates of *P*, *Q* and *R* are specified.

Carrying out the integration leads us through

$$[\tfrac{1}{3}ax^3+\tfrac{1}{2}bx^2+cx]_{-h}^{h}$$

to

$$\{\tfrac{2}{3}ah^3+2ch\}$$

or

$$A = \tfrac{1}{3}h(2ah^2+6c)$$

as an expression for the area A under the parabola.

Now if the original ordinates of *P*, *Q* and *R* were y_1, y_2 and y_3, respectively, we can say that the parabola $y = ax^2+bx+c$ must pass through the points $P(-h, y_1)$, $Q(0, y_2)$ and $R(h, y_3)$. The coordinates of these points must therefore satisfy the equation $y = ax^2+bx+c$, and writing down these conditions will define the values to be given to a, b and c. Thus

$$y_1 = ah^2-bh+c, \tag{1}$$

$$y_2 = c, \tag{2}$$

$$y_3 = ah^2+bh+c. \tag{3}$$

Adding (1) to (3) gives

$$y_1+y_3 = 2ah^2+2c,$$

and substituting into the expression for the area gives

$$A = \tfrac{1}{3}h(y_1+y_3+4c).$$

Now $y_2 = c$ (from (2)), and we can thus write that

$$A = \tfrac{1}{3}h(y_1+4y_2+y_3). \tag{9.2}$$

This, then, is the expression which enables us to calculate the approximate value of the area under the curve *PBQCR* of Fig. 9.26, and, as stated above, the value thus obtained will in general be more accurate than the similar value obtained by means of the trapezium rule.

If we require to evaluate an area bounded by ordinates rather far apart, or if we wish to improve upon the accuracy given by equation (9.2), we can divide the area in question into an even number of equally spaced strips, and apply equation (9.2) repeatedly to adjacent pairs of strips. Thus, if we require to find an approximate value for

$$\int_a^b f(x)\, dx$$

by this means, we imagine the area representing the value of the integral to be divided into n parts by n equally spaced ordinates (Fig. 9.28). The value chosen for n must, of course, be even. Applying equation (9.2) to the adjacent pairs of strips, we then obtain

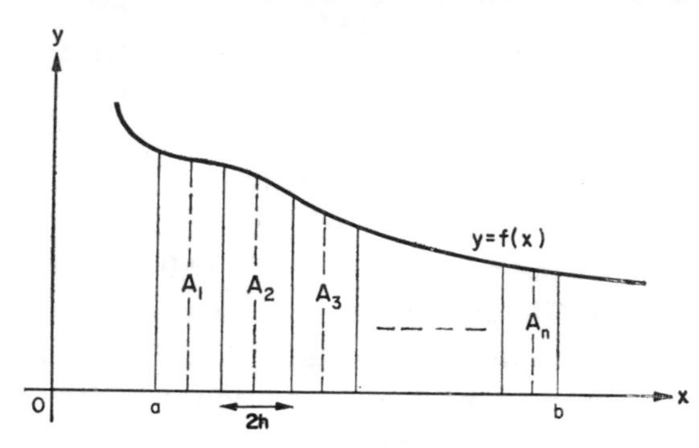

FIG. 9.28. To evaluate $\int_a^b f(x)\, dx$, imagine the area under the curve $y = f(x)$ to be divided into an even number of strips of equal width h.

$$A_1 = \tfrac{1}{3}h(y_1+4y_2+y_3),$$
$$A_2 = \tfrac{1}{3}h(y_3+4y_4+y_5),$$
$$A_3 = \tfrac{1}{3}h(y_5+4y_6+y_7)$$

and so on until the final pair, where

$$A_m = \tfrac{1}{3}h(y_{n+1}+4y_n+y_{n+1}),$$

and where $m = \frac{1}{2}n$.

Adding these expressions, we obtain the expression for the approximate value of the whole area between $x = a$ and $x = b$, i.e.

$$A = \tfrac{1}{3}h(y_1+4y_2+2y_3+4y_4+2y_5+\dots +4y_n+y_{n+1}).$$

This is one form of Simpson's rule for the area representing the value of

$$\int_a^b f(x)\, dx,$$

and, as with the trapezium method, the terms inside the main bracket are more conveniently evaluated when regrouped in the form

$$A = \tfrac{1}{3}h[(y_1+y_{n+1})+4(y_2+y_4+\dots+y_n) + 2(y_3+y_5+\dots+y_{n-1})]. \quad (9.3)$$

We can finally note that the trapezium rule and Simpson's rule are merely developments of the Gregory–Newton formula (p. 68). If, for example, we take the first two terms of that equation, namely $f(a+\theta h) \simeq f(a)+\theta\delta f(a)$, and integrate both sides with respect to θ, we obtain the trapezium rule on substituting the appropriate limits. In a manner of speaking, the trapezium rule is equivalent to replacing the integrand by the first two terms of the Gregory–Newton formula, and clearly this approximation is useful as long as $\delta^2 f(a)$ and higher order differences are small. Simpson's rule can be obtained in a similar manner by employing the first three terms of the Gregory–Newton formula, and is clearly more suitable when the second-order differences are not negligible but the higher order ones are. The detail of these identifications are left as an exercise for the reader (see questions 10–12 in Exercise 9e).

It goes almost without saying that the greater the value chosen for n, the greater will be the accuracy of the final answer. We need only remind ourselves finally that n must be even, for the whole method rests upon the assumption that the strips into which the area is divided are considered in pairs, and that once n has been chosen, h is defined as the value of $(b-a)/n$.

Example 9.12

Find an approximate value for $\int_1^2 (1/x)\,dx$ using Simpson's rule.

We will choose to divide the area representing the value of the integral into ten strips. Since the limits of integration are 1 and 2, this defines h as 0·1. Tabulating the quantities involved in the computation for convenience,

$y_1=\frac{1}{1}=1{\cdot}0000$; $y_2=\frac{1}{1{\cdot}1}=0{\cdot}9091$; $y_3=\frac{1}{1{\cdot}2}=0{\cdot}8333$;

$y_4=\frac{1}{1{\cdot}3}=0{\cdot}7692$; $y_5=\frac{1}{1{\cdot}4}=0{\cdot}7143$;

$y_6=\frac{1}{1{\cdot}5}=0{\cdot}6667$; $y_7=\frac{1}{1{\cdot}6}=0{\cdot}6250$;

$y_8=\frac{1}{1{\cdot}7}=0{\cdot}5882$; $y_9=\frac{1}{1{\cdot}8}=0{\cdot}5556$;

$y_{11}=\frac{1}{2}=0{\cdot}5000$; $y_{10}=\frac{1}{1{\cdot}9}=0{\cdot}5263$.

Adding,

$y_1+y_{11}=1{\cdot}5000$; $(y_2+y_4+y_6+y_8+y_{10})=3{\cdot}4595$; $(y_3+y_5+y_7+y_9)=2{\cdot}7282$.

Thus

$$\int_1^2 \frac{1}{x}\,dx = \frac{1}{3}(0{\cdot}1)\,[1{\cdot}5000+4(3{\cdot}4595)+2(2{\cdot}7282)]$$

$$= \frac{1}{3}(0{\cdot}1)\,[1{\cdot}5000+13{\cdot}8380+5{\cdot}4564]$$

$$= \frac{1}{3}(0{\cdot}1)\,[20{\cdot}7844]$$

$$0{\cdot}6928.$$

Exercise 9e

Evaluate the following integrals using Simpson's rule and the number or strips indicated:

1. $\int_0^1 \sqrt{(1+x^2)}\,dx$; 10. **2.** $\int_0^{\pi/3} \tan x\,dx$; 6.

3. $\int_0^8 \frac{dx}{\sqrt{(1+x^2)}}$; 8.

4. Compare the results given by the trapezium rule and Simpson's rule using six strips for the value of

$$\int_0^{\pi/6} \cos 2x\,dx.$$

Find the true value by direct integration.

5. Evaluate

$$\int_1^{11} \frac{1}{x}\,dx$$

using Simpson's rule and 4, 8, 10 and 20 strips in turn. Represent this graphically, and estimate the true value of the integral.

6. The depth (y m) of a stream varies with the distance (x m) from one bank as follows:

x (m)	0	2	4	6	8	10	12	14	16
y (m)	0	0·5	1·0	1·5	2·0	3·0	2·0	1·0	0

Calculate the cross-sectional area of the stream, and deduce the volume of water flowing past any point in one second if the velocity of the stream is taken as constant and equal to 3 m/s.

***7.** A force, constant in direction and variable in magnitude, acts on a particle of mass 4 kg initially at rest. The magnitude of the force (P newtons) and the distance travelled by the particle (s m) are given by the following table:

s	0	10	20	30	40	50	60	70	80
P	20	31	39	45	49	50	46	40	32

Draw a force-distance (force-displacement) graph and find, as accurately as your data permits and stating your units,

(i) the total work done by the force, when the particle has travelled 80 m;
(ii) the velocity of the particle at this time;
(iii) the power at which the force is then working. (SU)

8. During a recent official trial, the speed of a car starting from rest was noted at intervals of 5 seconds as shown by the table:

Time in seconds:	0	5	10	15	20	25	30
Speed in m/s:	0	12·7	19·6	24·5	27·9	30·3	32·8

Find the ratio of the average acceleration in the first 10 seconds to that in the last 10 seconds, the acceleration 15 seconds from the start, and the average speed during the 30 seconds. (SU)

†**9.** Find the area bounded by the curve $y = x\log_{10}x$, and the lines $x = 1$, $x = 5$ and $y = 0$.

†**10.** It can be shown that

$$\int_a^{a+nh} f(x)\,dx = h\int_0^n f(a+\theta h)\,d\theta.$$

Using the Gregory–Newton formula (p. 68) for $f(a+\theta h)$, find an expression for

$$h\int_0^n f(a+\theta h)\,d\theta$$

in terms of $\delta f(a)$, $\delta^2 f(a)$, etc.

Using your expression, write down its form when

(i) $n = 1$, and $\delta^2 f(a)$ and higher order differences are neglected;
(ii) $n = 2$, and $\delta^3 f(a)$ and higher order differences are neglected.

Replacing $\delta f(a)$ by $f(a+h)-f(a)$ and $\delta^2 f(a)$ by $f(a+2h)-2f(a+h)+f(a)$, rewrite your answers in terms of $f(a+2h)$, $f(a+h)$ and $f(a)$.

Comment on your results.

†**11.** Using the Gregory–Newton formula, find expressions for

$$\int_a^{a+nh} f(x)\,dx$$

if

(i) $n = 3$, and $\delta^4 f(a)$ and higher-order differences are ignored (Cote's three-eights rule);
(ii) $n = 6$ and $\delta^7 f(a)$ and higher-order differences are ignored (Weddle's rule). (See question 10.)

†**12.** By considering the magnitude of the first term ignored in the Gregory–Newton formula, obtain an expression for the approximate error obtained in using

(i) the trapezium rule, (ii) Simpson's rule.

Replace the difference expressions by their approximate equivalents in terms of h and the derivatives of $f(x)$, and use your results to estimate the errors given by the two "rules" in evaluating

(a) $\int_0^1 x^2\,dx$, (b) $\int_0^1 x^3\,dx$, (c) $\int_1^2 \frac{1}{x}\,dx$.

(Assume that two strips only are taken in each case, and remember that this will demand two applications of the trapezium rule.)

†***13.** Two variables x and y vary with each other as follows:

x	0	5	10	15	20
y	0	8·1	31·3	51·1	62·7

Obtain an approximate value of $\int_0^{20} y\,dx$.

What meaning can be attached to this value if

(i) x is the time in seconds, and y is
 (a) the force in N acting on a body,
 (b) the power in kW dissipated by an electrical machine,
 (c) the acceleration in m/s² of a body,
 (d) the velocity in m/s of a body,
 (e) the current in mA flowing onto the plates of capacitor;
(ii) x is the distance (in m) moved by a body acted on by a force of y N?

†**14.** The velocity v of a river varies with the distance x from a certain point A as follows:

Distance (m)	0	25	50	75	100	125	150
Velocity (m/s)	1·1	1·5	1·7	1·3	0·8	1·2	1·3

Show that the time taken to reach the point 150 m from A is

$$\int_0^{150} \frac{dx}{v},$$

and estimate its value.

†**15.** The probability density of a statistic x is given by

$$\frac{1}{\sigma\sqrt{(2\pi)}}\,e^{-(x-\bar{x})^2/2\sigma^2},$$

where $\bar{x}$ and σ are the mean and standard deviation of the statistic.

Explain why the proportion of the population in which a value of the statistic between x and $(x+\delta x)$ can be observed is approximately equal to

$$\frac{1}{\sigma\sqrt{2\pi}}\,e^{-(x-\bar{x})^2/2\sigma^2}\,\delta x$$

and deduce that the proportion displaying a score between α and β is given by 1

$$\frac{1}{\sigma\sqrt{2\pi}}\int_\alpha^\beta e^{-(x-\bar{x})^2/2\sigma^2}\,dx.$$

Hence show that between 60% and 70% of the population lie within σ of the mean.

9.6. MASS OF NON-UNIFORM BODIES

The mass of a body is usually calculable if its density is known. If the density is constant throughout the body, the problem of determining the mass of the body reduces to finding its surface area (for a lamina) or its volume. Methods of finding the surface area of a lamina have been discussed in Section 9.4, and it is only necessary to add here that the density of a lamina is often quoted in kg/m², i.e. in terms of the mass per unit area of lamina. If this is done, one has simply to find the surface area of the lamina, its thickness being included in the expression for the density of unit area.

Example 9.13

The density of a substance is 5 kg/m³. What is the area density of the lamina of the substance if its thickness is 0·001 m?

If we take 1 m² of lamina, its volume will be (1×0·001) m³, and its mass (0·001×50) or 0·05 kg. The superficial density is thus 0·05 kg/m².

If, however, the density of a body varies, the task of finding the total mass is not so simple. Methods involving integration are often useful, and these start by considering an element of the body so small and symmetrical that the density can be considered constant throughout it. By forming the sum of such elements, and considering the limiting value of this sum as the elements shrink to nothing (i.e. finding the value of the integral), the mass can be determined.

An important feature in calculations such as these is the shape and position of the element first taken, and a set of examples illustrating this point for bodies with commonly arising shapes is given beneath. Analytically the method can be described as follows. A small element of the body of volume δv has a mass of $\varrho\,\delta v$ where ϱ is a function of the position of the element, and δv is taken so small that ϱ can be considered constant throughout it. The total mass of the body can be written as approximately equal to the sum of the masses of such elements throughout the body, i.e. $\Sigma\,\varrho\,\delta v$. The exact mass of the body is the limit of this sum as $\delta v \to 0$, and thus the mass M can be expressed in the form

$$M = \lim_{\delta v=0} \Sigma\,\varrho\,\delta v = \int \varrho\,dv,$$

and where ϱ cannot be taken outside the integral as it is a function of position. In general, however, it will be possible to express ϱ and v as a function of a single space variable, and the integral is considerably simplified, as can be seen in the following examples.

Example 9.14

The density ϱ per unit length of a bar of length a varies with the distance x from one end according to the equation $\varrho = k(a+x)$. Find the total mass of the bar.

Consider an element of length δx situated at a distance x from the left-hand end as shown in Fig. 9.29. The mass of this element

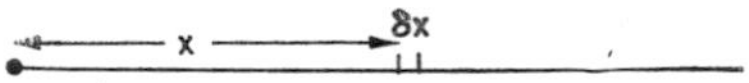

FIG. 9.29.

equals $\varrho\,\delta x$, where ϱ is the density of the unit length at a distance x from the left-hand end. The total mass thus equals

$$\lim_{\delta x=0} \Sigma\,\varrho\,dx = \int_0^a \varrho\,dx.$$

Now ϱ is not constant, and must therefore be replaced by $k(a+x)$ before integration. Thus the mass of the bar

$$\begin{aligned} M &= \int_0^a k(a+x)\,dx = k\int_0^a (a+x)\,dx \\ &= k[ax+\tfrac{1}{2}x^2]_0^a \\ &= k[(a^2+\tfrac{1}{2}a^2)-0] \\ &= \tfrac{3}{2}ka^2. \end{aligned}$$

Example 9.15

The density per unit area of a disc at any point is directly proportional to the distance of the point from the centre. Find the total mass of a disc of radius a.

Consider an annular element as shown in Fig. 9.30. If the inner radius of the annulus is r, and the width δr, the area of the annulus can be found by considering it cut along the radius and opened out (Fig. 9.31). A trapezium will be formed, and the area of the trapezium is equal to $\frac{1}{2}\{2\pi r+2\pi(r+\delta r)\}\ \delta r$, or,

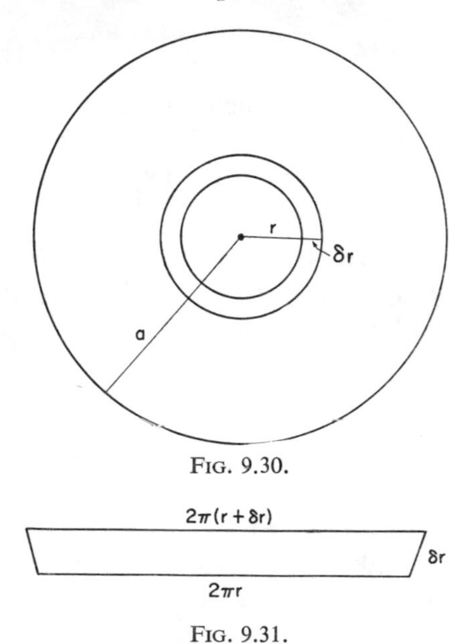

FIG. 9.30.

FIG. 9.31.

multiplied out, $\{2\pi r\ \delta r+\pi(\delta r)^2\}$. Neglecting the term in $(\delta r)^2$ (which disappears when the limiting value of the sum at $\delta r = 0$ is replaced by the integral*), this expression for the area becomes $2\pi r\ \delta r$. (This can alternatively be obtained by considering the area or the hole in the centre of the annulus and subtracting it from the area from the centre to the outer edge: thus

$$\begin{aligned} A &= \pi(r+\delta r)^2-\pi r^2 \\ &= 2\pi r\ \delta r, \quad \text{neglecting the } (\delta r)^2 \text{ term.)} \end{aligned}$$

The mass of the annular element is therefore $\{\varrho x(2\pi r\ \delta r)\}$, where ϱ is the density per unit area of the annulus (and since all parts of the annulus are equidistant from the centre of the disc, the value of ϱ will be constant throughout the annulus). Whence the total mass of the disc is equal to

$$\lim_{\delta r=0} \sum (\varrho\,.\,2\pi r\,.\,\delta r),$$

and this is the same as

$$\int_0^a 2\varrho\pi r\,\mathrm{d}r.$$

Since the density ϱ is a function of r, the factor ϱ in the integrand must be replaced before integration. The density is in fact proportional to the distance from the centre, so $\varrho = kr$, and thus the total mass of the disc is given by

$$\begin{aligned} M &= \int_0^a kr\,.\,2\pi r\,.\,\mathrm{d}r \\ &= 2\pi k \int_0^a r^2\mathrm{d}r \\ &= 2\pi k \left[\frac{r^3}{3}\right]_0^a \\ &= \tfrac{2}{3}\pi k a^3. \end{aligned}$$

Example 9.16

The curve $y = \sqrt{x}$ between $x = 0$ and $x = 4$, the x-axis and the line $x = 4$ define the shape of a lamina whose density varies as the square root of the distance from the x-axis. Find the total mass of the lamina.

Since the density is a function of the distance from the x-axis, the density will be constant throughout an element drawn parallel to the x-axis (see Fig. 9.32). The length of such element is given by the expression $(4-x)$,

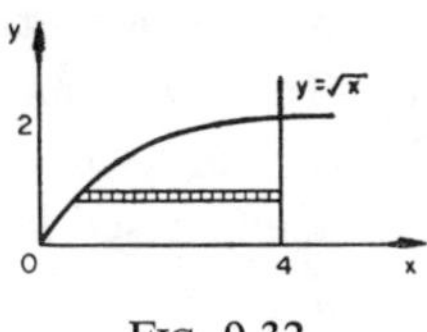

FIG. 9.32.

* It is left to the reader to show that $\lim\limits_{\delta r=0} \sum \{2\pi r\ \delta r+\pi\ (\delta r)^2\} = \int 2\pi r\,\mathrm{d}r$.

where x is the abscissa of the part of the curve situated at the left-hand end of the element. The mass of the element can be written as $\varrho(4-x)\,\delta y$, and replacing ϱ by $ky^{\frac{1}{2}}$, the total mass of the lamina is represented by the term

$$\lim_{\delta y=0} \sum \left[ky^{\frac{1}{2}}(4-x)\,\delta y\right], \quad \text{or} \quad \int_0^2 ky^{\frac{1}{2}}(4-x)\,dy.$$

Replacing the x in the integrand by the y^2 which it equals (from the equation $y = \sqrt{x}$), the total mass M can be found as

$$M = \int_0^2 ky^{\frac{1}{2}}(4-y^2)\,dy$$

$$= k\int_0^2 (4y^{\frac{1}{2}}-y^{\frac{5}{2}})\,dy$$

$$= k\,[4.\tfrac{2}{3}\,y^{\frac{3}{2}}-\tfrac{2}{7}.y^{\frac{7}{2}}]_0^2$$

$$= k\,[\tfrac{8}{3}.2\sqrt{2}-\tfrac{2}{7}.8\sqrt{2}]$$

$$= 16\sqrt{2}k[\tfrac{1}{3}-\tfrac{1}{7}]$$

$$= \frac{64\sqrt{2}}{21}k.$$

Exercise 9f

1. Find the mass of the laminae bounded by the following curves and lines if the density of each lamina is 3 g/cm²:

(i) $y = x^{\frac{1}{4}}$, $y = 0$, $x = 16$;

(ii) $y = x$, $y = 3x$, $y = 4-3x$;

(iii) $y = 4x^2+1$, $x = 0$, $y = 0$, $x = 1$;

(iv) $y = 3x$, $x = 0$, $y = 9$.

2. Repeat question 1 for the following laminae of density 2·5 g/cm²:

(i) $y = x\sqrt{(1-x^2)}$, $y = 0$, $x = 0$;

(ii) $y = 2-x^2$, $x = 0$, $y = 1$;

(iii) $y = \dfrac{1}{\sqrt{(4-x^2)}}$, $x = 0$, $x = 1$, $y = 0$.

3. Find the density per unit length of a bar of circular cross-section and density 4 g/cm³ if its radius is

(i) 0·1, (ii) 0·5, (iii) 2·5, (iv) a cm.

4. The density ϱ per unit length of a bar of length a varies with the distance x from one end according to the equation $\varrho = k\sqrt{x}$. Find the total mass of the bar.

5. Find the mass of a bar of length 8 cm if its density at a point x from one end is

(i) $2+\sqrt{x}$, (ii) $2\sqrt{(16-x)}$, g/cm.

6. Find the mass of a bar of length 5 cm if its density at a plane x cm from the centre is

(i) $(3-x)$, (ii) $\cos 0{\cdot}1x$, g/cm.

(Assume that x takes negative values to the left of the centre and positive values to the right of the centre.)

7. The density per unit area of a disc at any point is equal to the square root of the distance of that point from the centre. Find the total mass of a disc of radius a.

8. Find the mass of a disc of radius a whose density at any point x from the centre is equal to $k \sin(\pi x^2/2a^2)$ g/cm², where k is a constant.

9. Find the mass of a lamina bounded by $y = x^2$, $y = 0$ and $x = 2$, given that its density at any point is directly proportional to the distance of that point from the x-axis.

10. Repeat question 9 for the lamina bounded by $y = x^2$, $x = 0$, $y = 4$, given that the density at any point is directly proportional to the square root of the distance of that point from the y-axis.

11. The line density of a non-uniform rod of length L at a distance x from one end is $(a+2bx)$, where a and b are constants. Prove that the mass of the rod is $L(a+bL)$.

9.7. CENTRES OF GRAVITY

It is convenient to think of any solid as a collection of point masses. When such a body is placed in a gravitational field, each point mass will experience a force, and if we imagine the body to be situated on the surface of the earth, we can regard the collection of the operative gravitational forces as a system of parallel forces (Fig. 9.33a). Such a collection will, in general, have a resultant moment about any axis drawn through the body, but there will be a set of three mutually perpendicular axes about which the gravitational forces as a whole have no moment. The intersection of these three axes is called the centre of gravity or the centroid of the body (Fig. 9.33b).

A convenient consequence of this situation enables us to replace the multitude of gravi-

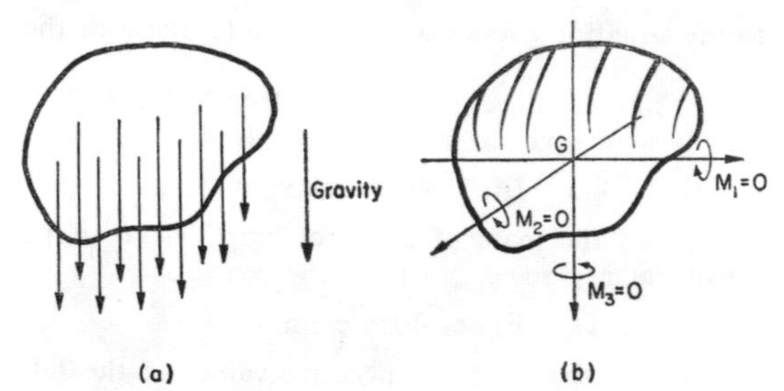

FIG. 9.33. (a) The forces of gravity on a collection of point masses can be regarded as parallel. (b) The intersection of the zero moment axes is the centre of gravity.

tational forces by a single force whose magnitude is equal to the sum of the forces acting on all the point masses (the so-called "weight" of the body), and whose line of action passes through the centre of gravity. If we then equate the sum of the moments of the individual point mass forces about any axis in or beyond the body with the moment of the total weight acting at the unknown position of the centre of gravity, we shall be able to locate the position of the centre of gravity from the resulting equation, there being only one unknown. Such a procedure forms the basis of the method by which we shall find the centres of gravity of a number of regular bodies.

In the general three-dimensional case, the method will demand that we find three axes about which the total point mass forces have no moment. To simplify the situation to start with, we shall confine our attention to two-dimensional (laminar) bodies and one-dimensional rods or bars.

To find an axis about which the forces of gravity have no moment requires that we choose a *direction* for the gravitational field. This choice will be a function of the orientation of the body, although in the simplest cases considerations of symmetry may enable us to write down the zero moment axis immediately. In the cases where the body possesses no useful symmetry, we shall have to choose two mutually perpendicular directions in turn, and find the equations or locations of two perpendicular zero moment axes. The intersection of such axes will then give us the required position of the centre of gravity. To illustrate the procedure, a number of examples are given below: the general principle concerning the direction of the gravitational field is that it is advisedly chosen to act in a direction perpendicular to any axis or axes of symmetry, and the force on a small element is considered first.

Expressed in mathematical terms, the total turning effect or moment of the weight of a laminar body about an axis perpendicular to its plane through a point O can be evaluated by first considering the moment of the weight of a small element (see Fig. 9.34a). If the mass of the element is $\varrho\,\delta v$, where ϱ is the density (assumed constant throughout the element's volume δv), the force of gravity on it will be $(\varrho\,\delta v)$ in gravitational units in the direction of the field. If the direction of the gravitational field is that indicated by the arrow to the right of the body in Fig. 9.34a, the weight of the element will have a moment about the axis through O equal to $(\varrho\,\delta v)x$, where x is the perpendicular distance from the line of action of the gravitational force to the axis. If we set

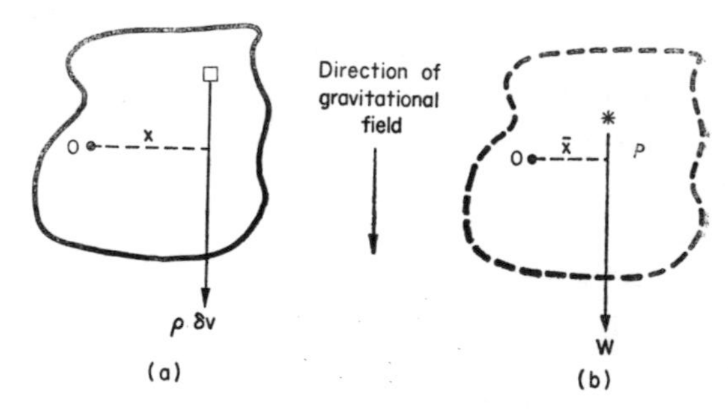

FIG. 9.34a. The turning effect of the small element of volume δV and density ϱ about O is $\varrho x\,\delta V$.

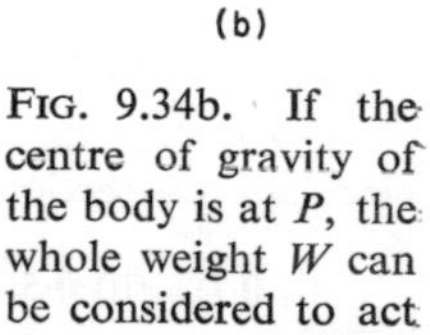

FIG. 9.34b. If the centre of gravity of the body is at P, the whole weight W can be considered to act there, and its moment about O is $W\bar{x}$.

up an x-coordinate axis with its origin at O, in the plane of the body, and in a direction perpendicular to the direction of the gravitational field, then the total moment of the weight of the body can be written as approximately equal to $\sum(\varrho\delta v)x$, where the value of x is the value of the x-coordinate of the element to be considered (and x will be measured in units such as m, cm, or mm), and where the sum is taken across the whole plane of the body. Taking the limiting value of this sum as the element size is reduced to zero, we obtain the equation

$$\text{total moment of the weight of the body about } O = \int \varrho x \, dv.$$

If now the centre of gravity is situated at a point P (see Fig. 9.34b), the whole weight W of the body can be considered to act along PW through P. (Note that we have arranged to have the plane of the body lying in an orientation to include the direction of the gravitational field.)

If this is so, the moment of the weight W about O equals $(W\bar{x})$, where $\bar{x}$ is the x coordinate of the centre of gravity. This must equal $\int \varrho x \, dv$, and thus

$$W\bar{x} = \int \varrho x \, dv.$$

Now the weight of the body W can be expressed as $\int \varrho \, dv$ (cf. Section 9.6), so

$$\bar{x} = \frac{\int \varrho x \, dv}{\int \varrho \, dv}, \qquad (9.4)$$

where the factors ϱ can be taken out of the integrals and cancelled only if the density is constant throughout the body.

The other coordinates of the centre of gravity with respect to some arbitrary origin such as the point O can be found by changing the direction of the gravitational field by 90°, but as mentioned before, considerations of symmetry will in general enable one or two of the coordinates to be written down without calculation. If, however, only one coordinate could be deduced in this way, the procedure to find the second or third coordinate of the centre of gravity (say $\bar{y}$) would be similar to that given above, and the equation

$$\bar{y} = \frac{\int \varrho y \, dv}{\int \varrho \, dv}, \qquad (9.5)$$

used in conjunction with Fig. 9.35, would enable the value of y to be found. There is,

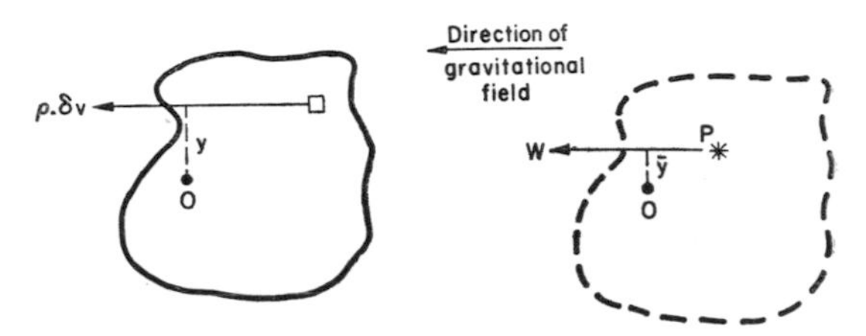

FIG. 9.35. By imagining the direction of gravity to have rotated through 90°, the y-coordinate of the centre of gravity can be found.

of course, no need to evaluate the denominator of this fraction again; as before, it represents the total weight of the body. Notice that the direction of the gravitational field is assumed perpendicular to the y-axis if $\bar{y}$ is required, and perpendicular to the x-axis if $\bar{x}$ is required.

As equations (9.4) and (9.5) stand, the expressions for the coordinates of the centre of gravity of a body are given in terms of integrals of a volume variable v. In cases involving rods or bars or laminae the expressions can be simplified by making the variable of integration x or y. This is achieved by replacing the $\varrho \, \delta v$ term for the mass of the element by an expression of the form $\varrho A \, \delta x$ in the case of a bar, where A is the area of cross-section of an element of length δx, or by one of the form $\varrho y t \, \delta x$ or $\varrho x t \, \delta y$ in the case of a lamina of thickness t, the length of the element being y or x (depending on which way it is drawn relative to the coordinate axes), and the breadth δx or δy (Fig. 9.36). A study of the worked examples following this section will make the procedure clear, and it is only necessary to add that, in general, the assumed direction of gravity will be along the length of the element. It is only necessary to take other steps if this leads to geometrical difficulties, or if the density of the lamina is not constant along the length of the element.

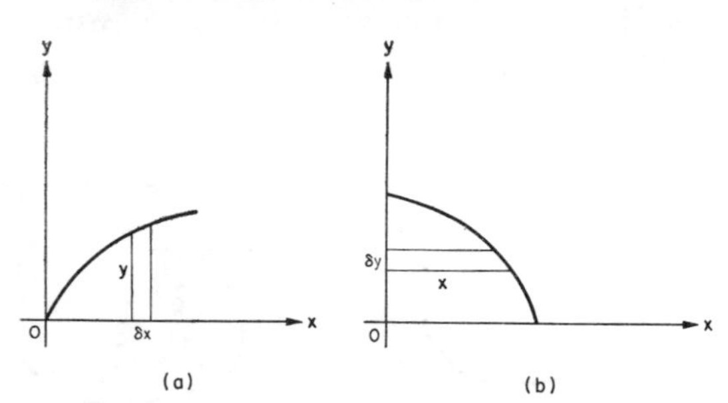

FIG. 9.36. The mass of the element can be written in the form (a) $\varrho yt\,\delta x$, or (b) $\varrho xt\,\delta y$.

It is useful to remember that since the position of the centre of gravity of a body can be expressed completely in terms of distances from a set of fixed points, any answer which does not dimensionally reduce to that of a length must be incorrect. This statement implies that the answer should be of the form al_1, $a\sqrt{(l_1^2+l_2^2)}$, $al_1^{\frac{3}{2}}/l_2^{\frac{1}{2}}$ or some similar expression where, if l_1 and l_2 are quantities possessing the unit or "dimension" of length (e.g. cm), and a is a constant without unit or dimension, the unit of the expression as a whole reduces to cm. Or again, if the numerator of a fraction expressing one coordinate of the centre of gravity possesses a mass unit (for example, grammes), the denominator must also possess a mass unit to cancel with it, leaving the whole expression with the unit or dimension of length only.

Example 9.17

Find the centre of gravity of a bar of uniform cross-section and of length l whose density at any point is directly proportional to the distance of that point from one end.

The problem of finding the centre of gravity of a bar of uniform cross-section is a one-dimensional problem, and the only sta-

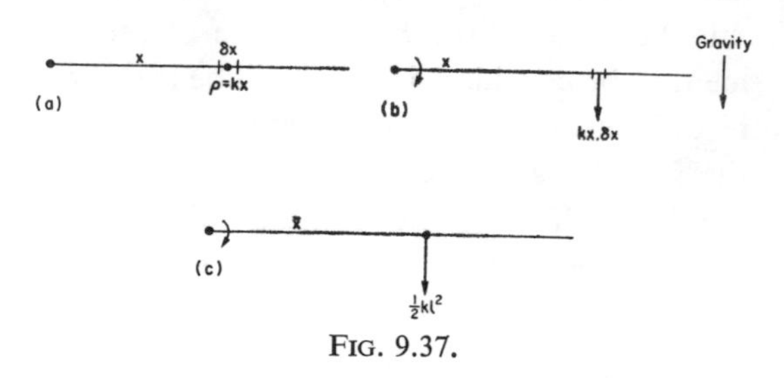

FIG. 9.37.

tistic to be found is the distance of the centre of gravity from one end. If we consider the density of the bar at a distance x from the left-hand end (Fig. 9.37a) to be kx per cm, where k is a constant, the mass of an element of length δx will be $kx\,\delta x$, assuming the density to be constant throughout the element. If we then assume the direction of gravity to be at right angles to the length of the bar or x-axis (i.e. down the page), the clockwise-turning effect of the weight of this element about the left-hand end will be $(kx\,\delta x)\,x$ (Fig. 9.37b), and so the total moment of the whole bar about the left-hand end will be given by the sum $\sum(kx\,\delta x)\,x$. Removing the approximation that the density is constant throughout an element of length δx by taking the limiting value of this sum at $\delta x = 0$, we obtain the total moment of the whole bar about the left-hand end as

$$\lim_{\delta x=0} \sum kx^2\,\delta x$$

or

$$\int_0^l kx^2\,dx.$$

Integrating, we find the total moment P about the left-hand end to be

$$\left[k\frac{x^3}{3}\right]_0^l, \quad \text{or} \quad \tfrac{1}{3}\,kl^3.$$

Now the total mass M of the bar is given by the value of

$$\lim_{\delta x=0} \sum kx\,\delta x \quad \text{or} \quad \int_0^l kx\,dx.$$

Integrating again, we find that $M = \frac{1}{2}kl^2$, and thus if the weight of this mass can be considered to act at a point $\bar{x}$ from the left-hand end (Fig. 9.37c), we can equate its turning effect about the left hand with P. Thus

$$\tfrac{1}{2}\,kl^2\,\bar{x} = \tfrac{1}{3}kl^3,$$

or

$$\bar{x} = \tfrac{2}{3}l,$$

and we can say that the centre of gravity of the bar is two-thirds of the way along the bar towards the denser end.

Example 9.18

Find the centre of gravity of the uniform lamina bounded by $x = 4$ and $y^2 = 16x$.

A sketch of the outline of the lamina is shown in Fig. 9.38: it should be recalled that the equation $y^2 - 16x$ defines two functions ($y = +4\sqrt{x}$ and $y = -4\sqrt{x}$), and thus both branches should be included.

By symmetry we can say that the centre of

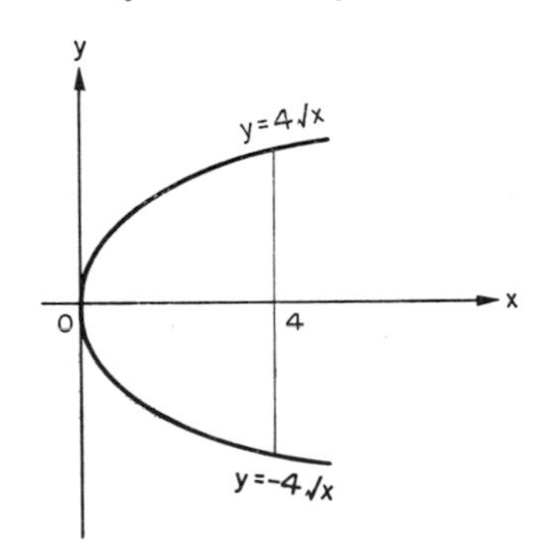

FIG. 9.38.

gravity lies on the x-axis, and thus $\bar{y}$, the y-coordinate of the centre of gravity, is 0.

Since we require $\bar{x}$, we draw an element perpendicular to the x-axis (Fig. 9.39), and write its mass as $\varrho\,.\,2y\,.\,\delta x$, remembering that

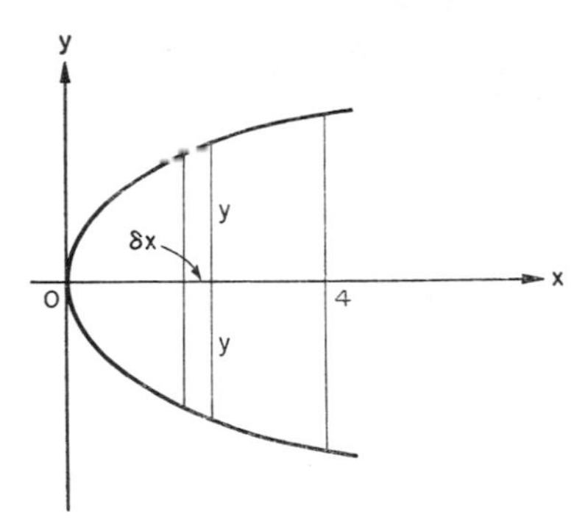

FIG. 9.39.

its length (up the page) is twice the y-coordinate of its mid-point. Then, assuming the direction of gravity to be perpendicular to the x-axis and along the length of the element, we can write that the weight of the element gives a clockwise moment about 0 equal to $\varrho.2y.\delta x.x$. Thus the total moment

$$P = \lim_{\delta x=0} \sum \varrho\,.\,2y.\delta x\,.\,x = \int_0^4 \varrho\,.\,2xy\,.\,dx,$$

adding the limits of integration as 0 and 4 since x (the variable with respect to which the integration is to be carried out) varies between 0 and 4 across the lamina.

Now the total weight $W = \lim_{\delta x=0} \sum \varrho\,.\,2y.\delta x$

$$= \int_0^4 2\varrho y\,\mathrm{d}x,$$

and equating the moment this produces about 0, assuming its force to be concentrated at a point $\bar{x}$ from 0 along the x-axis, we obtain the equation

$$\bar{x}\int_0^4 2\varrho y\,\mathrm{d}x = \int_0^4 2\varrho xy\,\mathrm{d}x.$$

We now replace y as a function of x in order to perform the integration: and since the density is constant throughout the lamina, we can take the 2ϱ factors outside the integral sign. Thus

$$\bar{x}.2\varrho \int_0^4 4\sqrt{x}\,\mathrm{d}x = 2\varrho \int_0^4 x\,.\,4\sqrt{x}\,\mathrm{d}x,$$

replacing y by the positive root $4\sqrt{x}$, remembering that we have allowed for the duplicity of the function by including the factor 2. (This expression should be compared with equation (9.4), when it can be seen that the δv term has been replaced by the expression $2y\delta x$ or $2.4\sqrt{x}\,\delta x$ in both integrals.) We can rewrite this equation in the form

$$\bar{x}\int_0^4 x^{\frac{1}{2}}\,\mathrm{d}x = \int_0^4 x^{\frac{3}{2}}\,\mathrm{d}x,$$

and, integrating,

$$\bar{x}\left[\tfrac{2}{3}x^{\frac{3}{2}}\right]_0^4 = \left[\tfrac{2}{5}x^{\frac{5}{2}}\right]_0^4.$$

Thus

$$\bar{x}\,.\,\tfrac{2}{3}(8) = \tfrac{2}{5}(32),$$

whence

$$\bar{x} = \tfrac{12}{5}.$$

The centre of gravity of the lamina is thus (2.4, 0).

Example 9.19

Find the centroid of a lamina of uniform density bounded by the lines $y = 2x$, $y = 0$ and $x = 5$.

The lamina is triangular in shape (Fig. 9.40a), and in order to find the x-coordinate of the centre of gravity, we first consider an element perpendicular to the x-axis (Fig.

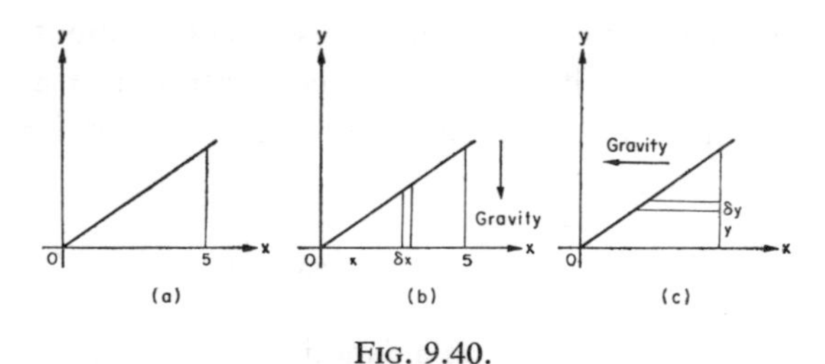

FIG. 9.40.

9.40b). If we assume the direction of gravity to be down the page, the moment of this element about the origin is $(\varrho y\delta x)\,x$. The total moment P of the whole lamina about 0 is thus

$$\lim_{\delta x=0}\sum \varrho yx\,\delta x \quad \text{or} \quad \varrho\int_0^5 yx\,\mathrm{d}x,$$

and replacing y by $2x$ before integrating, we find that

$$P = \varrho\int_0^5 2x^2\,\mathrm{d}x = 2\varrho\left[\tfrac{1}{3}x^3\right]_0^5 = \tfrac{250}{3}\varrho.$$

Now the mass of the lamina is readily obtained by multiplying its area by its density: i.e. $M = \frac{1}{2}.5.10.\varrho = 25\varrho$, and so, equating the moment about O of the weight of the whole lamina (assumed concentrated at $(\bar{x}, \bar{y})$) with P, we obtain the equation

$$25\varrho\bar{x} = \tfrac{250}{3}\varrho.$$

Thus

$$\bar{x} = \tfrac{10}{3}.$$

To obtain $\bar{y}$ we take an element perpendicular to the y-axis and assume the direction of gravity to be also perpendicular to the y-axis (Fig. 9.40c). Then, writing the length of the element as $(5-x)$, and its mass as $\varrho(5-x)\,\delta y$, we can proceed to write an equation for y similar to that above, viz.

$$\begin{aligned}25\varrho\bar{y} &= \varrho\int_0^{10}(5-x)y\,\mathrm{d}y\\ &= \varrho\int_0^{10}(5y-\tfrac{1}{2}y^2)\,\mathrm{d}y\\ &= \varrho\left[\tfrac{5}{2}y^2-\tfrac{1}{6}y^3\right]_0^{10}\\ &= 250/3\varrho.\end{aligned}$$

Thus $\bar{y} = \frac{10}{3}$, and the centroid is the point $(\frac{10}{3}, \frac{10}{3})$.

EXERCISE 9g

1. Find the position of the centre of gravity of a bar of uniform cross-section and length l cm if the density/cm at any point is proportional to

(i) the square root,
(ii) the square of the distance of that point from one end.

2. Repeat question 1 for a bar whose density per cm at a point x cm from one end is

(i) $k(2l-x)$,
(ii) $k/\sqrt{(4l^2-x^2)}$,

where k is a constant.

3. The density of a bar of uniform cross-section decreases steadily from 4 g/cm at one end to 3 g/cm at the mid-point and then increases steadily to 5 g/cm at the other end. Where is its centre of gravity?

Find the centres of gravity of the laminae of uniform density bounded by the following lines and curves:

4. $y^2 = 4x$, $x = 4$.
5. $y^2 = x$, $x = 25$.
6. $y = 9x^2$, $y = 9$.
7. $25y = x^2$, $y = 4$.
8. $y = 3x$, $y = -3x$, $x = 4$.
9. $y = 5x$, $y = -5x$, $y=2$.
10. $y^2 = 9-x$, the y-axis.
11. $x = 3\sqrt{y}$, $y = 1$, $x = 0$.
12. $y = x^3$, $y = 0$, $x = 2$.
13. $y = x^2$, $x = 1$, $x = 3$.

14. Find the position of the centre of gravity of a uniform triangular lamina.

15. Find the centre of gravity of a uniform semicircular lamina of constant density.

16. Find the position of the centre of gravity of a lamina in the shape of a quadrant of a circle, given that the density is uniform.

17. The density at any point in the lamina described by the curve $y^2 = 9x$ and the line $x = 4$ is proportional to the distance of that point from the y-axis. Does this mean that the density at each point in an element perpendicular to the y-axis is the same?

18. The constituency of the triangular lamina defined by the lines $y = x$, $y=-x$ and $x = 5$ varies in such a manner that the density at any point is proportional to the square root of the distance of that point from the y-axis. Where is the centre of gravity of the lamina?

19. The density at any point (x, y) in a lamina described by the curve $y = 2x^2$ and the line $y = 2$ is equal to $(2+y)$. Find the centre of gravity of the lamina.

20. Find the centre of gravity of a uniform lamina defined by the lines $y = 5x$, $y = 0$ and $x = 5$.

21. Three point masses are placed along a uniform bar 3 cm, 5 cm and 7 cm from one end. If the length of the bar is 10 cm, and the masses are 5 g, 8 g and 2 g, respectively, find the centre of gravity of the system, given that the mass of the bar is 5 g.

22. Find the area of the segment of the parabola $y^2 = 4x$ cut off by the line $x = 9$.

Find also the position of the centroid of this segment. (OC)

23. Sketch the graph of $y = x^2(6-x)$.

Find the area between the x-axis and the portion of the curve between $x = 0$ and $x = 6$.

Find also the distance from the y-axis of the centroid of this area. (OC)

24. Show that the area of an element between the curve $y = 2\sqrt{ax}$ and $y = 2x$ bounded by ordinates at x_1 and $(x_1+\delta x)$ is approximately equal to $(y_1-y_2)\,\delta x$ where $y_1 = 2\sqrt{(ax_1)}$ and $y_2 = 2x_1$. Deduce that the centroid of the element approaches $\{x_1, \frac{1}{2}(y_1+y_2)\}$ as $\delta x \to 0$, and hence find the centroid of the lamina bounded by the curve $y^2 = 4ax$ and the line $y = 2x$.

25. A uniform lamina is bounded by the curve $y^2 = 4x$ and the line $y = 2x$. Find its area and the coordinates of its centre of gravity. (OC)

†26. Find the centre of gravity of a lamina bounded by $y = \sqrt{x}$, $x = 4$, $y = 0$, given that the density at any point is proportional to the distance of that point from the x-axis.

†27. The function $f(x)$ is continuous at all values of x between 0 and a. Write down an expression for the centre of gravity of the lamina bounded by $y = f(x)$, $y = 0$, $x = 0$ and $x = a$, given that $f(x) \geqslant 0$ for all $x \geqslant 0$. The density of the lamina is constant.

†28. Find the centre of gravity of the lamina bounded by $y = 1/x$, $x = 1$, $x = 5$ and $y = 0$, given that its density is uniform.

†29. The points (0, 1), (0·5, 2·2), (1·0, 3·3), (1·5, 3·9), (2·0, 4·1), (2·5, 3·2), (3·0, 1·2) are joined by a smooth curve $y = f(x)$.

Find the approximate value of

$$\text{(i)} \int_0^3 f(x)\,dx; \qquad \text{(ii)} \int_0^3 xy\,dx,$$

and hence find the x coordinate of the centre of gravity of the area enclosed by the curve, $x = 0$, $x = 3$ and $y = 0$.

9.8. VOLUMES BY ROTATION

Solid bodies with circular symmetry can be described in terms of the symmetry they show. A sphere, for example, can be described as the solid form traced out when a semicircular lamina is rotated about a diameter (see Fig. 9.41). If the equation of the arc to be rotated is expressible in an algebraic form, the

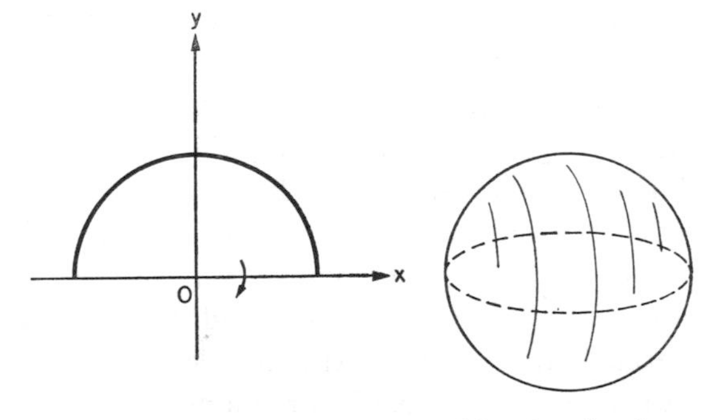

FIG. 9.41. A sphere can be considered to be the solid form described when a semicircular lamina is rotated about a diameter.

volume of the solid formed by its rotation can often be obtained by integration.

Let us consider the solid described by the rotation of an element drawn, as in Fig. 9.42, perpendicular to the axis of rotation. In one complete revolution this element describes a disc, and setting up coordinate axes, the volume of the disc can be written as approximately equal to $\pi y^2 \, \delta x$, where y is the ordinate length of the element and therefore the radius of the disc, and δx its thickness. The volume of the whole body can then be written as approximately equal to the sum of such elementary discs along the length of the arc, and exactly equal to the limit of this sum. Thus the volume V of the whole body is given by the expression

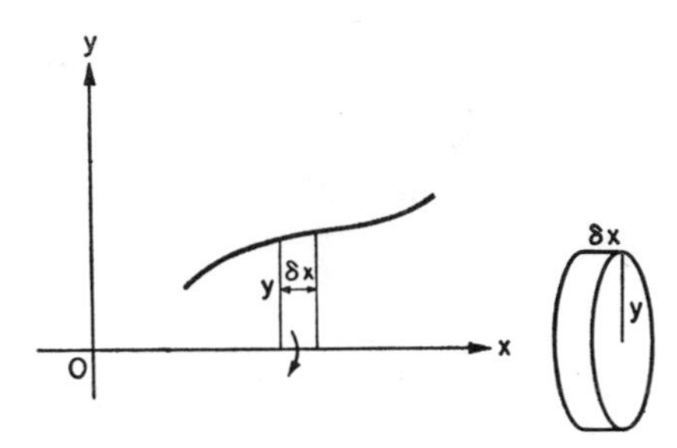

FIG. 9.42. The volume described by rotating an element of length y, perpendicular to the axis of rotation and of thickness δx, through one revolution is $\pi y^2 \, \delta x$.

$$V = \lim_{\delta x = 0} \sum \pi y^2 \, \delta x = \int \pi y^2 \, dx.$$

Knowing y as a function of x (the equation of the arc to be rotated), if the integral can be determined, the volume can be found.

It is sometimes necessary to rotate the arc describing the surface of the solid about some line other than the x-axis. Rotating the curve $y = x^2$ about the y-axis gives an entirely different shape from that obtained by rotating it about the x-axis (Fig. 9.43), and if it is to be rotated about the y-axis, the element to be considered first must be drawn perpendicular to the y-axis if a disc is to be obtained (Fig. 9.44). The radius of the disc obtained by rotation can be seen to equal the value of the abscissa of the point on the curve at its surface (i.e. x), and denoting the thickness by δy we can write that the total volume is numerically equal to the limiting value of $\sum(\pi x^2 \, \delta y)$ for all such elementary discs, i.e. $\pi \int x^2 \, dy$. The calculation then proceeds as before, replacing x^2 by the function of y to which it is equal (i.e. y in this case).

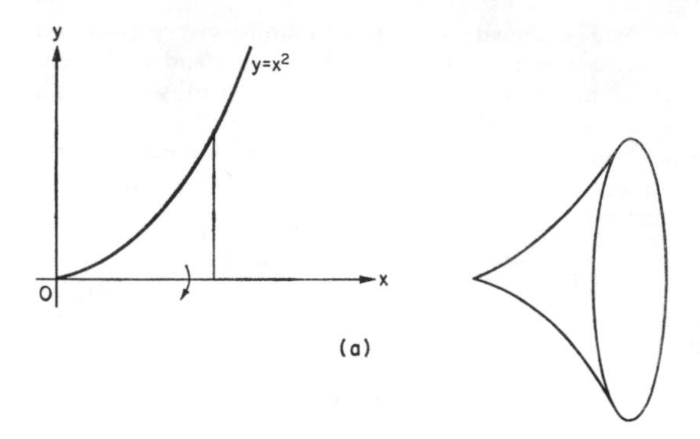

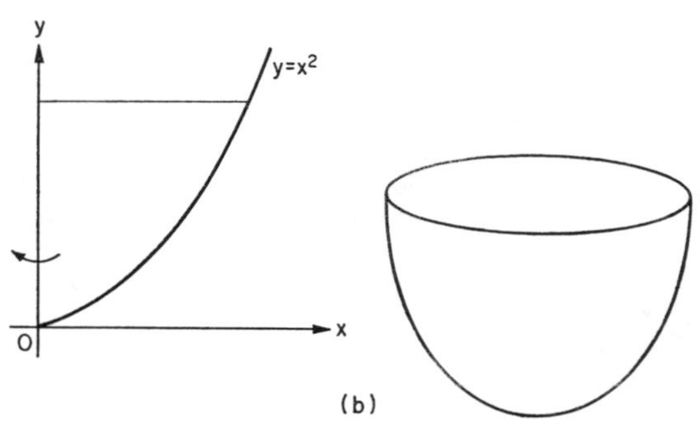

FIG. 9.43. The solid form resulting from the rotation of a curve about an axis depends very much on the disposition of the axis of rotation relative to the curve.

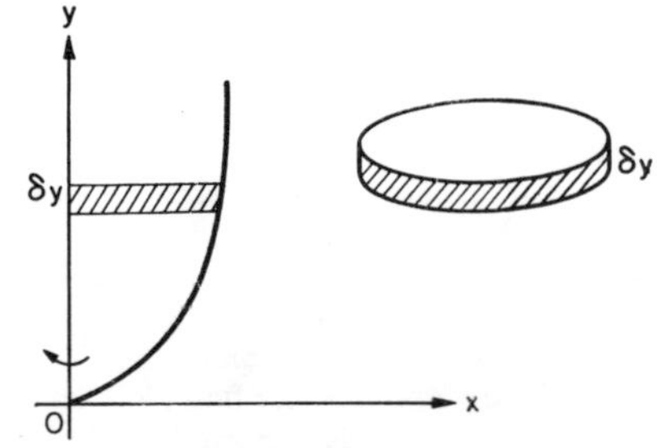

FIG. 9.44. For rotations about the y-axis, consider an element of length of x perpendicular to the y-axis.

Another situation arises when an arc is rotated about an axis such as $x = 1$ (see Fig. 9.45). The element drawn to describe a disc within the body now has its base on the line $x = 1$ and its end on the curve, and as a result its radius is neither simply x or y, but $(1-x)$. The volume of the elementary disc thus equals $\pi(1-x)^2\,\delta y$, and the total volume obtained is $\int \pi(1-x)^2\,dy$.

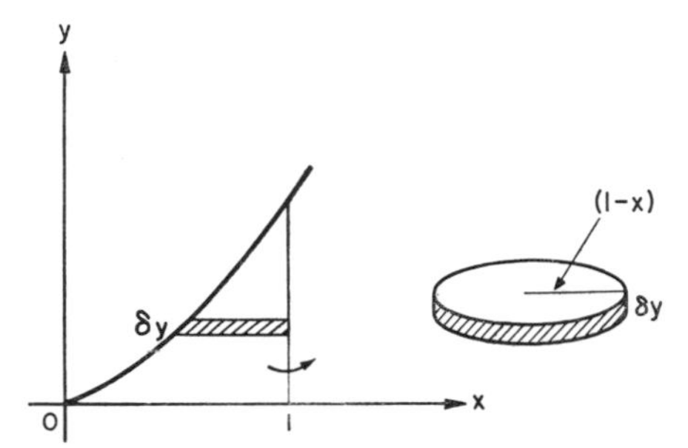

FIG. 9.45. If the axis of rotation is $x = 1$, the volume of the solid perpendicular to it is $\pi(1-x)^2\,\delta y$.

The mass of such solids is simply obtained if the density is constant throughout the body: if it is not, the variation of the density must be expressed as a function of the variable with respect to which the integration is being performed before integrating. Sometimes it may be necessary to use an element parallel to the axis of rotation to achieve an elementary solid with constant density throughout (see Example 9.21). In these cases, the volume of the hollow cylinder formed by the rotated element is given by an expression of the form $2\pi y(1-x)\,\delta y$, thinking of the cylinder cut along one side and opened out (see Fig. 9.46). Alternatively, the volume of the cylinder can be written as $\{\pi(y+\delta y)^2-\pi y^2\}(1-x)$, and expanding the $(y+\delta y)^2$ term to $\{y^2+2y\,\delta y+(\delta y)^2\}$ and neglecting the $(\delta y)^2$ term leads directly to the same result.

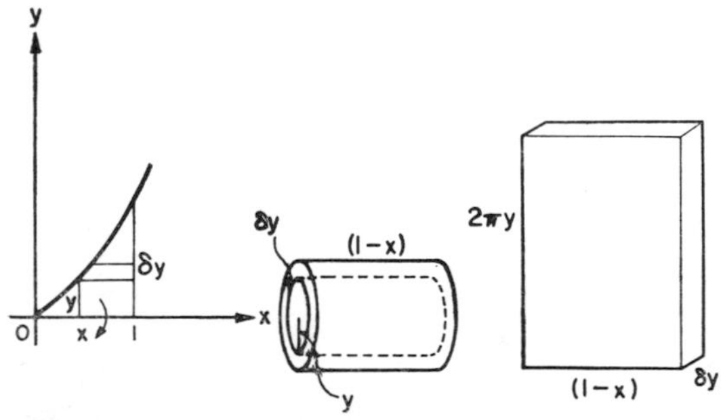

FIG. 9.46. If an element is taken parallel to the axis of rotation, a hollow cylinder will result, and its volume will be given by an expression of the approximate form $2\pi y(1-x)\delta y$.

Example 9.20

Find the volume obtained when the area bounded by the curve $y = x^{\frac{3}{2}}$, the x-axis and the line $x = 4$ is rotated about the x-axis.

Considering an element perpendicular to the axis of rotation (Fig. 9.47), we can say that as the curve is rotated, the element traces out a disc. The volume of this disc is $\pi y^2\,\delta x$, and we can say that the total volume obtained by rotating the given area about the x-axis is

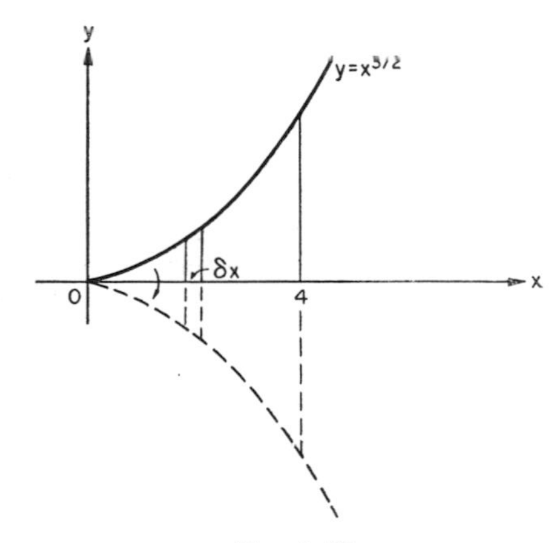

FIG. 9.47.

the value of

$$\lim_{\delta x=0} \sum \pi y^2\,\delta x.$$

Replacing the limiting value of the sum by the integral, we can write that

$$V = \int_0^4 \pi y^2\,dx,$$

and replacing y^2 by x^3,

$$V = \pi \int_0^4 x^3 \, dx$$

$$= \left[\frac{\pi x^4}{4}\right]_0^4$$

$$= 64\pi.$$

The volume formed is thus 64π.

Example 9.21

A solid of uniform density ϱ is formed by rotating the area bounded by the curve $y = x^3$, the x-axis and the line $x = 2$ about the line $x = 2$. Find (i) its mass, (ii) the position of its centre of gravity.

(i) Considering, as usual, an element perpendicular to the axis of rotation (Fig. 9.48a),

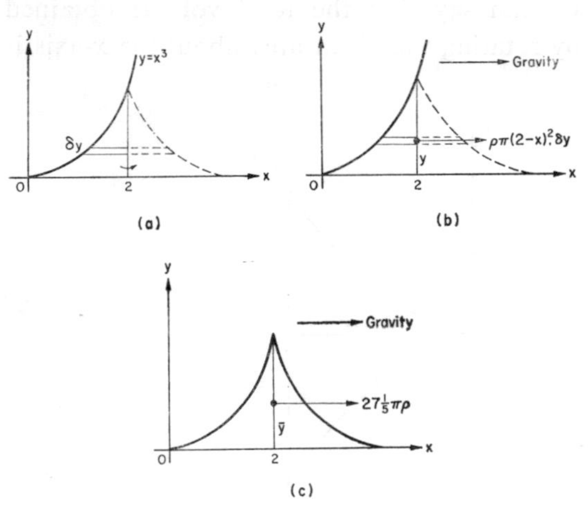

FIG. 9.48.

we can write that the mass of the disc formed by the rotation is $\varrho\pi(2-x)^2 \, \delta y$. Thus the total mass is given by the value of

$$\int_0^8 \varrho\pi(2-x)^2 \, dy,$$

remembering for the limits of integration that y varies from 0 to 8 if x goes from 0 to 2.

Thus

$$M = \pi\varrho \int_0^8 (2-x)^2 \, dy$$

$$= \pi\varrho \int_0^8 [4-4y^{\frac{1}{3}}+y^{\frac{2}{3}}] \, dy,$$

expanding the integrand and replacing x by $y^{\frac{1}{3}}$ (since $y = x^3$). Integrating,

$$M = \pi\varrho[4y-4.\tfrac{3}{4}y^{\frac{4}{3}}+\tfrac{3}{5}y^{\frac{5}{3}}]_0^8$$

$$= \pi\varrho[32-4.\tfrac{3}{4}.16+\tfrac{3}{5}.32]$$

$$= \pi\varrho[32-48+19\tfrac{1}{5}]$$

$$= 3\tfrac{1}{5}\pi\varrho.$$

(ii) By symmetry, $\bar{x} = 2$.

By imagining the direction of gravity to be parallel to the x-axis, the moment of the weight of an elementary disc about the x-axis can be written as $\varrho.\pi(2-x)^2.\delta y.y$ (its centre of gravity being at its centre) (Fig. 9.48b). Thus the total moment of the solid about the x-axis is

$$\varrho = \int_0^8 \pi\varrho y(2-x)^2 \, dy$$

$$= \pi\varrho \int \{4y-4y^{\frac{4}{3}}+y^{\frac{5}{3}}\} \, dy$$

$$= \pi\varrho[2y^2-4.\tfrac{3}{7}y^{\frac{7}{3}}+\tfrac{3}{8}y^{\frac{8}{3}}]_0^8$$

$$= \pi\varrho[2{\cdot}54-4.\tfrac{3}{7}.128+\tfrac{3}{8}.256]$$

$$= \pi\varrho\{128-219\tfrac{3}{7}+96\}$$

$$= 4\tfrac{4}{7}\pi\varrho.$$

Placing the centre of gravity at the point $(2, \bar{y})$ enables us to equate the moment of the total weight about the x-axis with this total moment (Fig. 9.48c):

i.e. $\quad 3\tfrac{1}{5}\pi\varrho\bar{y} = 4\tfrac{4}{7}\pi\varrho$

or $\quad \tfrac{16}{5}\bar{y} = \tfrac{32}{7}$

and $\quad \bar{y} = \tfrac{10}{7}.$

Thus the total mass is $3\tfrac{1}{5}\pi\varrho$, and the centre of gravity is $(2, \tfrac{10}{7}.)$

Example 9.22

A solid is formed by rotating the part of the curve $y = x^{\frac{3}{2}}$ for $0 \leq x \leq 4$ about the y-axis. If the density at any point in the solid is one-half of the ordinate of that point, find the total mass of the solid.

If we imagine an element parallel to the x-axis and of thickness δy, we can say that the density at all points in the element is the same. The mass of the disc formed by rotating this element about the y-axis is $\pi x^2 \varrho\, \delta y$ (Fig. 9.49),

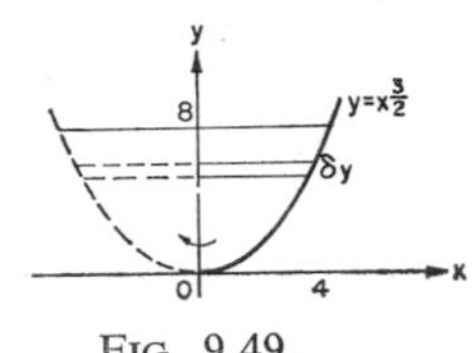

FIG. 9.49.

where $\varrho = \frac{1}{2}y$, and so the total mass of the solid is

$$\int_0^8 \pi x^2 \varrho\, \delta y = \pi \int_0^8 y^{\frac{4}{3}} \cdot \tfrac{1}{2} y\, dy$$

$$= \frac{\pi}{2}\left[\frac{3}{10} y^{\frac{10}{3}}\right]_0^8$$

$$= \frac{3\pi}{20} \cdot (1024)$$

$$= 151 \cdot 6\pi.$$

EXERCISE 9h

Find the volumes obtained when the areas bounded by the following curves and lines and the x-axis are rotated about the x-axis, and sketch the shape of the resulting solid in each case:

1. $y = \sqrt{x}$, $x = 4$. **3.** $y = 2/x$, $x = 1$, $x = 4$.
2. $y = x^2$, $x = 3$. **4.** $y = \sqrt{(1-x^2)}$, $x = 0$.

Find the volumes obtained when the areas bounded by the following curves and lines and the y-axis are rotated about the y-axis:

5. $y^2 = x+1$, $x = 0$. **6.** $y = x^2$, $y = 4$.
7. $y^2 = 1-1/x^4$, $x = 0$, $y = 0$, $y = \frac{1}{2}$.

8. By finding the volume obtained when the quadrant of $x^2+y^2 = a^2$ between $y = 0$ and $x = 0$ is rotated about the x-axis, show that the volume of a sphere of radius a is $\frac{4}{3}\pi a^3$.

9. By considering a cone to be the solid obtained when the line $y = (r/h)x$ between $x = 0$ and $x = h$ is rotated about the x-axis, show that the volume of a cone of height h and base radius r is $\frac{1}{3}\pi r^2 h$.

10. The size and shape of a pillar are determined by revolving about the x-axis that part of the curve $20y^2 = 80 - x$ which lies between the limits $x = 2$ m and $x = 6$ m. Find by integration the volume of the pillar, taking π to be $\frac{22}{7}$. (OC)

11. Find the volume obtained when the area between $y = 1-x^2$ and the x-axis is rotated about the x-axis.

12. Find the volume of a solid obtained by rotating the area contained by $y = 4-x^2$ and $y = 2$ about the line $y = 2$.

13. Find the volume obtained when the area between $y = \sqrt{x}$, the x-axis and $x = 9$ is rotated about the line $x = 9$.

14. A bowl is in the shape of a sphere of inside diameter 20 cm. Calculate the volume of water it contains when the maximum depth is (i) 2·5 cm; (ii) 3·75 cm; (iii) 5 cm.

15. A circle whose radius is 12 cm is revolved about a diameter so as to describe a sphere. Find by integration (giving your answer as a multiple of π) the volume of that portion of the sphere which is contained between two parallel planes which are 4 cm and 10 cm from the centre and on the same side of it.
Find also what fraction of the whole sphere is contained in this portion. (If a formula is used for the whole volume, it should be proved.) (OC)

16. A portion of the curve $y = x^2$ is cut off between the lines $x = 2$ and $x = 3$. Find the area bounded by these lines, the x-axis, and the curve.
This area is rotated through four right angles about the x-axis. Find the volume thus formed. (Give the answer as a multiple of π.) (OC)

17. Find the points of intersection of the curve $y = x(4-x)$ and the line $y = 2x$.
Find the volume generated when the area enclosed between the curve and the line makes one complete revolution about the x-axis. (OC)

18. Find the position of the centre of gravity of a uniform solid obtained by rotating the curve $y = 5\sqrt{x}$ between $x = 0$ and $x = 4$ about the x-axis.

19. Find the position of the centroid of the solid of uniform density obtained by rotating the curve $x = \sqrt{(2-y)}$ about the x-axis.

20. Find by integration
(i) the distance of the mean centre (centre of gravity) of a semicircular lamina from the bounding diameter,
(ii) the distance of the mean centre of a solid hemisphere from the bounding plane.

Deduce without further integration the positions of the mean centres of a lamina in the shape of a quadrant of a circle, and a solid in the shape of an octant of a sphere. (Take the radius to be r in each case.) (SU)

21. Find the position of the centre of gravity of a uniform cone.

22. Find the position of the centre of gravity of a cap of a hemisphere, given that the hemisphere from which it is cut is of radius 8 cm and the depth from the top of the cap to the mid-point of its base is 5 cm.

23. A bar of uniform density has a square cross-section. The side of the cross-section at one end is 5 cm, and this tapers steadily to 3 cm at the other, the length of the bar being 10 cm. Where is the centre of gravity of the bar?

24. The density at a point in the solid obtained by rotating the portion of the curve $y = 1-x^3$ between $x = 0$ and $x = 1$ about the x-axis is proportional to the distance of that point from the y-axis. Find the position of the centroid of the body.

25. The density at any point in a cone is proportional to the square root of the distance of that point from the base. Find the position of the centroid of the cone.

26. A vessel has the shape formed by rotating the curve $4y = x^2$ through 2π radians about the axis of y which is vertical. Find the volume of liquid which will fill the vessel to a depth of h cm.

If a hole at the lowest point of the vessel allows liquid to escape at the rate of 4π cm³/s, prove that the rate at which the level falls when the depth is h cm is $1/h$ cm/s. Find the time taken for the depth to decrease from 8 cm to 6 cm. (OC)

27. The area bounded by the parabola $y^2 = 20x$ and the lines $x = 0$ and $y = 10$ is rotated about the y-axis. Show that the volume of the solid obtained is 50π, and find the volume of the solid obtained by rotating the same area about the x-axis. (OC, part question)

28. The curve $\sqrt{x}+\sqrt{y} = \sqrt{a}$ touches the axes of x and y at A and B, respectively. Prove that the area enclosed by OA, OB (where O is the origin) and the arc AB is $a^2/6$, and find the coordinates of the mean centre of this area. Find also the volume generated if this area is rotated through four right angles about the axis of x. (SU)

29. Find the coordinates of the points at which the curve $y = 4x^2$ and the straight line $y = 8x$ intersect and calculate the area enclosed between the line and the curve. Find also the volume of the solid of revolution formed when this area is revolved through four right angles about the x-axis. (L)

30. Sketch the curve $y = x^2/2a+a$, where a is a positive number. Find the area bounded by the curve, the line $x = 4a$, and the coordinate axes.

Show also that the volume of the solid of revolution generated when this area is rotated about the y-axis is $80\pi a^3$ cubic units. (NUJMB)

31. Find the area of the portion of the plane enclosed by the curve $y = 1+\sin x$, the axis of y, and the axis of x from 0 to $\frac{3}{2}\pi$.

Find also the volume of the solid obtained by rotating this area about the axis of x. (Do not substitute for π in your answers.) (SU)

32. The part of the curve $y = 10^x$ between $x = 0$ and $x = 1$ is rotated about the x-axis. Show that the volume so formed has a magnitude given by the value of

$$\int_0^1 10^{2x}\,dx.$$

Using Simpson's rule, find its approximate value.

33. Find the approximate value of the volume formed by rotating the area bounded by $y = 1/\sqrt{x}$, $x = 1$, $x = 4$ and the x-axis about the x-axis.

†34. The cross-section of a bar length 8 cm tapers uniformly from a circle of radius 2 cm to one of radius 1 cm. Where is its centre of gravity if the density

(i) is uniform,

(ii) at any point is directly proportional to the distance of that point from the larger end?

†35. Sketch the curve $y^2 = x^4-x^5$, showing clearly the form at the origin. (The turning values need not be calculated.) Find the area of the loop and the volume described when the loop is rotated about the axis of x. (SU)

SUMMARY

The value of a definite integral such as

$$\int_a^b f(x)\,dx$$

can be found by obtaining the indefinite integral $F(x)$ of $f(x)$, and evaluating $\{F(b)-F(a)\}$. The differential equation $dy/dx = f(x)$ can be solved by integrating both sides with respect to x, giving $y = \int f(x)\,dx+c$.

The mean value of a function over a range is the magnitude of the constant value function whose integral over the same range is equal to the integral of the function in question. If $\bar{x}$ denotes the mean value of $f(x)$ for $a \leqslant x \leqslant b$, then

$$\bar{x} = \frac{\int_a^b f(x)\,dx}{b-a}.$$

The integral can be regarded as the limiting value of a sum:

$$\lim_{\delta x=0} \sum_{x=a}^{x=b} f(x)\,\delta x = \int_a^b f(x)\,\mathrm{d}x.$$

This concept has applications in finding the areas of laminae bounded by given lines and curves, the masses of non-uniform laminae, the centres of gravity of various bodies, and the volumes of bodies with circular symmetry.

The x-coordinate of the centre of gravity of a body bounded by $y = f(x)$, the x-axis and the ordinates $x = a$ and $x = b$ is given by the equation

$$\bar{x} = \frac{\int_a^b \varrho x y\,\mathrm{d}x}{\int_a^b \varrho y\,\mathrm{d}x}.$$

The volume V of a body obtained by rotating about the x-axis that part of $y = f(x)$ which lies between $x = a$ and $x = b$ is given by

$$V = \int_a^b \pi y^2\,\mathrm{d}x.$$

Miscellaneous Exercise 9

Write down the values of:

1. $[x]_3^5$.

2. $[x]_{-3}^5$.

3. $[x^2]_{-4}^3$.

4. $[7x-1]_{-5}^{-2}$.

5. $[1-3x]_{-2}^{-1}$.

6. $[\tan x]_0^{\pi/4}$.

7. $[\cot^2 x]_0^{\pi/4}$.

8. $[1-\sin x]_0^{\pi/3}$.

9. $\left[\dfrac{1}{x-7}\right]_8^{10}$.

10. $\left[\dfrac{x}{x+1}\right]_0^1$.

11. $[\cos x]_{-\pi/4}^0$.

12. $\left[\dfrac{7-x}{8-3x}\right]_{-2}^{-1}$.

13. $[x \sin x]_0^{\pi/2}$.

14. $\left[\dfrac{\sin x}{x}\right]_{-\pi/2}^{\pi/2}$.

15. $[\tan^{-1} x/4]_0^4$.

16. $[\sin^{-1} \tfrac{1}{2}x]_0^1$.

17. $\left[\dfrac{1}{x^2}\right]_{-3}^{-2}$.

18. $\left[\dfrac{2}{\sqrt{x}}\right]_4^9$.

19. $[(1-3x)^3]_{-1}^2$.

Evaluate the following definite integrals:

20. $\int_1^5 8x\,\mathrm{d}x$.

21. $\int_0^4 p^2\,\mathrm{d}p$.

22. $\int_0^1 \dfrac{\mathrm{d}t}{\sqrt{(1-t^2)}}$.

23. $\int_{-1}^0 (2z-5)\,\mathrm{d}z$.

24. $\int_{-4}^{-2} (5-x)\,\mathrm{d}x$.

25. $\int_{-6}^{-2} (2x+3)\,\mathrm{d}x$.

26. $\int_0^{\frac{1}{6}} \dfrac{\mathrm{d}x}{\sqrt{(1-9x^2)}}$.

27. $\int_{-2}^{-1} (4-p)\,\mathrm{d}p$.

28. $\int_{-2}^{-1} \dfrac{p^2-4}{p+2}\,\mathrm{d}p$.

29. $\int_0^1 x\sqrt{(1-x^2)}\,\mathrm{d}x$.

30. $\int_{-1/2}^{-1/4} \dfrac{\mathrm{d}t}{t^3}$.

31. $\int_0^1 p(1+p^2)^3\,\mathrm{d}p$.

32. $\int_0^4 \dfrac{\mathrm{d}p}{16+p^2}$.

33. $\int_0^{\frac{4}{3}} \dfrac{\mathrm{d}p}{16+9p^2}$.

34. $\int_0^1 \dfrac{p\,\mathrm{d}p}{\sqrt{(1-p^2)}}$.

35. $\int_0^{\frac{1}{3}} \dfrac{x}{\sqrt{(4-9x^2)}}\,\mathrm{d}x$.

36. $\int_0^{\frac{1}{2}} \dfrac{\mathrm{d}\theta}{(1+\theta)^4}$.

37. $\int_{-3}^{-2} (5-2q)\,\mathrm{d}q$.

38. $\int_{-3}^{-2} (5-2q)^2\,\mathrm{d}q$.

39. $\int_0^{\pi/2} x \sin x^2\,\mathrm{d}x$.

40. $\int_0^{\pi/4} \sec^2 x \tan x\,\mathrm{d}x$.

41. $\int_0^{\pi/4} \operatorname{cosec}^2 x \cot^4 x\,\mathrm{d}x$.

42. $\int_0^1 \dfrac{\theta\,\mathrm{d}\theta}{\sqrt{(9-4\theta^2)}}$.

43. $\int_{\pi/6}^{\pi/3} \dfrac{\sec^2 p}{\tan^3 p}\,\mathrm{d}p$.

44. $\int_0^{\pi/6} \sec x \tan x\,\mathrm{d}x$.

45. $\int_0^1 q\sqrt{(9-5q^2)}\,\mathrm{d}q$.

46. $\int_0^{\pi/12} \sin 3x \cos^9 3x\,\mathrm{d}x$.

47. $\int_0^{\pi/32} \dfrac{\sin 8x}{\cos^7 8x}\,\mathrm{d}x$.

48. $\int_{\pi/2}^{\pi} \sin (x-\tfrac{1}{2}\pi)\,\mathrm{d}x$.

49. $\int_0^{\frac{\pi}{4}} \cos^2 \theta\,\mathrm{d}\theta$.

50. $\int_0^{\frac{2}{3}} \dfrac{\mathrm{d}x}{(16-9x^2)}$.

51. $\int_0^{\frac{2}{3}} \dfrac{x\,\mathrm{d}x}{\sqrt{(16-9x^2)}}$.

52. $\int_{\pi/30}^{\pi/15} \operatorname{cosec} 5\theta \cot 5\theta \, d\theta$.

53. $\int_{0}^{\pi/32} \sec^2 8p \tan^2 8p \, dp$.

54. $\int_{0}^{\pi/8} \sec^2 2q \tan 2q \, dq$. **55.** $\int_{0}^{5/2} \frac{dz}{25+z^2}$.

56. $\int_{\pi/30}^{\pi/20} \frac{\cos 5x}{\sin^8 5x} \, dx$. **57.** $\int_{0}^{\pi/2} \frac{\cos x}{(1+\sin x)^2} \, dx$.

58. $\int_{0}^{4/5} \frac{3x \, dx}{\sqrt{(16-25x^2)}}$.

Find the mean values of the following functions over the ranges indicated:

59. $6x^5$, $0 \leqslant x \leqslant 2$. **60.** $(9x^2-7)$, $0 \leqslant x \leqslant 2$.

61. $(x-4)(3x-2)$, $1 \leqslant x \leqslant 3$.

62. $x\sqrt{(17-9x^2)}$, $\frac{1}{3} \leqslant x \leqslant 1$.

63. $\frac{1}{\sqrt{(16-9x^2)}}$, $0 \leqslant x \leqslant \frac{4}{3}$.

64. $\sec^2 x \tan x$, $0 \leqslant x \leqslant \frac{1}{4}\pi$.

65. $\sin x (1+\cos x)^4$, $0 \leqslant x \leqslant \frac{1}{3}\pi$.

66. $\frac{x^2}{(2-x^3)^4}$, $0 \leqslant x \leqslant 1$.

67. $(1-8x)^7$, $0 \leqslant x \leqslant \frac{1}{8}$.

68. $\cos^2 x$, $-\pi/2 \leqslant x \leqslant \pi/2$.

69. $(5-x)$, $-3 \leqslant x \leqslant 9$.

70. $\frac{\sec^2 x}{1+\tan^2 x}$, $0 \leqslant x \leqslant \pi/4$.

71. $\sec^2 x \tan^5 x$, $-\pi/4 \leqslant x \leqslant \pi/4$.

72. Find the areas enclosed by the x-axis and the following curves and lines:

(i) $y = ax^4+b$, $x = 0$, $x = 2$;

(ii) $y = (7x+1)^{\frac{1}{3}}$, $x = 0$, $x = 1$;

(iii) $y = \sin^2 x$, $x = \frac{1}{2}\pi$;

(iv) $y - \sec^2 x$, $x = 0$ and $x = \pi/20$.

73. Find the areas enclosed by the following curves and lines:

(i) $y = (3-2x)^2$, $y = 9$;

(ii) $y = \sqrt{(x+1)}$, $x = 0$, $y = 0$;

(iii) $y^3 = 8x^2$, $y = 2$, $x = 0$.

74. The part of the curve $y^2 = 4-x^2$ lying between $x = 1$ and $x = 2$ is rotated through four right angles about the x-axis. Find the volume generated. Supposing this volume to be filled with matter of uniform density, find the position of its centre of gravity. (OC)

75. A flat, thin plate of uniform density is bounded by the two curves $y = x^2$, $y = -x^2$ and the line $x = 2$. Find its area. Find also the coordinates of its centre of gravity. (OC)

76. Referring to Fig. 9.50, write down the values of the following integrals in terms of the areas P, Q, R and S:

(i) $\int_0^a [g(x)-f(x)] \, dx$,

(ii) $\int_a^b [g(x)-f(x)] \, dx$,

(iii) $\int_0^b [g(x)-f(x)] \, dx$,

(iv) $\int_0^b |g(x)-f(x)| \, dx$.

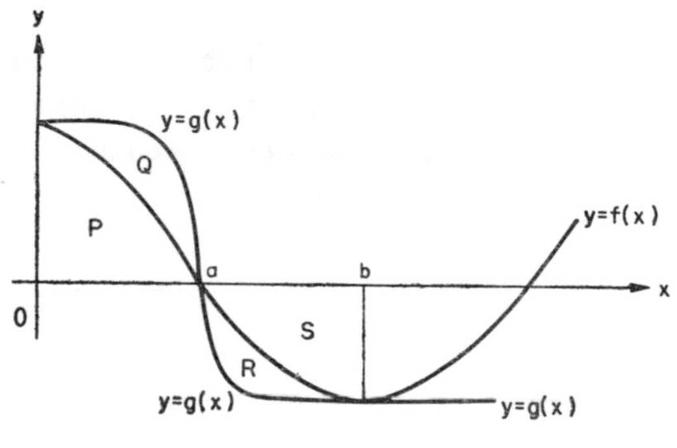

FIG. 9.50.

77. Find (approximately) the mean value of x^x between $x = 0$ and $x = 5$.

78. Find (approximately) the x coordinate of the centre of gravity of the uniform lamina bounded by $y = 1/\sqrt{(x-1)}$, $x = 4$, $x = 9$ and $y = 0$.

79. A solid has a circular base described by the equation $x^2+y^2 = a^4$; its section perpendicular to the z-axis at any point is a circle of radius (a^2+z^2). If its height is a, find its volume.

80. A cylindrical hole of radius b is bored through the centre of a sphere. Find the volume of the remainder of the sphere, given that its radius is a.

81. From a solid sphere of radius a, a minor segment is cut off by a plane distant $\frac{1}{2}a$ from the centre. Find the volume of the segment.

Prove that the centre of gravity of the segment is at a distance $7a/40$ from its plane face. (SU)

82. A segment is cut off from the curve $y^2 = 4ax$ by the line $y = x$. Find the volume of the solid obtained by rotating this segment about the x-axis. (L)

83. Find the volume of the solid formed when the half ellipse $x^2/a^2 + y^2/b^2 = 1$ for $0 \leqslant x \leqslant a$ is rotated about the x-axis. Find also the centroid.

84. The area between the circle $x^2 + y^2 = 9$ and the ellipse $x^2/25 + y^2/9 = 1$ is rotated through four right angles about the x-axis. What is the volume of the solid so formed?

85. The cross-sectional area of a solid in a plane perpendicular to the x-axis at a distance x from the origin is $S(x)$. Show that the volume δV of an element of the solid between two planes at x and $(x+\delta x)$ satisfies the inequality

$$S(x)\,\delta x \leqslant \delta V \leqslant S(x+\delta x)\,\delta x$$

if

$$S(x+\delta x) > S(x),$$

or

$$S(x)\,\delta x \geqslant \delta V \geqslant S(x+\delta x)\,\delta x$$

if

$$S(x+\delta x) < S(x).$$

Hence show that $\mathrm{d}V/\mathrm{d}x = S$, and state the conditions under which this equation holds.

Find the areas of the laminae bounded by the following lines and curves:

The x-axis and:

86. $y = 7x^2 + x$, $x = 3$, $y = 0$.

87. $y = 5\sqrt{x}$, $x = 9$.

88. $y = 1/3x^2$, $x = \frac{1}{9}$, $x = 1$.

89. $y = \sqrt{(8x+9)}$, $x = 0$, $x = 2$.

90. $y = 1/(1-2x)^2$, $x = 0$, $x = \frac{1}{4}$.

91. $y = \cos^2 2x$, $x = 0$, $x = \frac{1}{3}\pi$.

92. $y = 1/\sqrt{(9-4x^2)}$, $x = 0$, $x = \frac{3}{4}$.

93. $y = x\sqrt{(9-16x^2)}$, $x = 0$, $x = \frac{1}{4}$.

The y-axis and:

94. $y = x^2$, $y = 9$.

95. $y = \sqrt{x}$, $y = 2$.

96. $y = 128x^7$, $y = 128$.

97. $y = 1/x^2$, $y = 3$, $y = 4$.

98. $y^2 = x-4$, $y = 1$, $y = 0$.

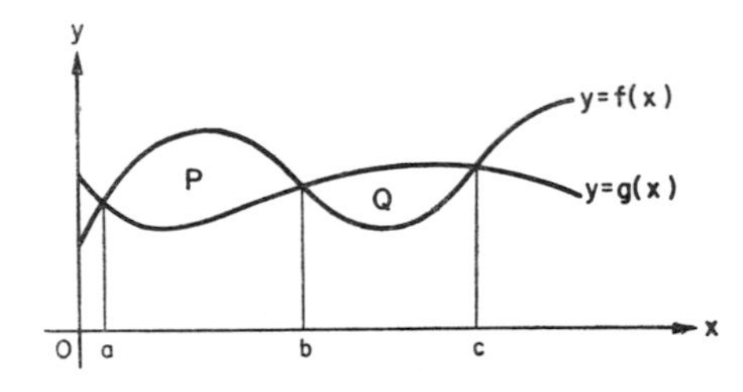

FIG. 9.51.

99. With reference to Fig. 9.51, describe in terms of P and Q the values of the following integrals:

(i) $\int_a^b [f(x)-g(x)]\,\mathrm{d}x$,

(ii) $\int_b^c [f(x)-g(x)]\,\mathrm{d}x$,

(iii) $\int_a^c [f(x)-g(x)]\,\mathrm{d}x$,

(iv) $\left|\int_a^c [f(x)-g(x)]\,\mathrm{d}x\right|$,

(v) $\int_a^c |[f(x)-g(x)]|\,\mathrm{d}x$.

100. Find the position of the centre of gravity of a frustum of a cone of height h, the radii of its faces being a and b $(a < b)$.

Appendix

TO SHOW THAT $\lim_{\theta \to 0+} \left(\frac{\sin \theta}{\theta} \right) = 1$

CONSIDER a circle of radius r whose centre is O (Fig. X.1). Construct two radii OP and OR so that $P\hat{O}R = \theta$ (radians), and produce OR so that it meets the tangent to the circle at P. Let the point of intersection of the tangent and the produced radius be S.

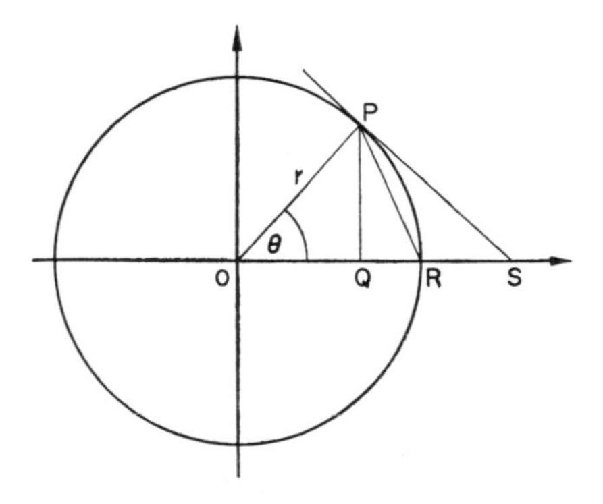

FIG. X.1.

Draw the perpendicular from P to OR, and let this meet OR at Q.

Now considering the triangle OPR, the sector OPR, and the triangle OPS, we can write that

$$\Delta OPR < \text{sector } OPR < \Delta OPS,$$

denoting the area of a triangle by the prefix Δ. We can also write that

(i) $\Delta OPR = \frac{1}{2} OR.PQ = \frac{1}{2} r.r \sin \theta$;

(ii) sector $OPR = \frac{1}{2} r^2 \theta$
(since the area of a sector subtending an angle θ at the centre of a circle of radius r is $\frac{1}{2} r^2 \theta$);

(iii) $\Delta OPS = \frac{1}{2} OP.PS = \frac{1}{2} r.r \tan \theta$
(OPS being a right angle). Thus
$\frac{1}{2} r^2 \sin \theta < \frac{1}{2} r^2 \theta < \frac{1}{2} r^2 \tan \theta$.

Dividing this inequality throughout by $r^2 \sin \theta$, we obtain

$$1 < \frac{\theta}{\sin \theta} < \sec \theta.$$

Now this inequality holds provided $\theta \neq 0$ (in which case we could not have divided throughout by $\sin \theta$), and we can see that since $\cos \theta$ approaches 1 as θ approaches 0 from above,* the value of $\theta/\sin \theta$ becomes "sandwiched" between 1 and something just

* This result can be derived in graphical terms by considering the continuity of $\cos \theta$ in the neighbourhood of $\theta = 0$. An analytical proof is rather beyond the scope of this book, and it will suffice to mention that the trigonometrical functions $\sin x$ and $\cos x$ can be defined as the series

$$\left(x - \frac{x^3}{3.2.1} + \frac{x^5}{5.4.3.2.1} - \ldots \right)$$

and

$$\left(1 - \frac{x^2}{2.1} + \frac{x^4}{4.3.2.1} - \ldots \right)$$

respectively. All the usual trigonometrical relationships can be derived from these two series, and differentiation, in a term-by-term fashion, establishes the derivatives of p. 109.

above 1 (if $\cos\theta \to 1-$, $1/\cos\theta \to 1+$). Thus the value of $\theta/\sin\theta$ must approach 1 as θ approaches 0. In this case its reciprocal $\sin\theta/\theta$ also approaches 1, and we can write that

$$\lim_{\theta\to 0+}\left(\frac{\sin\theta}{\theta}\right) = 1.$$

EXERCISE

1. Prove that

$$\lim_{\theta\to 0+}\left(\frac{\sin\theta}{\theta}\right) = 1,$$

using a circle of radius 1 (i.e. let $OP = 1$).

2. Draw sketches of the graphs of $y = \cos x$, $y = x/\sin x$ and $y = \sec x$ for $0 \leqslant x \leqslant \pi/2$, using only one pair of areas. For what value or values within the stated range are the functions undefined? Explain how your answer demonstrates that

$$\cos\theta \leqslant \theta/\sin\theta \leqslant 1/\cos\theta \quad \text{for} \quad 0 \leqslant \theta \leqslant \pi/2.$$

3. Prove that the area of a sector of a circle is $\frac{1}{2}r^2\theta$, where r is the radius and θ radians is the angle at the centre. (If the formula πr^2 for the area of a circle is used, it should be proved.)

If the bounding radii of the sector are OA and OB, and $\theta < \pi$, find a formula for the area of the minor segment of the circle cut off by the chord AB. If this segment has the same area as triangle AOB, find angle AOB to the nearest half degree. (A graphical method is recommended.) (SU)

Miscellaneous Exercises

MISCELLANEOUS EXERCISE A (CHAPTERS 1–3)

1. Draw on a set of graphs the boundaries defined by the following sets of inequalities, shading in the area common to all members in each case:

(a) $|x| < 2, \quad 0 < x < 6,$

(b) $|x-2| > 5, \quad x > 2,$

(c) $|x-3| < 2, \quad -2 < x < 2,$

(d) $|1-1/x| > \frac{1}{2}, \quad x > 7$ and $x < -3,$

(e) $|3-x| < 3, \quad 1 < x < 5,$

(f) $|x-2| < |x+2|, \quad x > 0$ and $x < -1,$

(g) $|x^2-1| < \sqrt{3}, \quad -2 < x < 2,$

(h) $|2-x| > 1, \quad x < 1$ or $x > 3.$

2. Prove that

(i) $||x|-|y|| \leqslant |x-y|$, (ii) $|x-y| \leqslant |x|+|y|$.

3. Show that the equation $ax^2+bx+c = 0$ can be expressed in the form

$$\left(x+\frac{b}{2a}\right)^2 = \frac{b^2-4ac}{4a^2}.$$

Hence show that $b^2-4ac \geqslant 0$ if x is real, and $ax^2+bx+c = 0$.

4. If $f(x) = x^2$, show that $f(x+h)-f(h) = 2xh+h^2$.

5. Show that the sum of two step functions is itself a step function.

6. Sketch graphs of the functions:

(a) $x+[x]$,

(b) $x-[x]$,

(c) $[-x]$,

(d) $[x+\frac{1}{4}]$,

(e) $[x]+[x+\frac{1}{2}]$.

7. Sketch the graph of $y = [1/x]$.

8. Sketch the graph of the function $f(x)$ defined by the equations $f(x) = \frac{1}{2}(3-x^2)$ for $1 \geqslant x$; $f(x) = 1/x, \ x \geqslant 1$.

9. Sketch the graph of $y = x+1/x$, and explain why it is symmetrical about the line $x = 1$.

10. Express (a) 0·44444˙; (b) 0·123 123 123˙; (c) 2·02 02 02˙ as an infinite series of fractions, and find the sum to n terms for each. Thus express the infinitely recurring decimals as a ratio of two numbers.

11. Show that for all integral values of n, the number $(2n+1)$ is odd, and the number $(2n+2)$ is even.

12. Explain why the equations $\theta(x) = 1+x$ for $x \geqslant 0$ and $\theta(x) = 3+x$ for $x \leqslant 0$ do not define a functional relationship.

13. Describe the region occupied by the points representing $\phi(n)$ on a graph of $\phi(n)$ against n, where the n's are positive integers, for each of the following cases:

(i) $\phi(n) = (-1)^n$,

(ii) $\phi(n) = 100/n+(-1)^n$,

(iii) $1/n+100(-1)^n$,

(iv) $n(1+(-1)^n)$.

Is there any horizontal line or lines which can be drawn so that all the points lie to one side of the bounded region?

14. Regarding a function ϕ as continuous at a point $x = a$ if

$$\lim_{x \to a-} \phi(x) = \lim_{x \to a+} \phi(x) = \phi(a),$$

discuss whether each of the following functions $\phi(x)$ are continuous at $x = a$ or not, explaining carefully why they are not in the cases exhibiting discontinuities [in each case the dot is the point $(a, \phi(a))$]:

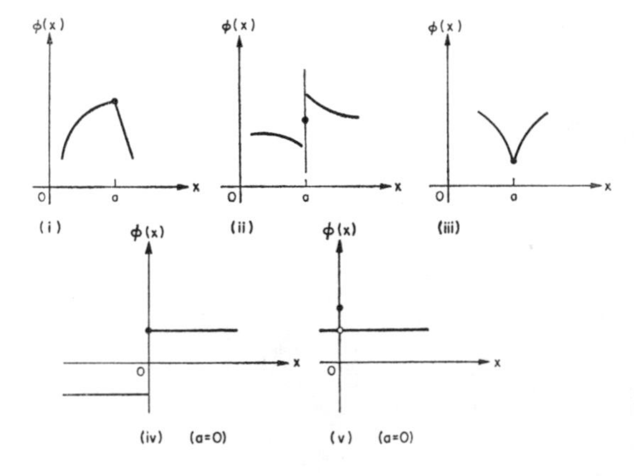

15. By rationalising the numerator, show that

(i) $\lim_{x=0}\left(\frac{1-\sqrt{(x+1)}}{x}\right) = -\frac{1}{2}$,

(ii) $\lim_{x=0}\sqrt{\frac{(1+x)-\sqrt{(1-x)}}{x}} = 1.$

16. Draw a sketch of the functions ϕ for which:

(a) $\lim_{x\to a+}\phi(x) = \lim_{x\to a-}\phi(x) \neq \phi(a)$,

(b) $\lim_{x\to a+}\phi(x) = \phi(a) \neq \lim_{x\to a-}\phi(x)$,

(c) $\lim_{x\to a+}\phi(x) \neq \lim_{x\to a-}\phi(x) \neq \phi(a)$.

There is no need to give algebraic equations which would describe the functions.

17. Give an algebraic expression of a function for which

$$\lim_{x\to 0+}\phi(x) = \lim_{x\to 0-}\phi(x)$$

and $\phi(0)$ is not defined. What can you say about

$$\lim_{x\to 0+}\phi(x),\ \lim_{x\to 0-}\phi(x) \text{ and } \phi(0)$$

if $\phi(x) \equiv 1/x^2$?

18. Draw a sketch graph of the function $[x]+[-x]$, and discuss its continuity.

19. Prove that

(i) $[2x] = [x]+[x+\frac{1}{2}]$, (ii) $[p+q] \geqslant [p]+[q]$.

20. Find out the world records for the 100, 200, 500, 1000, 1500 m and marathon races. Plot a graph of the average speed of the runners who set the records against the distance of the race, and discuss whether or not the value of this speed appears to approach a limit as the distance of the race increases without bound.

21. The following table gives pairs of values of the height and weight of 30 people. Plot a scatter diagram of the figures, and discuss whether there is any evidence of an association between height and weight. If there is, attempt to draw the best line you can through the cluster of points, and write down an equation relating the height to the weight.

Height (cm)	166	164	171	156	160	157
Weight (kg)	56	58	55	48	41	43
Height (cm)	162	165	160	173	175	157
Weight (kg)	45	46	47	44	55	77
Height (cm)	173	162	156	157	159	153
Weight (kg)	60	51	44	46	48	52
Height (cm)	173	178	147	155	162	150
Weight (kg)	60	66	40	42	48	42
Height (cm)	180	160	165	160	178	157
Weight (kg)	65	47	52	47	61	44

22. The following figures express the variation of the sensitivity of two microphones in a direction making an angle θ with their axes. Draw a polar diagram representing both sets of data, and comment on the results:

Sensitivity (arbitrary units)									
	1	50	35	0	35	50	35	0	35
	2	50	40	14	8	0	8	14	40
Direction ($\theta°$)		0	45	90	135	180	225	270	315

What sort of shape would you expect the polar diagram of the illumination of an electric lamp to have?

23. Show that the expression

$$\frac{2x^2+4x+7}{x^2+2x+5}$$

can be written in the form

$$a-\frac{b}{(x+x)^2+d}$$

where a, b, c and d are numbers. Hence or otherwise show that as x varies, the value of the function always lies between $1\frac{1}{4}$ and 2.

Draw a sketch of the graph of the function. (SU)

24. If $1/f = (\mu-1)(1/r_1+1/r_2)$, show that a small change $\delta\mu$ in μ produces a small change $\delta(1/f)$ in $1/f$, where $\delta(1/f) = (1/f)\,\delta\mu/(\mu-1)$.

25. If f_1 and f_2 are two functions of a variable x, show that the change in the value of the product $1/f_1f_2$ caused by a change δx in x is given approximately by

$$\frac{1}{f_1}\delta\left(\frac{1}{f_2}\right)+\frac{1}{f_2}\delta\left(\frac{1}{f_1}\right),$$

where $\delta(1/f_2)$ and $\delta(1/f_1)$ are the changes in the values of $1/f_2$ and $1/f_1$ caused by the change in x.

26. Prove that the number of primes is infinite.

27. (i) Explain why $2n+2$ and $2n+1$ are even and odd numbers respectively for all integral values of n.

(ii) Show that $4n^3+1$ is divisible by 3, given that n is an integer.

28. The hour hand of a clock has a lead of $\frac{1}{12}$ revolution over the minute hand at 1.0 p.m. When the minute hand is at 5 minutes past, the hour hand has moved $\frac{1}{12}\cdot\frac{1}{12}$ further: thus the total distance of the hour hand from the figure 12 is $(\frac{1}{12}+\frac{1}{12}\cdot\frac{1}{12})$ths of a revolution. When the minute hand has advanced to the second position, the hour hand has advanced to $(\frac{1}{12}+\frac{1}{12}\cdot\frac{1}{12}+$ $+\frac{1}{12}\cdot\frac{1}{12}\cdot\frac{1}{12})$ths of a revolution beyond 12. Find the limiting value of this process as the number of terms increases without bound, and hence deduce when the minute hand overtakes the hour hand.

29. Using the relationship $\sin(x+h)-\sin x = 2\cos(x+\frac{1}{2}h)\sin\frac{1}{2}h$, prove that the function $\sin x$ is continuous when $x = 0$, $x = \frac{1}{3}\pi$, and x has any value α.

MISCELLANEOUS EXERCISE B (CHAPTERS 4–8)

1. Find the rate of change of volume of a cube with respect to the length of its edge.

2. If $f(x) = (ax+b)\sin x+(cx+d)\cos x$, determine the values of a, b, c and d such that $f'(x) = x\cos x$.

3. Find the sum of the series $1+x+x^2+\ldots+x^n$, and hence, by differentiation, find the sum of the series $1+2x+3x^2+\ldots+nx^{n-1}$.

4. Sketch the graph of the functions $f(x)$ for which

(i) $f''(x) > 0$ for all values of x,

(ii) $f''(x) > 0$ for $0 \leqslant x < a$,
$f''(x) = 0$ for $a \leqslant x < b$,
$f''(x) < 0$ for $x \geqslant b$.

5. Sketch the graph of a function for which $f'(x) < 0$, and $f''(x) > 0$ for $0 \leqslant x \leqslant a$.

6. Differentiate with respect to x:

(i) $\sin(\cos x)$,
(ii) $\sin\{\sin(\sin x)\}$,
(iii) $\cos(\sin^2 x)$.

7. Find the derivative with respect to x of the expression $\sqrt{\{x+\sqrt{(x+\sqrt{x})}\}}$.

8. Find an expression for $f'(x)$ in terms of g' if:

(i) $f(x) = g(x^2)$,
(ii) $f(x) = g(\sin^2 x)$,
(iii) $f(x) = g\{g(x)\}$,
(iv) $f(x) = g(\sin^2 x)+g(\cos^2 x)$.

9. Find the approximate change of the volume of a cylinder of

(a) radius
(b) height

x, if x changes by δx.

Find also the change in its surface area for each of the two cases.

10. If $f(x) = g(x)h(x)$, find expressions for $f'(x)$ and $f''(x)$.

11. If $g(0) = 2$, and $g(x)$ is continuous for x, find the values of $f'(0)$ and $f''(0)$, where $f(x) = x^n g(x)$, for

(i) $n > 2$, (ii) $n = 2$.

12. If $f(x^2) = x$ for $x > 0$, find $f'(9)$.

13. Interpret the following theorems geometrically:

(i) *Mean value theorem:* if $a \neq b$ and $f(x)$ is a continuous function in the range $a \leqslant x \leqslant b$, then there is at least one value c between a and b for which $f(b)-f(a) = f'(c)(b-a)$, assuming that $f'(x)$ exists.

(ii) *Rolle's theorem:* if $f(x)$ is a continuous function for $a \leqslant x \leqslant b$, and if $f(a) = f(b)$, then there is at least one value c of x in the range $a \leqslant x \leqslant b$ for which $f'(c) = 0$.

14. If $f(x) = (x-a)^n g(x)$, the equation $f(x) = 0$ has n coincident roots at $x = a$. Deduce that $f'(x) = 0$ has $(n-1)$ coincident roots at $x = a$, and $f^r(x) = 0$ has $(n-r)$ coincident roots at $x = a$.

15. Show that the tangent to the parabola $y = x^2+px+q$ at the point whose abscissa is $(p+q)/2$ is parallel to the chord joining the points whose abscissae are p and q.

16. Discuss the continuity of the functions $f(x)$ and $f'(x)$ if $f(x) = \sqrt{x}$.

17. Find the shortest distance from the point $(0, 6)$ to the curve $x^2 = 4y$.

18. A window is in the shape of a rectangle surmounted by a semicircle whose diameter equals the width of the rectangle. The glass used in the rectangular part of the window transmits all the light incident upon it, but the glass used for the surmounting semicircle absorbs one half of the incident light. Find the dimensions of the window which will admit most light if the total perimeter (p) is fixed.

19. A lorry travels 100 km along a motorway at a speed of x km/h. The cost of the fuel it uses is 4p per dm³, and the rate at which it consumes the fuel is $4+x^2/100$ dm³/hr. If the driver is paid 50p per hour, find the speed at which the lorry should be driven so that the cost of the journey is least.

20. Show that the value of the ratio

$$\frac{f(x+\delta x)-f(x)}{\delta x}$$

for the functions $f(x)$ below can be expressed in the forms indicated in each case:

$f(x)$	$f\{(x+\delta x)-f(x)\}/\delta x$
$\frac{1}{x}$	$-\frac{1}{x(x+\delta x)}$
$\sqrt{x}$	$\frac{1}{\sqrt{(x+\delta x)}+\sqrt{x}}$
x^3	$3x^2+3x\,\delta x+(\delta x)^2$
x^{-2}	$\frac{-2}{x(x+\delta x)^2}-\frac{\delta x}{x^2(x+\delta x)^2}$

21. What is the gradient of the chord joining the points whose abscissae are 1 and 2 on the curve $y = x^2-1$?

22. Sketch the graphs of the sets of number pairs for which

(i) $\dfrac{dy}{dx}$ = a constant, (ii) $\dfrac{dy}{dx} = x$.

23. Discuss the continuity of $\phi(x)$ and $\phi'(x)$ at $x = 0$, a and b, where the graph of $\phi(x)$ against x is shown in the figure below.

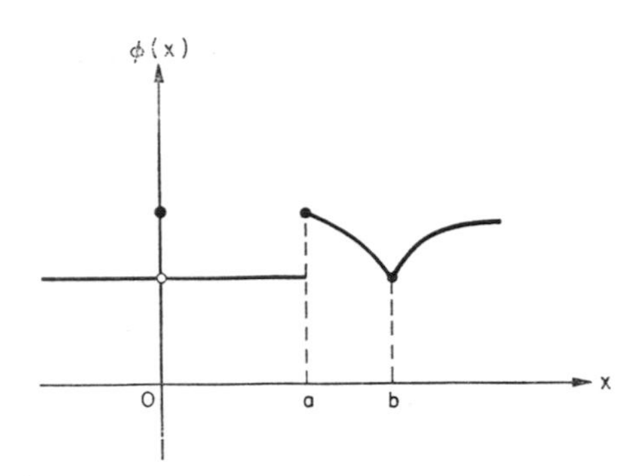

24. Explain why the gradient function of a curve with a kink in it at $x = a$ is discontinuous there.

25. Prove that the equation of the normal to the rectangular hyperbola $xy = c^2$ at the point $(ct, c/t)$ is $xt^3-yt = c(t^4-1)$. Four normals to the curve from a point meet the curve at P, Q, R, S. Prove that the pairs of lines PQ, RS; PR, QS; PS, QR are such that the lines in each pair are perpendicular to each other. (SU)

26. A curve is defined by parametric equations $x = a \cos^3 t$, $y = a \sin^3 t$. Prove that the equation of the tangent at the point given by the parameter t is

$$y \cos t+x \sin t = a \cos t \sin t.$$

Deduce that the tangents to the curve at the points given by parameters α and β meet at $R(X, Y)$, where

$$X = a \cos \alpha \cos \beta \frac{\cos \frac{1}{2}(\alpha+\beta)}{\cos \frac{1}{2}(\alpha-\beta)},$$

$$Y = a \sin \alpha \sin \beta \frac{\sin \frac{1}{2}(\alpha+\beta)}{\cos \frac{1}{2}(\alpha-\beta)}.$$

If $\alpha-\beta = \frac{1}{2}\pi$, evaluate X^2+Y^2 and Y/X in terms of α, and hence show that the locus of R as α varies satisfies the polar equation $2r^2 = a^2 \cos^2 2\theta$. (SU)

***27.** Two vertical conducting plates are connected together and held a distance b apart: a third plate is free to move laterally in a vertical plane between them, remaining vertical (see Fig. X.2). The outer plates are earthed, and the centre one is connected to a source of potential of magnitude V volts. The capacity (C) of the system when the centre plate is a distance x from one of the outer plates is

$$\frac{A}{4\pi x(b-x)},$$

and the energy E stored in the system is $(B-\frac{1}{2}CV^2)$, where B is a constant. Find an expression for the rate of change of E with respect to x, and by using the fact that the force F acting on the centre plate is $-dE/dx$ show that:

(i) if $x = \frac{1}{2}b$, there is no force on the centre plate;
(ii) if $x > \frac{1}{2}b$, the centre plate will move until it strikes the nearer outer plate;
(iii) if $x < \frac{1}{2}b$, the centre plate will move until it strikes the nearer outer plate.

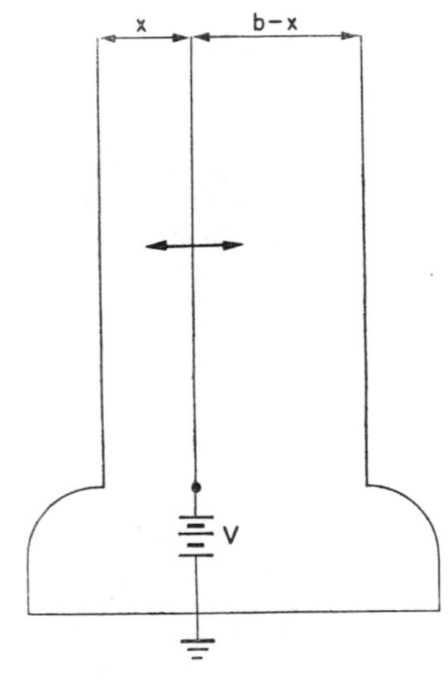

FIG. X.2.

***28.** If in question 27 the centre plate is disconnected from the source of potential, and the whole system is isolated, the energy E in the system is given by $N/2C$, where N is another constant. Show that there is no force on the centre plate if $b = \frac{1}{2}x$, and that it will again move to the nearer outer plate if it moves away from the centre at all.

***29.** A coil is to be wound in a hollow circular former of square cross-section. The mean radius of the coil is a, and the side of the square cross-section is b (Fig. X.3). Show that the number of turns of wire which can be wound on top of each other in the former is $b^2/4x^2$, where x is the radius of the wire used, and deduce the resistance of the wire if the resistivity is $\varrho\,\Omega$ cm. The coil is connected to a source of e.m.f. E and internal resistance r: find an expression for the field at the centre of the coil, and show that this is a maximum when $x^4 = \varrho ab^2/2r$.

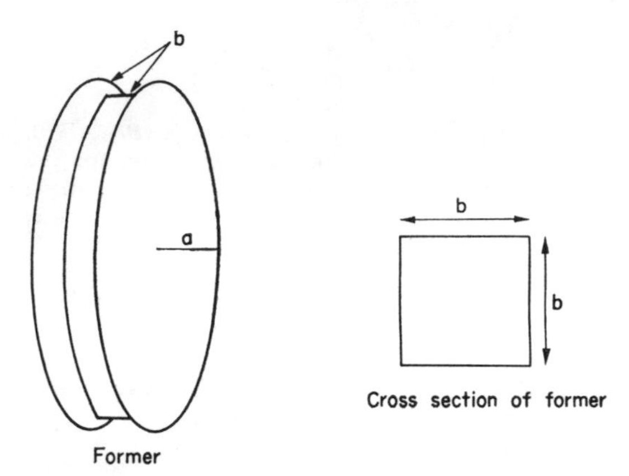

FIG. X.3.

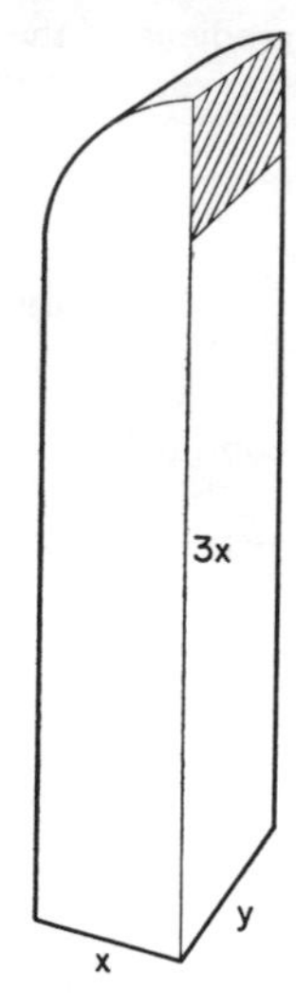

FIG. X.4.

30. Explain why a function must be continuous at a point in order to be differentiable at that point, and give an example to illustrate that the converse is not true. Sketch the graph of a function $f(x)$ such that $f(x) = x^2 \sin(1/x)$ for all x other than zero, and $f(0) = 0$, and write down the form of its derivative $f'(x)$. Explain why this derivative takes the value $+1$ or -1 at $x = 1/n\pi$ for any non-zero integral value of n, and explain why the value of

$$\lim_{x=0}\left\{\frac{f(x)-f(0)}{x}\right\} = \lim_{x=0}\left\{x \sin\frac{1}{x}\right\}$$

is zero.

Deduce that $\lim_{x\to 0} f'(x)$ does not exist, and hence that, although $f'(x)$ exists and $f(x)$ is differentiable at $x = 0$, the derivative $f'(x)$ is not continuous at $x = 0$.

Is it true to say that if a function is differentiable, its derived function must be continuous?

31. A curve is given by the parametric equations $x = a(t^2+1)$, $y = a/t$. Prove that the equation of the tangent at the point P, where the parameter is p, is $x+2p^3y = a(1+3p^2)$, and that it meets the curve again at the point Q, where the parameter is $-2p$. Prove that the least possible length of PQ is $(3\sqrt{3a})/2$. (SU)

32. A litter bin (see Fig. X.4) stands on a rectangular base of breadth x and length y. Each end is a rectangle of breadth x and height $3x$, surmounted by a quadrant of a circle of radius x. The bin is completely enclosed except for the rectangular opening at the top. It is made of thin metal, the total area of which (including the rectangular base and the curved top) is a given quantity S. If the total volume is to be as big as possible, prove that

$$\frac{y}{x} = \frac{2(12+\pi)}{14+\pi}. \qquad \text{(SU)}$$

33. Draw a graph of $\log_{10}x$, and use it to find the value of the rate of change of $\log_{10}x$ with x when $x = 2$. What is the value of $d(\log_{10}x)/dx$ when $x = 4$?

By finding the value of $d(\log_{10}x)/dx$ at three other points, draw a graph of $d(\log_{10}x)/dx$ against $1/x$, and guess an expression for $d(\log_{10}x)/dx$.

34. Explain why the use of the equation $dy/dx = (dy/dx)(dz/dx)$ fails at $x = 0$ if $y = z^2$, and $z = x^4 \sin(1/x)$ for $x \neq 0$, and $z = 0$ for $x = 0$.

Is the situation helped by defining $x \sin(1/x)$ and $x \cos(1/x)$ to be 0 at $x = 0$?

35. Draw sketches of the curves known as:

(i) the hyperbola, (iv) the caustic curve,
(ii) the catenary, (v) the equiangular spiral,
(iii) the cycloid, (vi) the normal curve,

and give a brief description of each, mentioning if possible one situation in which they arise in practice.

36. Prove that the function $1/(x-1)$ has no derivative at $x = 1$.

37. Sketch the graph of the function $|x|/x$.

MISCELLANEOUS EXERCISE C (CHAPTERS 7–9)

1. Write down the values of the following integrals:

(i) $\int_{-1}^{2} [x]\,dx$, (iii) $\int_{-1}^{2} [x+\frac{1}{2}]\,dx$,

(ii) $\int_{0}^{5} [2x]\,dx$, (iv) $\int_{-1}^{+1} [-x]\,dx$.

2. Show that:

(a) $\int_{a+c}^{b+c} f(x)\,dx = \int_{a}^{b} f(x+c)\,dx$,

(b) $\int_{ca}^{cb} f(x)\,dx = \int_{a}^{b} f(cx)\,dx$;

(c) $\int_{a}^{b} cf(x)dx = c\int_{a}^{b} f(x)\,dx$.

3. Show that $\int_{a}^{b} [x]\,dx + \int_{a}^{b} [-x]\,dx = a-b$.

4. Show that $|\int_{a}^{b} f(x)\,dx| \leqslant \int_{a}^{b} |f(x)\,dx|$ for all $f(x)$.

5. Write down the value of $\int_{0}^{1} x^n\,dx$, stating any restrictions on the value of n.

6. A function $f(x)$ is defined by the equations $f(x) = \frac{1}{2}x^2$ for $0 \leqslant x \leqslant 2$, and $f(x) = 4 - x$ for $2 < x \leqslant 4$. Write down the value of $\int_{0}^{4} f(x)\,dx$.

7. Show that $\int_{a}^{b} f(c-x)\,dx = \int_{c-a}^{c-b} f(x)\,dx$.

8. Given that $c \geqslant b$, show that $\int_{a}^{b} f(x)\,dx \leqslant \int_{a}^{c} f(x)\,dx$ only if $f(x) \geqslant 0$ for $a < x < c$.

9. Sketch the graph of $A(x)$, where $A(x) = \int_{0}^{x} f(t)\,dt$, and where $f(t)$ is equal to: (i) c; (ii) ct; (iii) ct^2; (iv) $[t]$, and where c is a constant.

10. Show that if $A(x)$ is the area of the cross-section of a solid made by a plane perpendicular to the x-axis at x, the volume of the solid between planes at $x = a$ and $x = b$ is

$$\int_{a}^{b} A(x)\,dx.$$

11. The cross-section of a solid at any point is a square perpendicular to the x-axis, and with its centre on the x-axis. The side of the square cut off at x has an edge whose length is $2x^2$, and the solid extends from $x = 0$ to $x = a$. Sketch the shape of the solid, and using the result of question 10, find its volume.

12. The area containing the points satisfying the inequalities $0 \leqslant x \leqslant 2$ and $\frac{1}{4}x^2 \leqslant y \leqslant 1$ is rotated about the following axes:

(i) $y = 0$, (ii) $x = 0$, (iii) $x = 2$, (iv) $y = 1$

Find the volumes of the resulting solids.

13. Discuss whether or not the value of

$$\int_{0}^{x} \sqrt[3]{|t|}\,dt$$

is more suitably written as $\frac{3}{4}x^{\frac{4}{3}}, \frac{3}{4}x^{\frac{4}{3}}, \frac{3}{4}\sqrt[3]{|x^4|}$, or $\frac{3}{4}|\sqrt[3]{x}|^4$.

14. If $f(1) = 3$, and $f'(x^2) = x$, find the value of $f(4)$.

15. Sketch the graph of the function

$$y = \frac{\sin x}{x+1}.$$

Show that the value of

$$\int_{0}^{a} \frac{\sin x}{x+1}\,dx \geqslant 0 \quad \text{for all} \quad a \geqslant 0.$$

16. Calculate the area enclosed on a unity scale graph by the ordinate $x = 1$, the x-axis, the curve $y = 1/x$ and the ordinate

(i) $x = 2$, (ii) $x = 3$, (iii) $x = 4$, (iv) $x = 5$,

using the trapezium rule and trapezia of base 0·2. Using 5 further trapezia calculate the area enclosed under the same curve between $x = 5$ and $x = 10$, and plot a graph showing how the values of the areas vary with the value of the abscissae of the upper boundary.

17. Show that the area of the triangle OAB in Fig. X.5 is $\frac{1}{2}r^2 \sin \delta\theta$, and that the area of the triangle

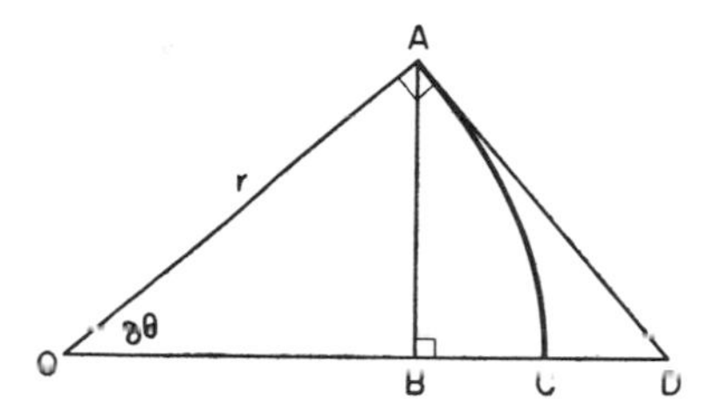

FIG. X.5.

OAD is $\frac{1}{2}r^2 \tan \delta\theta$. Hence deduce that $\frac{1}{2}r^2 \sin \delta\theta \leqslant A \leqslant \frac{1}{2}r^2 \tan \delta\theta$, where A = the area of the sector $OACB$ of the circle centre O and radius r, and by considering the behaviour of this inequality as $\delta\theta \to 0$, show that $dA/d\theta = \frac{1}{2}r^2$. Hence show that the area of a sector of a circle centre O and radius r, and which subtends an angle θ at the centre is given by the expression

$$\int_{0}^{\theta} \tfrac{1}{2}r^2\,d\theta,$$

and by using the equation of the circle in the form $r = a$ (where a is constant), show that the area of a quadrant of the circle is $\frac{1}{4}\pi r^2$.

18. Write down an expression for the value of

$$\sum_{r=1}^{n-1} r^2,$$

and hence show that

$$\int_{0}^{a} x^2\,dx = \lim_{n\to\infty} \left\{\frac{a^3}{2n} + \frac{a^3}{n^3}\,\frac{(n-1)n(2n-1)}{6}\right\} = \frac{1}{3}a^3,$$

explaining why the integral and the limit are equal.

19. Explain the meaning of the equation

$$A = \lim_{n \to \infty} \frac{5^3}{n^3} \sum_{r=2}^{n} r^2,$$

and deduce that

$$A = \frac{125}{3}.$$

***20.** Newton's Law of Gravitation states that the attraction between two particles of mass M and m situated a distance r apart is GMm/r^2, where G is a constant.

A uniform stick whose mass is 10 and whose length is 2 is placed with the nearer end at a distance 2 from a fixed point P. By considering a small element of length δx in the stick at a distance x from P, show that the gravitational field at P due to this element is $5G(\delta x)/x^2$. By integration find the total gravitational potential at P due to the whole stick.

21. Explain why a thin spherical shell of radius r and thickness δr has a volume approximately equal to $4\pi r^2 \delta r$. Hence show that the volume of a solid sphere of radius a is $\frac{4}{3}\pi a^3$.

***22.** The gravitational attraction at any point in space whose distance from the centre of the earth is r is inversely proportional to r^2. Given that the acceleration of gravity on the earth's surface is 10 m/s^2, and that the radius of the earth is 6400 km, calculate the least velocity of a rocket fired upwards from the earth's surface which is necessary if it is to escape completely from the earth's gravitational field.

The radius of the moon's orbit about the centre of the earth is 400,000 km. Calculate the least velocity necessary for a rocket to be fired from the earth's surface to the moon, neglecting the moon's gravitational field.

Answers

Chapter 1

Exercise 1a

2. Into. Proper subset. Subset; proper subset.
4. (i) D; (ii) C; (iii) E.
(a) Proper subset; (b) neither; (c) subset of A; (d) proper subset of B; (e) neither; (f) subset.
5. Into; 2.
7. X; onto.
8. Into; both.
11. (i), (ii), (iii), (v).
13. (i), (iv), (vi), (vii).
14. (i) lm; (ii) mn.
Yes; lmn.
15. (i) Yes; (ii) not necessarily; (iii) no.

Exercise 1b

1. (i) Yes; (ii) no; (iii) no; (iv) yes. (ii).
2. Yes; no.
3. Yes.
4. Yes; yes; yes.
5. Yes; yes.
6. (i) Yes; (ii) no.
9. (i) (a), (c), (d), (g); (ii) (b), (d), (h); (iii) (e), (f); (iv): (d); (v) (a), (c), (e), (f), (g). (e), (f).
10. Onto.

Exercise 1c

2. Positive integer, integer, positive integer, positive integer, rational number, rational number, non-terminating repeating decimal, repeating decimal, non-terminating repeating decimal, irrational number, irrational number, rational number, negative integer, (negative) rational number. $\frac{2}{3}$; $\frac{7}{99}$.
3. B is a function of A.
4. B is a function of A.
5. Yes; increasing downwards.
5. No in each case.
7. 4, 2, 10, −2.
8. −1; 1; 8; −9; −5; 1·101; −1·1; 0·1; 0·7; −0·3; n; $-n$; $-n+1$.
9. $(1-a/b)$ in all cases.
11. 1; 8; 7; 3; 0. 1; 3; 4; 10; n; $(n-2)$; $(n-1)$.
12. 5; 5; 10; 2.
13. 17; 8; $\frac{1}{2}$.
14. Yes; yes; yes; no; no.
15. (i) Yes, (ii) no.

18.

a	7	−2	0	2	1
b	48	3	−1	3	0

is a possible answer.

19. $\frac{9}{100}(1-(0{\cdot}09^n))/0{\cdot}91 = \frac{9}{91}(1-(0{\cdot}09)^n)$. Approaches $\frac{9}{91}$.
20. (i) 0·01, terminating; (ii) 0·00110011...; non-terminating repeating; (iii) 0·010101..., non-terminating, repeating; (iv) 0·001001001..., non-terminating, repeating; (v) 0·0001100110011..., non-terminating, repeating.
21. (i) Yes, (ii) no.

Exercise 1d

2. (i) time of day; (ii) sum of money invested; (iii) his age; (iv) speed of the car; (v) cost of butter; (vi) audience appreciation; (vii) length of service; (viii) time since it started.
5. t.
6. Time.
7. Upward.
8. (i) the years 1950–84; (ii) number of years elapsing; (iii) time of day; (iv) length of the pendulum; (v) number of cars on the road.
10. $H = ms$.
11. $l_1 = l_0(1+\alpha t)$.
12. $pv = k$.
13. $I = P(1+R/100)^T - P$.
14. $P = P_0(1{\cdot}02)^t$; 35 years.
15. (i) b, (ii) no.
16. £9; 26·8 cm, 19·0 cm.
19. 3·17 cm; 189 cm².
20. x^2+4xy; S; x^2y; $S = x^2+4V/x$.
22. $6t$ km/h; $t^2/600$ km.
23. $S = 8\pi r^2$ or $8\pi h^2/9$.

24. $\sqrt{(a^2+b^2+c^2)}$; $a\sqrt{3}$; $a = \sqrt{(D^2-b^2-c^2)}$; $a = \frac{1}{3}D\sqrt{3}$.
25. (i) Yes, (ii) yes, (iii) yes.
26. (i) $\frac{9}{16}$.
27. $y = kx^4$.
29. (i) 80 m, (ii) 1·23 or 4·05 seconds.
30. No.
32. Yes. By projecting onto two axes forming a plane perpendicular to the plane containing the points. A spherical surface.
34. (i) 0·45p, (ii) 0·70p, 6·4 cm.

Exercise 1e

1. (i) +3, increase; (ii) +11, increase; (iii) −2, decrease; (iv) −4, decrease; (v) +3, increase; (vi) +3, increase; (vii) −2, decrease; (viii) −14, decrease; (ix) −39, decrease; (x) +9, increase; (xi) −23, decrease; (xii) −22, decrease; (xiii) −8, decrease.
2. (i) +2, 0;
(ii) +2, 0;
(iii) 0, −1;
(iv) 0, +4;
(v) +3, −2;
(vi) +5, +1;
(vii) −4, −2;
(viii) +4, +2.
4. +3·3; +1·7; +11·0; +2·89; +5·61.
5. −2; −3; $-\frac{11}{10}$; $-\frac{3}{5}$; $\frac{3}{2}$.
6. (i) 291, (ii) 93, (iii) −72.
7. −3; +1; +16; +7; +9; −14: $+\frac{7}{12}$.
8. −9; −1; $-\frac{5}{3}$; $(49)^2$; $-\frac{9}{5}$; −11; $\frac{81}{25}$.

9.

y	0 1 4 9 16 25 36 49 64
δy	1 3 5 7 9 11 13 15
$\delta^2 y$	2 2 2 2 2 2 2
$\delta^3 y$	0 0 0 0 0 0

They are the squares of the first eight positive integers.

z	0 1 8 27 64 125 216 343 512
δz	1 7 19 37 61 91 127 169
$\delta^2 z$	6 12 18 24 30 36 42
$\delta^3 z$	6 6 6 6 6 6
$\delta^4 z$	0 0 0 0 0

(i) 5; (ii) 2; (iii) 19.

10. $(x+\delta x)^2$ or $x^2+2x\,\delta x+(\delta x)^2$; $2x\,\delta x+(\delta x)^2$.

Miscellaneous Exercise 1

1. Into. 2. Yes; no. No; no. 3. One to many. 4. Many to one. 5. Question 4 (the inverse relation). 6. Many to one. 8. One to many. 10. 1 : 1 correspondence; yes; yes. 11. (i) Yes. (ii) No.

16. (a) $A - \frac{3}{4}$; $B - 1$; $C - 2$; $D + 1\frac{1}{2}$; $E - 1$; $F + \frac{1}{4}$.
(b) $A+60$; $B+320$; $C+130$; $D-240$; $E-260$.
17. (i) 16·9; (ii) 16·5.
18. (i) 1, (ii) 3, (iii) 4, (iv) 2, (v) 3, (vi) 2, (vii) 2, (viii) 1, (ix) n.
19. (a) (i), (iv), (v), (vi). (b) All.
22. $x_n^2+5x_n-3$; $2x_n+4$.
29. (i) 249 mm, (ii) 62 mm.
30. (a) 0·61 dm³; (b) 0·25 dm³; 1·37 atm.
31. (a) £104, (b) £212·16, (c) £324·65, (d) $2600\{(1\cdot 04)^T-1\}$.
32. $(2+t)$, $(3+\frac{1}{2}t)$ cm; $\pi(5-t-\frac{3}{4}t^2)$ cm².
36. 36 cm³, 3 cm.
37. (12·6×12·6×6·3) in.
39. $a \simeq 6\cdot 15$, $b \simeq 1\cdot 06$.

Chapter 2

Exercise 2a

1. (a) 5 cm, 8 cm, 20 cm; (c) (i) 1 cm/s, (ii) 1·6 cm/s; (d) 1·9, 4·2, 11·0, 7·9, 2·4 cm; 0·95, 2·1, 2·75, 2·63, 2·4 cm/s.
2. (a) Descending steadily. (b) Descending steadily but less quickly than (a). (c) Ascending slowly with constant speed. (d) Remaining at the same height. (e) Ascending quickly with constant speed. (f) Staying steady at first, then descending, reaching a constant downward velocity. (g) Ascending with increasing speed. (h) Ascending with decreasing speed, reaching a maximum height and then decreasing with increasing speed.
3. When the speed of the lever is constant.
6. (i) 5·2 seconds; (ii) 33·8 metres; (iii) 3 seconds.
7. 1·47 p.m.; $28\frac{1}{2}$. 8. 10·31 a.m.

Exercise 2b

3. 3 km.
4. 11·8 km.
5. (a) 80 km; (b) 35 km.
6. 90 km.
7. (i) n km; (ii) n^2 km.
8. $4\frac{1}{6}$ km.
9. $q^2/60pr$ km.
10. 11.33 a.m.
11. $8\frac{2}{3}$ km.
12. 96 km/h.
14. (i) $7\frac{1}{8}$; (ii) $5\sqrt{3}/3$ seconds.

Exercise 2c

1. A 3·05 km N57°E; B 2·5 km S62°E; C 1·3 km S40°W.
2. $5\sqrt{2}$ km S45°E; 10 km.
3. 100 metres due E.
4. (i) 34 km/h; (ii) 30 km/h, N30°E.

5. 2·5 m/s; 2·5, 1·25, 0·63; when the graph is a standard scale graph.
6. 114 km/h; 120 km/h.
7. (i) 16·7 cm/s, −16·7 cm/s; (ii) 21·75 cm/s, −21·75 cm/s; (iii) 16·7 cm/s, −16·7 cm/s; (iv) 10·5 cm/s, −6·1 cm/s; (v) 16·3 cm/s, 13·7 cm/s; (vi) 18·75 cm/s, 0.
8. 100 km N36·9°E; 100 km/h N36·9°E.
9. 15 km due S; 143 km/h due S.
10. 10 knots at an angle of $\tan^{-1}(\frac{3}{4})$ with the downstream direction.

Exercise 2d

2. (6·37, 40·5), (0·63, 0·40).
4. (i) −2, −3; (ii) −4, −7; (iii) −7, +1; (iv) −6, −10; (v) +3, −6.
5. $Q; P; P; P; P; p$.
6. $2\frac{1}{3}$.
7. $y < \frac{3}{4}$; no; $x = \frac{1}{4}$; $y = \frac{3}{4}$.
8. 1·4; 1·26.
9. (i) $y = 1/3x^2$;
   (ii) $y = 7 - x^2$;
   (iii) $y = \sin^{-1} 5x$;
   (iv) $y = \frac{1}{4}x - \frac{7}{4}$;
10. (i) 9; (ii) 16; (iii) 3/2.
11. $\delta x(2x + \delta x)$.
12. (i) 25; (ii) 10; (iii) 1; (iv) 8.
13. (i) $u(u + 2x - 2y)$; (ii) $-2bx$.
14. $-\dfrac{1}{x(x+\delta x)}$
15. $y = +\sqrt{(a^2 - x^2)}$; $y = -\sqrt{(a^2 - x^2)}$;
   (i) $D\{x: -a \leqslant x \leqslant a\}$
   $R\{y: -a \leqslant y \leqslant a\}$
   (ii) $D\{x: -a \leqslant x \leqslant a\}$
   $R\{y: 0 \leqslant y \leqslant a\}$;
   $D\{x: -a \leqslant x \leqslant a\}$
   $R\{y: -a \leqslant y \leqslant 0\}$
16. 2; $D\{x: x \geqslant 1\}$, $R\{y: 0 \leqslant y\}$;
   $D\{x: x \geqslant 1\}$, $R\{y: y \leqslant 0\}$.
17. (i) Yes; (ii) no.
18. D\{all real numbers\}; $R\{0, 1\}$.
19. D\{all integers\}; $R\{2\}$.
21. $y_1 + \dfrac{(y_2 - y_1)}{(x_2 - x_1)}(a - x_1)$.
22. $y_1 - \dfrac{(y_2 - y_1)}{(x_2 - x_1)}(x_1 - a)$.
24. Reflection in $y = x$.
25. $x^2 + y^2 - 10x - 12y + 36 = 0$; $11x - 60y = 0$.
26. (i) $D\{x: x = 0\}$
   $R\{y: y = 0\}$
   (ii) $D\{x: 0 \leqslant x \leqslant 1\}$, $R\{y: \leqslant 0 \leqslant y \leqslant 1\}$.

Exercise 2e

5. (i) $x = 1$; (ii) $x = -1$.
6. (i) Continuous; (ii) discontinuous at $x = 2$.
7. (i) No; (ii) yes.
8. Yes.
9. $(2n+1)\pi/2$ where n is integral.
10. Yes. 14. No.
16. At the integral values of x.
16. Discontinuous at $a = 4$.
17. (i) $x = 3$; (ii) $x = -3\frac{1}{2}$; (iii) $x = (2n+1)\pi/2$; (iv) none; (v) $x = n\pi$; (vi) $x = \pm 4$; (vii) $x = n\pi$; (viii) $x = 0$.

Exercise 2f

1. (i) Decreases without bound; (ii) increases without bound; (iii) 2; (iv) 5; (v) $2a$.
2. (i) 1; (ii) 5; (iii) 0; (iv) 0.
3. 1.
4. $\dfrac{1 - (1/10)^n}{9}$; 1/9.
5. $\dfrac{1 - (1/10^6)^n}{7}$; 1/7.
6. 2. 8. 1/3.
9. (i) 2; (ii) 3; (iii) 2; (iv) $\frac{1}{2}$.
10. $-1/x^2$.
11. $|1/x|$ increases without bound in both cases; no.
15. (ii) $x = \pm 3$; (iii) $x = -1$.
16. (i) $\cos x$; (ii) $-\sin x$.
17. nx^{n-1}. 20. No.
21. $x = 1$; no limiting value for $x \leqslant -1$.
26. $s; q; r$.
29. Open interval $0 < x < 0{\cdot}14$.
30. Open interval (0, 1); no; no.

Exercise 2g

3. 0·2229.
5. $3x^2 + (3 + 2a)x + (1 + a + b)$; $6x + (6 + 2a)$; 6; 0.
9. 0·92575(5), 0·9500, 0·9731

Miscellaneous Exercise 2

2. 3·2 m, 0·8 s. 4. 15 m, 15 m, −25 m, 20 m.
5. (i) 2 kilometres, (ii) 4 kilometres, downstream from its starting-point in a direction making an angle of 36°50′ with the direction straight across; 20 km/h in the same direction. 5 kilometres in the same direction again.

6. (a) $\tan\theta = v$; (b) $2\tan\theta = v$; (c) $\tan\theta = 5v$; (d) $\tan\theta = v$.

7. (a) (1, 9); (b) (2, 15); (c) (−2, 15); (d) (0, 7); (e) +3 and −3; (f) +4 and −4.

8. A, C, E. 5.

9. Q.

10. A, B, E.

11. $a = -1$, $b = 0$, $c = 4$.

12. (i) $3x-y-11 = 0$; (ii) $2x-y-1 = 0$.

13. $x(d-b)+y(a-c)+a(c-d) = 0$.

14. (4, 2) and (16, −4)

15. (i) $3\frac{1}{2}$, −7; (ii) −5, $2\frac{1}{2}$; (iii) −1, −1; (iv) 0, 0; (v) ±2, −4; (vi) ±1, 1.

18. ($12\frac{1}{2}$, 0); $2\frac{1}{2}$; $4y-3x+25 = 0$.

19. $y^2 = 4ax$.

20. $y^2 = 4x$.

21. $x-y+3 = 0$, $x+y-3 = 0$.

24. D {all real numbers}, R{1, 2, 3}.

25. (i) −1; (ii) 0; (iii) −2/7; (iv) $\frac{1}{2}$; (v) $a^2/(a+b)$; (vi) −15/7; (vii) −4.

27. Area $= \frac{1}{2}bc\sin A$; $b = \dfrac{2(\text{area})}{c\sin A}$;

$$A = \sin^{-1}\left(\frac{2\text{ area}}{bc}\right).$$

Chapter 3

Exercise 3a

1. 206,300; 319,960; 239,360.

2. −21,558 p.a.; 4·46, 3·96; −0·5 p.a.; 3·46, 2·96.

3. (i) 1; (ii) 2; (iii) 4; (iv) $(2^b-2^a)/b-a$.

4. (i) 2; (ii) 4; (iii) a; (iv) $a+1$.

5. 33·8 crowns to the £; 0·6%.

6. 5%.

7. 1·25 m/s².

8. 1·39 m/s².

9. newton m⁻² s.

10. 0·0075 ohms per °C.

11. 0·0066 cm °C⁻¹; 0·005 cm °C⁻¹.

12. (i) 10; (ii) −20; (iii) −50 m/s⁻¹.

13. (i) 3; (ii) 5; (iii) 7.

16. 0·0026 per °C.

17. 0·00342 per °C; 272°C.

18. (i) 9·7 m/s; (ii) 11°32′.

Exercise 3b

1. (i) 25 m/s; (ii) 22·5 m/s; (iii) 21·25 m/s; $5(4+\delta t)$; 20·5 m/s; 20 m/s.

2. (i) $5(1+\delta t)$ m/s; (ii) $5(3+\delta t)$ m/s; (iii) $5(5+\delta t)$ m/s; 5, 15, 25 m/s.

3. (i) 40; (ii) 35; (iii) $32\frac{1}{2}$; (iv) $(30+5\delta t)$ m/s; 30 m/s.

4. (i) 45; (ii) $42\frac{1}{2}$; (iii) $41\frac{1}{4}$; (iv) $(40+5\delta t)$ m/s; 40 m/s.

5. $(50+5\delta t)$ m/s; 50 m/s.

6. (i) $(30-5\delta t)$ m/s; (ii) $(40-5\delta t)$ m/s; 30 m/s; 40 m/s.

7. (i) $(20-5\delta t)$; (ii) $(10-5\delta t)$ m/s; 10; 5 m/s.

8. $s = 5t^2$; 6 m/s.

9. $10t$ m/s; (i) 0·6 s; (ii) 1·5 s; (iii) 2·8 s.

10. (i) 50 m/s; (ii) 70 m/s; 252 km/h.

11. (i) 60 m/s; (ii) −60 m/s.

12. $(70-10t)$ m/s; (i) after 3·8 s; (ii) after 5·6 s.

13. (i) 2×10^4; (ii) 4×10^4; (iii) 6×10^4 m/s.

14. $s = 40t^2$; 80 m/s, 40 m.

15. (i) 5, (ii) 0.

16. (i) −20; (ii) −8.

17. (i) 5; (ii) −1.

18. (i) −14; (ii) −6.

19. (a) 4; (b) −2; (c) −3; (d) $-1+\delta x$; (e) −1.

20. (a) 9, (b) 1/9.

21. Average rate is less than the instantaneous rate at the upper end; the converse.

22. £105; £110·3; £115·8; £121·5; £127·6; 5·5%; 5%.

23. (a) 0·0045; (b) 0·0165; (c) 0·0015; (d) 0·0045; (e) 0·0315 cm °C⁻¹.

24. $b+2cx$.

Exercise 3c

1. (i) $\frac{1}{2}$; (ii) 5/3; (ii) $\frac{1}{3}$; (iv) −3/2; (v) 0.

2. −3/4; −1; −7/8.

3. 5; 3/4; −2/3.

4. 2; 63°26′; 2; 2; 75°58′; 4.

5. 3, 3, 3, 2; 1, −4, 0, 2/3.

6. (i) 5; (ii) −2; (iii) $\frac{1}{2}$; (iv) 3; (v) 5/2.

7. (i) $\frac{1}{3}$; (ii) −1; (iii) −3; (iv) 3/2; (v) $-\frac{1}{2}$.

8. (i) 1; (ii) −2/3; (iii) 5; (iv) 0; (v) −2/9.

9. 3·2, 3·6, 2·0 hm/min.

10. $-1\frac{1}{2}$, $-\frac{1}{2}$, 1; (i) 1; (ii) 1; (iii) 0.

11. 14·6 cm.

12. 10·2 m/s².

13. $a = 12$, $b = 2·8$.

14. 1·40.

16. 156 J °C⁻¹.

17. $\theta = 100-2·06L$.

19. (a) 0·01, 0°0·3′; (b) 20, 11°32′; (c) $12\frac{1}{2}$, 7°11′; (d) 33·3, 19°28′.

20. 0·2.

Exercise 3d

2. (i) 1; (ii) $-3\frac{1}{2}$; (iii) 3; (iv) −3.

3. 2, 6, −4.

4. 7, 4.
5. 0, -3, 5, 2, -4; $2y = x^2$.

MISCELLANEOUS EXERCISE 3

1. (i) 0·208, 0·52; (ii) 0·0142 per annum.
2. $-1{\cdot}21$ francs to the £ per annum.
3. 11·1 m/s²; $-4{\cdot}12$ m/s².
4. cm/s.
5. ohm m; 2×10^{-8} ohm m °C^{-1}.
6. 35·4 cm/s.
8. (i) 0·0112(5); (ii) 0·0066; (iii) 0·0062; (iv) 0·0035 °C^{-1}; 0·0058 °C^{-1}.
9. (i) 0·000195; (ii) 0·000384; (iii) 0·00059 °C^{-1}; 0·00083 °C^{-1}.
10. (i) 17; (ii) 14/13; (iii) 0; (iv) vertical; (v) 7/4.
11. $a = 1{\cdot}0$, $n = 0{\cdot}5$.
14. Chord resistance (i) $9{\cdot}1\times10^3\ \Omega$; (ii) $12{\cdot}5\times10^3\ \Omega$; incremental resistance (i) $7{\cdot}5\times10^3\ \Omega$; (ii) $9{\cdot}2\times10^3\ \Omega$. 8·02 mA.

Chapter 4

EXERCISE 4a

1. $\frac{1}{2}$; 2; 8: 3; 5; 6.
2. -15.
3. $2x-1+\delta x$; $2x-1$.
4. $6x+3\delta x$; 6·3, 6·03, 6·003, 6·0003.
5. 12; -4; 2.
6. 3·31, 3·0301, 3·003001: 3.
7. $9x^2$.
8. $\frac{1}{3}$, 3/16.
9. (i) Yes; (ii) no; (iii) no; (iv) no. $+1$, $x > 0$; -1, $x < 0$.
10. $3x^2+3x(\delta x)+(\delta x)^2$, $3x^2$.
11. $3(\sqrt{3}-1)/\pi$.

EXERCISE 4b

1. $10x$.
2. $8x^3$.
3. $2(x-4)$.
4. $-2/x^3$.
5. $\sqrt{(x+\delta x)}-\sqrt{x}$, $\dfrac{\delta x}{\sqrt{(x+\delta x)}-\sqrt{x}}$, $\dfrac{1}{2\sqrt{x}}$.

EXERCISE 4c

1. $8x^7$.
2. $40x^7$.
3. 1.
4. 6.
5. 0.
6. $\frac{1}{4}$.
7. 0.
8. -2.
9. 3.
10. $4x$.
11. $\frac{7}{6}x^6$.
12. 7.
13. x.
14. $20x^{19}$.
15. $14x+5$.
16. $30x^{29}-90x^{14}$.
17. $-4x^3$.
18. $50x^4$.
19. $3x^2+2x+7$.
20. $2(x+1)$.
21. $2x-3$.
22. $2p^2x$.
23. $-2x$.
24. $13x^{12}-7x^6$.
25. $20x^{19}$.
26. $26x^{25}$.
27. $11ax^{10}-7bx^2$.
28. 0.
29. 0.
30. $10x$.
31. $18x$.
32. $\frac{1}{2}x-\frac{1}{4}$.
33. $2x-a$.
34. $3x^2-\frac{1}{2}x$.
35. $(2n-1)x^{2n-2}$.
36. $6nx^{3n-1}$.
37. $2x-17$.
38. Increases uniformly from -24 to $+12$.
39. (i) -4; (ii) -4; (iii) -7.
40. $(1, -4)$.
41. (i) -3; (ii) $3t^2+17$; (iii) $10t$; (iv) $30at^{29}-60b^2t^{59}$.
42. (i) $2p-6p^2$; (ii) $4y-1$; (iii) $40t^4-270t^9$; (iv) 2π; (v) $8\pi r$; (vi) $4\pi r^2$.
43. 8.
44. (a) (i) $x > 1\frac{1}{2}$; (ii) $x > -5/8$; (b) $\frac{1}{3} < x < 3$.
45. $(\frac{1}{3}, 103/27)$; $(1, 3)$.
46. (i) $y = 4$; (ii) $24x+y+28 = 0$; (iii) $y = 0$.
49. $(3, -5)$, $(-1, 27)$.
51. $(0, 0)$, $(2+\sqrt{6}, 0)$, $(2-\sqrt{6}, 0)$, $2x+y = 0$, $4(3+\sqrt{6})x-y-4(12-5\sqrt{6}) = 0$, $4(3-\sqrt{6})x-y-4(12-5\sqrt{6}) = 0$.
52. 1.
53. 45°, 315°, 288°26′. 281°18′.

EXERCISE 4d

1. $-2/x^2$.
2. $-1/2x^2$.
3. $-3/2x^2$.
4. $1/x^2$.
5. $-2/x^3$.
6. $-3/x^4$.
7. $-4/x^5$.
8. $49/x^8$.
9. $-24x^{-5}$.
10. $4x^{-9}$.
11. $-2/x^7$.
12. $-10x^{-3}$.
13. $-2/(5x^3)$.
14. $-50/x^3$.
15. $45/x^4$.
16. $-24x^{-4}$.
17. $-6/x^7$.
18. $-36/x^5$.
19. $-6/x^7$.
20. $-7(3x+8)/x^9$.
21. $3x^2-1$.
22. $8x^2+21x^2$.
23. $-1/x^2$.
24. 1.
25. $2(x-1)$.
26. $6(2-x)/x^5$.
27. $-2/(3x^3)$.
28. $-3(9x^2+8x+2)/x^6$.
29. $1+8/x^2$.
30. $4x(x^2+1)$.
31. $3x^2-4x-3$.
32. $-6x^{-7}-48x^{-9}$
33. $-(5+x^4)/7x^6$.
34. $3x^2-4-4/x^2$.

36. (i) $2t-1$; (ii) $2y+1$; (iii) $3z^2+2z+1$.
37. (i) $3{\cdot}6\times10^{-5}$ cm/km; (ii) $6{\cdot}4\times10^{-5}$ cm/km; (iii) 12×10^{-7} cm h km^{-2}; (iv) 2×10^{-6} cm h km^{-2}.

39. $x = y^n$,

$$\frac{\delta y}{\delta x} = \frac{1}{ny^{n-1}+\frac{n(n-1)}{2!}y^{n-2}\delta y+\ldots+(\delta y)^{n-1}};$$

$$\frac{1}{n}x^{(1/n)-1}.\ \text{Yes.}$$

40. $$\frac{px^{p-1}+\frac{p(p-1)}{2!}x^{p-2}\delta x+\ldots+(\delta x)^{p-1}}{qy^{q-1}+\frac{q(q-1)}{2!}y^{q-2}\delta y+\ldots+(\delta y)^{q-1}}.$$

41. $(2, 0)$, $(-2, 0)$.
42. (i) $(1, -1)$, $(-1, 1)$; (ii) $(\sqrt{3}/3, -5\sqrt{3}/9)$, $(-\sqrt{3}/3, 5\sqrt{3}/9)$; (iii) $(1, -1)$, $(-1, 1)$.

Exercise 4e

1. $-14/x^8$; $112/x^9$.
2. $-2/x^4$; $8/x^5$.
3. $-4x^{-5}$; $20x^{-6}$.
4. $-3/2x^5$; $-15/2x^6$.
5. $-2/9x^3$; $2/3x^4$.
6. $-60x^{-13}$; $780x^{-14}$.
7. $14x^{-3}$; $-42x^{-4}$.
8. $(2-x)/x^3$; $2(x-3)/x^4$.
9. $-7/x^2$; $14/x^3$.
10. $-4x^3$; $-12x^2$.
11. $3x^2-5$; $6x$.
12. $4x+11$; 4.
13. $2a(ax+b)$, $2a^2$.
14. $9(3x-5)^2$; $54(3x-5)$
15. $2x-1$; 2.
16. $(x^2-3)/3x^4$; $\frac{2(6-x^2)}{3x^5}$.
17. $-2x^{-3}+3x^{-4}$; $6x^{-3}-12x^{-5}$.
18. $-3x^{-4}-3$; $12x^{-5}$.
19. $2x-6x^2+4x^3$; $2-12x+12x^2$.
20. $1-\frac{4}{x^2}$; $\frac{8}{x^3}$.
21. $-2x$; -2.
22. $(2p-1)x^{2(p-1)}$; $2(2p-1)(p-1)x^{2p-3}$.
23. Gradient $= 7\frac{1}{4}$, rate of change of gradient $= -1/8$.
24. $2x-1$; 2; $(2x-1)^2$.
25. $f(1) = 0$; $f'(1) = -3$; $f''(1) = 8$; $f'''(1) = -24$.
26. $-16x^{-3}$; $48x^{-4}$; $256x^{-6}$.
27. $-9x^{-10}$, $90x^{-11}$; $-28x^{-5}$; $140x^{-6}$; $-3/8x^4$, $3/2x^5$, $32/x^5$; $-160/x^6$; $4x^3-4x$, $12x^2-4$.
28. (i) $18t^5$, $90t^4$, $324t^{10}$; (ii) $2t-5$, 2, $(2t-5)^2$.
29. $6t$, $4t^3-4t$, 6, $12t^2-4$, $\frac{2}{3}(t^2-1)$.
32. (a) (i) $x < -a$; (ii) $c < x < e$; (iii) $b < x < d$; (iv) $x < -c$, $-c < x < b$, $x = d$, $x = f$, $g < x < h$, $h < x < k$, $x > g$; (v) $-c < x < b$, $x = c$, $x = e$, $g < x < h$, $x > k$.
(b) (i) $x < -a$, $e < x < f$; (ii) $d < x < f$, $f < x < g$; (iii) $-a < x < b$, $d < x < f$; (iv) $-b < x < 0$, $x > g$; (v) $-b \leqslant x \leqslant 0$, $x = d$, $g \leqslant x < h$, $x > k$.
34. $y = x^2-7x$.
35. $x+y-4 = 0$.
36. (a) > 0, increasing; (b) < 0, increasing; (c) > 0, decreasing; (d) < 0, increasing; (e) > 0, constant; (f) < 0, constant; (g) 0, decreasing; (h) 0, increasing. (a), (b), (d), (h) > 0; (c), (g), < 0; (e), (f) $= 0$.
37. $n(n-1)(n-2)\ldots(n-r+1)x^{n-r}$, no.
42. a^n2^x.
43. (a) (i) $x > 1$, $x < 1$; (ii) $x > 3$; (b) (i) none; (ii) $x < 3$; (c) (i) $x > 1$; (ii) $x > 2$, $x < 0$; (d) (i) $x < 1$; (ii) $0 < x < 2$.

Exercise 4f

8. All x for which $2n\pi < x < (2n+1)\pi$, where n is integral.
9. (a), (b).
10. $\frac{7}{4}x^{\frac{3}{4}}$.
11. $\frac{10}{3}x^{\frac{2}{3}}$.
12. $\frac{-9}{2}x^{\frac{1}{2}}$.
13. $-\frac{1}{2}x^{-\frac{3}{2}}$.
14. $-\frac{3}{2}x^{-\frac{3}{2}}$.
15. $-3x^{-\frac{5}{2}}$.
16. $\frac{3}{2}x^{-\frac{3}{2}}$.
17. $\frac{1}{6}x^{-\frac{5}{6}}$.
18. $-0{\cdot}1x^{-1{\cdot}1}$.
19. $0{\cdot}8x^{-0.9}$.
20. $\frac{3}{2}x^{-\frac{1}{2}}+\frac{7}{2}x^{-\frac{3}{2}}$.
21. $\frac{3}{2}x^{\frac{1}{2}}+\frac{1}{2}x^{-\frac{1}{2}}$.
22. $-x^{-\frac{3}{2}}+\frac{21}{2}x^{-\frac{5}{2}}$.
23. $1-\frac{1}{x^2}$.
24. $\cos x$.
25. $-3\cos x$.
26. $-\frac{1}{4}\sin x$.
27. $2x-\cos x$.
28. $5(\cos x-1)+1/(2\sqrt{x})$.
30. $-\sin x$, $-\cos x$, $\sin x$.
31. $(\cos x-\sin x)$, $-(\sin x+\cos x)$.
32. 3, 0, 0, $\frac{3}{2}$.
33. $\sec^2 x$.
34. $-5\sin 5x$.
35. (i) $\sec x\tan x$; (ii) $-\operatorname{cosec}^2 x$; (iii) $2\sin x\cos x$.
37. $\sin(x+r\pi/2)$.
39. (i) $2x+y+\delta x$; (ii) $x+4y+2\delta y$.
41. (i) True; (ii) untrue.

EXERCISE 4g

1. $\{\delta^2 f(a)-\delta^3 f(a)+\frac{11}{12}\delta^4 f(a)-\frac{5}{6}\delta^5 f(a)\ldots\}/h^2$.
2. (i) 0·0873; (ii) 0·0435; (iii) 0·0290; 0·2010, 0·1002, 0·668.
3. 0·195.
4. 1·001, 1·105, 1·223.
5. $2\frac{1}{2}$ m/s; $s = 2\frac{1}{2}t(t+1)$.
6. (i) 0·25; (ii) 0·75; (iii) 1·25; (iv) 1·75 m/s²; 0, 0·75, 3·75, 7·0 kW.

MISCELLANEOUS EXERCISE 4

1. $27x^{26}$.
2. $-19x^{-20}$.
3. $30ax^{29}$.
4. $-30x^{-31}$.
5. $54x^2$.
6. $135x^{14}$.
7. $-12x^{-3}$.
8. $-24x^{-9}$.
9. $\frac{1}{5}x^{-\frac{4}{5}}$.
10. $\frac{3}{4}x^{-\frac{1}{4}}$.
11. $\frac{1}{2}x^{-\frac{7}{8}}$.
12. $-\frac{7}{5}x^{-\frac{6}{5}}$.
13. $\frac{4}{3}\sqrt[3]{x}$.
14. $-16\frac{1}{2}x^{-2\frac{1}{2}}$.
15. $\frac{3}{4}x^{-\frac{1}{4}}$.
16. $-\frac{2}{3}x^{-\frac{5}{3}}$.
17. $-\frac{3}{2}x^{-\frac{5}{2}}$.
18. $-13\frac{1}{2}x^{-3\frac{1}{4}}$.
19. $5-3x^2$.
20. $24x^3-30x$.
21. $3x^2-16x+16$.
22. 1.
23. $-\frac{3}{4}x^{-\frac{5}{4}}-\frac{3}{4}x^{-\frac{1}{4}}$.
24. $-\frac{9}{2}x^{-\frac{3}{2}}-x^{-\frac{1}{2}}+\frac{3}{2}x^{\frac{1}{2}}$.
25. (i) $2at$; (ii) $2t-(a+b)$.
26. (i) $-(7/p^2)-3$; (ii) $4p^3$.
27. −1, 14, 2.
28. $-\frac{1}{4}$.
29. $-1 < x < 3$.
30. (i) $x < 0$, $x > 3$; (ii) $0 < x < 3$.
31. (a) $x > \frac{7}{3}$; (b) $x < \frac{7}{3}$.
33. (a) $x < -2$, $x > 1$; (b) $-2 < x < 1$.
35. All values of $x < 0$: always upward.
36. (i) $4x(x^2-3)$; (ii) $2x+2/x^2$.
37. $1+\frac{1}{2}x^{-\frac{1}{2}}-\frac{1}{2}x^{-\frac{3}{2}}-x^{-2}$.
38. $4x-8x^{-3}$, $1/(2\sqrt{x})$; $15\frac{7}{8}$, $\frac{1}{4}$.

Chapter 5

EXERCISE 5a

1. (i) $12t^2-7$; (ii) $-4t$; (iii) $-2\sin t$.
2. (i) $5/(2\sqrt{t})$; (ii) $-b$; (iii) $2pt^{p-1}$.
3. (i) $18t^2-8$; (ii) $10t^{\frac{3}{2}}$; (iii) $3\cos t+\sin t$.
4. (i) $\frac{5}{2}t^{\frac{3}{2}}$; (ii) $9t^{\frac{3}{2}}$.
5. (i) 88; (ii) 110; (iii) 176.
6. (i) 0·8; (ii) 0; (iii) 0·7; (iv) 0·43 cm/s.
7. $-0·04\sin t$: (i) 0; (ii) −35; (iii) −21·5; (iv) −40; (v) 0; (vi) 40: 0; −0·04; 0 cm.
8. (i) 48−50; (ii) 60; (iii) 84−86; (iv) 132−140 km/h.

EXERCISE 5b

1. 16·5 cm/s².
2. 2·5 m/s².
3. 16; 24; 32; 12 cm/s².
4. (i) 432; (ii) 3·3.
5. 9·80 m/s².
6. 980·8 cm/s².
7. (i) 47; (ii) $2/\sqrt{t}$; (iii) $-2\sin t$; (iv) $6(t-2)$.
8. 6 m/s² due east, 15 m/s N $\tan^{-1}(4/3)$E.
9. 30 m/s downwards. (i) 50 m/s downwards; (ii) 20 m/s downwards; (ii) 42·4 m/s at 45° downwards with the horizontal; (iv) 25 m/s downwards at $\tan^{-1}(3/4)$ with the horizontal; (v) 54 m/s downwards at 61° with the horizontal.
10. $9·8t$ m/s; 9·8 m/s².
11. (i) 10, (ii) 20, (iii) 30 m/s; 10 m/s².
12. 300 m/s; 30; 4500 m.
13. (a) 28; (b) 52; $1800-240t$, $t = 7·5$ min, 112·5 km/h.
14. (a) 100; (b) 10^4; (c) 10^7; 10^{10}.
15. $5\cos t$; (i) 5; (ii) 4·4; (iii) 3·0; (iv) 0; (v) −5; (vi) $1·4-5\sin t$.
16. (i) 3; 0; −3; (ii) 1·6; −2·5; −1·6; (iii) 0; −3; 0; (iv) 0; 3; 0.
18. 6; $3t^2$.
19. $3at^2+2bt+c$; $6at+2b$; $6a$; $b = c = 0$; $v = 120gt^2$ for $0 \leqslant t \leqslant 1/8$;

$v = 120gt^2 + 30g(t-1/8)$ for $t > 1/8$; $h = 40gt$ for $0 \leq t \leq 1/8$; $h = 40gt^3 + 15gt^2 - 15gt/4$ for $t > 1/8$.

20. (ii) $\sqrt{5}$ m; (iii) $-2{\cdot}19$ m/s.
21. $t = \tan^{-1} 3$; $-\sqrt{10}$ m/s²; $(2t\sqrt{10})$ m.
22. ± 12.
23. 10 m/s².
25. $\{2r\omega \sin(\frac{1}{2}\omega t)\}/t$; $r\omega^2$.

EXERCISE 5c

1. -12.
2. 1.
3. 0·9.
4. 1·8.
5. 5/16.
6. 4·2.
7. (i) $-\dfrac{\sqrt{2}}{2q^{\frac{2}{3}}}$; (ii) $-4/p^3$.
8. (i) 0·17; (ii) 0·10; (iii) 0·17.
10. (i) $2\pi r$; (ii) 2π; (iii) $4\pi r^2$.
11. (i) $\pi/\sqrt{(L/g)}$; (ii) $-\pi\sqrt{(L/g^3)}$.
12. (i) $C/2\pi$, (ii) $\frac{1}{2}\sqrt{(S/4\pi)}$.
13. pq cm²/s.
17. (i) 4×10^{-3}; (ii) $4{\cdot}4\times10^{-3}$; (iii) 5×10^{-3}; (iv) 6×10^{-3} per °C.
19. (i) k/x^2 away from P; (ii) 0; (iii) k towards P.
20. $\dfrac{2\pi Q^2}{A}$; $\dfrac{AV^2}{8\pi x^2}$: attraction.
21. $I = -2\pi fCE_0 \sin 2\pi ft$.

EXERCISE 5d

1. $0{\cdot}16\pi$ cm².
2. $0{\cdot}112\pi$ cm².
3. (i) $2\pi x\,\delta x$; (ii) $\cos x\,\delta x$; (iii) $-\frac{1}{2}\,\delta x$; (iv) $2(2x-1)\delta x$.
4. (i) 0·2; (ii) $-0{\cdot}1$; (iii) 0·025; (iv) $-0{\cdot}08$.
5. 5 million.
7. 22·4.
8. (i) 12·5; (ii) 2·5; (iii) $\frac{1}{4}$.
9. $1{\cdot}01\times10^{-3}$ seconds.
10. (i) $-\frac{1}{2}$; (ii) $-0{\cdot}005$.
11. np, only if $n > 0$.
12. 0·0015.
13. $\delta y = \delta x/(2\sqrt{x})$; 5·01, 7·014, 15·033.
14. 5·017.
15. $-2{\cdot}9$.
16. $-2{\cdot}4$.
17. $-0{\cdot}25$.
18. 0·2.
20. 0·739.
22. 0·38 cm³.
24. $\frac{2}{3}p$.

EXERCISE 5e

1. $2x-y-2 = 0$; $x+2y-1 = 0$.
2. $x+4y-24 = 0$; $4x-y-11 = 0$.
3. $5x-y-3 = 0$; $x+5y-11 = 0$.
4. $6\sqrt{3}x+12y+6-7\sqrt{3}\pi = 0$; $12x-6\sqrt{3}y-14\pi-3\sqrt{3} = 0$.
5. $70x+y-35 = 0$; $x-70y+12252 = 0$.
6. $x+16y-24 = 0$; $16x-y-127 = 0$.
7. $9x-2y+18 = 0$; $4x+18y+93 = 0$.
8. $4x+20y = 145$.
10. $9x+4y = 72$; $\frac{1}{2}$; $4x+y = 24$.
11. $6x-y-9 = 0$.
12. $4x-y-8 = 0$; $x-y+4 = 0$; 30°58′.
13. $a = -1$, $b = 2$; $(-4/3, -58/27)$; $(2, 6)$; $8x-y-10 = 0$; $(-11/7, -4/7)$.
14. 5.
15. -1, -1; $4x-y+4 = 0$; $x+y-1 = 0$.
16. $x+2y-3 = 0$; $(\frac{1}{2}, 5/4)$.

EXERCISE 5f

1. (i) 8·2; (ii) 8·7; (iii) $\sqrt[3]{(V/2\pi)}$.
2. (a) -3; (b) $-1{\cdot}5$; (c) 2; (d) 4; $-\frac{1}{2}a$.
3. (a) $-\frac{3}{4}$; $-3\frac{7}{8}$; (b) 1; -3; (c) 1/5; $16\frac{4}{5}$; (d) $\sqrt{3{\cdot}5}$; $-10\frac{1}{4}$.
4. (i) $x = 1$; (ii) $x = 1\frac{1}{2}$.
5. $d^2y/dx^2 > 0$.
6. 108 dm².
8. -3; -8; -13.
9. (i) $\frac{93}{16}$ new pence per minute; (ii) 5np.
10. Yes; no. Minimum.

EXERCISE 5g

3. (i) $(-1, -3)$ min; (ii) $(-1, 4)$ max.
5. (i) (0, 0) max; $(-2, -4)$ min; (ii) (2, 20) max; (4, 16) min; (iii) $(\frac{1}{3}, -17\frac{14}{27})$ max; $(3, -27)$ min.
7. 57·2 cm, 92·8 cm.

EXERCISE 5h

1. $(-1, 15)$ max; $(5, -93)$ min.
2. $(-1, 6)$ max; $(2, -21)$ min.
3. $(2, -48)$ min.
4. $(-1, -5)$ min.
5. $(-\frac{2}{3}\sqrt{3}, 4(16\sqrt{3}-1)/9)$ min; $(\frac{2}{3}\sqrt{3}, -4(16\sqrt{3}+1)/9)$ min.
6. $(-3, -72)$ min; (0, 9) max; $(3, -72)$ min.
7. $(-a, -a^4)$ min; (0, 0) max; $(a, -a^4)$ min.
9. $312\frac{1}{2}$ m².

10. 625 m^2.
11. (i) 56·6 m; (ii) 80 m.
12. 4·65 cm.
13. 12·6 $\times$12·6$\times$6·3 cm.
14. (i) 193; (ii) 217 cm^3.
15. 8·22 dm^3.
16. ± 216, 432.
17. 96.
18. 10; 10: $t = 1{\cdot}55$; $t = 0$: $s = 0$.
19. (a) $dy/dx < 0$, $d^2y/dx^2 > 0$; (b) $dy/dx > 0$, $d^2y/dx^2 > 0$; (c) $d^2y/dx^2 > 0$.
20. $C_2 = Vt^2(1-t)$.
22. 12, $4\frac{1}{4}$, 6.
27. 24 ft.
28. (i) Max.; (ii) min.
30. Less.
31. Continuous, $d^2y/dx^2 \geqslant 0$.
33. 8/27.
34. $S = \sum_{i=1}^{n} y_i^2 - 2m \sum_{i=1}^{n} x_i y_i + m^2 \sum_{i=1}^{n} x_i^2$;
$\sum_{i=1}^{n} x_i y_i / \sum_{i=1}^{n} x_i^2$; $y \sum_{i=1}^{n} x_i^2 = x \sum_{i=1}^{n} x_i y_i$.
38. $\frac{1}{2}, \frac{1}{3}, \frac{1}{4}$.

Exercise 5i

1. (1, 0), horizontal inflexion.
2. (-1, 0), horizontal inflexion.
3. (0, -7), min.
4. (0, -7), horizontal inflexion.
5. (0, -7), horizontal inflexion.
6. (-3, 59), max; (3, -49), min.
7. (-1, 11), max; (1, 3), min.
8. (2, 0), horizontal inflexion.
9. ($-\frac{1}{2}$, $-8\frac{1}{2}$), max; ($\frac{1}{2}$, $8\frac{1}{2}$), min.
11. Max $\frac{3}{4}$; min $-\frac{3}{4}$: $\frac{3}{4}$ at $x = 0$.
14. $a = 1$, $b = -3$.
15. (i) $(u-u_1)^2/2f$; (ii) $(2u-u_1)f/u$.
16. $c-b^2/4a$.
17. 0, $-\frac{2}{3}$; $2\frac{2}{3}$, $-3\frac{1}{3}$.
18. $1+4/x^2$; $4x^2+2/x^3$; $|x| < 1$.
20. $-2\pi, -\pi, \pi, 2\pi$.
21. 5; $\frac{1}{2}\pi - \tan^{-1}(\frac{4}{3})$.
22. $\sqrt{(a^2+b^2)}$.
23. (i) 83; (ii) 75.
24. $W - W^2r/E^2$; $E^2/4r$; E^2/r.
26. 4/7 cm.
27. $S = 8x(1-x)/3$; 4/3.
28. Yes.
29. $\dfrac{x(1-x)}{a}$; $\dfrac{1}{2(1+a^2)}$.
30. ($8 \sin \theta$) cm^2.
31. $-2 < x < 0$; -16, minimum; 0, maximum; 16, minimum.
32. (a) $f'(a) = 0, f''(a) < 0$; (b) $f'(a) = 0, f''(a) > 0$; $-1/8$ at $x = \pm 2$, minima.
34. At (2, 0); gradient $= -8$.
35. Max; min.
36. $\dfrac{b(2b^2-9ac)+2(b^2-3ac)\sqrt{(b^2-3ac)}}{27a^2}+d$.

Miscellaneous Exercise 5

1. (i) 30 cm/s; 42 cm/s^2; (ii) $6\frac{3}{4}$ cm.
2. (i) $\frac{3}{2}x^{-\frac{1}{2}}+\frac{5}{2}x^{-\frac{3}{2}}$; (ii) $+2$, max; -1, min.
3. $6x-y-20 = 0$, $x+6y-65 = 0$; (3, 2).
4. 375 km; 2·5 hours; 225 km/h.
6. 9.
7. x^2+16/x; 2.
9. (i) $\dfrac{64\sqrt{2}}{27}$ cm; (ii) 32/9 cm/s; (iii) 4 cm/s; (iv) $3\frac{2}{3}$ cm/s.
10. Positive; minimum.
11. $z = 2x^3-9x^2+12x-27$; -22, -23.
12. (i) $6(1+3x)$; $42x-y-35 = 0$; (ii) $\frac{2}{3}$ m/s.
13. (i) $3x^2+4x+3$; $2(x-9/x^3)$; (ii) (-2, 26) max; (1, -1) min.
16. (i) $(2\pm\sqrt{7})/3$; (ii) 2/3.
17. $x+y-4 = 0$.
18. (i) $20x^4+1+20x^{-5}$; (ii) 32 max; 0 min.
19. $13x-y-59 = 0$; $x-y-17 = 0$.
20. (i) (a) $-5+8x-3x^2$; (b) $6+4/x^2$; (ii) $-1 < x < 3$.
21. $1/8\pi$ cm.
22. 33·3 cm from the end where $x = 0$.
24. (i) (a) $4x(x^2-3)$; (b) $(2/x^3)+1$; (ii) $9x-y-16 = 0$; (-4, -52).
25. $a(12-a)$; 6.
26. $360t-270t^2$; $360-540t$: (i) 80 min, (ii) 320/3 km, (iii) 12 km/h.
27.

x	0	-1	-2
y	-4	-2	0
y'	0	-3	0
y''	6	0	-6

(i) Min; (ii) horizontal inflexion; (iii) max.
28. $x(a+h) = ah$.
29. 2; -1; 11: 6; 0.
30. $4\pi^2 Rr\,\delta r$.
31. 1·6.
32. 0·00728.
33. 2·0025.
34. 1·045; 1·02; 1·00005; 2·025; 5·03.

35. 0·5013, 0·5020.
36. 0·02%; 13·3 seconds.
37. $1{\cdot}6\,\pi$ cm³.
38. 1·999975, 10·0025.
39. $0{\cdot}0375\pi$ cm².
40. $a = 3$.
41. 1·53, 0·35, −1·88; 45°.

Chapter 6

Exercise 6a

1. $14(2x+3)^6$.
2. $-10(5-x)^9$.
3. $\dfrac{-12}{(4+3x)^5}$.
4. $\dfrac{-1}{2\sqrt{(2-x)}}$.
5. $3x(1+x^2)^{\frac{1}{2}}$.
6. $\frac{5}{4}(4x-1)(1-x+2x^2)^{\frac{1}{4}}$.
7. $18x(11-3x^2)^{-4}$.
8. $\dfrac{-21}{(8+7x)^4}$.
9. (i) $4t(1-t^2)^{-3}$; (ii) $\dfrac{-2}{(7+4t)^{\frac{3}{2}}}$; (iii) $16(4-t)(t-t^2)^7$.
10. $1+\dfrac{x}{\sqrt{(1+x^2)}}$.
11. $dy/dt = \cos x(dx/dt)$; 0·0707; 4.
12. (i) 0·2; (ii) −1·2; (iii) −0·1; (iv) 0·084; (v) −0·2.
16. 600; 12,200; 200π; 10.
17. (i) $-10{,}000\,\pi\sqrt{2}$; (ii) $-20{,}000\pi$; (iii) 0; 200.
18. $-0{\cdot}1\pi \sin \pi t$; $-0{\cdot}1\pi^2 \cos \pi t$.

Exercise 6b

1. $-4(3-x)^3$.
2. $-36x(1+3x^2)^{-7}$.
3. $10 \cos x\,(1+\sin x)^9$.
4. $24 \cos 8x$.
5. $-3 \sin 15x$.
6. $\frac{1}{4}\cos \frac{1}{4}x$.
7. $-3 \sin (3x-1)$.
8. $24(1+1/x^2)(x-1/x)^7$.
9. $\dfrac{-10x}{(1+x^2)^2}$. **10.** $\dfrac{5}{(9-5x)^2}$.
11. $2 \sec \frac{1}{2}x \tan \frac{1}{2}x$.
12. $7 \sin (3-7x)$.
13. $\cos (x-\frac{1}{4}\pi)$.
14. $2 \sin x \cos x$.
15. $\dfrac{7 \sin x}{(1+\cos x)^2}$.
16. $\dfrac{\cos x}{\sqrt{(\sin x)}}$.
17. $-\dfrac{1}{8}\sin \dfrac{(x-1)}{8}$.
18. $-\frac{1}{2}\pi \cos (4-2\pi x)$.
19. $-2 \cos 2x\,(\sin 2x)^{-2}$.
20. $15(2+3x)^4$.
21. $-7(1+1/x)^6/x^2$.
22. $20(\cos x-\sin x)(\sin x+\cos x)^{19}$.
23. $2(1+\sqrt{x})^3/\sqrt{x}$.
24. $\dfrac{4x}{(1-x^2)^3}$.
25. $\dfrac{3x}{\sqrt{(1+x^2)}}$.
26. $-12 \sin x(\cos x)^3$.
27. $\dfrac{\sin 2x}{(\cos 2x)^{\frac{3}{2}}}$.
28. $\dfrac{-3x^2}{(9+x^3)^2}$.
29. $\dfrac{x^3}{(14-x)^{\frac{5}{4}}}$.
30. $16 \sec^2 x \tan x$.
31. $-21 \sin^2 x \cos x$.
32. $-\frac{1}{2}\operatorname{cosec} \frac{1}{2}x \cot \frac{1}{2}x$.
33. $80x^3(1-x^4)^3$.
34. $\dfrac{15}{4\sqrt{(5x)}}\{1+\sqrt{(5x)}\}^{\frac{1}{2}}$.
35. $-14x \sin (7x^2+5)$.
36. $\dfrac{-1}{2\sqrt{x}}\sin \sqrt{x}$.
37. $\dfrac{3 \cos 3x}{2\sqrt{(\sin 3x)}}$.
38. (i) $-\frac{1}{7}p^{-\frac{8}{7}}$; (ii) $-20b(a-bp)^{19}$; (iii) $-56(2-7p)^7$; (iv) $3(2p+2/p^3)(p^2-1/p^2)^2$; (v) $15 \sec 3p \tan 3p$; (vi) $6p \cos (3p^2)$; (vii) $2/(1-2p)^2$; (viii) $-2p \sin (p^2)$; (ix) $-2 \sin p \cos p$; (x) $-\sqrt{p}\sin(\sqrt{p})/2$; (xi) $-\sin p/\sqrt{(\cos p)}$; (xii) $-12p^2(3+p^3)^{-5}$; (xiii) $2 \operatorname{cosec}^2 p \cot p$; (xiv) $12(1+4p)^2$; (xv) $6 \cos p/(1-\sin p)^2$.
39. (i) $18\frac{1}{2}x^{17\frac{1}{2}}$; (ii) $24x(3x^2)^3$; (iii) $-1/x^2$; (iv) $3 \sin^2 x \cos x$; (v) $2 \sec^2 x \tan x$; (vi) $4-24x^2+20x^4$; (vii) $6 \sec^2 3x \tan 3x$; (viii) $105(1-1/x^3)^6/x^4$; (ix) $3(2-3x)/2\sqrt{(1-x)}$; (x) $-35 \operatorname{cosec} 5x \cot 5x$; (xi) $4x \sin (x^2) \cos (x^2)$; (xii) $2(\sin x-\cos x)(\cos x+\sin x)^{-3}$; (xiii) $-2(3+x)/\sqrt{(7-6x-x^2)}$; (xiv) $20 \sin^3 5x \cos 5x$; (xv) $-x/\sqrt{(1-x^2)^2}$; (xvi) $-2(1+x)/\sqrt{(1-2x-x^2)}$; (xvii) $3(5+6x)(8+5x+3x^2)^2$.

40. (i) 2·51 m/h. (ii) $3.37\frac{1}{2}$ a.m.; 24 m; (iii) 67 minutes.

41. 10; 50; 250; $\pi/20$, $7\pi/20$, $\frac{1}{4}\pi$; -10, -50, -250.

43. $-3/2$.

45. $6x(x^4+2x^2+1) = 6x(x^2+1)^2$.

46. $\dfrac{3}{1+9x^2}$; $\dfrac{\cos x}{(1+\sin^2 x)}$; $\dfrac{2x}{(1+x^4)}$.

49. $\dfrac{x}{\sqrt{(a^2+x^2)}} - \dfrac{\mu(c-x)}{\sqrt{\{b^2+(c-x)^2\}}} = 0$.

Exercise 6c

1. (i) $2x+2y(\mathrm{d}y/\mathrm{d}x) = 0$; (ii) $1+22y(\mathrm{d}y/\mathrm{d}x) = 0$; (iii) $10x-18y(\mathrm{d}y/\mathrm{d}x) = 0$; (iv) $1+2(y-4)(\mathrm{d}y/\mathrm{d}x) = 0$; (v) $2x-3+2(y+1)(\mathrm{d}y/\mathrm{d}x) = 0$.

2. (i) $\dfrac{3x}{2y}$; (ii) $\dfrac{2x+1}{2y-1}$; (iii) $\dfrac{2(x+3)}{3-2y}$.

5. (i) $3x+4y+25 = 0$; (ii) $3x+2y = 1$; (iii) $5x-y+25 = 0$; (iv) $2x-9y-19 = 0$.

6. (i) $\dfrac{\cos x}{\sin t}$; (ii) $\dfrac{2x^3}{2t^3-t}$.

7. (i) $\dfrac{1}{0{\cdot}2\pi}$; (ii) $1/\pi$; (iii) $1/2\pi$ cm/s; $5/\pi$ cm.

8. 240π cm³/h.

9. $a^2+b^2 = 25$; $a(\mathrm{d}a/\mathrm{d}t) = -b(\mathrm{d}b/\mathrm{d}t)$; $-\frac{3}{4}$ cm/s.

10. $\frac{1}{5}\sqrt{10}$.

11. $-Q^2(\mathrm{d}C/\mathrm{d}x)/2C^2$; $-Q^2/2\varepsilon_0 A$; attraction.

12. $-v^2/u^2$.

13. 2 m/s.

14. 0·36.

15. 0·53 cm/s.

16. 1·5 m/s.

17. $\varepsilon_0 V^2 A/2x^2$; attraction.

18. $64\sqrt{6}/9$ cm/s.

19. $\frac{1}{2}$ cm/s.

22. $1{\cdot}2\,\pi$ cm²/s.

23. 2×10^{-3} cm/s.

24. $2\pi r^2 M/x^3$; $(-6\pi r^2 M/x^4)$ (velocity of magnet).

25. $8ab(a+bt)$.

28. 0·00050.

29. $k/\pi a^2$ cm/s.

30. (i) $\frac{25}{16}\pi$; (ii) $\frac{9}{4}\pi$; 1·44.

32. (i) $\sqrt[3]{4}$ m; (ii) $\dfrac{1}{12\sqrt[3]{16}}$ m/h.

34. $(2x+x^2)$ cm²; 2/3 cm/min; 8 cm²/min.

Exercise 6d

1. $x\cos x+\sin x$.

2. $3x^2\cos x - x^3\sin x$.

3. $3(\sin 2x+2x\cos 2x)$.

4. $x(2\sin\frac{1}{2}x+\frac{1}{2}\cos\frac{1}{2}x)$.

5. $\cos x(\cos^2 x-2\sin^2 x)$.

6. $\sin x(2\cos^2 x-\sin^2 x)$.

7. $(8\tan 4x\cos^2 4x-4\tan^{0} 4x)$ or $4\sin 4x(2\cos^2 4x-\sin^2 4x)$.

8. $(2-3x)/\sqrt{(1-x)}$.

9. $x^3(1-x^3)^6(4-25x^3)$.

10. $(1+x)^2(1+4x)$.

11. $2(5x+4)^5(3x-8)^3(75x-144)$.

12. $x\cos x(1+2\sin x)+\sin x-\cos^2 x$.

13. $3\sin^2 x\cos 3x\,(\cos x\cos 3x-2\sin x\sin^3 3x)$.

14. $\frac{1}{2}\sec^2\frac{1}{2}x$.

15. $-\sqrt{(\sin x-\frac{1}{2}\cos x\cot x)}(\operatorname{cosec} x)$.

16. $-6x^5$.

17. $a\cos 2ax$.

18. $\dfrac{3}{(1-5x^2)^{\frac{3}{2}}}$.

19. $2\sin x(x\sin x+(x^2-7)\cos x)$.

20. $x^2(1-7x)^9(3-91x)$.

21. $E^2(r-R)/(r+R)^3$; r.

23. $5(2\cos 2t\sin 3t+3\sin 2t\cos 3t)$; $5(12\cos 2t\cos 3t-13\sin 2t\sin 3t)$.

24. $2^8\,5^5/3^{18}$, max; 0, min; $2^8\,5^5/3^{18}$, max; 0, min.

25. $\sec^2 x$; $-\operatorname{cosec}^2 x$.

26. (i) $\omega(\cos^2\omega t-\sin^2\omega t)$; (ii) $\omega\cos 2\omega t$.

27. $3\cos 3\theta\cos 2\theta-2\sin 3\theta\sin 2\theta$; $5/2\cos 5\theta+\frac{1}{2}\cos\theta$.

28. $w^2(3-w^2)^3(9-11w^2)$.

29. $\theta = \cot\theta$; (49°18′); 0·561.

30. (a) $8/3\pi$ cm³/s; (b) $63\pi/\sqrt{29}$ cm²/s.

31. 4/9 cm/s.

32. $4a/25$, $3a/25$ km.

33. $\frac{1}{2}a^2$.

35. $\sin\frac{1}{2}\pi t+\frac{1}{2}\pi t\cos\frac{1}{2}\pi t$; $\pi(\cos\frac{1}{2}\pi t-\frac{1}{4}\pi t\sin\frac{1}{2}\pi t)$.

36. $(\sin x+x\cos x)/2\sqrt{(x\sin x)}$; (i) $+1$; (ii) $+1$; derivative $= 1$.

39. (i) $4x^3-3x^2+1$; (ii) $x\cos x$; (iii) $\dfrac{x(8-15x)}{2\sqrt{(2-3x)}}$.

41. 30° with the direction straight across; $\left(10\sqrt{3}+\dfrac{50}{3}\right)$ seconds;
(i) same direction, $\left(10\sqrt{3}+\dfrac{80}{3}\right)$ seconds;
(ii) same direction, $\left(10\sqrt{3}+\dfrac{25}{6}\right)$ seconds.

42. $\cos\{(60+D)/2\}$; 1635 radian/cm.

44. $\operatorname{cosec} x(1-x\cot x)$; 257–258°.

45. $4\pi a^2\sin\theta\cos^2\theta$.

46. Max 0, min -1; (i) $\frac{1}{2}d^2$; (ii) $\frac{1}{4}(d^2\sqrt{3})$.

48. $(\frac{1}{4}, \frac{27}{4})$ min; (1, 0) inflexion.

49. $uv(\mathrm{d}w/\mathrm{d}x)+uw(\mathrm{d}v/\mathrm{d}x)+vw(\mathrm{d}u/\mathrm{d}x)$; $(1+2x)^3(1-x^2)^4\{(1+10x-11x^2-30x^3)\}$.

50. $(gf'-fg')/g^2$.

51. r/n.

Exercise 6e

1. $x(2-x)/(1+x)^4$.
2. $(8x - x^2-4)/x^2(1-x)^2$.
3. $\dfrac{2x}{(3-x^2)^2}$.
4. $\dfrac{x(x-2)}{(x-1)^2}$.
5. $\dfrac{(2\cos x+\cos^2 x+2\sin^2 x)}{(2+\cos x)^4}$.
6. $\dfrac{-6x^2}{(1+x^3)^2}$.
7. $\dfrac{(\cos^2 x+3\sin^2 x)}{\cos^4 x}$.
8. $\dfrac{3\sin 3x}{\cos^2 3x}$.
9. $\dfrac{-x^2(3+2x^2)}{(2x^2-1)^3}$.
10. $(8-3x)/x^5$
11. -4 max; 0 min.
12. $V = -\frac{1}{6}\sqrt{3}$ at $t = -\sqrt{3}$, min; $V = \frac{1}{6}\sqrt{3}$ at $t=\sqrt{3}$, max.
13. (i) $\dfrac{1+x}{(1+x^2)}$; (ii) $\dfrac{x^2+2x-1}{x^2(1-x)^2}$.
15. $\sec x\tan x$; $-\operatorname{cosec} x\cot x$.
16. $\dfrac{1}{t}\sec^2(t/a)-\dfrac{a}{t^2}\tan(t/a)$.
17. $3x^2$; $\dfrac{3x^2}{(1-x^3)^6}$ 6; $\dfrac{2+3x}{2\sqrt{(1+x)}}$; $2\tan x\sec^2 x$.
18. $100/(100-x)^2$.
19. $R = r$.
22. $(-4, -\frac{5}{4})$ min; $(1, 5)$ max.
23. 10.
24. $\dfrac{1-x^2}{(x^2+1)^2}$; $\dfrac{x-2}{2x^2\sqrt{(1-x)}}$; $\sec x(\tan x+\sec x)$.
25. (i) $\dfrac{x(2+x)}{(1+x)^2}$; (ii) $\dfrac{-3}{(2x+1)^5}$; (iii) $x^2(3\tan x+x\sec^2 x)$.
28. $6\sqrt{10}$.
30. $10/3(6-t)^2$ m/s.
34. $\dfrac{B}{C}\left(\dfrac{100}{81}-\sqrt{5439}\right)\leqslant f^2\leqslant\dfrac{B}{C}\left(\dfrac{100}{81}+\sqrt{5439}\right)$.
35. 0; 2: $11x-8y+4-11\pi/4 = 0$.
37. 1, min; $\frac{1}{5}$, max.
39. $(-3, -\frac{1}{6})$ min; $(-1, -\frac{1}{2})$ max.

Exercise 6f

1. $2x-3y-(3x-2y)\,dy/dx = 0$.
2. $2x(2x^2+y^2)+2y(x^2y-16y^2)dy/dx = 0$.
3. $3x^2+10xy^2+(x^3+10x^2y)dy/dx = 0$.
4. $2y\sin x\cos x+\sin^2 x\,dy/dx = 0$.
5. $-\sin y\sin x+\tan y +(\cos x\cos y+x\sec^2 y)(dy/dx) = 0$.
6. (i) 1 or 2; (ii) 2 or 1.
7. $\dfrac{q(5p^2-1)}{p(1-5q^2)}$.
8. $x\,\delta y+y\,\delta x+\delta x\,\delta y$.
9. (i) $6q^3p+9q^2p^2$; (ii) $\omega x\cos\omega t+\sin\omega t(dx/dt)$.
10. (i) $\dfrac{x(x-2)}{(x-1)^2}$; (ii) $\dfrac{(4x-4x^2-2)}{\sqrt{(1+x^2)}}$.
11. $\dfrac{\sec^2 x}{\sqrt{(1-\tan^2 x)}}$. (ii) $\dfrac{y(a-y)}{x(x-a)}$.
13. 2.
14. (i) p; (ii) $8p$ or $1{\cdot}4p$.

Exercise 6g

1. $\dfrac{1}{\sqrt{(9-x^2)}}$.
2. $\dfrac{1}{\sqrt{(16-x^2)}}$.
3. $\dfrac{2}{\sqrt{(25-x^2)}}$.
4. $-\dfrac{1}{\sqrt{(16-x^2)}}$.
5. $\dfrac{-1}{\sqrt{(49-x^2)}}$.
6. $\dfrac{4}{16+x^2}$.
7. $\dfrac{-3}{\sqrt{(81-x^2)}}$.
8. $\dfrac{10}{4+x^2}$.
9. $2x\sin^{-1}x+\dfrac{x^2}{\sqrt{(1-x^2)}}$.
10. $\dfrac{1}{x(1+x^2)}-\dfrac{1}{x^2}\tan^{-1}x$.
11. $3\cos^{-1}x-\dfrac{(3x+7)}{(1-x^2)}$.
12. $\dfrac{1}{(9+x^2)(1+x^2)}-\dfrac{2x\tan^{-1}x}{(9+x^2)^2}$.
13. $\dfrac{1}{(4-x^2)}\left\{1+\dfrac{x\sin^{-1}x}{\sqrt{(4-x^2)}}\right\}$.
14. $-1-\dfrac{x\cos^{-1}(x/3)}{\sqrt{(9-x^2)}}$.
15. $\dfrac{2}{\sqrt{(1-4x^2)}}$.
16. (i) $\dfrac{5}{1+25t^2}$; (ii) $\dfrac{3}{\sqrt{(1-9t^2)}}$; (iii) $\dfrac{2}{\sqrt{(1-4t^2)}}$; (iv) $\dfrac{3}{\sqrt{(1-(3t+\pi)^2)}}$ or $\dfrac{3}{\sqrt{(1-9t^2)}}$; (v) $\dfrac{25}{1+(5t-\frac{1}{2}\pi)^2}$ or $\dfrac{25}{1+25t^2}$.
17. (i) $\dfrac{2\tan^{-1}x}{1+x^2}$; (ii) $\dfrac{\left\{\dfrac{2\cos^{-1}3x}{\sqrt{(1-4x^2)}}+\dfrac{+3\sin^{-1}2x}{\sqrt{(1-9x^2)}}\right\}}{(\cos^{-1}3x)^2}$; (iii) $\dfrac{-1}{x\sqrt{(1-16x^2)}}$; (iv) $-\dfrac{1}{1+x^2}$.

18. $\dfrac{dz}{dt} = \sin^{-1} t + \dfrac{t}{\sqrt{(1-t^2)}}$;
$\dfrac{d^2z}{dt^2} = \dfrac{2-t^2}{(1-t^2)^{\frac{3}{2}}}$.

19. $-4\sqrt{3}$.

20. $\dfrac{1}{\sqrt{(1-x^2)}}$ and $\dfrac{1}{1+x^2}$.

21. $\dfrac{2x}{1+x^4}$; 1.

22. $\dfrac{\phi'}{\sqrt{(1-\phi^2)}}$; $-\dfrac{\phi'}{(1-\phi^2)}$; $\dfrac{\phi'}{1+\phi^2}$.

24. 0.

25. $-\operatorname{cosec}^2 x$; $-\dfrac{1}{1+x^2}$.

26. (i) $\dfrac{2 \sin x \cos x}{1+\sin x}$; (ii) $\dfrac{-2x}{x^4+(1-x^2)^2}$.

Exercise 6h

1. $1/t$.

2. $a \cos t$.

3. $-b/a \tan t$.

4. $2\sqrt{t}$.

5. $-\cos^2 2t$.

6. -1.

7. $-(1/t^2) \cot^2 t$.

8. (i) $4ax = y^2$; (ii) $y = a \sin (x-a)$; (iii) $x^2/a^2+y^2/b^2 = 1$; (iv) $y = x^2-3$; (v) $xy = 1$; (vi) $\sin x = \cos y$; (vii) $xy = 1$.

9. (i) $\sqrt{(a/x)}$; (ii) $a \cos (x-a)$; (iii) $\dfrac{-bx}{a\sqrt{(a^2-x^2)}}$; (iv) $2x$; (v) $-1/x^2$; (vi) -1; (vii) $-1/x^2$.

10. $\dfrac{6t-1}{2(6t^2+1)}$; (i) $\dfrac{x(4-15x)}{2\sqrt{(1-3x)}}$; (ii) $\dfrac{x^2(3+4x)}{(1+2x)^2}$; (iii) $6 \sin^2 2x \cos 2x$.

12. (i) $\dfrac{-1}{2at^3}$; (ii) $\dfrac{-b}{a^2 \cos^3 3t}$; (iii) 0.

13. $4/9t^2$; $x^4 = y^3$.

14. $\dfrac{(3t-2)}{2t}$; $\dfrac{3\sqrt{(x/a)}-2}{2\sqrt{(x/a)}}$; $-8a/27$

15. 0·85 kg/mm.

16. (i) $(2x^2-3)/x^4\sqrt{(1-x^2)}$; (ii) $\dfrac{1}{(1+\cos x)}$; $\dfrac{-5t^{\frac{3}{2}}}{6(1+t^2)}$.

17. (i) $3(1+x)^2 \{\tan 3x+(1+x) \sec^2 3x\}$; $\frac{9}{4}$.

18. (i) $1+\dfrac{x}{\sqrt{(1+x^2)}}$; (ii) $\dfrac{-11 \sin x}{(3+5 \cos x)^2}$; $\dfrac{4\sqrt{2}}{3a}$.

21. (i) $\dfrac{-4 \sin \theta \cos \theta}{\sqrt{(3-4 \sin^2 \theta)}}$; (ii) $\dfrac{2(1+t^2)}{3t}$.

22. $3-(1-t)^2$; $1+(1+t)^2$; max 3, min 1; max 1, min -3.

23. $\dfrac{\sin \theta}{(1-\cos \theta)}$.

24. $-\sec^2 \theta$.

25. $-3 \cos^2 t \sin t$; $-3a \cos^2 t \sin t$; $3a \sin^2 t \cos t$; $x \tan t+y = a(\cos^3 t \tan t + \sin^3 t) = a \sin t$; $a(\cos t-\cos 2t)$; $3\sqrt{3}a/4$.

28. $\sqrt{(\dot{r}+r^2\omega^2)}$.

29. $\pm 3a\omega \sin \theta \cos \theta$.

Miscellaneous Exercise 6

1. $\dfrac{\mu \cos r}{\cos i}$; $\dfrac{\mu \cos r'}{\cos i'}$.

2. $1+\dfrac{di'}{di}$.

3. $\dfrac{\sqrt{2}}{(1-x^2)}$.

7. $y' = (x \sec^2 (x/a)-a \tan (x/a))x^2$;
$z = \dfrac{a(x \sec^2 (x/a)-a \tan (x/a))}{x^2+a^2 \tan^2 (x/a)}$.

9. (i) $\dfrac{\sin \sqrt{x}}{2\sqrt{x} \cos^2 \sqrt{x}}$; (ii) $\dfrac{x^2+2x-1}{x^2(1-x)^2}$; (iii) $\dfrac{a}{2(1+ax+b)\sqrt{(ax+b)}}$.

10. (i) $\dfrac{z^2+2z-3}{(z+1)^2}$; (ii) $2z\left(\tan^{-1} z^2+\dfrac{z^2}{1+z^4}\right)$; (iii) $\dfrac{z(2-z)}{(1+z)^4}$; (iv) $-\frac{1}{2}$.

11. (i) $\dfrac{(1+3x^2)}{2\sqrt{\{x(1+x^2)\}}}$; (ii) $\dfrac{2+2 \cos x-\cos^2 x}{(2+\cos x)^3}$.

13. $P = 2a(1+\cos \theta+2 \sin \frac{1}{2}\theta)$.

15. 5·75 metres; 2/39.

17. $\dfrac{Bt(\varepsilon-1)}{\varepsilon_0[ab+x(\varepsilon-1)]^2}$; $\varepsilon = 1$, no force; $\varepsilon < 1$, force repelling dielectric.

18. L.

19. $(100\pi/27)$ m/s; $\dfrac{100\pi^2}{729}$ m/s^2.

20. $(0, \frac{1}{4}\pi)$, $(\frac{1}{2}\pi, \frac{3}{4}\pi)$.

22. (i) $3x^2+2x+1$; (ii) $x \sin x$; (iii) $\dfrac{2x-5x^2}{\sqrt{(1-2x)}}$.

23. $2x$, $-1/x^2$; (i) $\dfrac{x}{\sqrt{(x^2+1)}}$; (ii) $\dfrac{2x-x^2}{(1-x)^2}$; (iii) $\cos 2x$.

25. (i) $1-1/x^2$, (ii) $\frac{3}{2}(1+3x)^{-4}$; (iii) $-2 \sin x \cos x$.

26. (1, 0) max; (2, -13) min.

29. 5 ohms.

31. 2 dm^3.

32. $tx-y-t^3 = 0$; $x+uy-u^2(3+2u) = 0$.

34. (i) $\dfrac{\sin t}{1-\cos t}$; (ii) $\dfrac{\sqrt{(2a-y)}}{\sqrt{y}}$.

36. $\dfrac{(1+x)}{(1+x^2)^{3/2}}$.

37. $F(a) < 0$.

39. $(b/a) \sin t$.

40. 0; 0; 2.

42. $A = \frac{1}{10}$, $B = -\frac{1}{5}$.

45. $\dfrac{1}{3\pi}$ cm/s.

46. 0·00025. **47.** $\dfrac{h+ut}{\sqrt{(a^2+v^2t^2)}}$.

49. $2\tan^2\theta(1-\tan^2\theta)$; $\tan^{-1}(1/\sqrt{2})$.

50. (i) $x \geqslant 0$; (ii) $|x| > 0$.

51. (i) Positive; (ii) negative; $-\frac{1}{2}$ min, 1 max, 5/3 min.

52. $5\sin 5\theta/\sin\theta$.

54. $(\pi/6, \sqrt{3})$ max; $(5\pi/6, -\sqrt{3})$ min.

55. 20×20×5 m.

57. (i) $50/\sqrt{3}$ metres downstream from the point directly opposite; (ii) $250/\sqrt{3}$ metres downstream.

59. $\sqrt{(a^2+x^2)}+\sqrt{(a^2+(b-x)^2)}$; $\frac{1}{2}b$.

60. $x = -2$; $x = 0, -3$.

Chapter 7

Exercise 7a

4. 0·4 cm; 0·63 cm; 0·895 cm; 1·005 cm; 1·27 cm; 1·41 cm; 1·55 cm; 1·67 cm; 1·79 cm; 1·92 cm; 2·0 cm; 5·68 cm; 4·88 cm.

5. (a) 8; (b) 20; (c) 2; (d) $4a$; (e) 8; (f) $2(2a+b)$ cm².

6. $2\pi a$; $2\pi(a+h)$.

7. 18·6; 12·7.

10. (i) $x > 4$, $x < -2$; (ii) $-1·24 < x < 3·24$; (iii) $x = 4·74$, $x = -2·74$; (iv) 11·6.

11. 20 N; $2\sqrt{(11/3)}g$ m/s.

12. (i) 3900; (ii) 39; (iii) $39n$; (iv) $39n^2$.

Exercise 7b

1. 19·2 and 16·2. **2.** 10·6 and 5·8.

3. 1·12 and 0·86. **4.** 1·83 and 1·43.

5. 1·53 and 0·54.

6. 4·375, 6·875 cm²; 3·75, 6·25 cm².

7. The distances are approximately 625 cm, 675 cm and 1300 cm.

8. 82 metres.

Exercise 7c

1. 2. **2.** 0·0348.

3. 82·5. **4.** 18·5; 19·6.

5. 340; 335; 333·8; 333·3. **6.** (a) 56·8; (b) 0·994.

7. 12 m, 9 m; 11 m. **8.** 0·25; 1·85; 3·70; 4·2.

11. 193 m.

12. (i) 48; (ii) 96; (iii) 174; (iv) 345 cm²; (a) £480; (b) £960; (c) £1740; (d) £3450.

13. 28, 30; 35, 51; $38\frac{3}{4}$, $46\frac{3}{4}$ cm²; 42·4–42·8 cm².

14. 21·3 cm².

Exercise 7d

1. 3. **9.** 18. **17.** $3x$.

2. 2. **10.** $22\frac{1}{2}$. **18.** $5x+c$.

3. 1. **11.** $31\frac{1}{2}$. **19.** $5x^2+c$.

4. 21. **12.** 16. **20.** 3/10.

5. $18a$. **13.** 1/2. **21.** 12.

6. 7200. **14.** 12. **22.** 80.

7. 2. **15.** 8. **26.** 16.

8. 20. **16.** 180.

27. (i) 2; (ii) 3; (iii) 13.

28. (i) 2; (ii) 4; (iii) 10·4; (iv) 12; (v) 13. No.

29. (i) 3; (ii) 2·18.

30. (i) $\frac{1}{2}(n-1)n$; (ii) $\frac{1}{2}m(2n-1-m)$ where $m = n]$.

34. (i) 46,750 J; (ii) 10·4 m/sec.

35. (i) 151·5 J; (ii) 31·1 m/s.

Exercise 7e

1. -4. **8.** 8. **15.** $25\pi/4$.

2. $-a^2$. **9.** -6. **16.** $9\pi/4$.

3. -4. **10.** $\frac{1}{2}x^2$. **17.** $3x$.

4. a^2. **11.** $\frac{1}{2}(x^2-1)$. **18.** $5x+c$.

5. $-\frac{1}{2}b^2$. **12.** 0. **19.** $\frac{1}{2}x^2+c$.

6. -4. **13.** $-10x$.

7. 0. **14.** 0.

20. $5x+c$; $-c/5$. (i) 5; (ii) 10; (iii) 0; (iv) $5a$.

21. (i) $\frac{1}{2}x^2$; (ii) $\frac{1}{2}(x^2-a^2)$; (iii) $\frac{1}{2}(x^2-a^2)$; (iv) $\frac{1}{2}(x^2-a^2)$; (v) $\frac{1}{2}(x^2-a^2)$.

27. (i) $n = 0, 2, 4, 6$, etc.; (ii) $n = 1, 4, 7$, etc.; (iii) $m = n, n+2, n+4$, etc.

28. (i) 8; (ii) 3; (iii) -1; (iv) 7.

29. (i) $\frac{1}{2}$; (ii) 0·26; (iii) $\frac{1}{4}+\frac{1}{4}\{a-2(a-b)\}^2$ where $b = [a]$.

33. 40; £40.

34. 3 mm.

Miscellaneous Exercise 7

1. 2·56, 5·76.

2. 3·625, 10·25.

5. 12 cm². (i) 3 cm²; (ii) 45 cm²; (iii) 0·785 cm.

6. $4\pi(a+\delta a)^2$, $4\pi a^2$, $4\pi a^2\delta a$.

7. 3·81%.

8. (i) 0·686; (ii) 1·867; (iii) 2·38.

9. 2 cm.

10. 101 m.

11. 1·52.

12. (i) 8; (ii) 81; (iii) 42·5; (iv) 2·5; (v) 0·5.

13. (i) 15; (ii) −15; (iii) $-\frac{1}{2}$; (iv) −4; (v) 4; (vi) 0.

14. (i) −1·5; (ii) −6; (iii) 3·5; (iv) 0.

15. (i) −16; (ii) −12; (iii) −10; (iv) −30; (v) −18; (vi) 7.

16. (i) 13·5; (ii) −4; (iii) 10; (iv) $-4b$; (v) 8; (vi) 0; (vii) −10; (viii) 26; (ix) 4; (x) $-\frac{1}{2}x^2$.

17. 11·5.

19. 7.

20. $112\frac{1}{2}$; $142\frac{1}{2}$; $52\frac{1}{2}$.

21. 1.

22. $\frac{1}{2}[a]\ \{1+2a-[a]\}$.

23. 796 m².

Chapter 8

Exercise 8a

4. (i) $-\frac{1}{2}\cos 2\theta$; (ii) $\frac{1}{2}\sin 2\theta$.

6. 0·366.

7. (i) 0; (ii) $(x-1)$.

Exercise 8b

An indeterminate constant c should be added to all answers up to and including question 32.

1. $\frac{1}{6}x^6$.
2. $\frac{1}{4}x^4$.
3. $\frac{1}{2}x^2$.
4. $4x^2$.
5. $3x$.
6. x^2.
7. $-x^{-1}$.
8. $-1/x^2$.
9. $-7/4x^4$.
10. $\frac{4}{5}x^{5/4}$.
11. $2\sqrt{x}$.
12. $3x^{1/3}$.
13. $\frac{32}{3}x^{3/4}$.
14. $-6x^{-1/2}$.
15. $2x^{1/3}$.
16. $\frac{1}{3}x^3-\frac{7}{2}x^2$.
17. $\frac{5}{2}x^2+15x$.
18. $\frac{1}{3}x^3+x^2+x$.
19. $\frac{2}{3}x^3-\frac{13}{2}x^2-7x$.
20. $\frac{2}{5}x^{5/2}+\frac{4}{3}x^{3/2}$.
21. ax.
22. $ax^8/8$.
23. $\frac{1}{3}x^3-7/x$.
24. $2x\sqrt{x}-4\sqrt{x}$.
25. $\frac{2}{5}x^{1/2}\ (x^2+5)$.
26. $-(7x+6)/42x^7$.
27. $\frac{1}{2}p(3p+2y)$.
28. $-(3q^2+1)/3q^3$.
29. $\frac{2}{5}z^{5/2}$.
30. $(a^2t^5)/5+\frac{2}{3}bat^3+b^2t$.
31. (i) $-\cos x$; (ii) $\sin^{-1}(x/3)$; (iii) $-\cos^3 2x$.
32. $-1/(1+x^2)^3$.
33. 512/9.
34. $\frac{1}{2}$; $\frac{2}{3}$.
35. $s=\frac{5}{2}t^2+7t+c$.
36. (a) $\frac{1}{6}(2x-3)^3+c$; (b) $\frac{2}{3}\sqrt{(x)^3}-2\sqrt{x}+c$.

Exercise 8c

An indeterminate constant c should be added to all answers.

1. $-\frac{1}{3}\cos 3x$.
2. $2\sin\frac{1}{2}x$.
3. $\frac{1}{5}(1+x^2)^5$.
4. $\frac{2}{3}(1+x^2)^{3/2}$.
5. $-\frac{1}{4}\cos^4 x$.
6. $-\frac{1}{3}\cot 3x$.
7. $-\frac{2}{3}\sqrt{(1-x^3)}$.
8. $\frac{1}{6}\tan^6 x$.
9. $\dfrac{1}{2\cos^2 x}$.
10. $\frac{1}{5}\sec 5x$.
11. $\dfrac{(5+4x)^{11}}{36}$.
12. $-\frac{1}{12}\cot^6 2x$.
13. $-\operatorname{cosec} x$.
14. $\sin^{-1}(x/3)$.
15. $\dfrac{-1}{2(16+x^2)}$.
16. $\frac{1}{2}\tan^{-1}(x/4)$.
17. $\dfrac{-1}{2(16+x^2)}$.
18. $\frac{3}{2}\sin^{-1}(x/2)$.
19. $\frac{1}{20}\tan^{-1}(4x/5)$.
20. $-\frac{1}{130}(1-2x^5)^{13}$.
21. $-\sqrt{(3-x^2)}$.
22. $-\frac{1}{9}\cos^9 x$.
23. $\dfrac{1}{\sqrt{35}}\tan^{-1}(\sqrt{5}x/\sqrt{7})$.
24. $\frac{1}{2}\tan^2 x$.
25. $\frac{1}{3}\cos x\ (\cos^2 x-3)$.
26. $\frac{1}{4}\sin 2x+\frac{1}{2}x$.
27. $2\tan x-x$.
28. $-\frac{1}{3}\sin 3x-\frac{1}{9}\sin^3 3x$.
29. $\sin^{-1} x-\sqrt{(1-x^2)}$.
30. $\cos x-\frac{2}{3}\cos^3 x$.
31. $\tan x-x$.
32. $\frac{3}{8}x-\frac{1}{4}\sin 2x+\frac{1}{32}\sin 4x$; or $\frac{1}{8}\sin x\cos x\,(5-2\sin^2 x)$.

Exercise 8d

1. 10.
2. 30.
3. 14.
4. 162.
5. 39.
6. 18.
7. 62.
8. 36.
9. 36.
10. 15·103.
11. 25·373.
12. $2^{12}-1$.
13. $4(2^{12}-1)$.
14. $\frac{1}{2}n[2a+(n-1)d]$.
15. $\dfrac{a(r^n-1)}{1-r}$.
16. 2.
17. 4/3.
18. 10/9.
19. 5/7.
20. 9/8.
21. (i) $\sin x+c$; (ii) $\frac{1}{6}x^6+c$; (iii) $-(1/y)+c$.
22. (i) $x(2-x)/2+c$; (ii) $\frac{1}{2}\tan 2x+c$; (iii) $\frac{2}{3}\sqrt{p^3}+c$.
23. (ii), (vi).

26. (i) $-\frac{1}{4}\cos^4 x+c$; (ii) $\frac{1}{3}\sqrt{(1+x^2)^3}+c$;
(iii) $\frac{1}{2}(\sin^{-1}x)^2+c$.

27. $\frac{1}{2}A\varrho h^2$.

32. $pa^3/3$.

MISCELLANEOUS EXERCISE 8

The constant of integration has been omitted; a fully correct answer should therefore have $+c$ at the end

1. $\frac{1}{9}x^9$.
2. $\frac{1}{5}x^5$.
3. $4x$.
4. $-\frac{1}{2}x^2$.
5. $-\frac{1}{3}x^3$.
6. $-\frac{7}{2}x^2$
7. $-\frac{8}{9}x^9$.
8. $7x$.
9. $7x-x^9$.
10. $2x-\frac{1}{2}x^2$.
11. $5x+x^2$.
12. $6x+x^2+3x^3$.
13. $3x^4-7x$.
14. $-2/\sqrt{x}$.
15. $\frac{2}{7}x^{7/2}$.
16. $x^{7/2}$.
17. $\frac{3}{4}x^{4/3}$.
18. $\frac{9}{8}x^{8/3}$.
19. $5x^{6/5}$.
20. $\frac{20}{9}x^{9/5}$.
21. $4x^{2\frac{1}{4}}$.
22. $-\frac{1}{2}x^{-2}$.
23. $-x^{-7}$.
24. $2\sqrt{x}$.
25. $\frac{9}{2}x^{2/3}$.
26. $-7x^{-1}-\frac{5}{2}x^{-2}$.
27. $-2x^{1/2}$.
28. $2x^{1/2}+2x^{-1}$.
29. $q/5x^5-p/4x^4$.
30. $-\dfrac{(7ax+6b)}{42x^7}$.
31. $5/4x^4-3/2x^2$.
32. $2(x+2)\sqrt{x}$.
33. $\frac{1}{6}(2x-5)^3$.
34. $\frac{1}{7}x$.
35. $\frac{1}{18}x^6-8x$.
36. $-\frac{1}{48}x^6$.
37. $-3/(5x^3)$.
38. $-12/(35x^5)$.
39. $\frac{11}{10}\sqrt{x}$.
40. $(7x-9)\sqrt{x}/12$.
41. $-4x^{-1}$.
42. $-2x^{-1}$.
43. $-\frac{4}{3}x^{-3}$.
44. $\frac{1}{3}x^3-x^2+x$.
45. $x^3-\frac{23}{2}x^2-8x$.
46. $7x+2x^2-x^3$.
47. $\dfrac{9x-6x^2-4}{6x^3}$.
48. $\frac{32}{7}x^{7/4}$.
49. $\frac{1}{2}x^{1/4}$.
50. $-\frac{1}{8}\cos 8x$.
51. $\frac{1}{5}\sin 5x$.
52. $\sin 2x$.
53. $-\frac{1}{3}\cos 9x$.
54. $-2\cos\frac{1}{2}x$.
55. $14\sin\frac{1}{2}x$.
56. $32\sin\frac{1}{4}x$.
57. $55\sin\frac{1}{5}x$.
58. $\frac{1}{8}(1+x^2)^8$.
59. $\frac{1}{14}(1+x^2)^7$.
60. $\frac{1}{3}(1+x^2)^{3/2}$.
61. $-\frac{1}{6}(1+x^3)^{-2}$.
62. $\frac{1}{14}(1-x^2)^{-7}$.
63. $\dfrac{7}{6(1-x^2)^3}$.
64. $-2\sqrt{(1-x^2)}$.
65. $\sin^{-1}x$.
66. $\sin^{-1}(x/3)$.
67. $\frac{5}{4}\sin^{-1}(4x)$.
68. $3\sin^{-1}(x/2\sqrt{2})$.
69. $2\tan^{-1}x$.
70. $-\dfrac{1}{7(1+x^2)^7}$.
71. $\frac{1}{2}\tan^{-1}(2x)$.
72. $\frac{1}{3}\tan^{-1}(x/3)$.
73. $\frac{7}{4}\tan^{-1}(x/4)$.
74. $2\tan^{-1}(4x)$.
75. $-\frac{5}{6}\tan^{-1}(2x/3)$.
76. $\tan^{-1}(3x/4)$.
77. $\tan^{-1}(3x/2)$.
78. $\frac{1}{2}\sin^{-1}(4x/5)$.
79. $-\frac{1}{8}\sqrt{(25-16x^2)}$.
80. $\frac{5}{2}\sin^{-1}(2x/3)$.
81. $-\frac{5}{4}\sqrt{(9-4x^2)}$.
82. $-\frac{1}{42}\sqrt{(36-49x^2)^3}$.
83. $-\frac{1}{7}\sqrt{(36-49x^2)}$.
84. $\frac{1}{6}\tan^{-1}(7x/6)$.
85. $-\frac{1}{3}\cos^3 3x$.
86. $-\frac{1}{10}\cos^{10} 3x$.
87. $\dfrac{1}{35\cos^7 5x}$.
88. $-\dfrac{1}{40\sin^{10} 4x}$.
89. $\frac{1}{90}\sin^9 10x$.
90. $\frac{1}{90}\tan^5 2x$.
91. $-\frac{1}{4}\cot^4\frac{1}{2}x$.
92. $7r+c$.
93. r^3-8r.
94. $\frac{2}{7}q^{7/2}$.
95. $\frac{20}{29}t^{29/20}$.
96. $\frac{2}{5}a^{5/2}$.
97. $-1/16s^4$.
98. $-1/2b^8$.
99. $\frac{1}{5}(1+x^2)^5$.
100. $-\dfrac{1}{2(1+y^2)^2}$.
101. $\frac{1}{6}\cos^6\theta$.
102. $\frac{1}{5}\sin^5\theta$.
103. $\dfrac{1}{2\cos^2\phi}$.
104. $\tan x$.
105. $\frac{1}{3}\sec^3 x$.
106. $\frac{1}{10}\tan^2 5x$.
107. $-\frac{1}{27}\cot^9 3x$.
108. $-\frac{2}{3}\sqrt{(\cos x)^3}$.
109. $-2\sqrt{(\cos x)}$.
110. $\frac{1}{4}\sin^2 2\theta$ or $-\frac{1}{4}\cos^2 2\theta$.
111. $\frac{8}{7}$.
112. $\frac{9}{8}$.
113. $\frac{1}{5}x^5+\frac{2}{3}x^3+x$.
114. (i) $-\frac{1}{2}(\cos^{-1}x)^2$;
(ii) $\frac{4}{3}\sin^{\frac{3}{4}}x$.
115. $\frac{1}{2}\varrho h^2$.
117. $GMmh/R(R+h)$, where $G=$ the universal constant of gravitation, M = mass of earth, m = mass of body, R = radius of earth.

Chapter 9

EXERCISE 9a

1. 1.
2. 24.
3. 56.
4. 6.
5. 1.
6. -8.
7. 2.
8. 3/4.
9. 6.
10. -24.
11. $1\frac{2}{7}$.
12. $16\frac{1}{4}$.
13. 0.
14. $2\frac{2}{3}$.
15. 4/5.
16. 7/96.
17. $\dfrac{\sqrt{2}-1}{2\sqrt{2}}$ or 0·1465.

18. 6.

19. 1/6.

20. −9.

21. 5/72.

22. −4/7.

23. −2/5.

24. 17.

25. $\pi/2$.

26. $\pi/2$.

27. $\pi/8$.

28. 1/3.

29. $3\frac{1}{5}$.

30. 3/16.

31. $\pi/12$.

32. $(3-\sqrt{5})/4$.

33. $(2\sqrt{3}-\sqrt{7})/9$.

34. 31/160.

35. $\frac{1}{2}$.

36. $\pi/6$.

37. $\frac{1}{3}$.

38. 1.

39. $\pi/12$.

40. $\pi/4$.

41. (i) $\frac{2}{3}-1/x+\frac{1}{3}x^3$; (ii) $\frac{1}{3}x^3-1/x$.

42. $\frac{1}{5}(\sin 5x-1)$.

43. $\frac{4}{3}(x_2-x_1)(x_2{}^2+x_1x_2+x_1{}^2+6)$.

44. $x/a+y/b = 1$; $\frac{1}{2}ab$.

45. $\frac{1}{3}h(y_1+4y_2+y_3)$.

46. 5.

47. 5·1 or 5·2.

48. 0·13.

49. (a) 252; (b) $6\frac{2}{3}$.

50. (i) $25x-15x^2+3x^3+c$; (ii) 5/3.

52. $2{\cdot}6\times 10^3$ J.

53. 2/3.

54. (a) $-\frac{1}{3}$; (b) $(\pi-2)/8$.

55. $3\frac{1}{4}$.

56. −1.

57. (a) 7/8; (b) 1/6.

58. 0·69.

59. $a > 0, b > 0$ or $a < 0, b < 0$.

61. (i) $(b^2-a^2)/2$; (ii) $(b^2-a^2)/2$; (iii) $(b^2+a^2)/2$.

62. (a) $\frac{1}{4}x^4-\frac{1}{2}x^2+c$; (b) 9.

63. (a) 0; (b) 1/2.

64. $1/\sqrt{3}$.

65. $e^{\sin x}+c$.

66. $[a]\{2a-[a]-1\}/2$.

67. 3/2.

69. $\pi/2$, 1, $(\pi/2-1)$.

70. $(\pi/2-1)$.

7. (i) $2y=3-\cos 2x$; (ii) $6y = 9x^2 \; 2x^3-12x+12$.

8. (i) $2y = 4+\tan^{-1}(x/2)$;
(ii) $2y = 10+\sin^{-1}(2x/3)$.

10. (i) $y = 6-\sqrt{(9-x^2)}$; (ii) $8y = (1+x^2)^8-1$.

11. (i) $20y = \sin^{10} 2x+10$; (ii) $25y = \tan^5 5x+50$.

12. (i) $2y = \sin x^2$; (ii) $4y = \sec^4 x-1$.

13. (i) $3y = 21x-10x^{\frac{3}{2}}$; (ii) $4y=2\sin(8x+\pi/6)+1$.

14. (i) $2x = b^2\sqrt{3}$; (ii) $6x = 1-\dfrac{1}{(1+2t)^3}$.

15. $6y = x(6-5x^2)$.

16. $16y = 12x-2\sin(4x+\pi/3)+(16+\sqrt{3})$.

17. 10 m.

18. 4/3.

19. (i) 0; (ii) $t = 1, t = 2$; (iii) $t = 1\frac{1}{2}$; (iv) $5\frac{1}{6}$ m.

20. 27 m/s; 486 m.

21. (−1/3, 32/27) max, (1, 0) min.

22. $7x-y-22=0$; $y = x^2-x-6$; $(\frac{1}{2}, -6\frac{1}{4})$.

23. 12 m; 14 m/s; 16 m/s².

24. $v^2 = 16-x^2$.

25. $y = x^3-2x^2+x+5$, $5\frac{4}{27}$.

26. (a) 100 m/s²; (b) 15 s; (c) 1225 m/s.

27. $3T/4$.

28. $x(1-y) = 2$.

29. $x(y-3) = 9$; $x+y-9 = 0$; $x-y+3 = 0$.

30. $m\ddot{x} = kx$.

31. $\ddot{x} = -n^2x$.

32. $m\ddot{x}-k\dot{x}+n^2x = 0$.

35. 201 m, 6 m/s².

36. (i) $3y = 3+\sin^{-1}(3x/5)$;

(ii) $9y=1-(1+\cos x)^9$; (iii) $y=\dfrac{1}{(1+\cos x)}-1$;

(iv) $6y = \dfrac{1}{(1+\cos^2 x)^3}-1$.

37. $5{\cdot}28\times 10^{12}t$ m/s; $2{\cdot}64\times 10^{12}t^2$ m; $y^2 = 606x$.

39. $dy/dx = v+x(dv/dx)$; $2y = x^3+cx$.

40. $3y = x(1+x^2)^{3/2}+cx$.

41. $y = W\{6lx^3-x^4-14l^3x+9l^4\}/12EIz$.

42. $d\omega/dt = -40$; $\omega = 200-40t$; $0 \leqslant t \leqslant 5$; 500°.

43. $I = \dfrac{E_0}{2\pi fL}\sin(2\pi ft)$.

Exercise 9b

1. $d\sqrt{(2m/Ve)}$; eV.

5. (i) $10a$; (ii) $5a$.

6. (i) $2y = 5x^2$; (ii) $y = 3x-4x^2$.

Exercise 9c

1. (i) 5; (ii) 8; (iii) $5\pi/4$.

2. (i) 10; (ii) 17; (iii) 16.

3. (i) $2/\pi$; (ii) $2/\pi$; (iii) $\frac{1}{2}$; (iv) $2/\pi$.
4. (i) $3\frac{1}{5}$; (ii) 2; (iii) $\pi/3$; (iv) $(2-\sqrt{3})/3$; (v) $\pi/4$.
5. (i) 3; (ii) 0; (iii) $\frac{1}{3}$.
6. $66\frac{2}{3}$ km/h; 22·2 km.
7. 25.
8. $\frac{E_0^2}{2Z}\cos\varepsilon$.
9. 5·9%.
10. 81·6 km/h; 8·2 km.
13. 3·9.
14. $-\frac{3}{4}$.
15. (i) $2i_0/\pi$; (ii) 0, $i_0^2/2$.
16. (i) 0; (ii) $2aw/\pi$; (iii) $ma^2w^2/4$.
17. 1.
19. (i) $\frac{4}{3}$; (ii) $-\frac{16}{7}$; (iii) $-3\frac{1}{4}$.
20. 0·5.
21. (i) $50/\pi$; (ii) $100/\pi$.
22. (i) $\frac{5}{6}$ V; (ii) 0·08 A; (iii) 0·07 W.
40. $\frac{(m-n)}{(m+1)(n+1)}$.
41. 8/3.
42. $3\pi/4$.

EXERCISE 9d

1. 8/3.
2. 26/3.
3. 4/3.
4. 16/3.
5. 96/5.
6. $\sqrt{3}/2$.
7. 15/4.
8. $8\sqrt{2}/3$.
9. 1/3.
10. $(2-\sqrt{2})/4$.
11. 325/4.
12. 1/6.
13. 1/3.
14. 1/2.
15. 1/6.
16. 1/20.
17. 8/3.
18. 6.
19. $1-\frac{1}{4}\pi$.
20. 1.
21. $16\sqrt{3}$.
22. (i) 9; (ii) 32/3.
23. $41\frac{2}{3}$.
24. 2/3, 1/3, 1/3.
25. 13·88, 35·12.
27. -9, (1, 2), $6\frac{3}{4}$.
28. $y = \frac{1}{3}x^3-2x^2+3x+1$, 10/3.
29. $y = 24-\frac{2}{3}x^2$.
30. $(\frac{1}{3}\sqrt{3}, \frac{3}{4})(-\frac{1}{3}\sqrt{3}, \frac{3}{4})$.
31. $32x-4y-31 = 0$; $32x+4y+31 = 0$.
32. 1/2.
33. (i) P; (ii) $P-Q$; (iii) $P-Q+R$; (iv) $P+Q+R$; (v) $|P-Q+R|$ or $(P+R) \sim Q$.
34. (i) -1; (ii) $-\frac{1}{2}$; (iii) 2.
35. (i) 4/3; (ii) 8/3.
36. (i) $-4/3$; (ii) 20/3; (iii) 4/3; (iv) 36.
37. 1/6.
38. 500/9.
39. (i) 4/3; (ii) 64/3; (iii) $\frac{(\pi-12+6\sqrt{3})}{12}$; (iv) 1/6; (v) 1/12.

EXERCISE 9e

1. 1·22.
2. 0·695.
3. 2·76.
4. 0·432; 0·433; 0·433.
5. 2·3979.
6. 68/3 m², 68 m³.
7. (i) 3280 J; (ii) 40·5 m/s; (iii) 1·30 W.
8. 4; 0·9 m/s²; 22·1 m/s.
9. 6·1.
11. (i) $3h(y_1+3y_2+3y_3+y_4)/8$;
    (ii) $3h\{y_1+5y_2+y_3+6y_4+y_5+5y_6+y_7\}/10$.
12. (i) The highest term ignored is $-\frac{1}{12}\delta^2 f(a)$, since $n = 1$. Since also $f''(a) \simeq \frac{1}{h^2}\delta^2 f(a)$ (see Section 4.12), the error is $\simeq -\frac{h^2 f''(a)}{12}$.
    (ii) The highest term ignored in Simpson's rule (where $n = 2$) is $\delta^3 f(a)/3!$. Since the coefficient of $\delta^3 f(a)/3!$ is zero if $n = 2$, this constitutes no error. The first contributing term is $-h\delta^4 f(a)/90$, and this is approximately $-h^5 f^{iv}(a)/90$.
    (a) $-1/12$, 0; (b) $-1/16$, 0; (c) $-35/648$, $-1/120$.
13. 603·5;
    (i) (*a*) The impulse experienced by the body in Ns; (*b*) the work done by the machine in kilojoules; (*c*) the gain in velocity in m/s experienced by the body; (*d*) the distance covered by the body in metres; (*e*) the charge in millicoulombs acquired by the capacitor.
    (ii) The work done by the force in joules.
14. 195 seconds.

EXERCISE 9f

1. (i) 76·8 g; (ii) 2 g; (iii) 7 g; (iv) 40·5 g.
2. (i) 0·8 g; (ii) 1·7 g; (iii) 1·31 g.
3. (i) 0·13 g/cm; (ii) 3·14 g/cm; (iii) 7·85 g/cm; (iv) $12·6a^2$ g/cm.
4. $2ka^{\frac{3}{2}}/3$.
5. (i) $16(3+2\sqrt{2})/3$ g; (ii) $64(4-\sqrt{2})/3$ g.
6. (i) 15 g; (ii) 4·8 g.
7. $4\pi r^{\frac{5}{2}}/5$.
8. $2\pi ka^2$.
9. 16/5.
10. 4.

EXERCISE 9g

1. (i) 31/5; (ii) 31/4.
2. (i) 41/9; (ii) $61(2-\sqrt{3})/\pi$.
3. 471/45 from the 4 g cm^{-1} end.
4. (2·4, 0).
5. (15, 0).
6. (0, 5·4).
7. (0, 2·4).
8. (8/3, 0).
9. (0, 4/3).
10. (18/5, 0).
11. (9/8, 3/5).
12. (8/5, 16/7).
13. (30/13, 363/130).
14. Intersection of medians.
15. $(0, 4a/3\pi)$.
16. $(2a/3\pi, 2a/3\pi)$.
17. Yes.
18. (4, 0).
19. (0, 9/7).
20. (10/3, 25/3).
21. 4·7 cm from the end nearest 3 g mass.
22. 72, (0, 27/5).
23. 108, 3·6.
24. $(2a/5, 1)$.
26. (8/3, 16/15).
27. $\left\{\dfrac{\int_0^a xf(x)\,dx}{\int_0^a f(x)\,dx}, \dfrac{\int_0^a (f(x))^2\,dx}{\int_0^a f(x)\,dx}\right\}$.
28. (3·1, 0·25).
29. (i) 9·03; (ii) 14·4; $\bar{x} = 1·59$.

EXERCISE 9h

1. 8π.
2. $243\pi/5$.
3. 3π.
4. $2\pi/3$.
5. $16\pi/15$.
6. 8π.
7. $\pi^2/6$.
10. 47·7 m^3.
11. $16\pi/15$.
12. $128\sqrt{2}\pi/15$.
13. $1296\pi/5$.
14. (a) 245π; (b) 358π; (c) 458π cm^3.
15. 552π cm^3: 23/96.
16. 19/3; $211\pi/5$.
17. $96\pi/15$.
18. (8/3, 0).
19. (0, 0).
20. (i) $4r/3\pi$; (ii) $3r/8$; $(4r/3\pi, 4r/3\pi)$; $(3r/8, 3r/8)$.
21. $(0, \frac{1}{4}h)$ (the y-coordinate is measured from the base).
22. 1·8 cm above the base.
23. 4·2 cm from the 5 cm end.
24. (40/81, 0).
25. $2h/5$ from the base.
26. $2\pi h^2$ cm^3; 14 seconds.
27. $(128\sqrt{2})/35$.
28. $(a/5, 0)$; $\pi a^3/15$.
29. (0, 0); (2, 16); 16/3; $1024\pi/15$.
30. $20a^2$.
31. $(3\pi/2+1)$; $\pi(9\pi+8)/4$.
32. 22·9.
33. 13·8.
34. (i) 22/7 cm; (ii) 4·65 cm from the larger end.
35. 1·83; $\pi/30$.

MISCELLANEOUS EXERCISE 9

1. 2.
2. 8.
3. −7.
4. 21.
5. −3.
6. 1.
7. 1.
8. $\sqrt{3}/2$ or 0·87.
9. −2/3.
10. 1/2.
11. $(2-\sqrt{2})/2$ or 0·293.
12. 13/154.
13. $\pi/2$.
14. 0.
15. $\pi/4$.
16. $\pi/6$.
17. 5/36.
18. −1/3.
19. 61.
20. 96.
21. 64/3.
22. $\pi/2$.
23. −6.
24. 16.
25. −20.
26. $\pi/18$.
27. $5\frac{1}{2}$.
28. $-3\frac{1}{2}$.
29. 1/3.
30. −6
31. $1\frac{7}{8}$.
32. $\pi/16$.
33. $\pi/48$.
34. 1.
35. $(2-\sqrt{3})/9$.
36. 19/81.
37. 10.
38. $100\frac{1}{3}$.
39. 0·59.
40. 1/2.
41. −1/5.
42. $(3-\sqrt{5})/4$.
43. 4/3.
44. $(2\sqrt{3}-3)/3$.
45. 19/15.
46. 31/960.
47. 7/48.
48. 1.
49. $(\pi+1)/4$.
50. $\pi/24$.
51. $2(2-\sqrt{3})/9$.
52. $\dfrac{2\sqrt{3}(\sqrt{3}-1)}{15}$.
53. 1/24.
54. 1/4.
55. 0·46.
56. $8(16-\sqrt{2})/35$.
57. 1/2.
58. 12/25.
59. 32.
60. 5.
61. −12.
62. $16(4-\sqrt{2})/27$.
63. $\pi/8$.
64. $2/\pi$.
65. $2343/5\pi$.
66. 7/144.
67. 1/8.

68. 1/2.

69. 2.

70. 1.

71. 0.

72. (i) $2(16a+5b)/5$; (ii) 60/21; (iii) $\pi/4$; (iv) 0·16.

73. (i) 27; (ii) 2/3; (iii) 4/5.

74. $5\pi/3$; (27/20, 0).

75. 16/3, (3/2, 0).

76. (i) Q; (ii) $-R$; (iii) $Q-R$; (iv) $Q+R$.

77. 370.

78. 6·2.

79. $28\pi a^5/15$.

80. $2\pi(2a^3-3a^2b+b^3)/3$.

81. $5\pi a^3/24$.

83. $32\pi a^3/3$.

84. $2\pi ab^2/3$; $(3a/8, 0)$.

85. $56{\cdot}5\pi$.

86. 67·5.

87. 90.

88. 8/3.

89. 49/6.

90. 1/2.

91. $(8\pi-3\sqrt{3})/48$.

92. $\pi/12$.

93. $(27-16\sqrt{2})/48$.

94. 18.

95. 8/3.

96. 112.

97. $2(2-\sqrt{3})$.

98. $4\frac{1}{3}$.

99. (i) P; (ii) $-Q$; (iii) $P-Q$; (iv) $|P-Q|$; (v) $P+Q$.

100. $h(a^2+2ab+3b^2)/(a^2+ab+b^2)$ from the smaller end and on the axis.

Appendix 1

2. $y = \cos x$, all values of x; $y = x/\sin x$, $x = 0$; $y = \sec x$; $x = \frac{1}{2}\pi$.

3. $\frac{1}{2}r^2(\theta-\sin\theta)$; $108\frac{1}{2}°$.

Miscellaneous Exercise A

10. (a) $4/10+4/100+4/1000+\ldots+4/10^n+\ldots$; 4/9.
(b) $123/1000+123/(1000)^2+123/(1000)^3+\ldots+123/(1000)^n+\ldots$; 123/999.
(c) $2+2/100+2/10{,}000+\ldots 2/(100)^{n-1}+\ldots$; 200/99.

18. Continuous except at integral values.

27. 5/11; $5\frac{5}{11}$ minutes past 1.

Miscellaneous Exercise B

1. $3l^2$, where l = length of side.

2. $a = 1$, $b = 0$, $c = 0$, $d = 1$.

3. $(x^{n+1}-1)/(x-1)$; $(nx^{n+1}-(n+1)x^n+1)/(x-1)^2$.

6. (i) $-\sin x \cos(\cos x)$;
(ii) $\cos x \cos(\sin x)\cos\{\sin(\sin x)\}$;
(iii) $-2\sin x\cos x \sin(\sin^2 x)$.

7. $\dfrac{2\sqrt{(x+\sqrt{x})}+2\sqrt{x}+1}{8\sqrt{x}\,\sqrt{(x+\sqrt{x})}\,\sqrt{\{x+\sqrt{(x+x)}\}}}$.

8. (i) $2xg'(x^2)$; (ii) $2\sin x\cos x\, g'(\sin^2 x)$;
(iii) $g'(x)g'\{g(x)\}$;
(iv) $\sin 2x\{g'(\sin^2 x)-g'(\cos^2 x)\}$.

9. (a) $2\pi xl\delta x$, x = radius, l = height;
(b) $\pi r^2\delta x$, x = height, r = radius.

10. $f'(x) = g(x)h'(x)+g'(x)h(x)$;
$f''(x) = g(x)h''(x)+2g'(x)h'(x)+g''(x)h(x)$.

11. (i) 0, 0; (ii) 0, 4.

12. 1/6.

16. $f(x)$ continuous for all $x \geqslant 0$; $f'(x)$ continuous for all $x > 0$.

17. $b < 2$, $2\sqrt{(b^2-3b+3)}$, $b \geqslant 2$, b.

18. Horizontally $4p/(8+3\pi)$;
vertically $(12+5\pi)p/(16+6\pi)$.

19. 40·6 km/h.

21. $\sqrt{3}-\sqrt{2}$.

27. $-\frac{1}{2}V^2 A/4\pi(x-tx^2)^2$.

29. Resistance $= \varrho ab^2/2x^4$;
field $= \pi b^2Ex^2/(2ax^4r+\varrho a^2b^2)$.

30. $2x\sin(1/x)-\cos(1/x)$.

33. 0·22; 0·11; $d(\log_{10} x)/dx = 0{\cdot}43/x$.

Miscellaneous Exercise C

1. (i) 2; (ii) 55; (ii) 3/2; (iv) -1.

5. $1/n$, $n \neq -1$.

6. $5\frac{1}{3}$.

11. $\frac{4}{5}a^5$.

12. (i) $1{\cdot}6\pi$; (ii) 2π; (iii) $\frac{10}{3}\pi$; (iv) $\frac{16}{5}\pi$.

14. $4\frac{1}{2}$.

16. (i) 0·696; (ii) 1·102; (iii) 1·390; (iv) 1·613; (v) 0·696.

18. $n(n-1)(2n-1)/6$.

20. $5G/4$.

22. 11,300 m/s; 11,250 m/s.

Index

Italic figures in parentheses after page references refer to exercise question numbers